Kishan G Mehrotra

# An Introduction to Multivariate Statistics

# An Introduction to Multivariate Statistics

M. S. Srivastava
Professor of Statistics
University of Toronto

C. G. Khatri
Professor of Statistics
Gujerat University

NORTH HOLLAND • NEW YORK
New York • Oxford

Elsevier North Holland, Inc.
52 Vanderbilt Avenue, New York, New York 10017

Distributors outside the United States and Canada:

Thomond Books
(A Division of Elsevier/North-Holland Scientific Publishers, Ltd.)
P.O. Box 85
Limerick, Ireland



Library of Congress Cataloging in Publication Data

Srivastava, Muni Shanker, 1936-
An introduction to multivariate statistics.

Bibliography: p.324
Includes index.
1. Multivariate analysis. I. Khatri, C. G., joint author. II. Title.
QA278.S69 519.5′3 78-21491
ISBN 0-444-00302-9

Manufactured in the United States of America

To Usha and Kanak

# Contents

# Preface

The content of this book was first prepared by M. S. Srivastava during 1973–1976 and is based on the lectures given to the first-year graduate students at the University of Toronto since 1963. A revision of this material took place in the summer of 1976 in collaboration with C. G. Khatri, who also added materials on S. N. Roy's union–intersection principle, confidence intervals, quadratic forms, and probability inequalities. A final revision took place during 1977–1978 and we added results on estimation in linear models (Chapter 5) without the assumption of normality. In particular, C. R. Rao's MINQUE theory was included.

The rest of the book gives an up-to-date account of the basic multivariate normal theory. Properties of tests such as monotonicity and unbiasedness are included (Chapter 10). Several new approaches and techniques are presented. For example, a unified approach is given for obtaining maximum likelihood estimates without differentiation, and new simple methods are explained for obtaining the asymptotic distribution of the roots (Chapter 9).

The basic emphasis of this book is on the maximum likelihood method in testing and on Roy's union–intersection principle in obtaining confidence intervals, although the latter method has been used to obtain tests as well. The material on matrix theory and the Jacobian of transformations is presented in Chapter 1; readers not familiar with matrix theory may find this chapter very useful. The reader, however, is expected to know the basic univariate results. Although this book is aimed at a two-semester course, it could be adopted as a one-semester course by choosing material appropriately. During 1977–1978 the first author tried it as a one-semester course at the University of Wisconsin at Madison and at the Indian Statistical Institute, New Delhi.

This book was made possible by generous help from several friends, colleagues, and students. Edward M. Carter gave his time generously and helped prepare the first three sections of Chapter 10. Discussion with C. R. Rao has been helpful in the preparation of Chapter 5. S. K. Mitra and J. Kleffe provided many references and suggestions for this chapter. I. Olkin and J. B. Selliah commented on the first draft. S. Minkin pointed out an error/omission in Chapter 2, M. Perlman in Chapter 4, C. Y. Leung in Chapter 7 and W. Oldford in Chapter 8. Of course, for a book of this nature, it is a formidable job to guarantee absolutely no errors or no omissions, but it is hoped that any that may occur would not mar the reading of the book. It will be greatly appreciated if errors and omissions are brought to our attention.

The outcome of this book was possible through the generous support of the National Research Council of Canada. The excellent typing was done by Mrs. June Hubert of Guelph, Ontario, to whom we express our sincerest thanks.

M. S. SRIVASTAVA
C. G. KHATRI

*January 1979*

# An Introduction to Multivariate Statistics

# 1

# Some Results on Matrices and Jacobians of Transformations

## 1.1 Notation and Definitions

Let $F$ be a field—an arbitrary collection of numbers within which the four operations of addition, subtraction, multiplication, and division by a nonzero number can always be carried out—of real or complex numbers. Suppose that $a_{11}, a_{12}, \ldots, a_{pq}$ is a collection of $pq$ elements of $F$. The rectangular array of these elements,

$$\begin{bmatrix} a_{11} & a_{12} & \cdots & a_{1q} \\ a_{21} & a_{22} & \cdots & a_{2q} \\ \vdots & \vdots & & \vdots \\ a_{p1} & a_{p2} & \cdots & a_{pq} \end{bmatrix},$$

consisting of $p$ rows and $q$ columns, is called a $p \times q$ matrix. We write $A = (a_{ij}) : p \times q$, where $a_{ij}$ is the element in the $i$th row and $j$th column. If all the elements of $A$ are zero, it is called a *zero* or *null* matrix and is denoted by $0_{pq}$, or simply 0 if there is no confusion. If $p = q$, $A$ is called a *square* matrix of order $p$. If $q = 1$, it is called a $p$ column vector or simply $p$ vector. If $p = 1$, it is called a $q$ row vector. A $p$ column vector of elements $a_1, \ldots, a_p$ will be denoted by $\mathbf{a}$. Similarly, the $p$ row vector of elements $a_1, \ldots, a_p$ will be denoted by $\mathbf{a}' = (a_1, \ldots, a_p)$.

A square matrix in which all elements above the main diagonal are zero is called a *lower triangular matrix* (denoted by $\underset{\sim}{A}$); one in which all the elements below the main diagonal are zero is called an *upper triangular matrix* (denoted by $\tilde{A}$). A square matrix $A : p \times p$ is said to be a *diagonal* matrix if all the off-diagonal elements are zero except the main diagonal elements. A diagonal matrix with diagonal elements $\alpha_i$ will be denoted by $D_\alpha$ or $\mathrm{diag}(\alpha_1,\ldots,\alpha_p)$. Note that some of the $\alpha$'s could be zero; if all the $\alpha$'s are zero, then it is a zero matrix of order $p$; if all the $\alpha_i$'s are equal to 1, it is called an *identity* matrix, denoted by $I_p$, where $p$ denotes the order of the matrix (or simply $I$, if the order is apparent). If the rows and columns of a matrix $A$ are interchanged, the resulting matrix is called the *transpose* of $A$ and is denoted by $A'$. Thus if $A=(a_{ij}):p\times q$, then $A'=(a_{ji}):q\times p$. If $A$ is a matrix of complex numbers, the *conjugate transpose* of $A$ is defined by $A^*=(a_{ji}^*)$, where $a_{ij}^*$ is the complex conjugate of $a_{ij}$. A square matrix $A:p\times p$ is said to be *symmetric* if $A=A'$, and *Hermitian* if $A=A^*$. Thus the symmetric matrix is a special case of the Hermitian matrix when all the elements are real. A square matrix $A$ is said to be *skew symmetric* (*skew Hermitian*) if $A=-A'$ $(A=-A^*)$. Clearly, if $A$ is a skew symmetric matrix, then $a_{ii}=0$.

A matrix $A:m\times n$ is said to be *semiorthogonal* if $AA'=I_m$ $(n\geqslant m)$, and it is said to be *orthogonal* when $n=m$. Notice that the inverse of an orthogonal matrix $A$ is its transpose $A'$. Similarly, a *semiunitary* matrix $A:m\times n$ satisfies the condition $AA^*=I_m$ $(n\geqslant m)$, and it is said to be *unitary* when $m=n$. The inverse of a unitary matrix $A$ is its conjugate transpose $A^*$. Also, if $AA^*=A^*A$, then $A$ is said to be *normal*.

*After this chapter all the elements of a matrix are taken to be real, unless otherwise stated.* This will avoid the use of the adjective "real" with every matrix.

## 1.2 Matrix Operations

Sometimes it is convenient to represent a matrix $A$ as

$$A=\left(\begin{array}{c|c} A_{11} & A_{12} \\ \hline A_{21} & A_{22} \end{array}\right),$$

where $A_{ij}$ is an $m_i\times m_j$ submatrix of $A$ $(i,j=1,2)$. A submatrix of $A$ is a matrix obtained from $A$ by deleting certain rows and columns. The above representation of $A$ is called a partitioned one. If there are more than two partitions of rows (or columns), we can write $A=(A_{ij})$, where $A_{ij}$ is the submatrix of $A$ in the $i$th row and $j$th column partition. Another type of partitioned matrix can be written as

$$A=\left[\begin{array}{c|c} A_{11} & \multirow{2}{*}{$A_2$} \\ \cline{1-1} A_{21} & \end{array}\right].$$

Let $A:m\times n$ and $B:p\times q$ be two matrices. Then, their *direct sum* is defined as the $(m+p)\times(n+q)$ matrix

$$A\oplus B=\begin{pmatrix} A & 0_{mq} \\ 0_{pn} & B \end{pmatrix}=\text{diag}(A,B),$$

while their *direct product* (or *Kronecker product*) is defined as the $mp\times nq$ matrix

$$A\otimes B=(a_{ij}B) \qquad \text{with } A=(a_{ij}).$$

Notice that $(A\oplus B)^*=A^*\oplus B^*$, $(A\otimes B)^*=A^*\otimes B^*$, and $I_n\otimes B=\text{diag}(B,B,\ldots,B):np\times nq$.

The usual addition of two matrices is defined only if both the matrices are of the same order. Thus, if $A=(a_{ij}):p\times q$ and $B=(b_{ij}):p\times q$, then their *sum* is defined by

$$A+B=(a_{ij}+b_{ij}):p\times q.$$

The usual multiplication of $A$ by $B$ is defined only if the number of columns of $A$ is equal to the number of rows of $B$. Thus, if $A=(a_{ij}):m\times n$ and $B=(b_{ij}):n\times q$, then the product of $A$ and $B$ (denoted by $AB$) is defined as an $m\times q$ matrix $AB=C=(c_{ij})$ where

$$c_{ij}=\sum_{k=1}^{n} a_{ik}b_{kj}, \qquad i=1,2,\ldots,m, \quad j=1,2,\ldots,q.$$

Notice that if $A$ and $B$ are two matrices of the same order and are partitioned in the same way [i.e., if $A=(A_{ij})$ and $B=(B_{ij})$, having the same orders for $A_{ij}$ and $B_{ij}$], then $A+B=(A_{ij}+B_{ij})$, while if two matrices $A:m\times n$ and $B:n\times q$ are partitioned as $A=(A_{ij})$ and $B=(B_{ij})$ with submatrices $A_{ij}:m_i\times n_j$ and $B_{ij}:n_i\times q_j$, then $AB=(\Sigma_k A_{ik}B_{kj})$. Further, $(A+B)^*=A^*+B^*$ and $(AB)^*=B^*A^*$.

## 1.3 Determinants

The determinant of a square matrix $A=(a_{ij}):n\times n$ is defined by

$$|A|\equiv\sum_{\boldsymbol{\alpha}}(-1)^{N(\boldsymbol{\alpha})}\prod_{j=1}^{n} a_{\alpha_j,j},$$

where $\Sigma_\alpha$ denotes the summation over the distinct permutations $\boldsymbol{\alpha}$ of the numbers $1,2,\ldots,n$, and $N(\boldsymbol{\alpha})$ is the total number of inversions of a permutation. An inversion of a permutation $(\alpha_1,\alpha_2,\ldots,\alpha_n)$ is an arrangement of two indices such that the larger index comes after the smaller index. For example, $N(2,1,4,3)=1+N(1,2,4,3)=2+N(1,2,3,4)=2$, because $N(1,2,3,4)=0$. Similarly, $N(4,3,1,2)=1+N(3,4,1,2)=3+$

$N(3,1,2,4)=5$. The value of the determinant is denoted as $|A|$ or det.$A$. If $|A|$ is real, then $|A|_+$ denotes the positive value of $|A|$. Notice that

$$|A'|=\sum_{\alpha}(-1)^{N(\alpha)}\prod a_{j,\alpha_j}=\sum_{\alpha}(-1)^{N(\alpha)}\prod a_{\alpha_j,j}=|A|.$$

The following are the immediate consequences and follow from the definition of $|A|$.

(i) If the $i$th row (or column) is multiplied by a constant $c$, the value of the determinant is multiplied by $c$. Hence $|cA|=c^n|A|$ if $A$ is an $n\times n$ matrix.
(ii) If any rows (or columns) of a matrix are interchanged, the sign of the determinant is changed. Hence, if two rows (or columns) of a matrix are identical, the value of the determinant is zero.
(iii) The value of the determinant is unchanged if to the $i$th row (or column), a multiple of the $j$th row (or column) is added. Hence, the value of the determinant is zero if a row (or column) is a linear combination of other rows (or columns).
(iv) $|I_n|=1$, $|D_a|=\prod a_i$, and $|\underset{\sim}{A}|=|\underset{\sim}{A}'|=\prod a_{ii}$.
(v) $|A^*|=(|A|)^*=$complex conjugate of $|A|$.

**Theorem 1.3.1** *If $A$ and $B$ are $n\times p$ and $p\times n$ matrices, then*

$$|AB|=\begin{cases}0 & \text{if } p<n,\\ \sum_{\mathbf{k}}{}_1|A_{(\mathbf{k})}|\,|B^{(\mathbf{k})}| & \text{if } p>n,\\ |A|\,|B| & \text{if } p=n,\end{cases}$$

*where $\sum_{\mathbf{k}}{}_1$ denotes the summation over $k_1,\dots,k_n$ such that $1\leqslant k_1<k_2<\cdots<k_n\leqslant p$, $A_{(\mathbf{k})}$ is a submatrix of $A$ obtained by taking columns $k_1,k_2,\dots,k_n$, and $B^{(\mathbf{k})}$ is a submatrix of $B$ obtained by taking rows $k_1,k_2,\dots,k_n$.*

PROOF. By definition,

$$|AB|=\sum_{\alpha}(-1)^{N(\alpha)}\prod_{j=1}^{n}\left(\sum_{k=1}^{p}a_{\alpha_j,k}b_{k,j}\right)$$
$$=\sum_{k_1=1}^{p}\cdots\sum_{k_n=1}^{p}\sum_{\alpha}(-1)^{N(\alpha)}\left(\prod_{j=1}^{n}a_{\alpha_j,k_j}\right)\left(\prod_{j=1}^{n}b_{k_j,j}\right)$$
$$=\sum_{k_1=1}^{p}\cdots\sum_{k_n=1}^{p}|A(k_1,\dots,k_n)|\left(\prod_{j=1}^{n}b_{k_j,j}\right),$$

where $A(k_1,\ldots,k_n)$ is an $n\times n$ matrix whose $j$th column is the $k_j$th column of $A$. By property (ii), we may note that if any two of $k_1,k_2,\ldots,k_n$ are equal, then $|A(k_1,\ldots,k_n)|=0$, while if all $k_1,k_2,\ldots,k_n$ are distinct integers, then $|A(k_1,\ldots,k_n)|=(-1)^{N(\boldsymbol{\beta})}|A(k_{\beta_1},\ldots,k_{\beta_n})|$, $\boldsymbol{\beta}$ being a permutation of $1,2,\ldots,n$ such that $1\leqslant k_{\beta_1}<k_{\beta_2}<\cdots<k_{\beta_n}\leqslant p$. Hence, if $p<n$,

$$|AB|=0,$$

while if $p\geqslant n$,

$$|AB|=\sum_{k_1\neq k_2\neq\cdots\neq k_n=1}^{p}\cdots\sum^{p}|A(k_1,\ldots,k_n)|\left(\prod_{j=1}^{n}b_{k_j,j}\right)$$

$$=\sum_{\mathbf{k}}{}_1\sum_{\boldsymbol{\beta}}(-1)^{N(\boldsymbol{\beta})}|A_{(\mathbf{k})}|\left(\prod_{j=1}^{n}b_{k_{\beta_j},j}\right)$$

$$=\sum_{\mathbf{k}}{}_1|A_{(\mathbf{k})}||B^{(\mathbf{k})}|,$$

where $\sum_{\mathbf{k}}{}_1$ is defined in Theorem 1.3.1 and $\sum_{\boldsymbol{\beta}}$ denotes the summation over the permutations $\boldsymbol{\beta}$ of $1,2,\ldots,n$. This proves Theorem 1.3.1. □

The following corollary follows from Theorem 1.3.1 and property (v):

**Corollary 1.3.1** $|AA^*|\geqslant 0$, *where $A$ is any $m\times n$ matrix.*

**Corollary 1.3.2** *Let $P$ and $R$ be square matrices. Then*

$$\begin{vmatrix} I_n & 0\\ Q & P\end{vmatrix}=|P|,\qquad \begin{vmatrix} R & S\\ 0 & I_p\end{vmatrix}=|R|,$$

*and*

$$\begin{vmatrix} R & S\\ 0 & P\end{vmatrix}=\begin{vmatrix} R & 0\\ Q & P\end{vmatrix}=|R||P|.$$

**Corollary 1.3.3** *If $A$ and $B$ are $p\times q$ and $q\times p$ matrices, then $|I_p+AB|=|I_q+BA|$.*

This follows from Theorem 1.3.1, Corollary 1.3.2 and the fact that

$$\left|\begin{pmatrix} I_p & A\\ -B & I_q\end{pmatrix}\begin{pmatrix} I_p & 0\\ B & I_q\end{pmatrix}\right|=\left|\begin{pmatrix} I_p & 0\\ B & I_q\end{pmatrix}\begin{pmatrix} I_p & A\\ -B & I_q\end{pmatrix}\right|=\begin{vmatrix} I_p & A\\ -B & I_q\end{vmatrix}.$$

### 1.3.1 Cofactors and Minors

Let $A_j^i$ be a submatrix of $A$, obtained by deleting the $i$th row and the $j$th column of $A$. Then $m_{ij}=|A_j^i|$ is known as the *minor* of $a_{ij}$, and $c_{ij}(A)=(-1)^{i+j}|A_j^i|=(-1)^{i+j}m_{ij}$ is known as the *cofactor* of $a_{ij}$. We note that

$$\sum_{j=1}^{n} c_{ij}(A)a_{kj}=\sum_{j=1}^{n}(-1)^{i+j}a_{kj}|A_j^i|$$

$$=\sum_{j=1}^{n}(-1)^{i+j}a_{kj}\sum_{\alpha}(-1)^{N(\alpha)}\prod_{t=1}^{n-1}a'_{\alpha_t,t},$$

where $A_j^i=(a'_{rs})$. Notice that $a'_{rs}=a_{rs}$ for $r<i$, $s<j$; $a'_{rs}=a_{r,s+1}$ for $r<i$, $s\geqslant j$; $a'_{rs}=a_{r+1,s}$ for $r\geqslant i$, $s<j$; and $a'_{r,s}=a_{r+1,s+1}$ for $r\geqslant i$, $s\geqslant j$. Hence,

$$\sum_{j=1}^{n} c_{ij}(A)a_{kj}=|A^i(k)|=\begin{cases}|A|, & i=k,\\ 0, & i\neq k,\end{cases}$$

where $A^i(k)$ is a matrix obtained from $A$ by replacing its $i$th row with its $k$th row. Similarly, we get

$$\sum_{i=1}^{n} c_{ij}(A)a_{ik}=|A_j(k)|=\begin{cases}|A|, & j=k,\\ 0, & j\neq k,\end{cases}$$

where $A_j(k)$ is a matrix obtained from $A$ by replacing the $j$th column with the $k$th column. Thus, we get

**Theorem 1.3.2** $|A|I=[c_{ij}(A)]A'=A[c_{ij}(A)]'$.

Let $A_{(j_1,\ldots,j_t)}^{(i_1,\ldots,i_t)}$ be a submatrix of $A$, consisting of rows $i_1,i_2,\ldots,i_t$ and columns $j_1,j_2,\ldots,j_t$. Then $|A_{(\alpha_1,\ldots,\alpha_t)}^{(\alpha_1,\ldots,\alpha_t)}|$ is known as the *principal minor* of order $t$, and

$$\operatorname{tr}_t A=(\text{sum of all the principal minors of order } t)$$

$$=\sum_{\alpha}|A_{(\alpha_1,\ldots,\alpha_t)}^{(\alpha_1,\ldots,\alpha_t)}|$$

is known as the $t$th trace of $A$. Thus when $t=1$, we have $\operatorname{tr}_1 A\equiv\operatorname{tr}A=\sum_{i=1}^{n}a_{ii}$.

By Theorem 1.3.2, it is easy to see that $|\lambda I_n+A|$ is a polynomial in $\lambda$ of degree $n$. Thus, we can write

$$|A+\lambda I_n|=c_0+c_1\lambda+c_2\lambda^2+\cdots+c_n\lambda^n \qquad \text{for all } \lambda,$$

where the $c_i$'s are constants depending on $i$. Differentiating this $i$ times with respect to $\lambda$, and then putting $\lambda=0$, we get

$$i!c_i=i!\operatorname{tr}_{n-i}A.$$

This gives the following theorem:

**Theorem 1.3.3** *Let $A$ be a square matrix of order $n$. Then*

$$|A+\lambda I_n| = \sum_{i=1}^{n} \lambda^i \operatorname{tr}_{n-i} A, \qquad \textit{where} \quad \operatorname{tr}_0 A \equiv 1.$$

**Corollary 1.3.4** *Let $A$ and $B$ be respectively $p \times q$ and $q \times p$ matrices with $p > q$. Then $\operatorname{tr}_t AB = \operatorname{tr}_t BA$ for $t \leqslant q$, and $\operatorname{tr}_t AB = 0$ for $t > q$.*

This follows from the fact that

$$|AB+\lambda I_p| = \lambda^{p-q}|BA+\lambda I_q| \qquad \text{for any } \lambda.$$

## 1.4 Inverse of a Nonsingular Matrix

A square matrix $A$ is said to be *nonsingular* if $|A| \neq 0$. For a nonsingular matrix $A$, let $B' = [c_{ij}(A)]/|A|$, where $c_{ij}(A)$ is the cofactor of $a_{ij}$ in $A$. Then by Theorem 1.3.2, $I = B'A' = AB$. Such a matrix $B$ will be said to be an inverse of $A$, and will be denoted by $B = A^{-1}$. Notice that $|B| = 1/|A| \neq 0$, and $B$ is a nonsingular matrix. Hence, there exists a matrix $C$ such that $BC = I_n$. Then $I = AB \Rightarrow C = ABC$, i.e., $A = C$. This shows that

$$(A^{-1})^{-1} = A, \qquad AA^{-1} = A^{-1}A = I, \quad \text{and} \quad A^{-1} \text{ is unique.}$$

Now we shall establish

**Theorem 1.4.1** *Let $P$ and $Q$ be nonsingular matrices with proper orders. Then*

(i) $[P \oplus Q]^{-1} = P^{-1} \oplus Q^{-1}$,

(ii) $(P \otimes Q)^{-1} = P^{-1} \otimes Q^{-1}$,

(iii) $(PQ)^{-1} = Q^{-1}P^{-1}$, *and*

(iv) *if $Q = P + UV$, then*

$$Q^{-1} = P^{-1} - P^{-1}U(I+VP^{-1}U)^{-1}VP^{-1}$$

*and*

$$Q^{-1}U = P^{-1}U(I+VP^{-1}U)^{-1}.$$

PROOF. (i), (ii), and (iii) follow from

$$(P^{-1} \oplus Q^{-1})(P \oplus Q) = I,$$
$$(P \otimes Q)(P^{-1} \otimes Q^{-1}) = I,$$

and

$$(Q^{-1}P^{-1})(PQ)=I.$$

For (iv), we note that $Q^{-1}P+Q^{-1}UV=I$ or $Q^{-1}+Q^{-1}UVP^{-1}=P^{-1}$. Hence, $Q^{-1}U(I+VP^{-1}U)=P^{-1}U$ or $Q^{-1}U=P^{-1}U(I+VP^{-1}U)^{-1}$. Using this in $Q^{-1}=P^{-1}-Q^{-1}UVP^{-1}$, we get the required result. □

**Corollary 1.4.1** *Let P and Q be nonsingular square matrices. Then*

$$\begin{pmatrix} P & 0 \\ R & Q \end{pmatrix}^{-1}=\begin{pmatrix} P^{-1} & 0 \\ -Q^{-1}RP^{-1} & Q^{-1} \end{pmatrix}$$

*and*

$$\begin{pmatrix} P & S \\ 0 & Q \end{pmatrix}^{-1}=\begin{pmatrix} P^{-1} & -P^{-1}SQ^{-1} \\ 0 & Q^{-1} \end{pmatrix}.$$

**Corollary 1.4.2** *Let P and Q be square matrices.*

(i) *If P is nonsingular, then*

$$\begin{pmatrix} P & S \\ R & Q \end{pmatrix}^{-1}=\begin{pmatrix} P^{-1} & 0 \\ 0 & 0 \end{pmatrix}+\begin{pmatrix} -P^{-1}S \\ I \end{pmatrix}(Q-RP^{-1}S)^{-1}(-RP^{-1},I).$$

(ii) *If P and Q are nonsingular matrices, then*

$$\begin{pmatrix} P & S \\ R & Q \end{pmatrix}^{-1}=\begin{bmatrix} (P-SQ^{-1}R)^{-1} & -P^{-1}S(Q-RP^{-1}S)^{-1} \\ -(Q-RP^{-1}S)^{-1}RP^{-1} & (Q-RP^{-1}S)^{-1} \end{bmatrix}.$$

PROOF. Since

$$\begin{pmatrix} P & S \\ R & Q \end{pmatrix}=\begin{pmatrix} P & 0 \\ R & Q-RP^{-1}S \end{pmatrix}+\begin{pmatrix} P \\ R \end{pmatrix}(0 \quad P^{-1}S),$$

the result follows from Theorem 1.4.1(iv) and Corollary 1.4.1. Part (ii) can be obtained on the same lines. □

## 1.5 Rank of a Matrix

An $n\times m$ matrix $A$ is said to be of rank $r$, and we write $\rho(A)=r$, if and only if (iff) there is at least one nonzero minor of order $r$ from $A$, and all the minors of order $r+1$ are zero. It is easy to establish the following:

(i) $\rho(A)=0$ iff $A=0$.

(ii) If $A$ is an $n\times m$ matrix and $A\neq 0$, then $1\leqslant\rho(A)\leqslant \mathrm{Min}(m,n)$. Further, $\rho(A)=\rho(A')=\rho(A^*)$.

(iii) $\mathrm{Max}[\rho(A),\rho(B)]\leqslant\rho(A \mid B)\leqslant\mathrm{Min}[n,\rho(A)+\rho(B)]$, where $n$ is the number of rows in $A$.

(iv) $\rho\begin{pmatrix} 0 & Q \\ R & 0 \end{pmatrix}=\rho\begin{pmatrix} Q & 0 \\ 0 & R \end{pmatrix}=\rho(\mathrm{R})+\rho(\mathrm{Q})$.

**Theorem 1.5.1** $\rho(AB)\leqslant\mathrm{Min}[\rho(A),\rho(B)]$.

PROOF. We may assume without loss of generality that $\rho(B)\geqslant\rho(A)=r$. Then consider any minor of order $r+1$ from $AB$ and expand it by Theorem 1.3.1. This expansion is a linear function in the $(r+1)$th minors of $A$, which are zero. Hence, all the $r+1$ minors of $AB$ are zero, which shows that $\rho(AB)\leqslant r=\rho(A)$. □

**Corollary 1.5.1** $\rho(AB)=\rho(A)$ *if* $B$ *is nonsingular.*

**Corollary 1.5.2**

$$\rho\begin{pmatrix} B & BC \\ AB & 0 \end{pmatrix}=\rho(B)+\rho(ABC).$$

This follows from Corollary 1.5.1, (iv), and

$$\begin{pmatrix} I & 0 \\ -A & I \end{pmatrix}\begin{pmatrix} B & BC \\ AB & 0 \end{pmatrix}\begin{pmatrix} I & -C \\ 0 & I \end{pmatrix}=\begin{pmatrix} B & 0 \\ 0 & -ABC \end{pmatrix}.$$

**Theorem 1.5.2** $\rho(A)=\rho(AA^*)=\rho(A^*A)$.

PROOF. Let $A$ be an $m\times n$ matrix of rank $r\leqslant\mathrm{Min}(m,n)$. Let $A_1: s\times n$ be a submatrix of $A$. Then by Theorem 1.3.1 and property (v) of Section 1.3,

$$|A_1A_1^*|=\sum_{\mathbf{k}}{}_1|A_{1(\mathbf{k})}|(|A_{1(\mathbf{k})}|)^*$$

where $\sum_{\mathbf{k}}{}_1$ denotes the summation over $k_1,\ldots,k_s$ such that $1\leqslant k_1<k_2<\cdots<k_s\leqslant n$, and $A_{1(\mathbf{k})}$ is a submatrix of $A_1$ obtained by taking columns $k_1,k_2,\ldots,k_s$ only. Notice that if $s>r$, then $|A_1A_1^*|=0$, while if $s=r$, then there is at least one nonzero minor of order $r$ in $A$. In the latter case, let $A_1$ contain only these $r$ rows that give a nonzero minor of order $r$. Then the above expression shows that $|A_1A_1^*|>0$. This shows that $\rho(A)=\rho(AA^*)$. Similarly, the other part can be established. □

**Corollary 1.5.3** *Let $\rho(B:p\times q)=p$. Then $\rho(AB)=\rho(A)=\rho(B^*A^*)$.*

**Corollary 1.5.4**

$$\rho\begin{pmatrix} P & Q \\ R & 0 \end{pmatrix} \geqslant \rho(Q)+\rho(R).$$

*The equality holds if there exist $B_1$ and $B_2$ such that $P=B_1R+QB_2$.*

PROOF. Let us assume that $P=B_1R+QB_2$. Then, by Corollary 1.5.1,

$$\rho\begin{pmatrix} P & Q \\ R & 0 \end{pmatrix}=\rho\left[\begin{pmatrix} I & B_1 \\ 0 & I \end{pmatrix}\begin{pmatrix} 0 & Q \\ R & 0 \end{pmatrix}\begin{pmatrix} I & 0 \\ B_2 & I \end{pmatrix}\right]=\rho\begin{pmatrix} 0 & Q \\ R & 0 \end{pmatrix}.$$

For the general case, let $BR=(R_1 \,\vdots\, R_2)$, where $B$ is nonsingular and $\rho(R)=(\text{number of columns of } R_1)=\rho(R_1)$. Then

$$\rho\begin{pmatrix} P & Q \\ R & 0 \end{pmatrix}=\rho\begin{pmatrix} BP & Q \\ BR & 0 \end{pmatrix}=\rho\begin{pmatrix} P_1 & P_2 & Q \\ R_1 & R_2 & 0 \end{pmatrix}\geqslant\rho\begin{pmatrix} P_1 & Q \\ R_1 & 0 \end{pmatrix}.$$

Noting that $P_1=(P_1(R_1^*R_1)^{-1}R_1^*)R_1$, we get

$$\rho\begin{pmatrix} P & Q \\ R & 0 \end{pmatrix}\geqslant\rho(R_1)+\rho(Q)=\rho(R)+\rho(Q).$$ □

**Corollary 1.5.5** $\rho(ABC)\geqslant\rho(AB)+\rho(BC)-\rho(B)$.

This is known as the Frobenius theorem on ranks and it follows from Corollaries 1.5.2 and 1.5.4.

**Corollary 1.5.6**

(i) *Let $A$ and $B$ be matrices of order $m\times n$ and $n\times m$ respectively. Then $\rho(A-ABA)=\rho(A)+\rho(I_n-BA)-n=\rho(A)+\rho(I_m-AB)-m$.*

(ii) *Let $C$ and $D$ be $r\times n$ and $m\times p$ matrices such that $R_1CB+BDR_2=B$ for some $R_1$ and $R_2$. Then $\rho(CBD)=\rho(CB)+\rho(BD)-\rho(B)$.*

This follows from Corollaries 1.5.2 and 1.5.4, using

$$P=\begin{pmatrix} I_n & I_n-BA \\ A & 0 \end{pmatrix} \quad\text{or}\quad P=\begin{pmatrix} I_m & A \\ I_m-AB & 0 \end{pmatrix} \quad\text{or}\quad P=\begin{pmatrix} B & BD \\ CB & 0 \end{pmatrix}.$$

**Theorem 1.5.3** *Let $A: m\times n$ be a matrix of rank $r$. Then there exist two matrices $F: m\times r$ and $G: r\times n$, each of rank $r$, such that $A=FG$.*

PROOF. Since $A$ is of rank $r$, there is at least one nonzero minor of order $r$ in $A$, and all the $(r+1)$th minors are zero. By changing rows and columns, we can write the changed matrix as

$$PAQ=\begin{pmatrix} A_{11} & A_{12} \\ A_{21} & A_{22} \end{pmatrix},$$

where $P: m\times m$, $Q: n\times n$, and $A_{11}: r\times r$ are nonsingular matrices. If the $j$th column of $A_{12}$ and the $i$th row of $A_{21}$ are $\mathbf{b}_j$ and $\mathbf{c}_i'$ respectively, and $A_{22}=(d_{ij})$, then using Problem 1.3(iii) and Theorem 1.3.1, we have

$$0=\begin{vmatrix} A_{11} & \mathbf{b}_j \\ \mathbf{c}_i' & d_{ij} \end{vmatrix}=|A_{11}|\left(d_{ij}-\mathbf{c}_i'A_{11}^{-1}\mathbf{b}_j\right),$$

i.e., $A_{22}=A_{21}A_{11}^{-1}A_{12}$. Letting

$$F=P^{-1}\begin{pmatrix} A_{11} \\ A_{21} \end{pmatrix} \quad \text{and} \quad G=(I_r, A_{11}^{-1}A_{12})Q^{-1},$$

we get the required theorem. □

**Corollary 1.5.7** *If $A^2=A$ (idempotent), then* $\operatorname{tr} A=\rho(A)$.

PROOF. Follows by taking $A=FG$ (see Theorem 1.5.3) and $GF=I_r$ on account of $A^2=A$. □

**Corollary 1.5.8** *Let $A$ be an $m\times n$ matrix of rank $m$ $(\leqslant n)$. Then there exists a matrix $G:(n-m)\times n$ of rank $n-m$ such that $AG^*=0$ and $(A^* \mid G^*)$ is nonsingular.*

PROOF. Let $B=I_n-A^*(AA^*)^{-1}A$. Note that $AB=0$, $B^2=B$, and $\rho(B)=\operatorname{tr} B=n-\operatorname{tr} A^*(AA^*)^{-1}A=n-m$. Then, by Theorem 1.5.3, we can write $B=G^*G_1$ where $G:(n-m)\times n$ and $G_1:(n-m)\times n$ are matrices each of rank $n-m$. Then $AG^*G_1=0$ gives $AG^*=0$, and $(A^*, G^*)$ is a nonsingular matrix. □

The converse of Corollary 1.5.8 is obvious.

## 1.6 Generalized Inverse of a Matrix

In Section 1.4, we have considered the inverse of a nonsingular matrix. Here, we define a $g$ inverse of a matrix $A: m\times n$ in such a way that it gives the inverse of $A$ if $A$ becomes a nonsingular matrix. Let $A$ be an $m\times n$ matrix of rank $r$, and write $A=FG$, where $F: m\times r$ and $G: r\times n$ are matrices of rank $r$. Define $B_0=G^*(GG^*)^{-1}(F^*F)^{-1}F^*$. Then $AB_0A=A$. Let $B=B_0+[I_n-G^*(GG^*)^{-1}G]T_1+T_2[I_m-F(F^*F)^{-1}F^*]$, where $T_1: n\times m$ and $T_2: n\times m$ are arbitrary matrices. It is easy to verify that $ABA=A$. Hence, we define a $g$ *inverse* of $A$ as a matrix $B$ such that $ABA=A$, and it will be denoted by $A^-$. Notice that $A^-$ is not unique and $\rho(A^-)\geqslant\rho(A)$. We observe that $AB_0A=A$, that $B_0AB_0=B_0$, and that $AB_0$ and $B_0A$ are Hermitian matrices, where $B_0$ is defined above. Suppose, there exists a matrix $C$ (other than $B_0$) such that $ACA=A$, $CAC=C$, and $AC$ and $CA$ are Hermitian matrices. Then from the given conditions, we have

$$B_0A=B_0(ACA)=B_0A(CA)=B_0AA^*C^*=A^*B_0^*A^*C^*=A^*C^*=CA.$$

Similarly, we can prove $AB_0=AC$. Hence

$$B_0=B_0AB_0=CAB_0=CAC=C.$$

Thus, there is only one matrix satisfying the above conditions. This unique matrix $C$ (or $B_0$) is called the *Moore–Penrose inverse* of $A$ and is denoted by $A^+$.

**Theorem 1.6.1** *Let $A$ be an $m\times n$ matrix. $B$ is a $g$ inverse of $A$ iff*

(i) $\rho(I_n-BA)=n-\rho(A)$ *or*

(ii) $\rho(BA)=\rho(A)$ *and $BA$ is idempotent or*

(iii) $\rho(AB)=\rho(A)$ *and $AB$ is idempotent.*

PROOF. By Corollary 1.5.6(i), we have

$$\rho(A-ABA)=\rho(I_n-BA)+\rho(A)-n.$$

From this, (i) follows immediately. Further, by Corollary 1.5.6(ii),

$$\rho\left[BA-(BA)^2\right]=\rho(I_n-BA)+\rho(BA)-n.$$

This shows that if (ii) is given, then using (i), $B$ is a $g$ inverse of $A$. Suppose $B$ is a $g$ inverse of $A$, i.e.,

$$A=ABA \quad \text{or} \quad BA=BABA,$$

i.e., $BA$ is idempotent. Now

$$\rho(A)=\rho(ABA)\leqslant\rho(BA)\leqslant\rho(A).$$

This proves the converse of (ii). Thus (ii) is proved. Similarly, (iii) can be established. □

**Corollary 1.6.1**

(i) $A^*(AA^*)^-$ *is a g inverse of* $A$, *and* $(AA^*)^-A$ *is a g inverse of* $A^*$,

(ii) $A^*(AA^*)^-AA^*=A^*$ *and* $AA^*(AA^*)^-A=A$, *and*

(iii) $A^*(AA^*)^-A$ *is Hermitian, idempotent of rank* $\rho(A)$, *and unique with respect to any g inverse* $(AA^*)^-$ *of* $AA^*$.

PROOF. (i) follows from Theorem 1.6.1 because $A[A^*(AA^*)^-]=(AA^*)(AA^*)^-$ is idempotent of rank $\rho(A)=\rho(AA^*)$, and $(AA^*)^-AA^*$ is idempotent of rank $\rho(A^*)=\rho(AA^*)$. (ii) follows from (i). In proving (iii), let $P=A^*(AA^*)^-A$. Then $P$ is idempotent of rank $\rho(A)$, because $A^*(AA^*)^-$ is a $g$ inverse of $A$. Let $P_i=A^*[(AA^*)^-]_iA$ for $i=1,2$, where $[(AA^*)^-]_1$ and $[(AA^*)^-]_2$ are two different $g$ inverses. Notice that

$$P_1=A^*\left[(AA^*)^-\right]_1A=A^*\left[(AA^*)^-\right]_1\left(AA^*\left[(AA^*)^-\right]_2A\right)$$
$$=\left(A^*\left[(AA^*)^-\right]_1AA^*\right)\left[(AA^*)^-\right]_2A=A^*\left[(AA^*)^-\right]_2A=P_2.$$

This shows that $P$ is unique with respect to any $g$ inverse of $(AA^*)$. Notice that $P^*=[A^*(AA^*)^-A]^*=A^*[(AA^*)^-]^*A=P$ for $[(AA^*)^-]^*$ is a $g$ inverse of $AA^*$. This proves Corollary 1.6.1. □

**Corollary 1.6.2** $C(ABC)^-A$ *is a g inverse of* $B$ *if* $\rho(ABC)=\rho(B)$.

PROOF. Since $\rho(ABC)=\rho(B)$, we get $\rho(B)=\rho(AB)=\rho(BC)=\rho(ABC)$. Using Corollary 1.5.5 in $0=A(BC)[I-(ABC)^-ABC]$, we have $0\geqslant\rho(ABC)+\rho[BC(I-(ABC)^-ABC)]-\rho(BC)$ and hence $BC[I-(ABC)^-ABC]=[I-BC(ABC)^-A]BC=0$. Using Corollary 1.5.5, we get the result. □

**Corollary 1.6.3** *If* $\rho(ABC)=\rho(A)=\rho(C)$, *then*

(i) $(ABC)^-AB$ *and* $CB(ABC)^-$ *are respectively g inverses of* $C$ *and* $A$, *and*

(ii) $C(ABC)^-A$ *is invariant with respect to any g inverse* $(ABC)^-$ *of* $ABC$, *and it is a Hermitian matrix if* $C=A^*$ *and* $B$ *is Hermitian.*

PROOF. (i) follows from Theorem 1.6.1 because $(ABC)^-ABC$ and $ABC(ABC)^-$ are idempotent of rank $\rho(ABC)=\rho(A)=\rho(C)$. (ii) can be established as in the proof of Corollary 1.6.1. This proves the required result. □

## 1.7 Idempotent Matrices

A square matrix $A$ will be called an *idempotent* matrix iff $A^2=A$. We have seen some of the properties of an idempotent matrix through Corollaries 1.5.6 and 1.5.7, and they are restated below:

(i) If $A$ is idempotent, then $\operatorname{tr} A=\rho(A)$, and

(ii) $A: n\times n$ is idempotent iff $\rho(A)+\rho(I-A)=n$.

**Theorem 1.7.1** *If $A: p\times p$ is Hermitian and idempotent of rank $r$, then there exists a matrix $G: p\times r$ of rank $r$ such that $A=G(G^*G)^{-1}G^*$.*

PROOF. By Theorem 1.5.6, we can find two matrices $G: p\times r$ and $F: r\times p$ of rank $r$ such that $A=GF$. Since $A$ is Hermitian idempotent, we get $GF=F^*G^*$ and $FG=I_r$. Hence, $G=GFG=F^*G^*G$, and $F^*=G(G^*G)^{-1}$, or $F=(G^*G)^{-1}G^*$. □

The following theorem was established by Khatri (1968a, 1971a).

**Theorem 1.7.2** *Let $A_1,\ldots,A_k$ be square matrices such that $A=\Sigma_{i=1}^{k}A_i$. Consider the following statements*:

(a) *$A_i$ $(i=1,2,\ldots,k)$ are idempotent.*

(b) *$A_iA_j=0$ for $i\neq j$ and $\rho(A_i^2)=\rho(A_i)$, $i,j=1,2,\ldots,k$.*

(c) *$A$ is idempotent.*

(d) *$\rho(A)=\Sigma_{i=1}^{k}\rho(A_i)$.*

*Then*

(i) (c) *and* (d) *imply* (a) *and* (b), *and*

(ii) *any two of* (a), (b), (c) *imply all conditions.*

PROOF. (i) Let (c) and (d) be given. Let $Q=\operatorname{diag}(A_1,A_2,\ldots,A_k)$ and $P=(I,I,\ldots,I)$. Then $A=PQP'$, and (d) gives $\rho(PQP')=\rho(Q)$. Notice that on account of (c), $I$ is a $g$ inverse of $PQP'$, and by Corollary 1.6.1(i), $P'P$ is a $g$ inverse of $Q$, i.e., $Q(P'P)Q=Q$, and this gives conditions (a) and (b).

(ii) Let (a) and (b) be given. Then $A^2=A$, and by Corollary 1.5.7 we get (d). Then (a) and (b) $\Rightarrow$ (c) and (d).

Let (a) and (c) be given. Then, by Corollary 1.5.7, (d) is obvious. Then, using (i), (c) and (d) $\Rightarrow$ (a) and then (b).

Let (b) and (c) be given. Then $A_i^2=AA_i=A(AA_i)=A_i^3$ for $i=1,2,\ldots,k$. $0=\rho(A_i^2-A_i^3)=\rho(A_i^2)+\rho(A_i-A_i^2)-\rho(A_i)$ by Corollary 1.5.6(ii), and hence $A_i=A_i^2$ for all $i$. Thus, (b) and (c) $\Rightarrow$ (a) and then (d). □

Note that condition $P(A_i^2)=P(A_i)$ for $i=1,2,\ldots,k$ in (b) can be replaced by $\operatorname{tr} A_i=P(A_i)$, $i=1,2,\ldots,k$, since it can be shown that $A^2=A^3$ and $\operatorname{tr} A=P(A)$ imply $A=A^2$.

**Corollary 1.7.1** *Let $G_i$ be an $m\times p_i$ matrix and $H_i$ be a g inverse of $G_i$ such that $G_iH_i$ is a Hermitian matrix for all $i=1,2,\ldots,k$. Let*

(i) $G_i^*G_j=0$ *for* $i\neq j=1,2,\ldots,k$ *and*

(ii) $\left(\sum_{i=1}^{k} G_iH_i\right)^2=\sum_{i=1}^{k} G_iH_i$.

*Then* (i) *and* (ii) *are equivalent.*

This follows from Theorem 1.7.2, since $(G_iH_i)(G_jH_j)=(H_i^*G_i^*)G_jH_j=H_i^*(G_i^*G_j)H_j=0$ for $i\neq j$.

**Theorem 1.7.3** *Let $A_1,A_2,\ldots,A_k,A$ be Hermitian matrices such that*

(i) $A,A_1,\ldots,A_{k-1}$ *are idempotent,*

(ii) $A=\sum_{i=1}^{k}A_i$, *and*

(iii) $A_k=XX^*$.

*Then $A_iA_j=0$ for $i\neq j$, and $A_k$ is idempotent.*

PROOF. Let $A_0=I-A$. Then $A_0$ is idempotent, and

$$I=A_0+A_1+\cdots+A_k,$$

and

$$A_j=A_j(I)A_j=A_j(A_0+A_1+\cdots+A_k)A_j$$

for $j=0,1,\ldots,k-1$. Let $P_j=(A_jA_0,\ldots,A_jA_{k-1},A_jX)$ except for $A_jA_j$ in the $(j+1)$th partition. Then, since $A$ and the $A_i$'s are Hermitian matrices,

$$P_jP_j^*=\sum_{\substack{i=0\\ i\neq j}}^{k} A_jA_iA_j=0.$$

By Theorem 1.5.2, this gives $P_j=0$ for $j=0,1,\ldots,k-1$. This proves $A_jA_i=0$ for $j\neq i$, $i,j=0,1,\ldots,k$. Now

$$A_k=\left(\sum_{i=0}^{k} A_i\right)A_k \quad \text{gives} \quad A_k=A_k^2. \qquad \square$$

## 1.8 Characteristic Roots and Characteristic Vectors

Let $A$ be an $n\times n$ square matrix, and let $\mathbf{x}$ be an $n\times 1$ nonnull vector such that $A\mathbf{x}=\lambda\mathbf{x}$. Then, $\lambda$ is called a characteristic (ch) root or an eigenvalue of $A$, and $\mathbf{x}$ is called a ch vector or an eigenvector corresponding to the ch root $\lambda$. Since $(A-\lambda I_n)\mathbf{x}=\mathbf{0}$ implies that $\rho(A-\lambda I_n)\leqslant n-1$, we get $|A-\lambda I|=0$. Conversely, $|A-\lambda I|=0$ implies the existence of an $\mathbf{x}$ such that $(A-\lambda I)\mathbf{x}=\mathbf{0}$. This shows that there always exist ch roots and ch vectors of a square matrix $A$. The ch roots are the solutions of a polynomial equation $|A-\lambda I|=0$ in $\lambda$ of degree $n$. The solutions of this polynomial equation may be real, complex or both. If

$$A=\begin{pmatrix} 0 & 1 \\ -1 & 0 \end{pmatrix},$$

then $|A-\lambda I|=\lambda^2+1=0$ has no real roots in $\lambda$. Thus, even though $A$ may be real, we may get complex roots. We may note that if the root $\lambda$ is repeated $r$ times ($r>0$), we may or may not have $r$ *linearly independent* ch vectors, where a set of vectors $\mathbf{x}_1,\ldots,\mathbf{x}_t$ is said to be linearly independent if $\rho(\mathbf{x}_1,\ldots,\mathbf{x}_t)=t$ or there does not exist a nonnull vector $\mathbf{y}'=(y_1,\ldots,y_t)$ such that $\Sigma_{i=1}^t y_i\mathbf{x}_i=\mathbf{0}$. For example, if

$$A=\begin{pmatrix} 1 & -1 \\ 1 & -1 \end{pmatrix},$$

then $|A-\lambda I|=\lambda^2=0$ gives $\lambda=0$, repeated twice, but there is only one nonnull ch vector, $(1,1)'$. The following theorem gives necessary and sufficient conditions for this.

**Theorem 1.8.1**

(i) *The number of nonzero ch roots of a square matrix $A$ is equal to its rank iff $\rho(A)=\rho(A^2)$.*

(ii) *If $\lambda$ is a ch root of $A$ repeated $r$ times, then there exist $r$ linearly independent ch vectors corresponding to $\lambda$ iff $\rho(A-\lambda I)=\rho(A-\lambda I)^2$.*

PROOF. (i) Let $\rho(A)$ be $r$. By Theorem 1.5.3, we can write $A=FG$, where $F:n\times r$ and $G:r\times n$ are matrices of rank $r$. Then $r=\rho(A^2)=\rho(FGFG)=\rho(GF)$, and so $GF$ is nonsingular. By Corollary 1.3.4, $|FG-\lambda I_n|=\lambda^{n-r}|GF-\lambda I_r|$ and $|GF-\lambda I_r|=0$ will give nonzero values of $\lambda$. Hence, $|A-\lambda I_n|=|FG-\lambda I_n|=0$ gives $r$ nonzero ch roots. Conversely, suppose there are $r$ nonzero ch roots. Then $|FG-\lambda I_n|=\lambda^{n-r}|GF-\lambda I_r|=0$ shows that $|GF-\lambda I_r|=0$ must give all nonzero solutions of $\lambda$, and so $|GF|\neq 0$, i.e., $GF$ is nonsingular. Therefore $r=\rho(GF)=\rho(FGFG)=\rho(A^2)$. This proves the first part.

(ii) Let $A_1 = A - \lambda I_n$. Then, from Corollary 1.8.1, $A_1$ has $n-r$ nonzero ch roots and if $\rho(A_1) = \rho(A_1^2)$, then by (i), $\rho(A_1) = n-r$. The number of linearly independent ch vectors corresponding to zero ch roots of $A_1$ is the number of linearly independent solutions of $A_1\mathbf{x} = \mathbf{0}$, and by Corollary 1.5.8, we get $r$ linearly independent solution vectors. The converse result can be obtained similarly. □

**Corollary 1.8.1** *Let $\lambda$ be a ch root of a normal matrix $A$ (i.e., $AA^* = A^*A$), and let its multiplicity be $r$. There are exactly $r$ linearly independent ch vectors corresponding to the root $\lambda$.*

PROOF. Let $\lambda$ be a ch root of $A$ repeated $r$ times. Then $|A - \omega I_n| = (\omega - \lambda)^r f(\omega)$, where $f(\lambda) \neq 0$. Let $B = A - \lambda I_n$. Then $B$ has exactly $n-r$ nonzero ch roots, because $|B - \omega I| = |A - (\lambda + \omega)I| = \omega^r f(\lambda + \omega)$ and $f(\lambda) \neq 0$. Note that $BB^* = (A - \lambda I_n)(A^* - \lambda^* I_n) = AA^* - \lambda A^* - \lambda^* A + \lambda\lambda^* I_n = A^*A - \lambda A^* - \lambda^* A + \lambda\lambda^* I_n = (A^* - \lambda^* I_n)(A - \lambda I_n) = B^*B$, and by Theorem 1.5.2,

$$\rho(B^2) = \rho(B^{*2}B^2) = \rho(B^*BB^*B) = \rho(B^*B) = \rho(B).$$

Hence, by Theorem 1.8.1, there are exactly $r$ linearly independent ch vectors corresponding to $\lambda$. This proves the required corollary. □

**Theorem 1.8.2** *For any normal matrix $A$, there exists a unitary matrix $U$ such that $UAU^*$ is a diagonal matrix, say $D_a$, i.e., $A = U^* D_a U$.*

PROOF. Let $\lambda$ be a ch root of $A$ and $\mathbf{x}$ be its ch vector. Then $B = A - \lambda I$ is a normal matrix, and $B\mathbf{x} = \mathbf{0}$ or $\mathbf{x}^* B^* B\mathbf{x} = 0$ or $\mathbf{x}^* BB^*\mathbf{x} = 0$ or $B^*\mathbf{x} = \mathbf{0}$. Let $\lambda_1$ and $\lambda_2$ be two distinct ch roots of $A$, and $\mathbf{x}_1$ and $\mathbf{x}_2$ be respectively the ch vectors corresponding to $\lambda_1$ and $\lambda_2$. Then

$$\mathbf{x}_1^* A\mathbf{x}_2 = (\mathbf{x}_1^* A)\mathbf{x}_2 = \lambda_1 \mathbf{x}_1^*\mathbf{x}_2 \quad \text{and} \quad \mathbf{x}_1^* A\mathbf{x}_2 = \mathbf{x}_1^*(A\mathbf{x}_2) = \lambda_2 \mathbf{x}_1^*\mathbf{x}_2.$$

Hence, $(\lambda_1 - \lambda_2)\mathbf{x}_1^*\mathbf{x}_2 = 0$ or $\mathbf{x}_1^*\mathbf{x}_2 = 0$. This shows that for distinct roots the vectors are orthogonal, and these vectors can be normalized, i.e., $\mathbf{x}_i^*\mathbf{x}_i = 1$, $(i = 1,2)$.

If $\lambda$ is repeated $r$ times, then by Corollary 1.8.1, there are exactly $r$ linearly independent ch vectors, $Y = (\mathbf{y}_1, \mathbf{y}_2, \ldots, \mathbf{y}_r)$, say. These vectors may not be orthogonal, but they can be orthogonalized by taking $Z = Y\underset{\sim}{T}^{*-1}$—where $Y^*Y = \underset{\sim}{T}\underset{\sim}{T}^*$, and $\underset{\sim}{T}$ is a nonsingular lower triangular matrix (see Problem 1.21)—for $Z^*Z = \underset{\sim}{T}^{-1}Y^*Y\underset{\sim}{T}^{*-1} = I_r$ and $AZ = \lambda Z$. Combining all these vectors, we get the required unitary matrix $U$. □

**Corollary 1.8.2** *For any Hermitian (or skew-Hermitian) matrix $A$, there exists a unitary matrix $U$ such that $UAU^*$ is a diagonal matrix with real (or pure imaginary) diagonal elements.*

PROOF. Notice that any Hermitian (or skew-Hermitian) matrix is a normal matrix. Hence, there exists a unitary matrix $U$ such that $UAU^* = D_\lambda$. Hence $D_\lambda = D_\lambda^*$ if $A$ is Hermitian, while $D_\lambda^* = -D_\lambda$ if $A$ is a skew-Hermitian. This proves Corollary 1.8.2. □

**Corollary 1.8.3** *For any real symmetric matrix $A$, there exists an orthogonal matrix with real elements such that $UAU'$ is a diagonal matrix.*

PROOF. By Corollary 1.8.2, the ch roots of $A$ will be real. Let $\lambda$ be any ch root of $A$. Then the ch vector $\mathbf{x}$ is a solution of $(A - \lambda I_n)\mathbf{x} = \mathbf{0}$, and this solution can be obtained on the real space. Hence, all the ch vectors will be real, and so the unitary matrix $U$ reduces to the orthogonal matrix on the real space. This proves Corollary 1.8.3. □

## 1.9 Hermitian and Positive Definite (or Semidefinite) Matrices

Let $A$ be an $n \times n$ Hermitian matrix and $\mathbf{x}$ be any $n \times 1$ vector. Then $\mathbf{x}^* A\mathbf{x}$ will be called a Hermitian (or quadratic) form of $A$, and it will be a real number. Notice that $\mathbf{x}^* A\mathbf{x}$ is a second-degree polynomial in the elements of $\mathbf{x}$.

$A$ will be said to be positive (semi)definite iff $\mathbf{x}^* A\mathbf{x} > 0$ ($\geqslant 0$) for all nonnull vectors $\mathbf{x}$, and negative (semi)definite iff $\mathbf{x}^* A\mathbf{x} < 0$ ($\leqslant 0$) for all nonnull vectors $\mathbf{x}$. Notice that definite (or semidefinite) matrices will be always Hermitian and we shall not mention this at all.

**Theorem 1.9.1** *If $A$ is positive (semi)definite* (p.d. *or* p.s.d.*), then all the* ch *roots are positive (nonnegative), and all the principal minors of $A$ are positive (nonnegative).*

PROOF. By Corollary 1.8.2, we can write $A = U^* D_\lambda U$, where $U$ is a unitary matrix and $D_\lambda = \text{diag}(\lambda_1, \ldots, \lambda_n)$, the $\lambda_i$'s being the ch roots of $A$. Then

$$\mathbf{x}^* A\mathbf{x} = \mathbf{x}^* U^* D_\lambda U\mathbf{x} = \mathbf{y}^* D_\lambda \mathbf{y} > 0 \quad (\text{or } \geqslant 0)$$

for all nonnull vectors $\mathbf{x}$ and $\mathbf{y} = U\mathbf{x}$. By choosing $\mathbf{y}' = (0, \ldots, 0, 1, 0, \ldots, 0)$ where 1 is at the $i$th place, we find that $\lambda_i > 0$ (or $\geqslant 0$). For the second part, we use Corollary 1.3.1 for the principal minors. Thus, Theorem 1.9.1 is proved. □

**Corollary 1.9.1** *A matrix $A$ is* p.s.d. *iff $A$ can be written as $XX^*$. $X$ can be taken as a nonsingular matrix if $A$ is* p.d.

PROOF. Let $A$ be p.s.d. Then, by Theorem 1.9.1, all the roots of $A$ are nonnegative, and by Corollary 1.8.2, we have $A = U^* D_\lambda U$, where $D_\lambda$ has nonnegative diagonal elements. Let $D_1 = \text{diag}(\sqrt{\lambda_1}, \ldots, \sqrt{\lambda_n})$, and $X = U^* D_1 V$ where $V$ is a matrix such that $VV^* = I_n$. Then $A = XX^*$. If $A$ is p.d., we have $U, D_1$ as nonsingular matrices and we can choose $V$ as a unitary matrix. The converse is obvious. □

**Corollary 1.9.2** *Let $S: p \times p$ be* p.d., *and $B: p \times m$ and $B_1: p \times n$ be two matrices such that $\rho(B_1) = p - \rho(B)$ and $B^* B_1 = 0$. Then $S^{-1} - S^{-1}B(B^*S^{-1}B)^- B^* S^{-1} = B_1(B_1^* S B_1)^- B_1^*$ is a* p.s.d. *matrix of rank $p - \rho(B)$.*

PROOF. By Corollary 1.9.1, we can write $S = XX^*$, where $X$ is nonsingular. Let $Y = X^{-1}B$ and $Y_1 = X^* B_1$. Then, if $Y_2 = (Y, Y_1)$, we have

$$\rho(Y_2) = p \quad \text{and} \quad Y_2^* Y_2 = \begin{pmatrix} Y^* Y & 0 \\ 0 & Y_1^* Y_1 \end{pmatrix}.$$

By Corollary 1.6.1(i) and Theorem 1.6.1, $I_p = Y_2(Y_2^* Y_2)^- Y_2^*$, and taking

$$(Y_2^* Y_2)^- = \begin{pmatrix} (Y^* Y)^- & 0 \\ 0 & (Y_1^* Y_1)^- \end{pmatrix},$$

we get the required result. □

**Theorem 1.9.2** *Let $A$ be a Hermitian matrix and $B$ be a* p.d. *matrix. Then there exists a nonsingular matrix $P$ such that $B = PP^*$ and $A = PD_\lambda P^*$, where the elements of the diagonal matrix $D_\lambda$ are the* ch *roots of $AB^{-1}$. If $A$ is* p.s.d., *then the $\lambda_i$'s are nonnegative.*

PROOF. By Corollary 1.9.1, we can write $B = XX^*$, where $X$ is a nonsingular matrix. Using Corollary 1.8.2, we can write $X^{-1}AX^{*-1} = UD_\lambda U^*$, where $U$ is unitary and the diagonal elements of $D_\lambda$ are the ch roots of $X^{-1}AX^{*-1}$ or $AB^{-1}$ (by Problem 1.13), and they will be nonnegative if $A$ is p.s.d. Taking $P = XU$, we get the required result. □

**Corollary 1.9.3** *Let $A$ be a positive definite matrix with distinct* ch *roots $\lambda_1 > \lambda_2 > \cdots > \lambda_p > 0$. Then there exists a unique unitary matrix $U$ such that $A = UD_\lambda U^*$, where the diagonal elements of $U$ are all positive.*

**Corollary 1.9.4** *Let $A$ be a* p.d. *matrix, and let*

$$A = \begin{pmatrix} A_{11} & A_{12} \\ A_{12}^* & A_{22} \end{pmatrix},$$

*where $A_{11}, A_{22}, A_{12}$ are respectively $p\times p$, $q\times q$, and $p\times q$ matrices and $q\geqslant p$. Then there exist nonsingular matrices $F: p\times p$ and $G: q\times q$ such that $A_{11}=FF^*$, $A_{22}=GG^*$, and $A_{12}=F(D_\lambda,0)G^*$, where the squares of the diagonal elements of $D_\lambda$ are the* ch *roots of $A_{11}^{-1}A_{12}A_{22}^{-1}A_{12}^*$.*

PROOF. By Theorem 1.9.2, we can find a nonsingular matrix $F$ such that $A_{11}=FF^*$ and $A_{12}A_{22}^{-1}A_{12}^*=FD_\omega F^*$. Notice that the diagonal elements of $D_\omega$ are nonnegative. By Corollary 1.9.1, we can write $A_{22}=XX^*$, where $X$ is nonsingular. Let $D_\lambda=\operatorname{diag}(\sqrt{\omega_1},\ldots,\sqrt{\omega_p})$ and $A_{12}=F(D_\lambda,0)VX^*$ (see Theorem 1.9.3 below), where $V$ is a $q\times q$ unitary matrix. Then it is easy to see that if $G=XV^*$, then $A_{22}=GG^*$, $A_{12}=F(D_\lambda,0)G^*$, and $A_{12}A_{22}^{-1}A_{12}^*=FD_\omega F^*$. This proves the required result. □

**Theorem 1.9.3** *Let $A: n\times p$ and $B: n\times q$ be matrices such that $q\geqslant p$ and $AA^*=BB^*$. Then there exists a unitary matrix $U: q\times q$ such that $B=(A,0)U$.*

PROOF. By Corollary 1.8.2 and Theorem 1.9.1, we can write

$$AA^*=BB^*=VD_\lambda V^*,$$

where $V$ is an $n\times n$ unitary matrix and the diagonal elements of $D_\lambda$ are nonnegative. By interchanging columns of $V$, we can always assume that $\lambda_1\geqslant\lambda_2\geqslant\cdots\geqslant\lambda_n\geqslant 0$. Let $D_1^{-1}=\operatorname{diag}(\sqrt{\lambda_1},\ldots,\sqrt{\lambda_r})$, $r=\rho(A)=\rho(B)$, $V=(V_1,V_2)$, $V_1: n\times r$, $F=A^*V_1D_1$, and $G^*=D_1V_1^*B$. Then $FG^*=A^*V_1D_1^2V_1^*B$, $(AA^*)^+=(BB^*)^+=V_1D_1^2V_1^*$, $FF^*=A^*(AA^*)^+A$, $GG^*=B^*(BB^*)^+B$, and $F^*F=G^*G=I_r$. Hence, by Corollary 1.6.1(ii),

$$B=BB^*(BB^*)^+B=AA^*V_1D_1^2B_1^*B=AFG^*.$$

By Corollary 1.5.8, we can find matrices $F_1: p\times(p-r)$ and $G_1: q\times(q-r)$ such that $F_2=(F,F_1)$ and $G_2=(G,G_1)$ are orthogonal matrices. Then $F^*F_1=0 \Leftrightarrow AF_1=0$ and $G^*G_1=0\Leftrightarrow BG_1=0$. If

$$U=\begin{pmatrix} F_2 & 0_{p,q-p} \\ 0_{q-p,p} & I_{p-q}\end{pmatrix}G_2^*,$$

then $(A,0)U=(AF_2,0_{n,q-p})G_2^*=(AF,0_{n,q-r})G_2^*=AFG^*=B$. This proves the required result. □

## 1.10 Some Inequalities

Any ch root of $A: n\times n$ will be denoted by ch($A$). If $A$ is a Hermitian matrix, then the $i$th largest ch root of $A$ will be denoted by $\operatorname{ch}_i(A)$ $(i=1,2,\ldots,n)$, i.e., $\operatorname{ch}_1(A)\geqslant\operatorname{ch}_2(A)\geqslant\cdots\geqslant\operatorname{ch}_n(A)$.

**Theorem 1.10.1** *Let $A$ be any $n\times n$ Hermitian matrix and $\mathbf{x}$ be any $n\times 1$ vector. Then*

$$\mathrm{ch}_n(A)\leqslant \mathbf{x}^*A\mathbf{x}/\mathbf{x}^*\mathbf{x}\leqslant \mathrm{ch}_1(A),$$

*and equality holds on the right side iff $\mathbf{x}$ is the* ch *vector corresponding to* $\mathrm{ch}_1(A)$, *whereas equality holds on the left side iff $\mathbf{x}$ is the* ch *vector corresponding to* $\mathrm{ch}_n(A)$.

PROOF. Since $A$ is Hermitian, by Corollary 1.8.2 we can write $A=UD_\lambda U^*$, where $\lambda_i=\mathrm{ch}_i(A)$ $(i=1,2,\ldots,n)$ and $U$ is unitary. Let $\mathbf{y}=U^*\mathbf{x}/\sqrt{\mathbf{x}^*\mathbf{x}}$ . Then $\mathbf{y}^*\mathbf{y}=1$ and

$$\frac{\mathbf{x}^*A\mathbf{x}}{\mathbf{x}^*\mathbf{x}}=\mathbf{y}^*D_\lambda\mathbf{y}=\sum_{i=1}^{n}\lambda_i|y_i|^2,$$

which is the weighted mean of $\lambda_1,\lambda_2,\ldots,\lambda_n$ with weights $|y_1|^2,|y_2|^2,\ldots,|y_n|^2$ such that $\sum_{i=1}^n|y_i|^2=1$. Hence

$$\lambda_n\leqslant\sum_{i=1}^{n}\lambda_i|y_i|^2\leqslant\lambda_1,$$

and the right-side equality holds iff $|y_1|^2=1$, $y_j=0$ $(j\neq 1)$, whereas the left-side equality holds iff $|y_n|^2=1$ and $y_j=0$ $(j=1,2,\ldots,n-1)$. This proves the required result. □

**Corollary 1.10.1** *Let $A:n\times n$ be Hermitian and $B:n\times n$ be* p.d. *Then*

$$\mathrm{ch}_n(B^{-1}A)\leqslant\frac{\mathbf{x}^*A\mathbf{x}}{\mathbf{x}^*B\mathbf{x}}\leqslant\mathrm{ch}_1(B^{-1}A).$$

*Equality holds on the right side iff $\mathbf{x}$ is the* ch *vector corresponding to* $\mathrm{ch}_1(B^{-1}A)$, *whereas equality holds on the left side iff $\mathbf{x}$ is the* ch *vector corresponding to* $\mathrm{ch}_n(B^{-1}A)$.

The proof follows from Theorem 1.10.1 by noting that

$$\frac{\mathbf{x}^*A\mathbf{x}}{\mathbf{x}^*B\mathbf{x}}=\frac{\mathbf{y}^*A_1\mathbf{y}}{\mathbf{y}^*\mathbf{y}},$$

where $B=PP^*$, $\mathbf{y}=P\mathbf{x}$, and $A_1=P^{*-1}AP^{-1}$.

**Corollary 1.10.2** $[\mathrm{mod}(\mathbf{x}^*\mathbf{y})]^2\leqslant(\mathbf{x}^*B\mathbf{x})(\mathbf{y}^*B^{-1}\mathbf{y})$, *where $B$ is any* p.d. *and* $\mathrm{mod}\,\mathbf{b}=(\mathbf{b}^*\mathbf{b})^{1/2}$. *The equality holds iff $\mathbf{x}=\lambda B^{-1}\mathbf{y}$, $\lambda$ being an arbitrary positive quantity.*

PROOF. Notice that

$$\frac{|\text{mod}(\mathbf{x}^*\mathbf{y})|^2}{(\mathbf{x}^*B\mathbf{x})}=\frac{\mathbf{x}^*(\mathbf{y}\mathbf{y}^*)\mathbf{x}}{\mathbf{x}^*B\mathbf{x}}.$$

There is only one nonzero ch root of $B^{-1}\mathbf{y}\mathbf{y}^*$, namely $\mathbf{y}^*B^{-1}\mathbf{y}$, and its ch vector will be $\lambda B^{-1}\mathbf{y}$, where $\lambda$ is a constant. Then by Corollary 1.10.1, we get the required result. □

**Theorem 1.10.2**

(i) *Let $A:n\times n$ be a Hermitian matrix with roots $\lambda_1\geqslant\lambda_2\geqslant\cdots\geqslant\lambda_n$. Then for any Hermitian idempotent matrix $B$ of rank $r$,*

$$\sum_{i=1}^{r}\lambda_{n-i+1}\leqslant \operatorname{tr}AB\leqslant\sum_{i=1}^{r}\lambda_i.$$

(ii) *Let $A:n\times n$ be a Hermitian matrix with roots $\lambda_1\geqslant\lambda_2\geqslant\cdots\geqslant\lambda_n$, and let $B:n\times n$ be a* p.d. *matrix with roots $\mu_1\geqslant\mu_2\geqslant\cdots\geqslant\mu_n>0$. Then*

$$\sum_{i=1}^{n}\frac{\lambda_i}{\mu_i}\leqslant\operatorname{tr}AB^{-1}\leqslant\sum_{i=1}^{n}\frac{\lambda_i}{\mu_{n-i+1}}.$$

PROOF. (i) By Corollary 1.8.2, there exists a unitary matrix $U$ such that $UAU^*=D_\lambda=\operatorname{diag}(\lambda_1,\lambda_2,\ldots,\lambda_n)$ is a diagonal matrix. Let $UBU^*=C$. Then $C=(c_{ij})$ is a Hermitian idempotent matrix of rank $r$, $\sum_{i=1}^{n}c_{ii}=r$, and $0\leqslant c_{ii}\leqslant 1$ for all $i$. Further,

$$\operatorname{tr}AB=\operatorname{tr}D_\lambda C=\sum_{i=1}^{n}c_{ii}\lambda_i.$$

Since $\sum_{i=1}^{n}c_{ii}=r$, and for $i=1,2,\ldots,n$, $0\leqslant c_{ii}\leqslant 1$, we have

$$\begin{aligned}\sum_{i=1}^{n}c_{ii}\lambda_i&=\sum_{i=1}^{n-1}c_{ii}\lambda_i+\left(r-\sum_{i=1}^{n-1}c_{ii}\right)\lambda_n\\&=\sum_{i=1}^{n-1}c_{ii}(\lambda_i-\lambda_n)+r\lambda_n\leqslant\sum_{i=1}^{n-1}c_{ii}^{(1)}(\lambda_i-\lambda_n)+r\lambda_n\end{aligned}$$

provided $1\geqslant c_{ii}^{(1)}\geqslant 0$ for $i=1,2,\ldots,n-1$ and $\sum_{i=1}^{n-1}c_{ii}^{(1)}=r$. Hence,

$$\sum_{i=1}^{n}c_{ii}\lambda_i\leqslant\sum_{i=1}^{n-1}c_{ii}^{(1)}\lambda_i.$$

This shows that

$$\operatorname*{Sup}_{B}\operatorname{tr}AB\leqslant\sum_{i=1}^{r}\lambda_i.$$

Similarly, it can be shown that

$$\operatorname*{Inf}_{B} \operatorname{tr} AB \geqslant \sum_{i=1}^{r} \lambda_{n-i+1}.$$

This will establish (i).

(ii) As in (i), let $A = U^* D_\lambda U$ and $E = UB^{-1}U^*$. Then

$$\operatorname{tr} AB^{-1} = \sum_{i=1}^{n} \lambda_i e_{ii}.$$

Let the diagonal elements $e_{11}, e_{22}, \dots, e_{nn}$ be ordered as $e_{(11)} \geqslant e_{(22)} \geqslant \cdots \geqslant e_{(nn)}$. Since the effect of changing any two coefficients, say $e_{ii}$ and $e_{jj}$, in $\sum_{i=1}^{n} \lambda_i e_{ii}$ is

$$(\lambda_i e_{ii} + \lambda_j e_{jj}) - (\lambda_i e_{jj} + \lambda_j e_{ii}) = (\lambda_i - \lambda_j)(e_{ii} - e_{jj}),$$

we get

$$\sum_{i=1}^{n} \lambda_i e_{(n-i+1,n-i+1)} \leqslant \operatorname{tr} AB^{-1} = \sum_{i=1}^{n} \lambda_i e_{ii}$$

$$\leqslant \sum_{i=1}^{n} \lambda_i e_{(ii)}.$$

Now, let $U_1 : r \times n$ be a submatrix obtained from $U$ by taking the rows corresponding to the diagonal elements $e_{(11)}, e_{(22)}, \dots, e_{(rr)}$. Hence

$$\sum_{i=1}^{r} e_{(ii)} = \operatorname{tr} U_1 B^{-1} U_1^* = \operatorname{tr} B^{-1} U_1^* U_1$$

and by (i),

$$\sum_{i=1}^{r} e_{(ii)} \leqslant \sum_{i=1}^{r} \frac{1}{\mu_{n-i+1}} \qquad \text{for} \quad r = 1, 2, \dots, n.$$

Similarly, it can be shown that

$$\sum_{i=1}^{r} e_{(n-i+1,n-i+1)} \geqslant \sum_{i=1}^{r} \frac{1}{\mu_i} \qquad \text{for} \quad r = 1, 2, \dots, n.$$

It is easy to verify that

$$\sum_{i=1}^{n} \lambda_i e_{(ii)} = \sum_{i=1}^{n-1} (\lambda_i - \lambda_{i+1}) \sum_{j=1}^{i} e_{(jj)},$$

and so

$$\sum_{i=1}^{n} \lambda_i e_{(ii)} \leqslant \sum_{i=1}^{n-1} (\lambda_i - \lambda_{i+1}) \sum_{j=1}^{i} \frac{1}{\mu_{n-i+1}} = \sum_{i=1}^{n} \frac{\lambda_i}{\mu_{n-i+1}}.$$

Similarly, it can be shown that

$$\sum_{i=1}^{n} \lambda_i e_{(n-i+1,n-i+1)} = \sum_{i=1}^{n-1} (\lambda_i - \lambda_{i+1}) \sum_{j=1}^{i} e_{(n-j+1,n-j+1)}$$

$$\geqslant \sum_{i=1}^{n-1} (\lambda_i - \lambda_{i+1}) \sum_{j=1}^{i} \frac{1}{\mu_j} = \sum_{i=1}^{n} \frac{\lambda_i}{\mu_i}.$$

Combining all the above results, we get (ii). Thus, Theorem 1.10.2 is established. □

**Theorem 1.10.3** *Let $X: p\times n$, $B: p\times q$, and $A: m\times n$ be matrices such that $(X-B\xi A)(X-B\xi A)^*$ is* p.d. *for every matrix $\xi: q\times m$. Then*

$$|(X-B\xi A)(X-B\xi A)^*| \geqslant |T| \qquad \textit{for all } \xi,$$

*where $T = S + [I - B(B^*S^{-1}B)^- B^*S^{-1}]S_1[I - B(B^*S^{-1}B)^- B^*S^{-1}]^*$, $S_1 = XA^*(AA^*)^- AX^*$, and $S = XX^* - S_1$. The equality holds iff $B\xi A = B(B^*S^{-1}B)^- B^*S^{-1}XA^*(AA^*)^- A$.*

PROOF. By Corollary 1.6.1(ii) and (iii), we notice that

$$\begin{aligned} P &\equiv (X-B\xi A)(X-B\xi A)^* \\ &= (X-B\xi A)(I-A^*(AA^*)^- A)(X-B\xi A)^* \\ &\quad + (X-B\xi A)A^*(AA^*)^- A(X-B\xi A)^* \\ &= S + (Y-B\xi)AA^*(Y-B\xi)^*, \end{aligned}$$

where $Y = XA^*(AA^*)^-$, $S_1 = YAA^*Y^*$, and $S = XX^* - S_1$. Since $P$ is nonsingular for every $\xi$, $S$ will be nonsingular also. Hence, we can rewrite $|P|$ as

$$\begin{aligned} |P| &= |S|\cdot|I_p + S^{-1}(Y-B\xi)AA^*(Y-B\xi)^*| \\ &= |S|\cdot|I_m + AA^*(Y-B\xi)^*S^{-1}(Y-B\xi)|. \end{aligned}$$

Taking

$$S^{-1} = \left[S^{-1} - S^{-1}B(B^*S^{-1}B)^- B^*S^{-1}\right] + \left[S^{-1}B(B^*S^{-1}B)^- B^*S^{-1}\right]$$

in $(Y-B\xi)^*S^{-1}(Y-B\xi)$, we get

$$(Y-B\xi)^*S^{-1}(Y-B\xi) = H + \eta^*(B^*S^{-1}B)\eta$$

where

$$H = Y^*\left[S^{-1} - S^{-1}B(B^*S^{-1}B)^- B^*S^{-1}\right]Y$$

and

$$\eta = (B^*S^{-1}B)^- B^*S^{-1}Y - \xi.$$

Then, we can write

$$|P|=|S|\cdot\left|(I_n+A^*HA)+\left[A^*\eta^*(B^*S^{-1}B)\eta A\right]\right|.$$

Notice that by Corollary 1.9.2, $H$ is p.d.s., and hence $I_n+A^*HA$ is p.d. Let the ch roots of $(I_n+A^*HA)^{-1}A^*\eta^*(B^*S^{-1}B)\eta A$ be $\lambda_i$ $(i=1,2,\ldots,n)$, which gives $\lambda_i\geqslant 0$ for all $i$ and

$$|P|=|S|\cdot|I_n+A^*HA|\prod_{i=1}^{n}(1+\lambda_i)=|T|\prod_{i=1}^{p}(1+\lambda_i).$$

The minimum value of $\prod_{i=1}^{p}(1+\lambda_i)$ under the variation of $\eta$ (i.e., $\xi$) is 1, and it can be attained iff $\lambda_1=\cdots=\lambda_n=0$, i.e., $A^*\eta^*(B^*S^{-1}B)\eta A=0$, i.e., $B\eta A=0$, i.e., $B\xi A=B(B^*S^{-1}B)^{-}B^*S^{-1}YA$. This proves the required Theorem 1.10.3, as well as the following. □

**Corollary 1.10.3** *If $S$ is positive definite,*

$$|S+(Y-B\xi)(Y-B\xi)^*|\geqslant|T_1|\qquad \textit{for all } \xi,$$

*where*

$$T_1=S+\left[I-B(B^*S^{-1}B)^{-}B^*S^{-1}\right]YY^*\left[I-B(B^*S^{-1}B)^{-}B^*S^{-1}\right]^*.$$

*Equality holds* iff $B\xi=B(B^*S^{-1}B)^{-}B^*S^{-1}Y.$

**Theorem 1.10.4** *Let $S:p\times p$ be a* p.d. *matrix. Then for $a>0$, $b>0$,*

$$|\Sigma|^{-b}\exp(-a\operatorname{tr}\Sigma^{-1}S)\leqslant|aS/b|^{-b}\exp(-pb)$$

*for all $p\times p$* p.d. *matrices $\Sigma$. Equality holds iff $\Sigma=aS/b$.*

PROOF. By Theorem 1.9.2, we can write $S=PP^*$ and $\Sigma=PD_\lambda P^*$, where $P$ is a nonsingular matrix and $D_\lambda$ is a diagonal matrix with $\lambda_i=\operatorname{ch}_i(\Sigma S^{-1})>0$ for all $i=1,2,\ldots,p$. Then, as $\Sigma$ varies, the values of $\lambda_i$ $(i=1,2,\ldots,p)$ will vary. We have

$$|\Sigma|^{-b}\exp(-a\operatorname{tr}\Sigma^{-1}S)=|S|^{-b}\prod_{i=1}^{p}\left[\lambda_i^{-b}\exp\left(-\frac{a}{\lambda_i}\right)\right].$$

First of all, we note that if $f(\lambda)=-b\log\lambda-a/\lambda$, then

$$f'(\lambda)=\frac{a-b\lambda}{\lambda^2}\gtreqqless 0\quad\text{according as}\quad\lambda\lesseqqgtr\frac{a}{b}.$$

This shows that

$$f(\lambda)\leqslant-b\log(a/b)-b$$

or

$$\lambda^{-b}\exp(-a/\lambda)\leqslant(a/b)^{-b}\exp(-b),$$

and equality holds iff $\lambda = a/b$. This shows that

$$|\Sigma|^{-b}\exp(-a\operatorname{tr}\Sigma^{-1}S) \leqslant |S|^{-b}(a/b)^{-pb}\exp(-bp)$$

and equality holds iff $\lambda_1 = \lambda_2 = \cdots = \lambda_p = a/b$, i.e., $\Sigma = aS/b$. □

**Theorem 1.10.5** *Let $F(\mathbf{x})$ be the distribution function of $\mathbf{x}$, and let $f(\mathbf{x})$ and $g(\mathbf{x})$ be two functions of $\mathbf{x}$ such that for any two points $\mathbf{x}_1$ and $\mathbf{x}_2$, either $\{f(\mathbf{x}_1) \geqslant f(\mathbf{x}_2), g(\mathbf{x}_1) \geqslant g(\mathbf{x}_2)\}$ or $\{f(\mathbf{x}_1) \leqslant f(\mathbf{x}_2), g(\mathbf{x}_1) \leqslant g(\mathbf{x}_2)\}$. Then*

$$\int f(\mathbf{x})g(\mathbf{x})\,dF(\mathbf{x}) \geqslant \left(\int f(\mathbf{x})\,dF(\mathbf{x})\right)\left(\int g(\mathbf{x})\,dF(\mathbf{x})\right).$$

PROOF. Let $\mathbf{x}$ and $\mathbf{y}$ be any two points. Then, under the given conditions,

$$[f(\mathbf{x}) - f(\mathbf{y})][g(\mathbf{x}) - g(\mathbf{y})] \geqslant 0 \qquad \text{for all } \mathbf{x} \text{ and } \mathbf{y}.$$

Hence

$$\iint [f(\mathbf{x}) - f(\mathbf{y})][g(\mathbf{x}) - g(\mathbf{y})]\,dF(\mathbf{x})\,dF(\mathbf{y}) \geqslant 0,$$

and this gives the required result, using $\int dF(\mathbf{x}) = 1$. □

## 1.11 Matrix Differentials and the Jacobian of Transformations

In this section, we shall consider elements and matrices on the real space. Let $f(\mathbf{x})$ be a function of a vector variable $\mathbf{x}' = (x_1, \ldots, x_n)$. Then, as usual, the partial derivative of $f(\mathbf{x})$ with respect to $x_i$ is obtained by treating $x_1, \ldots, x_{i-1}, x_{i+1}, \ldots, x_n$ as constants and taking the derivative with respect to $x_i$. This will be denoted by $\partial f(\mathbf{x})/\partial x_i$ $(i = 1, 2, \ldots, n)$. We shall denote

$$\text{(1.11.1)} \quad \frac{\partial f(\mathbf{x})}{\partial \mathbf{x}'} = \left(\frac{\partial f(\mathbf{x})}{\partial x_1}, \frac{\partial f(\mathbf{x})}{\partial x_2}, \ldots, \frac{\partial f(\mathbf{x})}{\partial x_n}\right) \quad \text{and} \quad \frac{\partial f(\mathbf{x})}{\partial \mathbf{x}} = \left(\frac{\partial f(\mathbf{x})}{\partial \mathbf{x}'}\right)'.$$

The differential of a function $f(\mathbf{x})$ is defined by

$$\text{(1.11.2)} \qquad df = \frac{\partial f(\mathbf{x})}{\partial \mathbf{x}'}\,d\mathbf{x}, \qquad \text{where} \quad d\mathbf{x} = (dx_1, dx_2, \ldots, dx_n)'.$$

$dx_i$ is called the differential of $x_i$ and $d\mathbf{x}$, the differential of $\mathbf{x}$. The differential of a matrix $F(\mathbf{x}) = (f_{ij}(\mathbf{x}))$ is defined by

$$\text{(1.11.3)} \qquad dF = dF(\mathbf{x}) = (df_{ij}(\mathbf{x})) = \left(\frac{\partial f_{ij}(\mathbf{x})}{\partial \mathbf{x}'}\,d\mathbf{x}\right) = \sum_{k=1}^{n} \frac{\partial F}{\partial x_k}\,dx_k,$$

where

$$\frac{\partial F}{\partial x_k} = \left( \frac{\partial f_{ij}(\mathbf{x})}{\partial x_k} \right).$$

This shows that if $F = GH$, which depends on $\mathbf{x}$, then with the help of

$$\frac{\partial f_{ij}}{\partial x_k} = \sum_t \frac{\partial g_{it}}{\partial x_k} h_{tj} + \sum_t g_{it} \frac{\partial h_{tj}}{\partial x_k}$$

or

$$\frac{\partial F}{\partial x_k} = \frac{\partial G}{\partial x_k} H + G \frac{\partial H}{\partial x_k},$$

we get

(1.11.4) $$dF = (dG)H + G(dH).$$

REMARKS. The above result is true even where $F$ is a function of a random matrix, because the elements of a matrix can be written as a vector containing all these elements. Hence, a separate treatment is not mentioned for matrix functions.

Equation (1.11.3) shows that we can obtain $\partial F / \partial x_k$ by putting $dx_k = 1$ and other $dx_j = 0$ in $dF$. This is useful for obtaining partial derivatives of complicated functions. For example, take $F = X^{-1}$, where $X = (x_{ij})$ is a symmetric matrix and the random elements are $x_{ij}$, $i \geqslant j = 1, \ldots, n$. To obtain $\partial F / \partial x_{ij}$, we first see that

$$dF = -F(dX)F, \qquad \text{since} \quad 0 = d(I) = d(XF) = (dX)F + X(dF).$$

Then, putting $dx_{ij} = 1$ and all other $dx_{\alpha\beta} = 0$, we get

(1.11.5) $$\frac{\partial F}{\partial x_{ij}} = -(\mathbf{f}_i \mathbf{f}_j' + \mathbf{f}_j \mathbf{f}_i'), \qquad i > j,$$

and

(1.11.6) $$\frac{\partial F}{\partial x_{ii}} = -\mathbf{f}_i \mathbf{f}_i',$$

where $F = X^{-1} \equiv (\mathbf{f}_1, \mathbf{f}_2, \ldots, \mathbf{f}_n)$. Thus, if $g(X) = \mathbf{a}' X^{-1} \mathbf{a}$,

(1.11.7) $$\frac{\partial g(X)}{\partial x_{ij}} = \begin{cases} -(\mathbf{a}'\mathbf{f}_i)^2 & \text{for} \quad i = j, \\ -2(\mathbf{a}'\mathbf{f}_i)(\mathbf{a}'\mathbf{f}_j) & \text{for} \quad i > j. \end{cases}$$

Hence, we get the following:

**Theorem 1.11.1**

(i) *If* $F = GH$, *then* $dF = (dG)H + G(dH)$.

(ii) *If* $F = X^{-1} = (\mathbf{f}_1, \ldots, \mathbf{f}_n)$, *where* $X$ *is a symmetric matrix, then*

$$dF = -F(dX)F$$

*and*

$$\frac{\partial F}{\partial x_{ij}} = \begin{cases} -(\mathbf{f}_i\mathbf{f}_j' + \mathbf{f}_j\mathbf{f}_i') & \text{for } i \neq j, \\ -\mathbf{f}_i\mathbf{f}_i' & \text{for } i = j. \end{cases}$$

*In particular,*

$$\frac{\partial(\mathbf{a}'X^{-1}\mathbf{a})}{\partial x_{ij}} = \begin{cases} -(\mathbf{a}'\mathbf{f}_i)^2 & \text{for } i = j \\ -2(\mathbf{a}'\mathbf{f}_i)(\mathbf{a}'\mathbf{f}_j) & \text{for } i \neq j. \end{cases}$$

**Corollary 1.11.1** *Let* $X$ *be any* $p \times n$ *random matrix. Then* $d(AX) = A(dX)$ *and* $d(X'BX) = (dX)'BX + X'B(dX)$ *where* $B$ *is a* $p \times p$ *symmetric matrix and* $A$ *is a* $q \times p$ *matrix. In particular, if* $n = 1$ *(i.e.,* $X = \mathbf{x}$*), then*

$$\frac{\partial(A\mathbf{x})}{\partial \mathbf{x}} = A \quad \text{and} \quad \frac{\partial(\mathbf{x}'B\mathbf{x})}{\partial \mathbf{x}} = 2B\mathbf{x}.$$

In some cases, the above concept will be useful in deriving some results on the Jacobians of the transformations. Let $n$ functionally independent variables $x_1, x_2, \ldots, x_n$ be transformed to new independent variables $y_1, y_2, \ldots, y_n$ through relations $y_i = f_i(x_1, \ldots, x_n)$, $i = 1, 2, \ldots, n$. These relations will be written as $\mathbf{y} = \mathbf{f}(\mathbf{x})$. Let us define a matrix of partial derivatives of $\mathbf{x}$ with respect to $\mathbf{y}$ as

$$\left(\frac{\partial x_i}{\partial y_j}\right) = \frac{\partial(x_1, \ldots, x_n)}{\partial(y_1, \ldots, y_n)} = \frac{\partial(\mathbf{x})}{\partial(\mathbf{y})}.$$

The absolute value of the determinant of $\partial(\mathbf{x})/\partial(\mathbf{y})$, denoted by

$$\left|\frac{\partial(\mathbf{x})}{\partial(\mathbf{y})}\right|_+ \quad \text{or} \quad \left|\frac{\partial(x_1, \ldots, x_n)}{\partial(y_1, \ldots, y_n)}\right|_+,$$

will be called the Jacobian of the transformation of $\mathbf{x}$ to $\mathbf{y}$. This Jacobian will be denoted by

$$J(\mathbf{x} \to \mathbf{y}) = \left|\frac{\partial(\mathbf{x})}{\partial(\mathbf{y})}\right|_+ = \left|\left(\frac{\partial x_i}{\partial y_j}\right)\right|_+. \tag{1.11.8}$$

From the definition of differentials, we have

(1.11.9) $$d\mathbf{y}=\frac{\partial\mathbf{f}(\mathbf{x})}{\partial\mathbf{x}'}d\mathbf{x},\qquad \frac{\partial\mathbf{f}(\mathbf{x})}{\partial\mathbf{x}'}=\frac{\partial(f_1,f_2,\ldots,f_n)}{\partial(x_1,x_2,\ldots,x_n)}.$$

Notice that the relation in the differentials is linear, and this shows that

(1.11.10) $$J(d\mathbf{y}\to d\mathbf{x})=J(\mathbf{y}\to\mathbf{x})=[J(\mathbf{x}\to\mathbf{y})]^{-1}.$$

Now, we shall establish some results on the Jacobian of transformations:

**Theorem 1.11.2** *Define the conditional transformations*

$$\mathbf{y}_1=\mathbf{f}_1(\mathbf{x}_1,\mathbf{x}_2,\ldots,\mathbf{x}_n),\qquad \mathbf{y}_2=\mathbf{f}_2(\mathbf{y}_1,\mathbf{x}_2,\ldots,\mathbf{x}_n),\ldots,$$

$$\mathbf{y}_i=\mathbf{f}_i(\mathbf{y}_1,\mathbf{y}_2,\ldots,\mathbf{y}_{i-1},\mathbf{x}_i,\ldots,\mathbf{x}_n)$$

*for* $i=1,2,\ldots,n$, *where* $\mathbf{y}_i$ *and* $\mathbf{x}_i$ *are* $q_i\times 1$ *vectors* $(i=1,2,\ldots,n)$. *Then*

$$J(\mathbf{y}_1,\ldots,\mathbf{y}_n\to\mathbf{x}_1,\ldots,\mathbf{x}_n)=\prod_{i=1}^{n}J(\mathbf{y}_i\to\mathbf{x}_i).$$

PROOF. Let $T_{ij}=\partial(\mathbf{f}_i)/\partial\mathbf{x}_j'$ for $i\leqslant j=1,2,\ldots,n$, and $T_{ij}=0$ for $i>j=1,2,\ldots,n$. Let $-G_{ij}=\partial\mathbf{f}_i/\partial\mathbf{y}_j'$ for $j<i$, $G_{ii}=I_{q_i}$, and $G_{ij}=0$ for $i>j$. Taking differentials, we have

$$d\mathbf{y}_i=-\sum_{j=1}^{i-1}G_{ij}d\mathbf{y}_j+\sum_{j=i}^{n}T_{ij}d\mathbf{x}_j\qquad\text{for}\quad i=1,2,\ldots,n.$$

Let us denote $G=(G_{ij})$, $T=(T_{ij})$, $(dy)'=((d\mathbf{y}_1)',\ldots,(d\mathbf{y}_n)')$, and $(dx)'=((d\mathbf{x}_1)',(d\mathbf{x}_2)',\ldots,(d\mathbf{x}_n)')$. Then the above set of equations can be written as

$$G(dy)=T(dx),\qquad\text{or}\quad (dy)=(G^{-1}T)(dx).$$

Hence

$$J(\mathbf{y}\to\mathbf{x})=J(d\mathbf{y}\to d\mathbf{x})=|G^{-1}T|_+=|G|^{-1}|T|_+=|T|_+$$

$$=\prod_{i=1}^{n}|T_{ii}|_+=\prod_{i=1}^{n}J(\mathbf{y}_i\to\mathbf{x}_i),$$

which proves the required result. □

For illustrative examples, see Problem 1.32 and the first part of Problem 1.29.

**Theorem 1.11.3** *Let* $\mathbf{y}=\mathbf{f}(\mathbf{x})$ *and* $\mathbf{x}=\mathbf{g}(\mathbf{z})$. *Then*

$$J(\mathbf{y}\to\mathbf{z})=J(\mathbf{y}\to\mathbf{x})J(\mathbf{x}\to\mathbf{z}).$$

PROOF. Follows by taking differentials,

$$d\mathbf{y}=\frac{\partial\mathbf{f}}{\partial\mathbf{x}'}d\mathbf{x}\quad\text{and}\quad d\mathbf{x}=\frac{\partial\mathbf{g}}{\partial\mathbf{z}'}(d\mathbf{z}).\qquad\square$$

**Corollary 1.11.2** *Let $Y$ and $X$ be $p\times n$ matrices with all entries functionally independent such that $Y=AXB$, where $A:p\times p$ and $B:n\times n$ are nonsingular matrices. Then $J(Y\to X)=|A|_+^n|B|_+^p$.*

The proof follows by taking $Y=AZ$ and $Z=XB$.

**Theorem 1.11.4** *Let $V$ and $W$ be two random symmetric (skew-symmetric) matrices of order $n$ such that $V=AWA'$, where $A:n\times n$ is nonsingular. Then*

$$J(V\to W)=|AA'|^{\frac{1}{2}(n+1)}\quad\left(=|AA'|^{\frac{1}{2}(n-1)}\right).$$

PROOF. Let $A$ be a lower triangular matrix $\underset{\sim}{T}$. Then we can write the transformations as

$$\begin{pmatrix} V_{11} & \mathbf{b}\\ \mathbf{b}' & v\end{pmatrix}=\begin{pmatrix} \underset{\sim}{T}_{11} & \mathbf{0}\\ \mathbf{t}' & t_1\end{pmatrix}\begin{pmatrix} W_{11} & \mathbf{c}\\ \mathbf{c}' & w\end{pmatrix}\begin{pmatrix} \underset{\sim}{T}'_{11} & \mathbf{t}\\ \mathbf{0} & t_1\end{pmatrix},$$

which gives

$$V_{11}=\underset{\sim}{T}_{11}W_{11}\underset{\sim}{T}'_{11},\qquad \mathbf{b}=\underset{\sim}{T}_{11}W_{11}\mathbf{t}+\underset{\sim}{T}_{11}\mathbf{c}t_1,$$

and

$$v=t_1^2w+2t_1\mathbf{c}'\mathbf{t}+\mathbf{t}'W_{11}\mathbf{t}.$$

Hence

$$J(V\to W)=J(V_{11}\to W_{11})J(\mathbf{b}\to\mathbf{c})J(v\to w)$$

$$=|\underset{\sim}{T}_{11}|_+(t_1)_+^{n+1}J(V_{11}\to W_{11}),$$

since $J(\mathbf{b}\to\mathbf{c})=|\underset{\sim}{T}_{11}|_+(t_1)_+^{n-1}$ and $J(v\to w)=t_1^2$. Hence by induction

$$J(V\to W)=|\underset{\sim}{T}|_+^{n+1}.$$

Similarly, if $A$ is an upper triangular matrix $\tilde{U}$, then $J(V\to W)=|\tilde{U}|_+^{n+1}$.

Since $A$ is nonsingular, rows and columns can be interchanged without affecting the Jacobian of the transformation, so that the leading principal minors of order $i$ $(i=1,2,\ldots,n)$ are nonzero. Then from Problem 1.20, we can write $A=\underset{\sim}{T}\tilde{U}$, where $\underset{\sim}{T}$ is lower triangular and $\tilde{U}$ is upper triangular. Hence,

$$V=\underset{\sim}{T}\tilde{U}W\tilde{U}'\underset{\sim}{T}'=\underset{\sim}{T}W_1\underset{\sim}{T},$$

where $W_1=\tilde{U}W\tilde{U}'$. Thus $J(V\to W)=J(V\to W_1)J(W_1\to W)=|\underset{\sim}{T}|_+^{n+1}|\tilde{U}|_+^{n+1}$, so that

$$\begin{aligned}J(V\to W)&=J(V\to W_1)J(W_1\to W)\\&=|\underset{\sim}{T}|_+^{n+1}|\tilde{U}|_+^{n+1}=|\underset{\sim}{T}\tilde{U}|_+^{n+1}=|A|_+^{n+1}\\&=|AA'|^{\frac{1}{2}(n+1)}.\end{aligned}$$

Similarly, the result for a skew-symmetric matrix can be established. This result was given by Hsu (see Deemer and Olkin 1951) and later by Jack (1965) and Olkin and Sampson (1972). The approach here is somewhat different. □

**Theorem 1.11.5** *Let $S$ be an $n\times n$ random symmetric matrix such that all the* ch *roots are distinct and nonzero. Let the transformation be $S=HD_\lambda H'$, where $D_\lambda=\operatorname{diag}(\lambda_1,\ldots,\lambda_n)$, $\lambda_1>\lambda_2>\cdots>\lambda_n$ $(\lambda_i\neq 0)$, and $H$ is an orthogonal matrix (whose elements in the first row are positive or whose all diagonal elements are positive). Then*

$$J(S\to H,D_\lambda)=\left(\prod_{i=1}^{n-1}\prod_{j=i+1}^{n}(\lambda_i-\lambda_j)\right)f_n(H),$$

$$f_n(H)=J(H'(dH)\to dH),$$

*and if the random elements in $H$ are $\{h_{12},\ldots,h_{1n},h_{23},\ldots,h_{2n},\ldots,h_{n-1,n}\}$, then*

$$f_n(H)=\frac{1}{\prod_{i=1}^{n}|H_{(i)}|_+},$$

*where $H_{(i)}=(h_{jk};\,j,k=1,2,\ldots,i)$.*

PROOF. Let $R=H'(dH)$. Since $I=H'H$ and $0=(dH')H+H'(dH)$, we get $R+R'=0$. Hence $R$ is a skew-symmetric matrix. Let

$$J(R\to dH)=f_n(H).$$

To obtain $f_n(H)$ explicitly, let the random set in $H$ be

$$\{h_{12},\ldots,h_{1n},h_{23},\ldots,h_{2n},\ldots,h_{n-1,n}\},$$

and $dH=HR$; i.e., taking $r_{ij}=-r_{ji}$ for $i<j$,

$$dh_{ij}=\sum_{k=1}^{j-1} h_{ik}r_{kj}-\sum_{k=j}^{n} h_{ik}r_{jk}, \qquad i<j=1,2,\ldots,n.$$

Notice that the transformations are conditional and hence

$$\text{(1.11.11)} \qquad J(dH\to R)=\prod_{j=2}^{n} J(dh_{1j},\ldots,dh_{j-1,j}\to r_{1j},r_{2j},\ldots,r_{j-1,j})$$
$$=\prod_{j=2}^{n} |H_{(j-1)}|_{+}=\prod_{i=1}^{n} |H_{(i)}|_{+}.$$

Now, taking the differentials of $S=HD_\lambda H'$, we have

$$dS=(dH)D_\lambda H'+H(dD_\lambda)H'+HD_\lambda(dH)'.$$

Hence,

$$H'(dS)H=RD_\lambda-D_\lambda R+dD_\lambda\equiv W \quad \text{(say)}.$$

Then $w_{ii}=d\lambda_i$ and $w_{ij}=(\lambda_i-\lambda_j)r_{ij}$ for $i<j=1,2,\ldots,n$. Hence,

$$J(S\to H,D_\lambda)=J(dS\to dH,dD_\lambda)=J(dS\to W)J(W\to dH,dD_\lambda)$$
$$=J(W\to R,dD_\lambda)J(R\to dH)$$
$$=\frac{\prod_{i<j}(\lambda_i-\lambda_j)}{J(dH\to R)}.$$

Using Eq. (1.11.11), we get the required result. □

## Problems

**1.1** Every square matrix $A$ can be expressed as the sum of a symmetric matrix $B$ and a skew-symmetric matrix $S$.

**1.2** Show the following

(i) $\operatorname{tr}_t I_n=\binom{n}{t}$.

(ii) $\operatorname{tr}(A+B)=\operatorname{tr}A+\operatorname{tr}B$.

(iii) $\operatorname{tr}_t \Gamma A\Gamma^*=\operatorname{tr}_t A$ if $\Gamma^*\Gamma=I$.

(iv) $\operatorname{tr}AB=\operatorname{tr}BA$.

**1.3** Establish the following:

(i) If $A$ is a skew-symmetric matrix with real elements, then

$$|A| \geqslant 0, \quad \text{and} \quad |I+A| > 0.$$

(ii) If $A$ and $B$ are $p \times p$ and $q \times q$ matrices, then

$$|A \otimes B| = |A|^q |B|^p.$$

(iii) $\begin{vmatrix} A & B \\ C & D \end{vmatrix} = |A| \cdot |D - CA^{-1}B|$ if $|A| \neq 0$.

**1.4** Find the determinant of $A = (a_{ij})$ where $a_{ij} = a_j^{i-1}$ for $i, j = 1, 2, \ldots, n$.

**1.5** Obtain the inverses of the following:

(i) $V = D_p - \mathbf{pp}'$, where $\mathbf{p}' = (p_1, \ldots, p_n)$, $\Sigma_{i=1}^n p_i < 1$, $p_i > 0$.

(ii) $V = (v_{ij})$, $v_{ii} = x$, and $v_{ij} = y$ for all $i \neq j$.

**1.6** Show that for any matrix $V$, $V^0 = I$, and for any finite positive integer $n$,

(i) $(I - tV)^{-1} = \sum_{i=0}^{n} t^i V^i + t^{n+1} V^{n+1} (I - tV)^{-1}$,

(ii) $(I - tV)^{-2} = \sum_{i=0}^{n} (i+1) t^i V^i - t^{n+1}[(n+2) V^{n+1} - (n+1) t V^{n+2}]$
$(I - tV)^{-2}$, and

(iii) $(I - tV)^{-k} = I + \{[1 - (1-t)^k]/(1-t)^k\} V$ if $V^2 = V$ and $I - tV$ is nonsingular.

**1.7** Establish

(i) $\rho(AB) = \rho(ABB^*) = \rho(A^*AB) = \rho(ABB^*A^*)$,

(ii) $\rho(AB) = \rho(A)$ if $\rho(B : p \times q) = p$, and

(iii) $|\rho(A) - \rho(B)| \leqslant \rho(A+B) \leqslant \rho(A) + \rho(B)$.

**1.8** Show that

(i) $\rho(A \otimes B) = \rho(A)\rho(B)$, and

(ii) $(A^- \otimes B^-)$ is a $g$ inverse of $(A \otimes B)$.

**1.9** A set of linear equations $AXB = C$ can be written as $(A \otimes B')\mathbf{x} = \mathbf{c}$, where $\mathbf{x}$ and $\mathbf{c}$ are vectors obtained by putting rows of $X$ and $C$ in columns. That is, if $X' = (\mathbf{x}_1, \mathbf{x}_2, \ldots, \mathbf{x}_p)$ and $C' = (\mathbf{c}_1, \mathbf{c}_2, \ldots, \mathbf{c}_q)$, then $\mathbf{x}' = (\mathbf{x}_1', \mathbf{x}_2', \ldots, \mathbf{x}_p')$ and $\mathbf{c}' = (\mathbf{c}_1', \mathbf{c}_2', \ldots, \mathbf{c}_q')$. Show that these equations will have a solution iff $\rho(A)\rho(B) = \rho(A \otimes B', \mathbf{c})$. The general solution for $X$ under this condition is $X = A^-ACBB^- + Z - A^-AZBB^-$ where $Z$ is an arbitrary matrix. [Hint: If there is a vector $\mathbf{d}$ such that $\mathbf{d}'(A \otimes B') = 0$ and $\mathbf{d}'\mathbf{c} \neq 0$, then there is no solution. Use this for solvability of the equations. The general solution will be the sum of a particular solution of $(A \otimes B')\mathbf{x} = \mathbf{c}$ and a general solution of $(A \otimes B')\mathbf{x} = \mathbf{0}$. A particular solution can be taken as $[A^-A \otimes (B^-)'B']\mathbf{c}$, and a general solution can be expressed as $[I - A^-A \otimes (B^-)'B']\mathbf{z}$, where $\mathbf{z}$ is an arbitrary vector.]

**1.10** If $\Gamma$ is an orthogonal matrix such that $|I+\Gamma|\neq 0$, then there is a one-to-one relation between $\Gamma$ and a skew-symmetric matrix $A$, namely,

$$\Gamma=2(I+A)^{-1}-I \quad \text{and} \quad A=2(I+\Gamma)^{-1}-I.$$

**1.11** If $V=R_1A$ and $U=AR_2$ for some matrices $R_1$ and $R_2$, then

(i) $A^- - A^-U(I+VA^-U)^-VA^-$ is a $g$ inverse of $A+UV$, and

(ii) $\begin{pmatrix} A^- & \vdots & 0 \\ \cdots & & \cdots \\ 0 & \vdots & 0 \end{pmatrix}+\begin{pmatrix} -A^-U \\ I \end{pmatrix}(B-VA^-U)^- \ (-VA^- \ \vdots \ I)$ is a $g$ inverse of $\begin{pmatrix} A & U \\ V & B \end{pmatrix}$.

**1.12** Show that $(A^+)^+=A$, $(AA^*)^+=(A^+)^*A^+$, $(A^*)^+=(A^+)^*$, and $(UAV)^+=V^*A^+U^*$ if $V$ and $U$ are unitary matrices. Show that if $U$ is a nonsingular matrix, then $A^+U^{-1}$ need not be a Moore–Penrose inverse of $UA$.

**1.13** Show that

(i) ch roots of $AB$ are equal to those of $BA$ except for some zero ch roots, and

(ii) if $\lambda$ is a ch root of $C$ and $f(C)$ is a finite degree polynomial in $C$, then $f(\lambda)$ is a ch root of $f(C)$.

**1.14** Let $B=(b_{ij})$, where $b_{ii}=2$ for $i=1,2,\ldots,p-1$, $b_{pp}=1$, $b_{i,i+1}=b_{i+1,i}=-1$ for $i=1,2,\ldots,p-1$, and other $b_{ij}=0$. Find

(i) $B^{-1}$,

(ii) the characteristic roots, and

(iii) the characteristic vectors.

**1.15** Let $A$ be a real skew-symmetric matrix. Show that there exists an orthogonal matrix $H$ such that

$$HAH'=\operatorname{diag}(E_1,E_2,\ldots,E_m,E_{m+1}),$$

where

$$E_j=\begin{pmatrix} 0 & \lambda_j \\ -\lambda_j & 0 \end{pmatrix}, \qquad j=1,2,\ldots,m, \quad \lambda_j\neq 0,$$

and $E_{m+1}=0$. Show that $(i\lambda_j, -i\lambda_j)$, $j=1,2,\ldots,m$ and $i=\sqrt{-1}$, are the nonzero ch roots of $A$. [Hint: Use Theorem 1.8.2 and Corollary 1.8.2. Let $A(\mathbf{x}_1-i\mathbf{x}_2)=i\lambda(\mathbf{x}_1-i\mathbf{x}_2)$. Then $\mathbf{x}_1'\mathbf{x}_2=0$ and

$$A\begin{pmatrix} \mathbf{x}_1 \\ \mathbf{x}_2 \end{pmatrix}=\begin{pmatrix} 0 & \lambda \\ -\lambda & 0 \end{pmatrix}\begin{pmatrix} \mathbf{x}_1 \\ \mathbf{x}_2 \end{pmatrix}.$$

Then normalize.]

**1.16** Let $A$ be a Hermitian matrix. Then show that

$$\mathop{\mathrm{Inf}}_{\substack{\mathbf{y}_1,\ldots,\mathbf{y}_i \\ \text{such that} \\ \mathbf{y}_j^*\mathbf{y}_{j'}=0, \\ j\neq j'}} \quad \mathop{\mathrm{Sup}}_{\substack{\mathbf{x} \\ \text{such that} \\ \mathbf{x}^*\mathbf{y}_j=0, \\ j=1,2,\ldots,i}} \left(\frac{\mathbf{x}^*A\mathbf{x}}{\mathbf{x}^*\mathbf{x}}\right)=\mathrm{ch}_{i+1}(A),$$

and

$$\mathop{\mathrm{Sup}}_{\substack{\mathbf{y}_1,\ldots,\mathbf{y}_i \\ \text{such that} \\ \mathbf{y}_j^*\mathbf{y}_{j'}=0, \\ j\neq j'}} \quad \mathop{\mathrm{Inf}}_{\substack{\mathbf{x} \\ \text{such that} \\ \mathbf{x}^*\mathbf{y}_j=0, \\ j=1,2,\ldots,i}} \left(\frac{\mathbf{x}^*A\mathbf{x}}{\mathbf{x}^*\mathbf{x}}\right)=\mathrm{ch}_{n-i}(A).$$

When $i=0$, the result follows from Theorem 1.10.1. Hence or otherwise, show that if $A(j_1,\ldots,j_t)$ is a submatrix of $A$, obtained by taking rows $j_1,\ldots,j_t$ and columns $j_1,j_2,\ldots,j_t$ $(t\geqslant i)$, then $\mathrm{ch}_{n-t+i}(A)\leqslant \mathrm{ch}_i[A(j_1,\ldots,j_t)]\leqslant \mathrm{ch}_i(A)$. Further, if $B$ is a Hermitian idempotent matrix of rank $t$, then $\mathrm{ch}_{n-t+i}(A)\leqslant \mathrm{ch}_i(AB)\leqslant \mathrm{ch}_i(A)$ for $i=1,2,\ldots,t$. [The ordering of the ch roots of $(AB)$ is done after omitting $n-t$ zero ch roots.] [Hint: Notice that it is sufficient to prove the result when $A$ is a diagonal matrix. Then choose one time the subspace of $\mathbf{x}$, and the other time the subspace of $\mathbf{y}_1,\ldots,\mathbf{y}_i$, to get two types of inequalities. Note that the Sup over a subspace is always less than or equal to the Sup over the space, and the inequality is reversed for Inf.]

**1.17** Let $A$ be an $n\times n$ Hermitian matrix, and $B$ an $n\times n$ p.s.d. matrix of rank $r$ $(\leqslant n)$. Then

$$\mathrm{ch}_i(AB)\leqslant \mathrm{Min}[\mathrm{ch}_1(A)\,\mathrm{ch}_i(B),\mathrm{ch}_i(A)\,\mathrm{ch}_1(B)]$$

and

$$\mathrm{ch}_i(AB)\geqslant \mathrm{Max}[\mathrm{ch}_{n-i+1}(A)\,\mathrm{ch}_r(B),\mathrm{ch}_{r-i+1}(B)\,\mathrm{ch}_n(A)]$$

for $i=1,2,\ldots,r$. (The ordering of the ch roots of $AB$ is done after omitting $n-r$ zero ch roots of $AB$.) [Hint: If $B$ is p.d., then note from Problem 1.15 that

$$\mathrm{ch}_i(AB^{-1})= \mathop{\mathrm{Inf}}_{\substack{\mathbf{y}_1,\ldots,\mathbf{y}_i \\ \text{such that} \\ \mathbf{y}_j^*B\mathbf{y}_{j'}=0, \\ j\neq j'}} \quad \mathop{\mathrm{Sup}}_{\substack{\mathbf{x} \\ \text{such that} \\ \mathbf{x}^*B\mathbf{y}_j=0, \\ j=1,2,\ldots,i}} \left(\frac{\mathbf{x}^*A\mathbf{x}}{\mathbf{x}^*B\mathbf{x}}\right).$$

Without loss of generality, first establish the result when $B$ is diagonal with positive elements. Then, for a general p.s.d. matrix of rank $r$, $\mathrm{ch}_i(AB)=\mathrm{ch}_i(\Gamma A\Gamma^* D_\omega)$, where $B=\Gamma^* D_\omega \Gamma$ and $\Gamma$ is semiunitary such that $\Gamma\Gamma^*=I_r$. Use the nonsingularity and the result of Problem 1.15.]

**1.18** (i) Let $A$ be any $n\times n$ square matrix. Then

$$\mathrm{ch}_n(AA^*)\leqslant |\mathrm{mod}\,\mathrm{ch}(A)|^2\leqslant \mathrm{ch}_1(AA^*)$$

(ii) For any $\mathbf{x}\neq 0$ and $\mathbf{y}\neq 0$, show that

$$\frac{(\mathrm{mod}\,\mathbf{x}^*C\mathbf{y})^2}{(\mathbf{x}^*A\mathbf{x})(\mathbf{y}^*B\mathbf{y})}\leqslant \mathrm{ch}_1(CB^{-1}C^*A^{-1}), \qquad A \text{ and } B \text{ p.d.}$$

**1.19** Let $A$ and $B$ be two Hermitian matrices such that $A-B$ is p.s.d. Then $\text{ch}_i(A)\geqslant\text{ch}_i(B)$ for all $i$. The converse is not true, i.e., if $\text{ch}_i(A)\geqslant\text{ch}_i(B)$ for all $i$, $A-B$ need not be p.s.d. [Hint: Use Problem 1.16 and $\mathbf{x}^*(A-B)\mathbf{x}\geqslant 0$ for all $\mathbf{x}$.]

**1.20** Let $A: n\times n$ be a nonsingular matrix.

(i) Show that if the principal minor of order $i$ is nonzero for $i=1,2,\ldots,n$, then there exists a representation

$$A=\underset{\sim}{T}\tilde{U}$$

where $\underset{\sim}{T}$ is a lower triangular matrix and $\tilde{U}$ is an upper triangular matrix with $u_{ii}=1$ for all $i=1,2,\ldots,n$. Notice that the representation is unique under the condition on the matrix $\tilde{U}$. The actual elements of $\underset{\sim}{T}$ and $\tilde{U}$ are

$$t_{ij}=\begin{cases}0 & \text{if } \ i<j,\\ \dfrac{\left|A\binom{1,2,\ldots,j-1,i}{1,2,\ldots,j}\right|}{\left|A\binom{1,2,\ldots,j-1}{1,2,\ldots,j-1}\right|} & \text{if } \ i\geqslant j,\end{cases}$$

and

$$u_{ij}=\begin{cases}0 & \text{if } \ i>j,\\ \dfrac{\left|A\binom{1,2,\ldots,i}{1,2,\ldots,i-1,j}\right|}{\left|A\binom{1,2,\ldots,i}{1,2,\ldots,i}\right|} & \text{if } \ i\leqslant j.\end{cases}$$

(ii) Show that $A=S_1S_2$ for some symmetric matrices $S_1$ and $S_2$. [Hint: Notice that

$$A\binom{1,2,\ldots,j-1,i}{1,2,\ldots,j}=\underset{\sim}{T}\binom{1,2,\ldots,j-1,i}{1,2,\ldots,j}\tilde{U}\binom{1,2,\ldots,j}{1,2,\ldots,j}\qquad\text{for }\ i\geqslant j$$

and

$$A\binom{1,2,\ldots,i}{1,2,\ldots,i-1,j}=\underset{\sim}{T}\binom{1,2,\ldots,i}{1,2,\ldots,i}\tilde{U}\binom{1,2,\ldots,i}{1,2,\ldots,i-1,j}\qquad\text{for }\ i<j.$$

Then take determinants.]

**1.21** Let $A$ be a p.d. matrix. Then there exist lower triangular matrices $\underset{\sim}{T}_1$ and $\underset{\sim}{T}_2$ with positive diagonal elements such that $A=\underset{\sim}{T}_1\underset{\sim}{T}_1^*=\underset{\sim}{T}_2^*\underset{\sim}{T}_2$. The representation is unique under the condition on the triangular matrices.

**1.22** Let $A$ be any $m\times n$ matrix of rank $r$ ($\leqslant m,n$).

(i) There exist two semiunitary matrices $U_1: r\times m$ and $U_2: r\times n$ such that $A=U_1^*D_\lambda U_2$, where $D_\lambda=\text{diag}(\lambda_1,\ldots,\lambda_r)$, with $\lambda_i^2$ the nonzero ch roots of $AA'$.

(ii) If $r=m$, there exist a lower triangular matrix $\underset{\sim}{T}$ with positive diagonal elements and a semiunitary matrix $U:r\times n$ such that $A=\underset{\sim}{T}U$.

**1.23** Let $A$ and $B$ be two p.d. matrices and $0\leqslant\lambda\leqslant 1$. Then

$$|\lambda A+(1-\lambda)B|\geqslant|A|^{\lambda}|B|^{1-\lambda}.$$

For $0<\lambda<1$, the equality holds iff $A=B$. The inequality is reversed if $\lambda>1$. [Hint: Show that $\lambda a+(1-\lambda)\geqslant a^{\lambda}$ for $a>0$, $0\leqslant\lambda\leqslant 1$, and that for $0<\lambda<1$ the equality holds iff $a=1$. Then use Theorem 1.9.2.]

**1.24** Let $A$ and $B$ be Hermitian matrices of order $n$, $0\leqslant\lambda\leqslant 1$, and $S_k(A)=\lambda_n+\lambda_{n-1}+\cdots+\lambda_{n-k+1}$, $\lambda_1\geqslant\lambda_2\geqslant\cdots\geqslant\lambda_n$ being the ch roots of $A$. Then show that

$$S_k(\lambda A+(1-\lambda)B)\geqslant\lambda S_k(A)+(1-\lambda)S_k(B)\qquad\text{for}\quad k=1,2,\ldots.$$

**1.25** Show that for a Hermitian matrix $A$ and for a sufficiently small value (indicate the choice of $t$),

(i) $\ln|I-tA|=-\sum_{r=1}^{\infty}(t^r/r)\operatorname{tr}A^r$, and

(ii) $(I-tA)^{-k}=\sum_{r=0}^{\infty}\binom{k+r-1}{r}t^rA^r$ with $A^0=I$.

**1.26** Let $X$ be any real random matrix, and denote

$$\frac{df(X)}{dX}=\left(\frac{\partial}{\partial x_{ij}}f(X)\right).$$

If $X$ is any real symmetric matrix, denote

$$\frac{df(X)}{dX}=\left(\varepsilon_{ij}\frac{\partial}{\partial x_{ij}}f(X)\right)$$

where $\varepsilon_{ii}=1$ and $\varepsilon_{ij}=\frac{1}{2}$ if $i\neq j$. With this notation, establish the following:

(i) $(d/dX)|X|^r=r|X|^r(X^{-1})$ for any symmetric $X$ or any random matrix $X$.

(ii) $(d/dX)\operatorname{tr}(A'X)=A$ for $A:q\times p$ and $X:q\times p$.

(iii) $\dfrac{d}{dX}\operatorname{tr}(X^{-1}A)=\begin{cases}-(X^{-1}AX^{-1})' & \text{if}\quad X \text{ is any matrix,}\\ -X^{-1}AS^{-1} & \text{if}\quad X \text{ and } A \text{ are symmetric.}\end{cases}$

(iv) $(d/dX)\operatorname{tr}X^2=2X$ if $X$ is symmetric.

**1.27** Let $\underset{\sim}{G}$ and $\underset{\sim}{T}$ be two real lower triangular random matrices of order $n$. Show that

(i) if $\underset{\sim}{G}=\underset{\sim}{A}\underset{\sim}{T}$, where $\underset{\sim}{A}$ is a real lower triangular matrix, then $J(\underset{\sim}{G}\to\underset{\sim}{T})=\prod_{i=1}^{n}(a_{ii})_{+}^{i}$,

(ii) if $\tilde{G}=\tilde{B}\tilde{T}$, where $\tilde{B}$ is a real upper triangular matrix, then $J(\tilde{G}\to\tilde{T})=\prod_{i=1}^{n}(b_{ii})_{+}^{n-i+1}$.

**1.28** Let $S$ be a $p\times p$ real p.d. matrix.

(i) If $S=\underset{\sim}{T}\underset{\sim}{T}'$, where $\underset{\sim}{T}$ is a lower triangular matrix with positive diagonal elements, then $J(S\to\underset{\sim}{T})=2^p\prod_{i=1}^p t_{ii}^{p-i+1}$.

(ii) If $S=\underset{\sim}{U}D_\lambda\underset{\sim}{U}'$, where $\underset{\sim}{U}$ is a lower triangular matrix with unit diagonal elements and $D_\lambda=\text{diag}(\lambda_1,\ldots,\lambda_p)$, then $J(S\to D_\lambda,\underset{\sim}{U})=\prod_{i=1}^p\lambda_i^{p-i}$.

**1.29** Let $V$ be a $p\times p$ random p.d. matrix on the complex field, and let $V=\underset{\sim}{T}\underset{\sim}{T}^*$, where $\underset{\sim}{T}$ is a lower triangular matrix with positive diagonal elements. Then $J(V\to\underset{\sim}{T})=2^p\prod_{i=1}^p t_{ii}^{2(p-i)+1}$. Now, if $V=\underset{\sim}{U}D_\lambda\underset{\sim}{U}^*$ where $\underset{\sim}{U}$ is a lower triangular matrix with unit diagonal elements, then $J(V\to\underset{\sim}{U},D_\lambda)=\prod_{i=1}^p\lambda_i^{2(p-i)}$.

**1.30** Let $S$ and $V$ be $p\times p$ random p.d. matrices on the complex field. Then show that the Jacobian of the transformation $V=ASA^*$ is $J(V\to S)=|AA^*|^p$.

**1.31** Let $x_j=\sum_{i=1}^p y_i^j$ for $j=1,2,\ldots,p-1$ and $x_p=y_1y_2\cdots y_p$. Then

$$J(x_1,\ldots,x_p\to y_1,y_2,\ldots,y_p)=(p-1)!\prod_{i=1}^{p-1}\prod_{j=i+1}^{p}(y_i-y_j)$$

if $y_1>y_2>\cdots>y_p$.

**1.32** Show that the Jacobian of the transformation

$$x_1=r\sin\theta_1,$$

$$x_i=r\cos\theta_1\cos\theta_2\cdots\cos\theta_{i-1}\sin\theta_i\qquad\text{for}\quad i=2,3,\ldots,p-1,$$

$$x_p=r\cos\theta_1\cos\theta_2\cdots\cos\theta_{p-1}$$

is

$$J(x_1,\ldots,x_p\to r,\theta_1,\ldots,\theta_{p-1})=r^{p-1}\prod_{i=1}^{p-1}\cos^{p-i-1}\theta_i.$$

**1.33** Let $X$ be a $p\times n$ real random matrix, $n\geqslant p$, and $X=\underset{\sim}{T}L$, where $\underset{\sim}{T}$ is a lower triangular matrix with positive diagonal elements and $L:p\times n$ is a semiorthogonal matrix. Then show that $J(X\to\underset{\sim}{T},L)=(\prod_{i=1}^p t_{ii}^{n-i})g_{n,p}(L)$, where $g_{n,p}(L)$ is a function of the elements of $L$. Notice that in this orthogonal matrix, we have no restrictions on the elements of $L$ except $LL'=I_p$, while in Theorem 1.11.5, we have $p$ restrictions on the elements of $L$ (when $p=n$). The Jacobian of the transformations $X=\underset{\sim}{T}H$ and $XX'=S=TT'$ is $J(X\to S,H)=|S|^{(n-p-1)/2}[g_{n,p}(H)/2^p]$. Further, show explicitly that $g_{n,p}(L)=(\prod_{i=1}^p|L_i|_+)^{-1}$ if the random elements of $L$ are $\{l_{12},\ldots,l_{1n},l_{23},\ldots,l_{2n},\ldots,l_{p,p+1},\ldots,l_{pn}\}$ and $L_i=(l_{jk}:j,k=1,2,\ldots,i)$.

**1.34** Let $S$ be a random p.d. matrix on the complex space, and let $S=HD_\lambda H^*$, where $H$ is a unitary matrix (with positive diagonal elements) and $D_\lambda=\text{diag}(\lambda_1,\ldots,\lambda_p)$, $\lambda_1>\lambda_2>\cdots>\lambda_p$. Then $J(S\to D_\lambda,H)=[\prod_{i=1}^{p-1}\prod_{j=i+1}^p(\lambda_i-\lambda_j)^2]f_p(H)$.

**1.35** (i) Let $S$ be a $p\times p$ symmetric p.d. matrix with distinct ch roots and $S=HD_\lambda H'$, where $H$ is an orthogonal matrix (with positive diagonal elements), $\lambda_1>\lambda_2>\cdots>\lambda_p>0$, and $D_\lambda$ is diagonal. Suppose $f(x)$ is a differentiable function of $x$ and $F(D_\lambda)=\text{diag}(f(\lambda_1),f(\lambda_2),\dots,f(\lambda_p))$. Define $F(S)=H[F(D_\lambda)]\mathbf{H}'$. Then the Jacobian of the transformation from $S$ to $F(S)$ is

$$J(S\to F(S))=\left\{\left[\prod_{i=1}^{p-1}\prod_{j=i+1}^{p}\frac{f(\lambda_i)-f(\lambda_j)}{\lambda_i-\lambda_j}\right]\prod_{i=1}^{p}\frac{df(\lambda_i)}{d\lambda_i}\right\}^{-1}.$$

If $F(D_\lambda)=\text{diag}(\sqrt{\lambda_1},\dots,\sqrt{\lambda_p})=(D_\lambda)^{1/2}$, then $S^{1/2}=H(D_\lambda)^{1/2}H'$ is called the symmetric square root of $S$. In later chapters, by a square root of $S$ we shall mean a symmetric square root of $S$, unless otherwise stated.

(ii) Let $S$ be a $p\times p$ symmetric matrix with all roots distinct. Let $Y=(n/2)^{1/2}\ln(n^{-1}S)$. That is,

$$S=ne^{(2/n)^{1/2}Y}=n\left[I+(2/n)^{1/2}Y+(2/n)Y^2/2!+\cdots\right].$$

Show that

$$J(S\to Y)=(2n)^{\frac{1}{4}p(p+1)}\left\{\exp\operatorname{tr}\left[\left(\frac{2}{n}\right)^{1/2}Y\right]\prod_{i<j}\frac{f(\lambda_i)-f(\lambda_j)}{\lambda_i-\lambda_j}\right\}$$

where $f(\lambda_i)=e^{\lambda_i}$, $\lambda_i=(2/n)^{1/2}\text{ch}_i(Y)$, and $\text{ch}_i(Y)$ denotes the $i$th ordered characteristic root of $Y$.

**1.36** Find the Jacobian of the transformation given by Problem 1.15, i.e., when $A$ is a $p\times p$ skew-symmetric matrix, $A=H'\text{diag}(E_1,\dots,E_m)H$ with $p=2m$, or $A=H'\text{diag}(E_1,E_2,\dots,E_m,0)H$ with $p=2m+1$ and

$$E_i=\begin{pmatrix}0 & \lambda_i\\ -\lambda_i & 0\end{pmatrix},\quad \lambda_i\neq 0,\qquad i=1,2,\dots,m.$$

**1.37** Let $V$ be a random p.d. matrix. Then show that the Jacobian of the transformation $W=V^{-1}$ is $J(W\to V)=|V|^{-(p+1)}$.

**1.38** Show that for all $\mathbf{x}\in R^p$ such that $C\mathbf{x}=\mathbf{0}$ for any $C:r\times p$ matrix of rank $r$,

$$\text{ch}_p(AQ)\leqslant\frac{\mathbf{x}'A\mathbf{x}}{\mathbf{x}'B\mathbf{x}}\leqslant\text{ch}_1(AQ),$$

where $A$ is Hermitian, $B$ is p.d., and

$$Q=B^{-1}-B^{-1}C^*(CB^{-1}C^*)^{-1}CB^{-1}.$$

In particular, if $A=\mathbf{y}\mathbf{y}'$ and $B^{-1}=\text{diag}(n_1,\dots,n_p)$, show that for all $\mathbf{x}\in R^p$ and $\sum_{i=1}^{p}x_i=0$

$$0\leqslant\frac{\mathbf{x}'A\mathbf{x}}{\mathbf{x}'B\mathbf{x}}\leqslant\sum_{i=1}^{p}n_i(y_i-\bar{y})^2,$$

where $\bar{y}=\sum_{i=1}^{p}n_iy_i/\sum_{i=1}^{p}n_i$. [Hint: Prove first for $B=I$. Also note that in this case $\mathbf{x}=[I_p-C^*(CC^*)^{-1}C]\mathbf{x}$, since $C\mathbf{x}=\mathbf{0}$.]

**1.39** Let $A$ and $B$ be $p\times p$ p.d. matrices, $\boldsymbol{\delta}_1$ and $\boldsymbol{\delta}_p$ be the ch vectors of $A$ corresponding to $\mathrm{ch}_1(A)$ and $\mathrm{ch}_p(A)$ respectively, and $\alpha_i=\boldsymbol{\delta}_i B\boldsymbol{\delta}_i$, $\beta_i=\boldsymbol{\delta}_i B^{-1}\boldsymbol{\delta}_i$, $i=1,p$. Let $A=(a_{ij})$ and $A^{-1}=(a^{ij})$. Then show that

(i) $\mathrm{ch}_p(AB^{-1})\leqslant[\mathrm{ch}_i(A)\beta_i, \mathrm{ch}_i(A)/\alpha_i,\ i=1,p]\leqslant \mathrm{ch}_1(AB^{-1})$,

(ii) $\mathrm{ch}_p(AB^{-1})\leqslant[a_{ii}\beta_p, \beta_p/a^{ii}, a_{ii}/\alpha_p, (\alpha_p a^{ii})^{-1},\ a_{ii}/\alpha_1,\ a_{ii}\alpha_1,\ (\alpha_1 a^{ii})^{-1},\ \beta_1/a_{ii}]\leqslant \mathrm{ch}_1(AB^{-1})$ for all $i=1,2,\ldots,p$.

**1.40** Let $A$ be a $p\times p$ p.d. matrix with roots $\lambda_1\geqslant\lambda_2\geqslant\cdots\geqslant\lambda_p$. Show that

(i) $1\leqslant(\mathbf{a}'A\mathbf{a})(\mathbf{a}'A^{-1}\mathbf{a})\leqslant(\lambda_1+\lambda_p)^2/4\lambda_1\lambda_p$, where $\mathbf{a}'\mathbf{a}=1$,

(ii) $1\leqslant|K_1'AK_1||K_1'A^{-1}K_1|\leqslant\prod_{i=1}^{q}(\lambda_i+\lambda_{p-i+1})^2/4\lambda_i\lambda_{p-i+1}$, where $K_1:p\times q$ is such that $K_1'K_1=I_q$.

# 2

# Multivariate Normal Distribution

## 2.1 Multivariate Normal Distribution

We begin with the definition of a unit (standard) normal random variable $u$ defined by its density

$$(2.1.1) \qquad (2\pi)^{-1/2}\exp\left(-\tfrac{1}{2}u^2\right),$$

and denoted as $u \sim N(0,1)$. A random variable $x$ has a normal distribution with mean $\theta$ and variance $\sigma^2 > 0$ if $x$ has the same distribution as

$$(2.1.2) \qquad \theta + \sigma u,$$

where $u \sim N(0,1)$. In this case its density is given by

$$(2.1.3) \qquad (2\pi\sigma^2)^{-1/2}\exp\left[-\frac{1}{2\sigma^2}(x-\theta)^2\right],$$

where $-\infty < \theta,\ x < \infty$, and $\sigma > 0$.

Let $\mathbf{u} = (u_1, \ldots, u_p)'$ be a vector of $p$ independent standard normal random variables $u_i \sim N(0,1)$, $i = 1, 2, \ldots, p$. Then $E(\mathbf{u}) \equiv (E(u_1), \ldots, E(u_n))' = \mathbf{0}$ and $E(\mathbf{uu}') \equiv (E(u_i u_j)) = I$. And the density of $\mathbf{u}$ is given by

$$(2.1.4) \qquad (2\pi)^{-\frac{1}{2}p}\exp\left(-\tfrac{1}{2}\mathbf{u}'\mathbf{u}\right)$$

and denoted by $\mathbf{u} \sim N_p(\mathbf{0}, I)$, where $\mathbf{0}$ denotes the mean of $\mathbf{u}$, $I$ the dispersion matrix of $\mathbf{u}$, and the subscript $p$ the dimensionality of the random vector. Analogously to the univariate case, a random vector $\mathbf{x}$ is said to have a nonsingular multivariate normal distribution with mean vector $\boldsymbol{\theta}$ and nonsingular $p \times p$ dispersion matrix $\Sigma$, denoted by $\mathbf{x} \sim$

$N_p(\boldsymbol{\theta},\Sigma)$, $\Sigma>0$, if $\mathbf{x}$ has the same distribution as

$$\boldsymbol{\theta}+A\mathbf{u}, \tag{2.1.5}$$

where $A$ is any nonsingular factorization of $\Sigma$ such that $\Sigma=AA'$, and where $\mathbf{u}\sim N_p(\mathbf{0},I)$. Since the Jacobian of the transformation is $J(\mathbf{u}\rightarrow\mathbf{x})=|A^{-1}|_+=|AA'|^{-1/2}=|\Sigma|^{-1/2}$, we find that the density of $\mathbf{x}$ is given by

$$(2\pi)^{-\frac{1}{2}p}|\Sigma|^{-1/2}\exp\left[-\tfrac{1}{2}(\mathbf{x}-\boldsymbol{\theta})'\Sigma^{-1}(\mathbf{x}-\boldsymbol{\theta})\right], \tag{2.1.6}$$

which is in conformity with Eq. (2.1.3).

From Eq. (2.1.5) the characteristic function (cf) of $\mathbf{x}\sim N_p(\boldsymbol{\theta},\Sigma)$ is given by

$$\begin{aligned} E(e^{i\mathbf{t}'\mathbf{x}}) &= E(e^{i\mathbf{t}'\boldsymbol{\theta}+i\mathbf{a}'\mathbf{u}}), \qquad \text{where } \mathbf{a}'=\mathbf{t}'A \text{ and } \Sigma=AA', \\ &= e^{i\mathbf{t}'\boldsymbol{\theta}}\prod_{r=1}^{p} Ee^{ia_ru_r}, \qquad \text{where } u_r\sim N(0,1), \\ &= e^{i\mathbf{t}'\boldsymbol{\theta}}\prod_{r=1}^{p} e^{-\frac{1}{2}a_r^2}=e^{i\mathbf{t}'\boldsymbol{\theta}}e^{-\frac{1}{2}\Sigma_{r=1}^{p}a_r^2}, \\ &= e^{i\mathbf{t}'\boldsymbol{\theta}-\frac{1}{2}\mathbf{t}'\Sigma\mathbf{t}}. \end{aligned} \tag{2.1.7}$$

The cf (2.1.7) is obtained when $\Sigma$ is p.d. Notice that Eq. (2.1.7) does exist when $\Sigma\geqslant 0$ [i.e., $\Sigma$ is positive semidefinite (p.s.d.)]. Let $\Sigma$ be of rank $r$. Then, by Corollary 1.8.3, we can find an orthogonal matrix $\Delta$ such that

$$\Sigma=\Delta\begin{pmatrix} D_\lambda & 0 \\ 0 & 0 \end{pmatrix}\Delta', \qquad \Delta=(\Delta_1,\Delta_2),$$

where $\Delta_1$ is a $p\times r$ matrix and $D_\lambda$ is an $r\times r$ nonsingular diagonal matrix. Let $\boldsymbol{\mu}=\Delta'\boldsymbol{\theta}$ and $\boldsymbol{\gamma}=\Delta'\mathbf{t}$. Then

$$\mathbf{t}'\boldsymbol{\theta}=\boldsymbol{\gamma}'\boldsymbol{\mu}=\boldsymbol{\gamma}_1'\boldsymbol{\mu}_1+\boldsymbol{\gamma}_2'\boldsymbol{\mu}_2 \quad \text{and} \quad \mathbf{t}'\Sigma\mathbf{t}=\boldsymbol{\gamma}'\begin{pmatrix} D_\lambda & 0 \\ 0 & 0 \end{pmatrix}\boldsymbol{\gamma}=\boldsymbol{\gamma}_1'D_\lambda\boldsymbol{\gamma}_1,$$

where $\boldsymbol{\gamma}'=(\boldsymbol{\gamma}_1',\boldsymbol{\gamma}_2')$ and $\boldsymbol{\mu}'=(\boldsymbol{\mu}_1',\boldsymbol{\mu}_2')$. Hence, (2.1.7) can be written as

$$\exp\left(i\boldsymbol{\gamma}_1'\boldsymbol{\mu}_1-\tfrac{1}{2}\boldsymbol{\gamma}_1'D_\lambda\boldsymbol{\gamma}_1+i\boldsymbol{\gamma}_2'\boldsymbol{\mu}_2\right)=E\exp(i\mathbf{t}'\Delta\Delta'\mathbf{x}), \tag{2.1.8}$$

which is valid for all $\boldsymbol{\gamma}$. Hence, if

$$\Delta'\mathbf{x}=\mathbf{y} \quad \text{and} \quad \mathbf{y}'=(\mathbf{y}_1',\mathbf{y}_2'),$$

then from the uniqueness property of cf, Eq. (2.1.8) shows that the distribution of $\mathbf{y}_2$ is degenerate $[P(\mathbf{y}_2=\boldsymbol{\mu}_2)=1]$ and the distribution of $\mathbf{y}_1$ is $N_r(\boldsymbol{\mu}_1,D_\lambda)$. The random vectors $\mathbf{y}_1$ and $\mathbf{y}_2$ are obviously independent. Hence, the natural definition of multivariate normal distribution is given as follows:

**Definition 2.1.1** A $p$-dimensional random vector $\mathbf{x}$ is said to have a multivariate normal distribution $N_p(\boldsymbol{\theta},\Sigma)$ if $\mathbf{x}$ has the same distribution as $\boldsymbol{\theta}+B\mathbf{u}$, where $B$ is a $p\times r$ matrix of rank $r$, $\Sigma=BB'$, and $\mathbf{u}\sim N_r(\mathbf{0},I_r)$. $r$ is called the rank of the distribution.

Obviously, the cf of $\mathbf{x}$ is given by Eq. (2.1.7) where $\Sigma\geqslant 0$. The pdf of $\mathbf{x}$ is given by Eq. (2.1.6) for $r=p$. When $r<p$, Khatri (1968a) has given one representation for the pdf:

$$(2.1.9)\qquad (2\pi)^{-r/2}\left(\prod_{i=1}^{r}\lambda_i\right)^{-1/2}\exp\left[-\tfrac{1}{2}(\mathbf{x}-\boldsymbol{\theta})'\Sigma^{-}(\mathbf{x}-\boldsymbol{\theta})\right],$$

where $\Sigma^{-}$ is a symmetric $g$ inverse of $\Sigma$ and $\lambda_1,\ldots,\lambda_r$ are the nonzero ch roots of $\Sigma$. We may note that the above representation of the pdf of $\mathbf{x}$ is not unique, because it can be shown that Eq. (2.1.9) is the pdf of $\mathbf{y}_1$ using

$$\Delta'\Sigma^{-}\Delta=\Delta'(\Delta_1 D_\lambda\Delta_1')^{-}\Delta=\begin{pmatrix} D_\lambda^{-1} & 0\\ 0 & 0\end{pmatrix}.$$

In Definition 2.1.1, the matrix $B$ need not be unique. This is true even when $r=p$, since $B$ can be replaced by $B\Gamma$, where $\Gamma$ is any $r\times r$ orthogonal matrix. This nonunique representation, however, characterizes the normal distribution [see Rao (1966, 1969)] in the following way.

Suppose $\mathbf{x}$ has two different representations:

$$\mathbf{x}=\boldsymbol{\theta}+B\mathbf{u}=\boldsymbol{\theta}+A\mathbf{v},$$

where the elements of $\mathbf{u}$ (and $\mathbf{v}$) are independent and no column vector of $B$ is proportional to any column vector of $A$. Then $\mathbf{x}$ is normally distributed. The proof of this can be obtained with the help of Theorem 2.2.8 given later.

Next we give an alternative definition of normality. Let $\mathbf{t}=s\mathbf{b}$ and $\mathbf{b}'\Sigma\mathbf{b}\neq 0$ in Eq. (2.1.7). Then it is easy to see that

$$\mathbf{b}'\mathbf{x}\sim N(\mathbf{b}'\boldsymbol{\theta},\mathbf{b}'\Sigma\mathbf{b}).$$

When $\mathbf{b}'\Sigma\mathbf{b}=0$, then $\mathbf{b}'\mathbf{x}=\mathbf{b}'\boldsymbol{\theta}$ with probability one. Writing this degenerate distribution as $N(\mathbf{b}'\boldsymbol{\theta},0)$, we find that every linear combination of $\mathbf{x}$ is normally distributed. The converse of this result is also true [see Cramèr (1946)]. This gives the following alternative

**Definition 2.1.2** A $p$-dimensional random vector $\mathbf{x}$ is said to have a multivariate normal distribution iff *every linear* combination of $\mathbf{x}$ has a univariate normal distribution (including the degenerate normal distribution).

Note that in the above definition, we require that every linear function is normally distributed. Then a natural question is: Can $\mathbf{x}$ be normally distributed if a finite or a countably infinite set of linear functions are normally distributed? For this purpose, linear functions $\mathbf{b}'\mathbf{x}$ and $\alpha(\mathbf{b}'\mathbf{x})$ are not considered as different. The answer is *no* for the finite case and is *yes* for the countably infinite case if $p=2$, as shown by Hamedani and Tata (1975); see Problem 2.3.

## 2.2 Some Properties of Multivariate Normal Distributions

We begin with the properties on linear transformations in the following

**Theorem 2.2.1** *Let* $\mathbf{x} \sim N_p(\boldsymbol{\theta}, \Sigma)$. *Then for any* $r \times p$ *matrix* $C$, $C\mathbf{x} \sim N_r(C\boldsymbol{\theta}, C\Sigma C')$.

PROOF. Let $\mathbf{t}$ be any nonnull $r$ vector, and let $\mathbf{b}' = \mathbf{t}'C$. Then the characteristic function of $C\mathbf{x}$ is given by

$$\begin{aligned} E(e^{i\mathbf{t}'C\mathbf{x}}) &\equiv E(e^{i\mathbf{b}'\mathbf{x}}) \\ &= e^{i\mathbf{b}'\boldsymbol{\theta} - \frac{1}{2}\mathbf{b}'\Sigma\mathbf{b}} \\ &= e^{i\mathbf{t}'(C\boldsymbol{\theta}) - \frac{1}{2}\mathbf{t}'(C\Sigma C')\mathbf{t}}. \end{aligned}$$

Hence, $C\mathbf{x} \sim N_r(C\boldsymbol{\theta}, C\Sigma C')$. □

An alternative proof of this theorem can also be obtained from Definition 2.1.2. This remark applies to all the theorems given below.

**Corollary 2.2.1** *Let* $\mathbf{x}' = (\mathbf{x}_1', \mathbf{x}_2')$. *Suppose*

$$\boldsymbol{\theta}' = (\boldsymbol{\theta}_1', \boldsymbol{\theta}_2') \quad \textit{and} \quad \Sigma = \begin{pmatrix} \Sigma_{11} & \Sigma_{12} \\ \Sigma_{12}' & \Sigma_{22} \end{pmatrix},$$

*where* $\mathbf{x}$ *and* $\boldsymbol{\theta}$ *are* $p$-*vectors and* $\mathbf{x}_1$ *and* $\boldsymbol{\theta}_1$ *are* $k$-*vectors, and* $\Sigma_{11}$ *is* $k \times k$, $k \leqslant p$. *Then* $\mathbf{x}_1 \sim N_k(\boldsymbol{\theta}_1, \Sigma_{11})$.

PROOF. This follows from Theorem 2.2.1 by taking $C = (I_k, 0)$. □

Since the selection of the subvector $\mathbf{x}_1$ of $\mathbf{x}$ is arbitrary, we get the following

**Theorem 2.2.2** *Let* $\mathbf{x} \sim N_p(\boldsymbol{\theta}, \Sigma)$. *Then any subvector of* $\mathbf{x}$ *is also normally distributed with mean equal to the corresponding subvector of* $\boldsymbol{\theta}$ *and dispersion matrix equal to the corresponding submatrix of* $\Sigma$.

**Corollary 2.2.2** *If* $\mathbf{x} \sim N_p(\boldsymbol{\theta}, \Sigma)$, *the marginal distribution of each component* $x_i$ *is* $N_1(\theta_i, \sigma_{ii})$, *where* $\mathbf{x} = (x_1, \ldots, x_p)'$, $\boldsymbol{\theta} = (\theta_1, \ldots, \theta_p)'$, *and* $\Sigma = (\sigma_{ij})$.

It should be pointed out that if the marginal distribution of each $x_i$ is normal, this does not necessarily imply that the joint distribution of $(x_1, \ldots, x_p)$ is also normal. A counterexample is as follows.

Let the joint density of two random variables $x_1$ and $x_2$ be given by

$$\text{(2.2.1)} \qquad \alpha\left[2\pi(1-\rho^2)^{1/2}\right]^{-1}\exp\left(-\frac{1}{2(1-\rho^2)}(x_1^2 - 2\rho x_1 x_2 + x_2^2)\right)$$

$$+(1-\alpha)\left[2\pi(1-\zeta^2)^{1/2}\right]^{-1}\exp\left(-\frac{1}{2(1-\zeta^2)}(x_1^2 - 2\zeta x_1 x_2 + x_2^2)\right),$$

where $0 < \alpha < 1$, $-1 < \rho, \zeta < 1$, $-\infty < x_1, x_2 < \infty$. Then it can easily be checked that the marginal densities of $x_1$ and $x_2$ are both normal, while obviously Eq. (2.2.1) is not a normal density.

REMARK 2.2.1. We have shown in Theorem 2.2.1 that if $\mathbf{x}$ is normal, then every linear function of $\mathbf{x}$ is normally distributed. Now, suppose $\mathbf{x}$ is normal and $f(\mathbf{x})$, a function of $\mathbf{x}$, is normal. Can we say that $f(\mathbf{x})$ must be essentially linear? In general, the answer to this question is no. Consider a univariate $x \sim N(0,1)$ and $y = \phi(x)|x|$, where $\phi(x)$ is an odd function of $x$ [i.e., $\phi(x) = -\phi(-x)$ for all $x$]. Then, if $\{\phi(x)\}^2 = 1$, we have $y \sim N(0,1)$, while $y = \phi(x)|x|$ is not a linear function of $x$. Now we enlarge the class of normal distributions for which the normality of $x$ and $f(x)$ holds; namely, suppose $x \sim N(\mu, 1)$ and $f(x) \sim N(\nu, \eta)$ for all $\mu \in R$, where $\nu$ and $\eta$ are functions of $\mu$. Suppose $f(x)$ is a 1-1 function of $x$; then $f(x)$ must be essentially linear. This result was established by Basu and Khatri (1969). Its multivariate version is given by

**Theorem 2.2.3** *Let* $\mathbf{x} \sim N_p(\boldsymbol{\mu}, \Sigma)$, $\Sigma > 0$. *Let* $f_j(\mathbf{x})$ *be a real-valued function of* $\mathbf{x}$, $j = 1, 2, \ldots, k$, $k \geqslant 1$. *If* $\mathbf{f}' = (f_1(\mathbf{x}), \ldots, f_k(\mathbf{x})) \sim N_k(\boldsymbol{\nu}, \boldsymbol{\psi})$ *for every* $\boldsymbol{\mu} \in R^p$ *and* $\Sigma > 0$, *then* $\mathbf{f} = A\mathbf{x} + \mathbf{b}$ *almost everywhere (a.e.), where* $A$ *and* $\mathbf{b}$ *do not depend on* $\boldsymbol{\mu}$ *and* $\Sigma$.

For the proof, see Basu and Khatri (1969). Ghosh (1969) and Mase (1977) have obtained the above results under much milder conditions. The conditions on the multivariate version can also be relaxed.

**Theorem 2.2.4** *Let* $\mathbf{x}_r$, $r=1,2,\ldots,k$, *be independently distributed as* $N_p(\boldsymbol{\theta}_r,\Sigma_r)$. *Then for fixed matrices* $A_i : m\times p$,

$$\sum_{r=1}^{k} A_r\mathbf{x}_r \sim N_m\left(\sum_{r=1}^{k} A_r\boldsymbol{\theta}_r,\ \sum_{r=1}^{k}(A_r\Sigma_r A_r')\right).$$

PROOF. The characteristic function of $\Sigma_{r=1}^{k}A_r\mathbf{x}_r$ is given by

$$E(e^{i\mathbf{t}'\Sigma_{r=1}^{k}A_r\mathbf{x}_r}) = \prod_{r=1}^{k} Ee^{i\mathbf{t}'A_r\mathbf{x}_r}$$

$$= \prod_{r=1}^{k} e^{i\mathbf{t}'A_r\boldsymbol{\theta}_r - \frac{1}{2}\mathbf{t}'(A_r\Sigma_r A_r')\mathbf{t}},$$

from which the result follows. □

**Corollary 2.2.3** *If* $\boldsymbol{\theta}_r\equiv\boldsymbol{\theta}$ *and* $\Sigma_r\equiv\Sigma$, $r=1,2,\ldots,k$, *then*

$$\bar{\mathbf{x}} = k^{-1}\sum_{r=1}^{k}\mathbf{x}_r \sim N_p(\boldsymbol{\theta},k^{-1}\Sigma).$$

This follows by taking $m=p$ and $A_r\equiv k^{-1}I_p$, $r=1,2,\ldots,k$.

**Corollary 2.2.4** *If* $\mathbf{x}_1$ *and* $\mathbf{x}_2$ *are independent* $N_p(\boldsymbol{\theta}_i,\Sigma_i)$, *then* $\mathbf{x}_1+\mathbf{x}_2\sim N_p(\boldsymbol{\theta}_1+\boldsymbol{\theta}_2,\Sigma_1+\Sigma_2)$.

The question now arises whether the converse of the above corollary is true. This is given in the following theorem due to Cramèr (1937).

**Theorem 2.2.5** *Let* $\mathbf{x}$ *and* $\mathbf{y}$ *be independent p-vectors such that* $\mathbf{x}+\mathbf{y}$ *is normally distributed. Then* $\mathbf{x}$ *and* $\mathbf{y}$ *are normally distributed.*

Cramèr (1937) has given the proof for $p=1$. The general result follows from Definition 2.1.2 by taking the linear function $\mathbf{a}'(\mathbf{x}+\mathbf{y})$ for any nonnull vector $\mathbf{a}$.

**Theorem 2.2.6** *Let* $\mathbf{x}=(\mathbf{x}_1',\mathbf{x}_2')'\sim N_p(\boldsymbol{\theta},\Sigma)$, *where* $\mathbf{x}_1$ *and* $\mathbf{x}_2$ *are r- and s-vectors respectively,* $r+s=p$. *Suppose the corresponding partition of* $\boldsymbol{\theta}$ *and* $\Sigma$ *are respectively given by*

$$\boldsymbol{\theta}=\begin{pmatrix}\boldsymbol{\theta}_1\\ \boldsymbol{\theta}_2\end{pmatrix}\quad \text{and}\quad \Sigma=\begin{pmatrix}\Sigma_{11} & \Sigma_{12}\\ \Sigma_{12}' & \Sigma_{22}\end{pmatrix}.$$

*Then* $\mathbf{x}_1$ *and* $\mathbf{x}_2$ *are independently distributed if and only if* $\Sigma_{12}=0$.

PROOF. We need only show that if $\Sigma_{12}=0$ then $\mathbf{x}_1$ and $\mathbf{x}_2$ are independently distributed (since independence implies that the covariance is zero). From the characteristic function of $\mathbf{x}$ we get

$$E(e^{i\mathbf{t}'\mathbf{x}})=\left(e^{i\mathbf{t}_1'\boldsymbol{\theta}_1-\frac{1}{2}\mathbf{t}_1'\Sigma_{11}\mathbf{t}_1}\right)\left(e^{i\mathbf{t}_2'\boldsymbol{\theta}_2-\frac{1}{2}\mathbf{t}_2'\Sigma_{22}\mathbf{t}_2}\right),$$

where $\mathbf{t}'=(\mathbf{t}_1',\mathbf{t}_2')$. Hence $\mathbf{x}_1$ and $\mathbf{x}_2$ are independently distributed.

Note that if $\mathbf{x}_1$ and $\mathbf{x}_2$ have (marginal) normal distributions but are not jointly normally distributed, then the zero correlation does not imply independence. A counterexample can be obtained from Eq. (2.2.1) by choosing $\zeta=-\rho$ and $\alpha=\frac{1}{2}$. □

**Theorem 2.2.7** *Let $\mathbf{x}=(\mathbf{x}_1',\mathbf{x}_2')'\sim N_p(\boldsymbol{\theta},\Sigma)$, $\Sigma>0$, where $\mathbf{x}_1$ and $\mathbf{x}_2$ are $r$- and $s$-vectors, respectively, $r+s=p$, and the partitioning of $\boldsymbol{\theta}$ and $\Sigma$ is similar to that in Theorem 2.2.6. Then the conditional distribution of $\mathbf{x}_1$ given $\mathbf{x}_2$ is*

$$N_r(\boldsymbol{\theta}_1+\Sigma_{12}\Sigma_{22}^{-1}(\mathbf{x}_2-\boldsymbol{\theta}_2),\Sigma_{1\cdot 2}),$$

*where*

$$\Sigma_{1\cdot 2}=\Sigma_{11}-\Sigma_{12}\Sigma_{22}^{-1}\Sigma_{12}'.$$

PROOF. Making a nonsingular linear transformation

$$\mathbf{w}=\begin{pmatrix} I & -\Sigma_{12}\Sigma_{22}^{-1}\\ 0 & I\end{pmatrix}\begin{pmatrix}\mathbf{x}_1\\ \mathbf{x}_2\end{pmatrix},$$

we find that

$$\begin{pmatrix}\mathbf{x}_1-\Sigma_{12}\Sigma_{22}^{-1}\mathbf{x}_2\\ \mathbf{x}_2\end{pmatrix}\sim N_p\left(\begin{pmatrix}\boldsymbol{\theta}_1-\Sigma_{12}\Sigma_{22}^{-1}\boldsymbol{\theta}_2\\ \boldsymbol{\theta}_2\end{pmatrix},\begin{pmatrix}\Sigma_{1\cdot 2} & 0\\ 0 & \Sigma_{22}\end{pmatrix}\right).$$

Hence, from Theorem 2.2.6, $\mathbf{x}_1-\Sigma_{12}\Sigma_{22}^{-1}\mathbf{x}_2$ and $\mathbf{x}_2$ are independently distributed, and

$$\mathbf{x}_1-\Sigma_{12}\Sigma_{22}^{-1}\mathbf{x}_2\sim N_r(\boldsymbol{\theta}_1-\Sigma_{12}\Sigma_{22}^{-1}\boldsymbol{\theta}_2,\Sigma_{1\cdot 2}).$$

Hence, the distribution of $\mathbf{x}_1$ given $\mathbf{x}_2$ is

$$N_r(\boldsymbol{\theta}_1+\Sigma_{12}\Sigma_{22}^{-1}(\mathbf{x}_2-\boldsymbol{\theta}_2),\Sigma_{1\cdot 2}).$$ □

REMARK 2.2.2. The above theorem can easily be generalized to the case when $\rho(\Sigma)<p$, by using generalized inverse of $\Sigma_{22}$ in place of $\Sigma_{22}^{-1}$.

**Corollary 2.2.5** *If $\mathbf{x}=(\mathbf{x}_1',\mathbf{x}_2')'\sim N_p(\boldsymbol{\theta},\Sigma)$, then*

(i) $\mathbf{x}_1-\Sigma_{12}\Sigma_{22}^{-1}\mathbf{x}_2$ *and* $\mathbf{x}_2$ *are independently normally distributed, and*

(ii) $\mathbf{x}_2-\Sigma_{12}'\Sigma_{11}^{-1}\mathbf{x}_1$ *and* $\mathbf{x}_1$ *are independently normally distributed.*

The above corollary raises a natural question. Is the converse of Corollary 2.2.5 true? That is, if (i) $\mathbf{x}_1 - A\mathbf{x}_2$ and $\mathbf{x}_2$ are independent and (ii) $\mathbf{x}_1 - A\mathbf{x}_2$ and $\mathbf{x}_2 - B\mathbf{x}_1$ are normally distributed, then can we say that $\mathbf{x} = (\mathbf{x}_1', \mathbf{x}_2')'$ is normally distributed? The answer is yes if $I - AB$ (or $I - BA$) is nonsingular [see Khatri (1975)]. This follows easily from Cramèr's Theorem 2.2.5, since

$$\mathbf{x}_2 - B\mathbf{x}_1 = (I - BA)\mathbf{x}_2 - B(\mathbf{x}_1 - A\mathbf{x}_2)$$

and since $\mathbf{x}_2$ and $\mathbf{x}_1 - A\mathbf{x}_2$ are independent.

Rao (1974) and Khatri and Rao (1976) give the following results and extensions: Let $\mathbf{x}_1 - A\mathbf{x}_2$ and $\mathbf{x}_2$ be independent, and let $\mathbf{x}_2 - B\mathbf{x}_1$ and $\mathbf{x}_1$ be independent. Then $B\mathbf{x}_1$ and $A\mathbf{x}_2$ have a joint multivariate normal distribution provided $I - AB$ (or $I - BA$) is nonsingular, or $\mathbf{x}_1 - A\mathbf{x}_2$ (or $\mathbf{x}_2 - B\mathbf{x}_1$) has a nonsingular distribution. For, we note that

$$\mathbf{x}_1 = (\mathbf{x}_1 - A\mathbf{x}_2) + A\mathbf{x}_2 \quad \text{and} \quad (\mathbf{x}_2 - B\mathbf{x}_1) = -B(\mathbf{x}_1 - A\mathbf{x}_2) + (I - BA)\mathbf{x}_2$$

are independent in linear functions of independent variables. Using Theorem 2.2.8 (stated at the end of this section), we find that $A\mathbf{x}_2$ and $B(\mathbf{x}_1 - A\mathbf{x}_2)$ are normally distributed, since $I - BA$ is nonsingular. Hence $B\mathbf{x}_1$ and $A\mathbf{x}_2$ are jointly normally distributed.

**Theorem 2.2.8** *Let $\mathbf{y}_i$ $(i = 1, 2, \ldots, k)$ be independent $p_i$-vectors. Suppose*

$$A_{i1}\mathbf{y}_1 + A_{i2}\mathbf{y}_2 + \cdots A_{ik}\mathbf{y}_k, \qquad i = 1, 2, \ldots, q \quad (q \geqslant 2)$$

*are independently distributed. Denote $A_j' = (A_{1j}', \ldots, A_{kj}')$, and by $A_{j(i)}'$ the matrix obtained from $A_j'$ by deleting the ith partition. If $\rho(A_{j(i)}') = p_j$ for all $i$ and $j$, then the $\mathbf{y}_i$'s are normally distributed.*

This theorem is due to Khatri and Rao (1972) and is a generalization of one due to Ghurye and Olkin (1962).

## 2.3 Some Probability Inequalities

Let $\mathbf{x} \sim N_p(\mathbf{0}, \Sigma)$ and $\Sigma > 0$, and let

$$\Phi(\mathbf{y}|\Sigma) = P(x_i \leqslant y_i,\ i = 1, 2, \ldots, p|\Sigma). \tag{2.3.1}$$

Note that if $\phi(\mathbf{x}|\Sigma)$ denotes the pdf of $\mathbf{x}$ as given in Eq. (2.1.6) with $\boldsymbol{\theta} = \mathbf{0}$, then

$$\frac{\partial^2 \phi(\mathbf{x}|\Sigma)}{\partial x_i \partial x_j} = \begin{cases} \dfrac{\partial \phi(\mathbf{x}|\Sigma)}{\partial \sigma_{ij}} & \text{if} \quad i \neq j, \\[2ex] 2\dfrac{\partial \phi(\mathbf{x}|\Sigma)}{\partial \sigma_{ii}} & \text{if} \quad i = j. \end{cases} \tag{2.3.2}$$

For the proof, we use the following results:

(a)
$$\frac{\partial|\Sigma|}{\partial\sigma_{ij}} = \begin{cases} 2\sigma^{ij}|\Sigma| & \text{for } i\neq j, \\ \sigma^{ii}|\Sigma| & \text{for } i=j, \end{cases}$$

(b)
$$\frac{\partial\Sigma^{-1}}{\partial\sigma_{ij}} = \begin{cases} -(\boldsymbol{\sigma}^{(i)}\boldsymbol{\sigma}^{(j)\prime}+\boldsymbol{\sigma}^{(j)}\boldsymbol{\sigma}^{(i)\prime}) & \text{for } i\neq j, \\ -\boldsymbol{\sigma}^{(i)}\boldsymbol{\sigma}^{(i)\prime} & \text{for } i=j, \end{cases}$$

and

(c)
$$\frac{\partial(\mathbf{x}'\Sigma^{-1}\mathbf{x})}{\partial x_i} = 2\boldsymbol{\sigma}^{(i)\prime}\mathbf{x}, \qquad \frac{\partial^2(\mathbf{x}'\Sigma^{-1}\mathbf{x})}{\partial x_i\,\partial x_j} = 2\sigma^{ij},$$

where $\Sigma^{-1}=(\sigma^{ij})=(\boldsymbol{\sigma}^{(1)},\boldsymbol{\sigma}^{(2)},\ldots,\boldsymbol{\sigma}^{(p)})$. For the proofs of the above results, see Problem 1.26(i), Theorem 1.11.1(ii), and Corollary 1.11.1.

With the help of Eq. (2.3.2), we have

$$\begin{aligned}\frac{\partial\Phi(\mathbf{y}|\Sigma)}{\partial\sigma_{12}} &= \int_{-\infty}^{y_1}\cdots\int_{-\infty}^{y_p}\frac{\partial\phi(\mathbf{x}|\Sigma)}{\partial\sigma_{12}}\,d\mathbf{x} \\ &= \int_{-\infty}^{y_1}\cdots\int_{-\infty}^{y_p}\frac{\partial^2\phi(\mathbf{x}|\Sigma)}{\partial x_1\,\partial x_2}\,d\mathbf{x} \\ &= \int_{-\infty}^{y_3}\cdots\int_{-\infty}^{y_p}\phi(y_1,y_2,x_3,\ldots,x_p|\Sigma)\,dx_3\cdots dx_p,\end{aligned}$$

which is nonnegative. Similarly, we can show that

(2.3.3)
$$\frac{\partial\Phi(\mathbf{y}|\Sigma)}{\partial\sigma_{ij}} \geqslant 0 \qquad \text{for } i\neq j.$$

Equation (2.3.3) establishes the following

**Theorem 2.3.1** *Let $\Sigma_1=(\sigma_{ij(1)})$ and $\Sigma=(\sigma_{ij})$ be two p.d. matrices such that $\sigma_{ii(1)}=\sigma_{ii}$ for all $i$ and $\sigma_{ij(1)}\leqslant\sigma_{ij}$. Then $\Phi(\mathbf{y}|\Sigma_1)\leqslant\Phi(\mathbf{y}|\Sigma)$ for all $\mathbf{y}$.*

REMARK 2.3.1. Theorem 2.3.1 is true even when $\Sigma_1$ and $\Sigma$ are singular matrices. First establish the result for p.d. matrices $\Sigma_1+\varepsilon I$ and $\Sigma+\varepsilon I$ by the choice of small $\varepsilon$. Then take the limit as $\varepsilon\to 0$.

**Corollary 2.3.1** *Let $\mathbf{x}\sim N(\mathbf{0},\Sigma)$, where $\Sigma=(\sigma_{ij})$, $\sigma_{ij}=\rho(\sigma_{ii}\sigma_{jj})^{1/2}$ for $i\neq j$. Then $\Phi(\mathbf{y}|\rho)$ is a monotonic increasing function of $\rho$, and in particular*

(2.3.4)
$$\Phi(\mathbf{y}|\rho) \gtreqless \prod_{i=1}^{p} P(x_i\leqslant y_i) \qquad \text{according as } \rho \gtreqless 0.$$

**Corollary 2.3.2** *Let $\mathbf{x}\sim N(\mathbf{0},\Sigma)$, where $\Sigma=(\sigma_{ij})$ and $\sigma_{ij}\geqslant 0$ for all $i\neq j$. Then $\Phi(\mathbf{y}|\Sigma)\geqslant\prod_{i=1}^{p}P(x_i\leqslant y_i)$.*

**Theorem 2.3.2** *Let* $\mathbf{x} \sim N_p(\mathbf{0}, \Sigma)$. *Then*

$$\text{(i)} \qquad P(|x_i| \leqslant c_i \text{ for } i=1,2,\ldots,p) \leqslant \prod_{i=1}^{p} P(|x_i| \leqslant c_i),$$

$$\text{(ii)} \qquad P(|x_i| \geqslant c_i \text{ for } i=1,2,\ldots,p) \geqslant \prod_{i=1}^{p} P(|x_i| \geqslant c_i),$$

*if* $\sigma_{ij} = l_i l_j (\sigma_{ii}\sigma_{jj})^{1/2}$ *for all* $i \neq j$ *and* $l_i^2 \leqslant 1$ *for all* $i$.

The inequality (i) was conjectured by Dunn (1958). It and its extensions were established by Sidak (1967a, b, 1968, 1971) and Khatri (1967, 1976a). Scott (1967) also gave a proof, but it is incorrect [see Sidak (1975)]. The inequality (ii) was given by Khatri (1967). These results are useful in simultaneous confidence bounds. Khatri (1976a) generalized (i) in

**Theorem 2.3.3** *Let* $\mathbf{x} \sim N_p(\mathbf{0}, \Sigma)$, $\mathbf{x}' = (x_1, \mathbf{x}_2')$, *and* $D(\mathbf{x}_2)$ *be a convex region in* $\mathbf{x}_2$ *symmetric about the origin. Then*

$$P[\,|x_1| \leqslant y_1,\ \mathbf{x}_2 \in D(\mathbf{x}_2)\,]$$

*is a monotonic nondecreasing function of all* $|\lambda|$ *for which*

$$\Sigma = \begin{pmatrix} \sigma_{11} & \lambda\boldsymbol{\sigma}' \\ \lambda\boldsymbol{\sigma} & \Sigma_{11} \end{pmatrix} \quad \textit{and} \quad \begin{pmatrix} \sigma_{11} & \boldsymbol{\sigma}' \\ \boldsymbol{\sigma} & \Sigma_{11} \end{pmatrix}$$

*are* p.s.d. *and fixed. In particular,*

$$P[\,|x_1| \leqslant y_1,\ \mathbf{x}_2 \in D(\mathbf{x}_2)\,] \geqslant P[\,|x_1| \leqslant y_1\,]\,P[\,\mathbf{x}_2 \in D(\mathbf{x}_2)\,]. \tag{2.3.5}$$

REMARK 2.3.2. A set $K$ will be said to be *convex* if for every $\mathbf{x}$ and $\mathbf{y} \in K$ and for every $\alpha$ lying in $[0,1]$, $\alpha\mathbf{x} + (1-\alpha)\mathbf{y} \in K$.

For example, let $K = \{\mathbf{x} : \mathbf{x}^* A \mathbf{x} \leqslant c\}$, where $A$ is p.s.d. and $0 \leqslant c < \infty$. Then $K$ is a convex set, because if $\mathbf{z} = \alpha\mathbf{x} + (1-\alpha)\mathbf{y}$ with $\mathbf{x} \in K$, $\mathbf{y} \in K$, and $0 \leqslant \alpha \leqslant 1$, then

$$\begin{aligned} \mathbf{z}^* A \mathbf{z} &= \alpha^2 \mathbf{x}^* A \mathbf{x} + (1-\alpha)^2 \mathbf{y}^* A \mathbf{y} + \alpha(1-\alpha)(\mathbf{x}^* A \mathbf{y} + \mathbf{y}^* A \mathbf{x}) \\ &\leqslant \left[\alpha(\mathbf{x}^* A \mathbf{x})^{1/2} + (1-\alpha)(\mathbf{y}^* A \mathbf{y})^{1/2}\right]^2 \leqslant c, \end{aligned}$$

since by Corollary 1.10.2, $\operatorname{Re}(\mathbf{x}^* A \mathbf{y}) \leqslant \operatorname{mod}(\mathbf{x}^* A \mathbf{y}) \leqslant [(\mathbf{x}^* A \mathbf{x})(\mathbf{y}^* A \mathbf{y})]^{1/2}$, where Re means *real part*.

## 2.4 Multiple Correlation

Let

$$\mathbf{z} = \begin{pmatrix} y \\ \mathbf{x} \end{pmatrix}$$

be a $p$-component random vector, where $\mathbf{x}$ is a $p-1$ vector. Let

$$E(\mathbf{z})=\mathbf{0} \quad \text{and} \quad \operatorname{Cov}(\mathbf{z})=\begin{pmatrix} \sigma_{11} & \boldsymbol{\sigma}_{12}' \\ \boldsymbol{\sigma}_{12} & \Sigma_{22} \end{pmatrix}.$$

Let us consider a linear combination of $\mathbf{x}$, say $\boldsymbol{\alpha}'\mathbf{x}$, such that

$$(2.4.1) \quad E(y-\boldsymbol{\alpha}'\mathbf{x})^2=\sigma_{11}-2\boldsymbol{\alpha}'\boldsymbol{\sigma}_{12}+\boldsymbol{\alpha}'\Sigma_{22}\boldsymbol{\alpha}$$
$$=(\boldsymbol{\alpha}-\Sigma_{22}^{-1}\boldsymbol{\sigma}_{12})'\Sigma_{22}(\boldsymbol{\alpha}-\Sigma_{22}^{-1}\boldsymbol{\sigma}_{12})+\sigma_{11}-\boldsymbol{\sigma}_{12}'\Sigma_{22}^{-1}\boldsymbol{\sigma}_{12}$$

has a minimum. It can easily be seen that the minimum occurs at

$$(2.4.2) \qquad \hat{\boldsymbol{\alpha}}\equiv\boldsymbol{\beta}=\Sigma_{22}^{-1}\boldsymbol{\sigma}_{12}.$$

Since

$$E(y-\boldsymbol{\alpha}'\mathbf{x})^2=E\left\{E[y-E(y|\mathbf{x})]^2|\mathbf{x}\right\}+E\left\{[E(y|\mathbf{x})-\boldsymbol{\alpha}'\mathbf{x}]^2|\mathbf{x}\right\},$$

it follows that $E(y-\boldsymbol{\alpha}'\mathbf{x})^2$ will be at a minimum iff

$$E(y|\mathbf{x})=\boldsymbol{\sigma}_{12}'\Sigma_{22}^{-1}\mathbf{x},$$

which is what we obtained in the previous section under the assumption of normality. Thus $\boldsymbol{\sigma}_{12}'\Sigma_{22}^{-1}\mathbf{x}$ may be considered as the best linear estimate of $y$ in terms of $\mathbf{x}$ in the sense that it minimizes Eq. (2.4.1) with or without the assumption of normality. Hence, for any constant $c$ and $\boldsymbol{\alpha}$, we have at this point

$$E(y-\boldsymbol{\beta}'\mathbf{x})^2\leqslant E(y-c\boldsymbol{\alpha}'\mathbf{x})^2.$$

That is,

$$-2\boldsymbol{\beta}'\boldsymbol{\sigma}_{12}+\boldsymbol{\beta}'\Sigma_{22}\boldsymbol{\beta}\leqslant-2c\boldsymbol{\alpha}'\boldsymbol{\sigma}_{12}+c^2\boldsymbol{\alpha}'\Sigma_{22}\boldsymbol{\alpha}.$$

Choosing

$$c^2=\frac{\boldsymbol{\beta}'\Sigma_{22}\boldsymbol{\beta}}{\boldsymbol{\alpha}'\Sigma_{22}\boldsymbol{\alpha}},$$

we find that

$$(2.4.3) \qquad \rho_{y\mathbf{x}}(\alpha)\equiv\frac{\boldsymbol{\alpha}'\boldsymbol{\sigma}_{12}}{\sigma_{11}^{1/2}(\boldsymbol{\alpha}'\Sigma_{22}\boldsymbol{\alpha})^{1/2}}\leqslant\frac{\boldsymbol{\beta}'\boldsymbol{\sigma}_{12}}{\sigma_{11}^{1/2}(\boldsymbol{\beta}'\Sigma_{22}\boldsymbol{\beta})^{1/2}}$$
$$=\frac{(\boldsymbol{\sigma}_{12}'\Sigma_{22}^{-1}\boldsymbol{\sigma}_{12})^{1/2}}{\sigma_{11}^{1/2}}=\rho_{y\mathbf{x}}.$$

It can be seen that the left side is the correlation between $y$ and $\boldsymbol{\alpha}'\mathbf{x}$, and the right side is the correlation between $y$ and $\boldsymbol{\beta}'\mathbf{x}$. Thus the right side is the maximum correlation between $y$ and linear combinations of $\mathbf{x}$. This maximum correlation is known as the *multiple* correlation and lies between 0 and 1 (see above). Another useful formula is

$$1-\rho_{y\mathbf{x}}^2 = 1-\frac{\boldsymbol{\sigma}_{12}'\Sigma_{22}^{-1}\boldsymbol{\sigma}_{12}}{\sigma_{11}}$$

$$=\frac{\sigma_{11}-\boldsymbol{\sigma}_{12}'\Sigma_{22}^{-1}\boldsymbol{\sigma}_{12}}{\sigma_{11}}=\frac{1}{\sigma^{11}\sigma_{11}}.$$

That is,

$$\rho_{y\mathbf{x}}^2 = 1-\frac{1}{\sigma_{11}\sigma^{11}}. \tag{2.4.4}$$

## 2.5 Canonical Correlations

The multiple correlation measures association between a random variable and a random vector variable. We will now extend it to the case of two random vector variables, say, an $r$ vector $\mathbf{x}$ and an $s$ vector $\mathbf{y}$. Without loss of generality, we assume that $r \leqslant s$. Since we are only interested in variances and covariances, we may assume without loss of generality that $E(\mathbf{x})=\mathbf{0}$ and $E(\mathbf{y})=\mathbf{0}$. Let the covariance matrix of the $r+s$ vector $(\mathbf{x}',\mathbf{y}')'$ be partitioned as

$$\Sigma = \begin{matrix} & r & s \\ r & \Sigma_{11} & \Sigma_{12} \\ s & \Sigma_{12}' & \Sigma_{22} \end{matrix},$$

where $\Sigma$ is positive definite. As in the previous section, let us consider arbitrary linear functions of $\mathbf{x}$ and $\mathbf{y}$, say, $\boldsymbol{\alpha}'\mathbf{x}$ and $\boldsymbol{\gamma}'\mathbf{y}$, respectively. The correlation between these two random variables, $\boldsymbol{\alpha}'\mathbf{x}$ and $\boldsymbol{\gamma}'\mathbf{y}$, is given by

$$\rho_{\mathbf{x},\mathbf{y}}(\boldsymbol{\alpha},\boldsymbol{\gamma})=\frac{\boldsymbol{\alpha}'\Sigma_{12}\boldsymbol{\gamma}}{(\boldsymbol{\alpha}'\Sigma_{11}\boldsymbol{\alpha})^{1/2}(\boldsymbol{\gamma}'\Sigma_{22}\boldsymbol{\gamma})^{1/2}}. \tag{2.5.1}$$

From Eq. (2.4.3), it follows that

$$\rho_{\mathbf{x},\mathbf{y}}(\boldsymbol{\alpha},\boldsymbol{\gamma}) \leqslant \frac{\boldsymbol{\alpha}'\Sigma_{12}\Sigma_{22}^{-1}\Sigma_{12}'\boldsymbol{\alpha}}{(\boldsymbol{\alpha}'\Sigma_{11}\boldsymbol{\alpha})^{1/2}(\boldsymbol{\alpha}'\Sigma_{12}\Sigma_{22}^{-1}\Sigma_{12}'\boldsymbol{\alpha})^{1/2}} \tag{2.5.2}$$

$$=\frac{(\boldsymbol{\alpha}'\Sigma_{12}\Sigma_{22}^{-1}\Sigma_{12}'\boldsymbol{\alpha})^{1/2}}{(\boldsymbol{\alpha}'\Sigma_{11}\boldsymbol{\alpha})^{1/2}}.$$

Since $\Sigma_{11}^{-1}\Sigma_{12}\Sigma_{22}^{-1}\Sigma_{12}'$ has at most $r$ nonzero roots ($r \leqslant s$), it follows that any of these $r$ roots will measure association between these two vectors $\mathbf{x}$ and $\mathbf{y}$. These $r$ roots are called canonical correlations.

When $\Sigma$ is singular, some of the canonical correlations will be unities, and the modified results are given by Khatri (1976b).

## 2.6 Partial Correlation

Let $\mathbf{x}$ be a $p$-component vector, where $\mathbf{x}'=(x_1,x_2,\mathbf{x}_3')$, and $\mathbf{x}_3$ is a $p-2$ vector. Let

$$\boldsymbol{\mu}=\begin{matrix}1\\1\\p-2\end{matrix}\begin{pmatrix}\mu_1\\\mu_2\\\boldsymbol{\mu}_3\end{pmatrix},$$

$$\Sigma=\begin{matrix}\\1\\1\\p-2\end{matrix}\begin{matrix}\begin{matrix}1 & 2 & p-2\end{matrix}\\\begin{pmatrix}\sigma_{11} & \sigma_{12} & \boldsymbol{\sigma}_{13}'\\\sigma_{12} & \sigma_{22} & \boldsymbol{\sigma}_{23}'\\\boldsymbol{\sigma}_{13} & \boldsymbol{\sigma}_{23} & \Sigma_{33}\end{pmatrix}\end{matrix}>0,$$

and

$$\Sigma^{-1}=\begin{pmatrix}\sigma^{11} & \sigma^{12} & -\\\sigma^{12} & \sigma^{22} & -\\- & - & -\end{pmatrix}.$$

Let $\mathbf{a}'\mathbf{x}_3$ be the best linear estimator of $x_1$ (that is, $\mathbf{a}$ is chosen so that the correlation between $x_1$ and $\mathbf{a}'\mathbf{x}_3$ is a maximum). Hence

$$\mathbf{a}=\Sigma_{33}^{-1}\boldsymbol{\sigma}_{13}.$$

Similarly, let $\mathbf{b}'\mathbf{x}_3$ be the best linear estimator of $x_2$. Then

$$\mathbf{b}=\Sigma_{33}^{-1}\boldsymbol{\sigma}_{23}.$$

The partial correlation between $x_1$ and $x_2$, denoted by $\rho_{12.34\cdots p}$ is defined to be the simple correlation between

$$x_1-\boldsymbol{\sigma}_{13}'\Sigma_{33}^{-1}\mathbf{x}_3 \quad \text{and} \quad x_2-\boldsymbol{\sigma}_{23}'\Sigma_{33}^{-1}\mathbf{x}_3,$$

that is, it is the correlation between $x_1$ and $x_2$ after eliminating the best linear effects of $\mathbf{x}_3$ from both variables. Hence

$$\rho_{12.3\cdots p}=\frac{\sigma_{12}-\boldsymbol{\sigma}_{13}'\Sigma_{33}^{-1}\boldsymbol{\sigma}_{23}}{\left(\sigma_{11}-\boldsymbol{\sigma}_{13}'\Sigma_{33}^{-1}\boldsymbol{\sigma}_{13}\right)^{1/2}\left(\sigma_{22}-\boldsymbol{\sigma}_{23}'\Sigma_{33}^{-1}\boldsymbol{\sigma}_{23}\right)^{1/2}}$$

$$=-\sigma^{12}/(\sigma^{11}\sigma^{22})^{1/2}.$$

Note that if $\mathbf{x}\sim N_p(0,\Sigma)$, $\Sigma>0$, the conditional distribution of $\begin{pmatrix}x_1\\x_2\end{pmatrix}$ given $\mathbf{x}_3$ is

$$N_2\left(\begin{pmatrix}\boldsymbol{\sigma}_{13}'\\\boldsymbol{\sigma}_{23}'\end{pmatrix}\Sigma_{33}^{-1}\mathbf{x}_3,\Delta\right),$$

where

$$\Delta=\begin{pmatrix} \sigma_{11}-\boldsymbol{\sigma}_{13}'\Sigma_{33}^{-1}\boldsymbol{\sigma}_{13} & \sigma_{12}-\boldsymbol{\sigma}_{13}'\Sigma_{33}^{-1}\boldsymbol{\sigma}_{23} \\ \sigma_{12}-\boldsymbol{\sigma}_{13}'\Sigma_{33}^{-1}\boldsymbol{\sigma}_{23} & \sigma_{22}-\boldsymbol{\sigma}_{23}'\Sigma_{33}^{-1}\boldsymbol{\sigma}_{23} \end{pmatrix}.$$

Hence we find that in this case, the partial correlation between $x_1$ and $x_2$ is nothing but the conditional correlation between $x_1$ and $x_2$ given $\mathbf{x}_3$:

$$\rho_{12.34\cdots p}=\frac{\operatorname{Cov}(x_1,x_2|\mathbf{x}_3)}{\left[\operatorname{Var}(x_1|\mathbf{x}_3)\operatorname{Var}(x_2|\mathbf{x}_3)\right]^{1/2}},$$

but this need *not* be true when $\mathbf{x}$ is not normally distributed.

## 2.7 Random Sample from $N_p(\boldsymbol{\theta},\Sigma)$

Let $\mathbf{x}\sim N_p(\boldsymbol{\theta},\Sigma)$ and $\Sigma>0$. Let $\mathbf{x}_1,\mathbf{x}_2,\dots,\mathbf{x}_N$ be independent samples on $\mathbf{x}$. We will write $X$ for the observation matrix

$$X\equiv(\mathbf{x}_1,\dots,\mathbf{x}_N)\equiv\begin{bmatrix} x_{11} & x_{12} & \cdots & x_{1N} \\ \vdots & \vdots & & \vdots \\ x_{p1} & x_{p2} & \cdots & x_{pN} \end{bmatrix}. \tag{2.7.1}$$

Noting that the trace of a scalar is scalar and $\operatorname{tr}(\mathbf{a}_1'\mathbf{a}_1+\mathbf{a}_2'\mathbf{a}_2+\mathbf{a}_3'\mathbf{a}_3)=\operatorname{tr}(\mathbf{a}_1\mathbf{a}_1'+\mathbf{a}_2\mathbf{a}_2'+\mathbf{a}_3\mathbf{a}_3')=\operatorname{tr}(\mathbf{a}_1,\mathbf{a}_2,\mathbf{a}_3)(\mathbf{a}_1,\mathbf{a}_2,\mathbf{a}_3)'$, we find that the pdf of $X$ is given by

$$\begin{aligned} p(X)&=\left[(2\pi)^p|\Sigma|\right]^{-\frac{1}{2}N}\exp\left(-\frac{1}{2}\sum_{i=1}^{N}(\mathbf{x}_i-\boldsymbol{\theta})'\Sigma^{-1}(\mathbf{x}_i-\boldsymbol{\theta})\right) \\ &=\left[(2\pi)^p|\Sigma|\right]^{-\frac{1}{2}N}\operatorname{etr}\left[-\tfrac{1}{2}\Sigma^{-1}(X-\boldsymbol{\theta}\mathbf{e}')(X-\boldsymbol{\theta}\mathbf{e}')'\right], \end{aligned} \tag{2.7.2}$$

where etr denotes the exponential of a trace of a matrix, $\mathbf{e}'=(1,1,\dots,1)$ (an $N$ row vector of ones), and

$$E(X)=\begin{bmatrix} \theta_1 & \theta_1 & \cdots & \theta_1 \\ \theta_2 & \theta_2 & \cdots & \theta_2 \\ \vdots & \vdots & & \vdots \\ \theta_p & \theta_p & \cdots & \theta_p \end{bmatrix}=\boldsymbol{\theta}\mathbf{e}'.$$

For convenience of notation, we will write

$$X\sim N_{p,N}(\boldsymbol{\theta}\mathbf{e}',\Sigma,I_N) \tag{2.7.3}$$

if $X$ has the pdf given by Eq. (2.7.2). If instead of Eq. (2.7.2), $X$ has the pdf given by

$$p(X)=\left[(2\pi)^p|\Sigma|\right]^{-\frac{1}{2}N}|A|^{-\frac{1}{2}p}\operatorname{etr}\left[-\tfrac{1}{2}\Sigma^{-1}(X-\eta)A^{-1}(X-\eta)'\right], \tag{2.7.4}$$

then we shall write

$$X \sim N_{p,N}(\eta, \Sigma, A). \tag{2.7.5}$$

This convention implies that the observations $\mathbf{x}_1, \ldots, \mathbf{x}_N$ are not independent. We shall use this notation even when $\Sigma \geqslant 0$ and $A \geqslant 0$, because the cf of $X$ is given by

$$E[\operatorname{etr}(iT'X)] = \operatorname{etr}\left(iT'\eta - \tfrac{1}{2}T'\Sigma TA\right) \tag{2.7.6}$$

where $T$ is any $p \times N$ real matrix.

## 2.8 Estimation of $\theta$ and $\Sigma$ if $X \sim N_{p,N}(\theta e', \Sigma, I_N)$

Let a random sample of size $N$ be given from $N_p(\theta, \Sigma)$. Then $X \sim N_{p,N}(\theta \mathbf{e}', \Sigma, I_N)$. Since $\mathbf{e}'X' = N\bar{\mathbf{x}}' = N(\bar{x}_1, \ldots, \bar{x}_p)$, where $\bar{x}_\alpha = N^{-1}\Sigma_{j=1}^N x_{\alpha j}$, $\alpha = 1, 2, \ldots, p$, and since

$$\begin{aligned}(X - \theta\mathbf{e}')(X - \theta\mathbf{e}')' &= XX' - N\theta\bar{\mathbf{x}}' - N\bar{\mathbf{x}}\theta' + N\theta\theta' \\ &= XX' - N\bar{\mathbf{x}}\bar{\mathbf{x}}' + N(\bar{\mathbf{x}} - \theta)(\bar{\mathbf{x}} - \theta)' \\ &\equiv V + N(\bar{\mathbf{x}} - \theta)(\bar{\mathbf{x}} - \theta)',\end{aligned}$$

where $V = XX' - N\bar{\mathbf{x}}\bar{\mathbf{x}}'$, we get

$$p(X) = \left[(2\pi)^p|\Sigma|\right]^{-\frac{1}{2}N} \operatorname{etr}\left\{-\tfrac{1}{2}\Sigma^{-1}\left[V + N(\bar{\mathbf{x}} - \theta)(\bar{\mathbf{x}} - \theta)'\right]\right\}.$$

Hence, from Neyman's factorization criterion [see, e.g., Lehmann (1959)], $(V, \bar{\mathbf{x}})$ is sufficient for $(\Sigma, \theta)$. Now we shall show that not only is $(V, \bar{\mathbf{x}})$ sufficient for $(\Sigma, \theta)$, but its distribution is *complete* as well. For this, we need to show that there does not exist a nontrivial unbiased estimate of 0 with respect to the distribution of $(V, \bar{\mathbf{x}})$ except for zero itself. That is, we will show that if

$$Eh(\bar{\mathbf{x}}, V) \equiv 0$$

for all $\theta$ and $\Sigma$, then $h(\bar{\mathbf{x}}, V) = 0$ almost everywhere, where $h(\bar{\mathbf{x}}, V)$ is any function of $\bar{\mathbf{x}}$ and $V$. Thus, we are given that

$$\int \cdots \int |V|^{\frac{1}{2}(n-p-1)} h(\bar{\mathbf{x}}, V) \operatorname{etr}\left\{-\tfrac{1}{2}\Sigma^{-1}\left[V + N(\bar{\mathbf{x}} - \theta)(\bar{\mathbf{x}} - \theta)'\right]\right\} d\bar{\mathbf{x}}\, dV \equiv 0,$$

where the pdfs of $V$ and $\bar{\mathbf{x}}$ are given in Chapter 3. Writing $\Sigma^{-1} = 2A$ and $\theta = (2A)^{-1}\mu$, we get

$$\int \cdots \int g\left[\bar{\mathbf{x}}, (V + N\bar{\mathbf{x}}\bar{\mathbf{x}}') - N\bar{\mathbf{x}}\bar{\mathbf{x}}'\right] \operatorname{etr}\left[-A(V + N\bar{\mathbf{x}}\bar{\mathbf{x}}') + N\mu\bar{\mathbf{x}}'\right] d\bar{\mathbf{x}}\, dV \equiv 0,$$

for all $A > 0$ and $\mu$, where

$$g\left[\bar{\mathbf{x}}, (V + N\bar{\mathbf{x}}\bar{\mathbf{x}}') - N\bar{\mathbf{x}}\bar{\mathbf{x}}'\right] = |V|^{\frac{1}{2}(n-p-1)} h(\bar{\mathbf{x}}, V),$$

which is the Laplace transform with respect to the variables $N\bar{\mathbf{x}}$ and $V+N\bar{\mathbf{x}}\bar{\mathbf{x}}'$. Hence

$$h(\bar{\mathbf{x}},V)=0$$

except for a null set. Hence $h(\bar{\mathbf{x}},V)=0$ a.e.

### *2.8.1 Maximum-Likelihood Estimates*

In this subsection we show that the maximum-likelihood estimates of $\Sigma$ and $\mu$ are respectively given by $N^{-1}V$ and $\bar{\mathbf{x}}$. For this, the likelihood function is

$$\begin{aligned} L(\boldsymbol{\theta},\Sigma) &\equiv (2\pi)^{-\frac{1}{2}pN}|\Sigma|^{-\frac{1}{2}N}\operatorname{etr}\left(-\tfrac{1}{2}\Sigma^{-1}V\right) \\ &\quad \times \exp\left[-\tfrac{1}{2}N(\bar{\mathbf{x}}-\boldsymbol{\theta})'\Sigma^{-1}(\bar{\mathbf{x}}-\boldsymbol{\theta})\right] \\ &\leqslant (2\pi)^{-\frac{1}{2}pN}|\Sigma|^{-\frac{1}{2}N}\operatorname{etr}\left(-\tfrac{1}{2}\Sigma^{-1}V\right) \\ &\leqslant (2\pi)^{-\frac{1}{2}pN}|V/N|^{-\frac{1}{2}N}\operatorname{etr}\left(-\tfrac{1}{2}pN\right) \\ &= L(\bar{\mathbf{x}},V/N), \end{aligned}$$

where the first inequality holds for $(\bar{\mathbf{x}}-\boldsymbol{\theta})'\Sigma^{-1}(\bar{\mathbf{x}}-\boldsymbol{\theta})\geqslant 0$, and the second inequality follows using Theorem 1.10.4, namely,

$$|\Sigma|^{-a}\operatorname{etr}(-b\Sigma^{-1}V)\leqslant|bV/a|^{-a}e^{-pa},$$

and the equality holds iff $\boldsymbol{\theta}=\bar{\mathbf{x}}$ and $\Sigma=V/N$. This proves the required result. For an alternative proof, see Watson (1964).

The following result is useful in obtaining maximum-likelihood estimates of any one-to-one function of the parameter.

**Lemma 2.8.1** *Let $\theta\in\Omega$, and let $L(\theta)$ denote the likelihood function. Let there exist a $\hat{\theta}\in\Omega$ such that $L(\hat{\theta})\geqslant L(\theta)$ for all $\theta\in\Omega$ (i.e., $\hat{\boldsymbol{\theta}}$ is* MLE*). Let $U$ be a one-to-one transformation from $\Omega$ to $\Lambda$. Then a maximum-likelihood estimator of $U(\theta)$ is $U(\hat{\theta})$.*

The proof is trivial.

## 2.9 Complex Multivariate Normal Distribution

Let $\mathbf{z}=\mathbf{x}+i\mathbf{y}$. Then $\mathbf{z}$ is said to be a complex random $p$ vector if $\mathbf{x}$ (the real part of $\mathbf{z}$) and $\mathbf{y}$ (the imaginary part of $\mathbf{z}$) are both random $p$ vectors. The mean and covariance matrix of $\mathbf{z}=(z_1,\ldots,z_p)'$ are, respectively, given by

$$\boldsymbol{\theta}\equiv E(\mathbf{z})=E(\mathbf{x})+iE(\mathbf{y})\equiv\boldsymbol{\theta}_1+i\boldsymbol{\theta}_2,$$

and

$$Q = \mathrm{Cov}(\mathbf{z}) = E[(\mathbf{z}-\boldsymbol{\theta})(\mathbf{z}-\boldsymbol{\theta})^*].$$

Since $Q$ is at least positive semidefinite Hermitian, we can write

$$Q = \Sigma_1 + i\Sigma_2,$$

where $\Sigma_1$ is at least positive semidefinite and $\Sigma_2$ is skew-symmetric.

For any complex vector $\mathbf{a}$, a linear combination of $\mathbf{z}$ is $\mathbf{a}^*\mathbf{z} = (\mathbf{a}_1' - i\mathbf{a}_2')(\mathbf{x}+i\mathbf{y}) = (\mathbf{a}_1'\mathbf{x}+\mathbf{a}_2'\mathbf{y}) + i(\mathbf{a}_1'\mathbf{y} - \mathbf{a}_2'\mathbf{x})$. The real part (Re) of this linear combination is given by $\mathbf{a}_1'\mathbf{x}+\mathbf{a}_2'\mathbf{y}$, and the imaginary part (Im) by $\mathbf{a}_1'\mathbf{y}-\mathbf{a}_2'\mathbf{x}$ [i.e., $\mathrm{Re}(\mathbf{a}^*\mathbf{z}) = \mathbf{a}_1'\mathbf{x}+\mathbf{a}_2'\mathbf{y}$ and $\mathrm{Im}(\mathbf{a}^*\mathbf{z}) = \mathbf{a}_1'\mathbf{x}-\mathbf{a}_2'\mathbf{y}$].

On the analogy of Definition 2.1.2, in the real case, we now give a definition of a complex multivariate normal distribution, denoted by $CN_p(\boldsymbol{\theta}, Q)$.

**Definition 2.9.1** Let $\mathbf{z}$ be a complex $p$ random vector with mean $\boldsymbol{\theta}$ and covariance matrix $Q$ such that the variances of the real and imaginary parts of every linear combination of $\mathbf{z}$ are equal (in this case given by $\frac{1}{2}\mathbf{a}^*Q\mathbf{a}$ for any $\mathbf{a}$). Then $\mathbf{z} \sim CN_p(\boldsymbol{\theta}, Q)$ if the real part (or the imaginary part) of every linear combination of $\mathbf{z}$ is normally distributed.

Some of the consequences of the above definition are given below. Let

$$\Sigma_{11} = \mathrm{Cov}(\mathbf{x}), \qquad \Sigma_{22} = \mathrm{Cov}(\mathbf{y}), \qquad \Sigma_{12} = \mathrm{Cov}(\mathbf{x}, \mathbf{y}).$$

Then, from the above definition (equality of two variances) we have for *every* $\mathbf{a}_1$ and $\mathbf{a}_2$,

$$\mathbf{a}_1'\Sigma_{11}\mathbf{a}_1 + \mathbf{a}_2'\Sigma_{22}\mathbf{a}_2 + 2\mathbf{a}_1'\Sigma_{12}\mathbf{a}_2 = \mathbf{a}_2'\Sigma_{11}\mathbf{a}_2 + \mathbf{a}_1'\Sigma_{22}\mathbf{a}_1 - 2\mathbf{a}_2'\Sigma_{12}\mathbf{a}_1.$$

Hence, taking $\mathbf{a}_1 \equiv \mathbf{a}_2$, we find that

$$\mathbf{a}_1'\Sigma_{12}\mathbf{a}_1 = 0.$$

Thus $\Sigma_{12}$ is a skew-symmetric matrix, and

$$\mathbf{a}_1'(\Sigma_{11}-\Sigma_{22})\mathbf{a}_1 - \mathbf{a}_2'(\Sigma_{11}-\Sigma_{22})\mathbf{a}_2 = 0$$

for every $\mathbf{a}_1$ and $\mathbf{a}_2$. This gives $\Sigma_{11} = \Sigma_{22}$. Hence if $\mathbf{z} \sim CN_p(\boldsymbol{\theta}, Q)$, where $Q = \Sigma_1 + i\Sigma_2$, then

$$\tfrac{1}{2}\Sigma_1 = \Sigma_{11} = \Sigma_{22},$$

and

$$\tfrac{1}{2}\Sigma_2 = \Sigma_{12}'.$$

(This follows from comparing the variance $\frac{1}{2}\mathbf{a}^*Q\mathbf{a}$ with $\mathbf{a}_1'\Sigma_{11}\mathbf{a}_1 + \mathbf{a}_2'\Sigma_{22}\mathbf{a}_2 + 2\mathbf{a}_1'\Sigma_{12}\mathbf{a}_2$ for every $\mathbf{a}_1$ and $\mathbf{a}_2$.) Thus we find that the covariance of $(\mathbf{x}', \mathbf{y}')'$ is given by

$$\Sigma \equiv \frac{1}{2}\begin{pmatrix} \Sigma_1 & -\Sigma_2 \\ \Sigma_2 & \Sigma_1 \end{pmatrix}.$$

Since the real part of any linear combination of $\mathbf{z}$ is

$$u \equiv \mathbf{a}_1'\mathbf{x} + \mathbf{a}_2'\mathbf{y} = (\mathbf{a}_1', \mathbf{a}_2')\begin{pmatrix} \mathbf{x} \\ \mathbf{y} \end{pmatrix},$$

which by Definition 2.9.1 is normally distributed for every $\mathbf{a}_1, \mathbf{a}_2$, it follows that

$$\begin{pmatrix} \mathbf{x} \\ \mathbf{y} \end{pmatrix} \sim N_{2p}\left(\begin{pmatrix} \boldsymbol{\theta}_1 \\ \boldsymbol{\theta}_2 \end{pmatrix}, \Sigma\right).$$

Hence,

$$v = (-\mathbf{a}_2', \mathbf{a}_1')\begin{pmatrix} \mathbf{x} \\ \mathbf{y} \end{pmatrix} \quad \text{is normally distributed.}$$

That is, the imaginary part of every linear combination of $\mathbf{z}$ is also normally distributed if the real part is normally distributed. It can easily be shown that the covariance between $u$ and $v$ is zero. Hence $u$ and $v$ are not only normally but also independently distributed.

For some applications of complex normal distribution, see Goodman (1963).

Thus we have another definition of a complex normal distribution.

**Definition 2.9.2** Let $\mathbf{z} = \mathbf{x} + i\mathbf{y}$ with mean $\boldsymbol{\theta}$ and covariance matrix $Q = (\Sigma_1 + i\Sigma_2)$. Then $z \sim CN_p(\boldsymbol{\theta}, Q)$ iff

$$\begin{pmatrix} \mathbf{x} \\ \mathbf{y} \end{pmatrix} \sim N_{2p}\left(\begin{pmatrix} \boldsymbol{\theta}_1 \\ \boldsymbol{\theta}_2 \end{pmatrix}, \Sigma\right),$$

where

$$\Sigma = \frac{1}{2}\begin{pmatrix} \Sigma_1 & -\Sigma_2 \\ \Sigma_2 & \Sigma_1 \end{pmatrix}, \qquad \theta = \boldsymbol{\theta}_1 + i\boldsymbol{\theta}_2.$$

If we assume that there are no linear dependences among the components of $\mathbf{x}$ and $\mathbf{y}$, that is, $\Sigma$ is positive definite, or equivalently $Q$ is positive definite, the density of $\mathbf{z}$ can be written down. Since

$$\begin{aligned} |Q|^2 &= |\Sigma_1 + i\Sigma_2|^2 \\ &= |\Sigma_1 + i\Sigma_2|\,|\Sigma_1 - i\Sigma_2| = |\Sigma_1|^2|I + i\Sigma_1^{-1}\Sigma_2|\,|I - i\Sigma_1^{-1}\Sigma_2| \\ &= |\Sigma_1|^2|I + \Sigma_1^{-1}\Sigma_2\Sigma_1^{-1}\Sigma_2| \\ &= \left|\begin{array}{c:c} \Sigma_1 & -\Sigma_2 \\ \hdashline \Sigma_2 & \Sigma_1 \end{array}\right| = |2\Sigma| \end{aligned}$$

and

$$Q^{-1}=(\Sigma_1+i\Sigma_2)^{-1}$$
$$=(\Sigma_1+\Sigma_2\Sigma_1^{-1}\Sigma_2)^{-1}-i\Sigma_1^{-1}\Sigma_2(\Sigma_1+\Sigma_2\Sigma_1^{-1}\Sigma_2)^{-1},$$

the density of $\begin{pmatrix}\mathbf{x}\\ \mathbf{y}\end{pmatrix}$, given by

$$(2\pi)^{-p}|\Sigma|^{-1/2}\exp\left[-\tfrac{1}{2}(\mathbf{x}'-\boldsymbol{\theta}_1',\mathbf{y}'-\boldsymbol{\theta}_2')\Sigma^{-1}(\mathbf{x}'-\boldsymbol{\theta}_1',\mathbf{y}'-\boldsymbol{\theta}_2')'\right],$$

can be written in the complex form as

$$(\pi)^{-p}|Q|^{-1}\exp\left[-(\bar{\mathbf{z}}-\bar{\boldsymbol{\theta}})'Q^{-1}(\mathbf{z}-\boldsymbol{\theta})\right],$$

and the cf of $\mathbf{z}$ is

$$E\left[\exp(i\,\mathrm{Re}(\mathbf{t}^*\mathbf{z}))\right]=\exp(i\,\mathrm{Re}(\mathbf{t}^*\boldsymbol{\theta})-\mathbf{t}^*Q\mathbf{t}),$$

where $\mathbf{t}=\mathbf{t}_1+i\mathbf{t}_2$.

## 2.10 Asymptotic Distributions

In this section we give some results on asymptotic distributions.

**Theorem 2.10.1 (Multivariate central limit theorem)** *Let the p-component vectors $\mathbf{x}_1,\mathbf{x}_2,\ldots$ be independently and identically distributed with means $E(\mathbf{x})=\boldsymbol{\mu}$ and covariance matrices $E[(\mathbf{x}-\boldsymbol{\mu})(\mathbf{x}-\boldsymbol{\mu})']=\Sigma$. Then as $n\to\infty$*

$$\mathcal{L}\left(n^{-1/2}\sum_{\alpha=1}^{n}(\mathbf{x}_\alpha-\boldsymbol{\mu})\right)\to N(\mathbf{0},\Sigma),$$

*where $\mathcal{L}(\mathbf{u})$ denotes the distribution of* $\mathbf{u}$.

This can be proved essentially on the same lines as in the univariate case by Cramèr (1946) by considering a linear combination of $\mathbf{x}$, say $\mathbf{l}'\mathbf{x}$.

**Theorem 2.10.2** *Let $\mathbf{u}_n$ be a p-component vector and $\boldsymbol{\theta}$ a fixed vector. Assume $\mathcal{L}(n^{1/2}(\mathbf{u}_n-\boldsymbol{\theta}))\to N_p(\mathbf{0},\Sigma)$ as $n\to\infty$. Let $\omega=f(\mathbf{u})$ be a real valued function of a vector $\mathbf{u}$ with first and second derivatives existing in a neighborhood of $\mathbf{u}=\boldsymbol{\theta}$. Then as $n\to\infty$*

$$\mathcal{L}(n^{1/2}[f(\mathbf{u}_n)-f(\boldsymbol{\theta})])\to N(0,\boldsymbol{\phi}_{\boldsymbol{\theta}}'\Sigma\boldsymbol{\phi}_{\boldsymbol{\theta}}),$$

*where $\boldsymbol{\phi}_Q$ is a p-component vector with ith element given by*

$$\left.\frac{\partial f(\mathbf{u})}{\partial u_i}\right|_{\mathbf{u}=\boldsymbol{\theta}}.$$

For proof, refer to Cramèr (1946, p. 366).

## 2.11 Results on Quadratic Forms

In this section, we give results on quadratic forms. Subsection 2.11.1 is devoted to the distribution, and Subsection 2.11.2 to the necessary and sufficient condition for independence and chi square.

### *2.11.1 Distribution of Quadratic Forms*

Let $\mathbf{x}$ be distributed as $N_p(\boldsymbol{\theta}, I)$. Then the distribution of $\mathbf{x}'\mathbf{x}=y$ is said to be noncentral chi square with $p$ degrees of freedom (d.f.) and noncentrality parameter $\lambda=\boldsymbol{\theta}'\boldsymbol{\theta}$. This will be denoted by $y=\mathbf{x}'\mathbf{x}\sim\chi_p^2(\lambda)$. When $\lambda=0$, then $y$ will be distributed as central chi square and it will be denoted by $y\sim\chi_p^2$. The cf of $y$ is given by

$$(2.11.1)\qquad E\left[\exp(ity)\right]=\prod_{j=1}^{p} E\left[\exp\left(itx_j^2\right)\right]$$

$$=\prod_{j=1}^{p}\left[(1-2it)^{-1/2}\exp\left(\frac{it\theta_j^2}{1-2it}\right)\right]$$

$$=(1-2it)^{-\frac{1}{2}p}\exp\left(\frac{it\lambda}{1-2it}\right).$$

From this, the distribution of $y$ is given by

$$(2.11.2)\qquad \sum_{j=0}^{\infty}\omega_j(\lambda)g(y\,|\,p+2j),$$

where

$$(2.11.3)\qquad g(y\,|\,\alpha)=\left\{2\Gamma\left(\tfrac{1}{2}\alpha\right)\right\}^{-1}\left(\tfrac{1}{2}y\right)^{\frac{1}{2}\alpha-1}\exp\left(-\tfrac{1}{2}y\right)$$

and

$$(2.11.4)\qquad \omega_j(\lambda)=\frac{\left(\frac{1}{2}\lambda\right)^j\exp\left(-\frac{1}{2}\lambda\right)}{j!}.$$

Let $\mathbf{x}\sim N_p(\boldsymbol{\theta}, I)$, and the quadratic function be given by $q=\mathbf{x}'A\mathbf{x}$. By Corollary 1.8.3, we can write $A=\Delta' D_\lambda\Delta$, where $\Delta$ is a semiorthogonal matrix such that $\Delta\Delta'=I_r$, $r=\rho(A)$, and $D_\lambda=\operatorname{diag}(\lambda_1,\ldots,\lambda_r)$ is nonsingular. Then, by Theorem 2.2.1, $\mathbf{y}=\Delta\mathbf{x}\sim N_r(\boldsymbol{\mu}, I_r)$, $\boldsymbol{\mu}=\Delta\boldsymbol{\theta}$, and the quadratic form

$q=\Sigma_{j=1}^{r}\lambda_j y_j^2$. Hence, the cf of $q$ is given by

$$E[\exp(itq)]=\prod_{j=1}^{r}E[\exp(it\lambda_j y_j^2)]$$

$$(2.11.5)\qquad =\prod_{j=1}^{r}\left\{(1-2it\lambda_j)^{-1/2}\exp\left[it\mu_j^2\lambda_j(1-2it\lambda_j)^{-1}\right]\right\}$$

$$=|I-2itD_\lambda|^{-1/2}\exp(it\boldsymbol{\mu}'D_\delta\boldsymbol{\mu})$$

$$=|I-2itA|^{-1/2}\exp\left[it\boldsymbol{\theta}'A(I-2itA)^{-1}\boldsymbol{\theta}\right],$$

where $\delta_j=\lambda_j(1-2it\lambda_j)^{-1}$ $(j=1,2,\ldots,r)$, $D_\delta=\operatorname{diag}(\delta_1,\ldots,\delta_r)$, $\Delta'D_\delta\Delta=A(I-2itA)^{-1}$, and $|I-2itD_\lambda|=|I-2itA|$.

The distribution of the quadratic form $q=\mathbf{x}'A\mathbf{x}$, when $q$ is p.d., has been considered by various authors in different situations and in different forms; for example, see Gurland (1955), Shah and Khatri (1961), Shah (1963), Ruben (1962, 1963), Johnson and Kotz (1967a, b) and its multivariate extension by Khatri (1971b). We shall present only one representation; for other representations, one should refer to Johnson and Kotz (1967b). For this purpose, let us denote

$$P=I-\lambda A^{-1}\quad\text{and}\quad z=(1-2it\lambda)^{-1}.$$

Then,

$$(I-2itA)=(\lambda^{-1}A)\left[I(1-2it\lambda)-(I-\lambda A^{-1})\right]=(I-P)^{-1}(I-Pz)z^{-1}$$

and

$$2itA(I-2itA)^{-1}=(I-2itA)^{-1}-I$$

$$=z(I-Pz)^{-1}(I-P)-I=(z-1)(I-Pz)^{-1}.$$

Hence, Eq. (2.11.5) can be rewritten as

$$(2.11.6)\qquad E[\exp(itq)]=a_0z^{p/2}|I-Pz|^{-1/2}\exp\left[\tfrac{1}{2}\boldsymbol{\theta}'\boldsymbol{\theta}+\tfrac{1}{2}(z-1)\boldsymbol{\theta}'(I-Pz)^{-1}\boldsymbol{\theta}\right],$$

where $a_0=|I-P|^{1/2}\exp(-\frac{1}{2}\boldsymbol{\theta}'\boldsymbol{\theta})$. Using Problem 1.25, we get

$$(2.11.7)\qquad |I-Pz|^{-1/2}\exp\left[\tfrac{1}{2}\boldsymbol{\theta}'\boldsymbol{\theta}+\tfrac{1}{2}(z-1)\boldsymbol{\theta}'(I-Pz)^{-1}\boldsymbol{\theta}\right]$$

$$=\exp\left(\sum_{j=1}^{\infty}z^jc_j\right)=\sum_{j=0}^{\infty}\omega_jz^j\quad\text{(say)},$$

where

$$(2.11.8)\qquad 2c_j=\lambda\boldsymbol{\theta}'A^{-1}P^{j-1}\boldsymbol{\theta}+\frac{\operatorname{tr}P^j}{j},\qquad P^0=I,\qquad j=1,2\ldots.$$

Differentiating Eq. (2.11.7) with respect to $z$, we get

$$\left(\sum_{j=1}^{\infty} jz^{j-1}c_j\right)\exp\left(\sum_{j=1}^{\infty} z^j c_j\right)=\sum_{j=1}^{\infty} j\omega_j z^{j-1},$$

or using Eq. (2.11.7),

$$\left(\sum_{j=1}^{\infty} jz^{j-1}c_j\right)\left(\sum_{j=0}^{\infty} \omega_j z^j\right)=\sum_{j=1}^{\infty} j\omega_j z^{j-1}.$$

Equating the coefficients of $z^k$, we get the recursive relation for $w_k$ as

$$(2.11.9)\qquad (k+1)\omega_{k+1}=\sum_{j=0}^{k}(j+1)\omega_{k-j}c_{j+1},$$

$$\omega_0=1 \quad\text{and}\quad \omega_1=c_1 \qquad \text{for}\quad k=0,1,\ldots.$$

Using Eqs. (2.11.7) and (2.11.6), we get the density function of $q$ as

$$(2.11.10)\qquad \sum_{j=0}^{\infty} a_0\omega_j\frac{g(q/\lambda|p+2j)}{\lambda},$$

where $g(y|\alpha)$ is defined in Eq. (2.11.3). The series given by Eq. (2.11.10) is absolutely convergent, and the best choice of $\lambda$ is $\lambda = 2\,\mathrm{ch}_1(A)\mathrm{ch}_p(A)/[\mathrm{ch}_1(A)+\mathrm{ch}_p(A)]$. For the proof, one can refer to Johnson and Kotz (1967a, b).

To obtain the distribution of any quadratic function, we should know the distribution of the difference of two noncentral chi-square variates. Let $y_j \sim \chi^2_{r_j}(\lambda_j), j=1,2$, be independently distributed, and let $y=ay_1-by_2$, with $a>0$, $b>0$. Then, using Eq. (2.11.2), the joint density of $y$ and $y_2$ is given by

$$(2.11.11)\quad \sum_{j_1=0}^{\infty}\sum_{j_2=0}^{\infty}\omega_{j_1}(\lambda_1)\omega_{j_2}(\lambda_2)g\left(\frac{y+by_2}{a}\middle| r_1+2j_1\right)\frac{g(y_2|r_2+2j_2)}{a}$$

for $[y+by_2>0,\ y_2>0]$ or $[-\infty<y<\infty$ and $y_2\geqslant \mathrm{Max}(-y/b,0)]$, where $\omega_j(\lambda)$ and $g(y|\alpha)$ are respectively defined by Eqs. (2.11.4) and (2.11.3). Integrating over $y_2$, we get the density of $y=a\chi^2_{r_1}(\lambda_1)-b\chi^2_{r_2}(\lambda_2)$ as

$$(2.11.12)\qquad \sum_{j_1=0}^{\infty}\sum_{j_2=0}^{\infty}\omega_{j_1}(\lambda_1)\omega_{j_2}(\lambda_2)\frac{f(y|r_1+2j_1,r_2+2j_2)}{a},$$

where

$$(2.11.13)\qquad f(y|\alpha,\beta)=\int_{\mathrm{Max}(-y/b,0)}^{\infty} g\left(\frac{y+bx}{a}\middle|\alpha\right)g(x|\beta)\,dx.$$

This gives the following:

**Theorem 2.11.1** *Let $y_1$ and $y_2$ be two independent noncentral chi squares $\chi^2_{r_1}(\lambda_1)$ and $\chi^2_{r_2}(\lambda_2)$. Then the density function of $y = ay_1 - by_2$ $(a>0, b>0)$ is given by Eq.* (2.11.12).

With the help of Eqs. (2.11.12) and (2.11.10), we can obtain the density function of any quadratic form, but this is left to the reader.

### *2.11.2 Necessary and Sufficient Conditions for Chi Square and Independence*

Let us consider the second degree polynomial, namely,

$$y = \mathbf{x}'A\mathbf{x} + 2\mathbf{b}'\mathbf{x} + c$$

where $\mathbf{x} \sim N_p(\mathbf{0}, I)$, $A$ is any $p \times p$ symmetric matrix, $\mathbf{b}$ is a $p \times 1$ column vector, and $c$ is a constant. By Corollary 1.8.3, we can write $A = \Delta' D_\lambda \Delta$, $\Delta_0' = (\Delta', \Delta_1')$ is an orthogonal matrix, $D_\lambda = \text{diag}(\lambda_1, \ldots, \lambda_r)$ is nonsingular, and $\rho(A) = r$. Let $\Delta_0 \mathbf{x} = \mathbf{z}$ and $\Delta_0 \mathbf{b} = \boldsymbol{\theta}$. Then $\mathbf{z} \sim N_p(\mathbf{0}, I)$, and the cf of $y$ is given by

$$(2.11.14)\quad E[\exp(ity)] = E\left[\exp\left(it\sum_{j=1}^{r}\lambda_j z_j^2 + 2it\sum_{j=1}^{p}\theta_j z_j + itc\right)\right]$$

$$= \exp(itc)\prod_{j=1}^{r} E\left[\exp\left(it\lambda_j z_j^2 + 2it\theta_j z_j\right)\right]\prod_{j=r+1}^{p} E\left[\exp(2it\theta_j z_j)\right]$$

$$= \left(\prod_{j=1}^{r}(1-2it\lambda_j)^{-1/2}\right)\exp\left(itc - 2t^2\sum_{j=r+1}^{p}\theta_j^2 - 2t^2\sum_{j=1}^{r}\frac{\theta_j^2}{1-2it\lambda_j}\right),$$

which can be rewritten as

$$(2.11.15)\quad E[\exp(ity)] = |I - 2itA|^{-1/2}\exp\left[itc - 2t^2\mathbf{b}'(I_p - 2itA)^{-1}\mathbf{b}\right].$$

The necessary and sufficient condition for $y = \mathbf{x}'A\mathbf{x} + 2\mathbf{b}'\mathbf{x} + c$ to be distributed as $\chi^2_s(\lambda)$ is that the cfs of $\chi^2_s(\lambda)$ and $y$ should be identical for all real $t$, i.e., for all real $t$,

$$(1-2it)^{-\frac{1}{2}s}\exp\left(it\frac{\lambda}{1-2it}\right) = |I - 2itA|^{-1/2}\exp\left[itc - 2t^2\mathbf{b}'(I - 2itA)^{-1}\mathbf{b}\right],$$

or for all real $t$

$$(2.11.16)\quad \frac{|I-2itA|}{(1-2it)^s} = \exp\left[-2it\lambda(1-2it)^{-1} + 2itc - 4t^2\mathbf{b}'(I-2itA)^{-1}\mathbf{b}\right].$$

Notice that the left side of this equation is a ratio of polynomials of finite degree in $t$, while the right side is the exponential of a ratio of polynomials of finite degree in $t$. This can happen iff for all real $t$ [see, for example, Laha (1956)],

$$|I-2itA|=(1-2it)^{s} \tag{2.11.17}$$

and

$$2it\lambda(1-2it)^{-1}-2itc+4t^{2}\mathbf{b}'(I-2itA)^{-1}\mathbf{b}=0. \tag{2.11.18}$$

Putting $\theta=1-2it$ in Eq. (2.11.17), we find

$$\prod_{j=1}^{r}\lambda_j\left[\theta+\left(\lambda_j^{-1}-1\right)\right]=\theta^{s},$$

where $r=\rho(A)$, and the $\lambda_j$'s are the nonzero ch roots of $A$. This shows that $r=s$ and $\lambda_j^{-1}=1$ for all $j=1,2,\ldots,s$. Further, putting this in Eq. (2.11.18), we find that

$$(I-A)\mathbf{b}=\mathbf{0}, \qquad \mathbf{b}'\mathbf{b}=\mathbf{b}'A\mathbf{b}=c=\lambda. \tag{2.11.19}$$

This proves

**Theorem 2.11.2** *Let* $\mathbf{x}\sim N_p(\mathbf{0},I)$ *and* $y=\mathbf{x}'A\mathbf{x}+2\mathbf{b}'\mathbf{x}+c$. *Then,* $y$ *is distributed as* $\chi_s^2(\lambda)$ iff $s=\rho(A)$, $A^2=A$, $\mathbf{b}=A\mathbf{b}$, *and* $\lambda=c=\mathbf{b}'\mathbf{b}=\mathbf{b}'A\mathbf{b}$.

**Corollary 2.11.1** *Let* $\mathbf{x}\sim N_p(\boldsymbol{\theta},\Sigma)$ *and* $y=\mathbf{x}'A\mathbf{x}+2\mathbf{b}'\mathbf{x}+c$. *Then* $y$ *is distributed as* $\chi_s^2(\lambda)$ *iff* $s=\operatorname{tr}A\Sigma=\rho(\Sigma A\Sigma)$, $\Sigma A\Sigma A\Sigma=\Sigma A\Sigma$, $\Sigma(\mathbf{b}+A\boldsymbol{\theta})=\Sigma A\Sigma(\mathbf{b}+A\boldsymbol{\theta})$, *and* $\lambda=c+2\mathbf{b}'\boldsymbol{\theta}+\boldsymbol{\theta}'A\boldsymbol{\theta}=(\mathbf{b}+A\boldsymbol{\theta})'\Sigma(\mathbf{b}+A\boldsymbol{\theta})$.

This follows from Theorem 2.11.2 by using Definition 2.1.1, i.e., $\mathbf{x}=\boldsymbol{\theta}+B\mathbf{u}$, $\mathbf{u}\sim N_r(\mathbf{0},I)$, $r=\rho(\Sigma)$, and $\Sigma=BB'$. This result is due to Khatri (1962b, 1963a).

**Corollary 2.11.2** *In Corollary* 2.11.1, *let* $\Sigma>0$ *and let* $\mathbf{y}=\mathbf{x}'A\mathbf{x}$. *Then a necessary and sufficient condition for* $y\sim\chi_s^2(\lambda)$ *is* $A=A\Sigma A$ *with* $s=\rho(A)$ *(or* $A\Sigma$ *is idempotent of rank* $s$*) where* $\lambda=\boldsymbol{\theta}'A\boldsymbol{\theta}$.

**Theorem 2.11.3** *Let* $\mathbf{x}\sim N_p(\mathbf{0},I)$. *Then* $y_j=\mathbf{x}'A_j\mathbf{x}+2\mathbf{b}_j'\mathbf{x}+c_j$, $j=1,2$, *are independently distributed iff* $A_1A_2=0$, $A_1\mathbf{b}_2=\mathbf{0}$, $A_2\mathbf{b}_1=\mathbf{0}$, *and* $\mathbf{b}_1'\mathbf{b}_2=0$.

PROOF. A necessary and sufficient condition for independence of $y_1$ and $y_2$ is

$$E[\exp(it_1y_1+it_2y_2)]=E[\exp(it_1y_1)]E[\exp(it_2y_2)] \tag{2.11.20}$$

for all real values of $t_1$ and $t_2$. Using Eq. (2.11.15), Eq. (2.11.20) can be written as

$$|I-2it_1A_1-2it_2A_2|^{-1/2}$$
$$\times\exp\left[-2(t_1\mathbf{b}_1+t_2\mathbf{b}_2)'(I-2it_1A_1-2it_2A_2)^{-1}(t_1\mathbf{b}_1+t_2\mathbf{b}_2)\right]$$
$$=\prod_{j=1}^{2}|I-2it_1A_1|^{-1/2}\exp\left(-2\sum_{j=1}^{2}t_j^2\mathbf{b}_j'(I-2it_jA_j)^{-1}\mathbf{b}_j\right).$$

As in the proof of Theorem 2.11.2, the necessary and sufficient condition reduces to

$$|I-2it_1A_1-2it_2A_2|=|I-2it_1A_1|\,|I-2it_2A_2| \tag{2.11.21}$$

and

$$(t_1\mathbf{b}_1+t_2\mathbf{b}_2)'(I-2it_1A_1-2it_2A_2)^{-1}(t_1\mathbf{b}_1+t_2\mathbf{b}_2)$$
$$=t_1^2\mathbf{b}_1'(I-2it_1A_1)^{-1}\mathbf{b}_1+t_2^2\mathbf{b}_2'(I-2it_2A_2)^{-1}\mathbf{b}_2 \tag{2.11.22}$$

for all real values of $t_1$ and $t_2$. The differential of $|I-2itQ|$ with respect to $t$ is given by

$$d(|I-2itQ|)=-2i|I-2itQ|\operatorname{tr}(I-2itQ)^{-1}Q\,dt.$$

Hence, taking differentials of Eq. (2.11.21) and equating the coefficients of $dt_1$ and $dt_2$, we get for all real values of $t_1$ and $t_2$

$$\operatorname{tr}\left[(I-2it_1A_1-2it_2A_2)^{-1}A_j\right]=\operatorname{tr}\left[(I-2it_jA_j)^{-1}A_j\right] \tag{2.11.23}$$

for $j=1,2$.

Subtracting $\operatorname{tr}A_j$ from both sides, we get

$$\operatorname{tr}\left[(I-2it_jA_j)^{-1}A_j^2\right]=\operatorname{tr}\left[(I-2it_1A_1-2it_2A_2)^{-1}(t_kA_kt_j^{-1}+A_j)A_j\right]$$

for ($j=1$, $k=2$) or ($j=2$, $k=1$), or

$$\operatorname{tr}\left\{\left[(I-2it_jA_j)^{-1}-(I-2it_1A_1-2it_2A_2)^{-1}\right]A_j^2\right\}$$
$$=\frac{t_k}{t_j}\operatorname{tr}\left[(I-2it_1A_1-2it_2A_2)^{-1}A_kA_j\right]. \tag{2.11.24}$$

Since the limit of the left side of this equation exists for $t_j\to 0$ when $t_k$ is given, the limit on the right side must exist, and this will be so if and only if $A_kA_j=0$, i.e., $A_1A_2=A_2A_1=0$. Putting this in Eq. (2.11.22), we get

$$\mathbf{b}_1'\mathbf{b}_2=0,\qquad A_1\mathbf{b}_2=\mathbf{0}\quad\text{and}\quad A_2\mathbf{b}_1=\mathbf{0}.$$

Thus, Theorem 2.11.3 is established. □

*Note 2.11.1.* In solving Eq. (2.11.21), various authors [for example, Craig (1943), Hotelling (1944), Ogawa (1950), Lancaster (1954)] have given different approaches, but we hope that the approach given here is a simple one.

**Corollary 2.11.3** *Let* $\mathbf{x} \sim N_p(\boldsymbol{\theta}, \Sigma)$. *Then* $y_j = \mathbf{x}'A_j\mathbf{x} + 2\mathbf{b}_j'\mathbf{x} + c_j$ $(j=1,2)$ *are independently distributed iff* $\Sigma A_1 \Sigma A_2 \Sigma = 0$, $\Sigma A_1 \Sigma(\mathbf{b}_2 + A_1\boldsymbol{\theta}) = \mathbf{0}$, $\Sigma A_2 \Sigma (\mathbf{b}_1 + A_2\boldsymbol{\theta}) = \mathbf{0}$, *and* $(\mathbf{b}_1 + A_1\boldsymbol{\theta})'\Sigma(\mathbf{b}_2 + A_2\boldsymbol{\theta}) = 0$.

This follows from Theorem 2.11.2 by using Definition 2.1.1, namely, $\mathbf{x} = \boldsymbol{\theta} + B\mathbf{u}, \mathbf{u} \sim N_r(\mathbf{0}, I)$, $r = \rho(\Sigma)$, and $\Sigma = BB'$. A multivariate generalization of this result is given by Khatri (1962b, 1963a).

**Corollary 2.11.4** *Let* $\mathbf{x} \sim N_p(\boldsymbol{\theta}, \Sigma)$ *and* $\Sigma > 0$. *Then*, $y_j = \mathbf{x}'A_j\mathbf{x} + 2\mathbf{b}_j'\mathbf{x} + c_j$, $j = 1,2$ *are independently distributed iff* $A_1\Sigma A_2 = 0$, $A_2\Sigma\mathbf{b}_1 = \mathbf{0}$, $A_1\Sigma\mathbf{b}_2 = \mathbf{0}$, *and* $\mathbf{b}_1'\Sigma\mathbf{b}_2 = 0$.

**Theorem 2.11.4** *Let* $\mathbf{x} \sim N_p(\boldsymbol{\theta}, \Sigma)$, $\Sigma > 0$, *and let* $q_j = \mathbf{x}'A_j\mathbf{x}$ $(j = 1,2,\ldots,k)$ *and* $\mathbf{x}'A\mathbf{x} = q = \Sigma_{j=1}^k q_j$.

(i) $q_j \sim \chi^2_{r_j}(\lambda_j)$, $j = 1,2,\ldots,k$.

(ii) $q_1, q_2, \ldots, q_k$ *are independently distributed.*

(iii) $q \sim \chi^2_r(\lambda)$.

(iv) $\rho(A) = \Sigma_{j=1}^k \rho(A_j)$.

*Then* (a) *any two of* (i), (ii), (iii) *imply all four, and* (b) (iii) *and* (iv) *imply* (i) *and* (ii).

PROOF. Using Corollaries 2.11.2 and 2.11.4, the four conditions of the Theorem 2.11.4 are equivalent to the following four conditions:

(i′) $A_j\Sigma$, $j = 1,2,\ldots,k$, are idempotent,

(ii′) $(A_j\Sigma)(A_i\Sigma) = 0$ for all $j \neq i$,

(iii′) $A\Sigma$ is idempotent and $A = \sum_{j=1}^{k} A_j$, and

(iv′) $\rho(A\Sigma) = \Sigma_{j=1}^k \rho(A_j\Sigma)$.

Then, by applying Theorem 1.7.2, we get the required result. □

In the same way, if we use Theorem 1.7.3, we get the following theorem, which is given by Hogg (1963):

**Theorem 2.11.5** *Let $\mathbf{x} \sim N_p(\boldsymbol{\theta}, \Sigma)$ and $\Sigma > 0$. Let $q = \mathbf{x}' A \mathbf{x} = \sum_{j=1}^{k} q_j$, $q_j = \mathbf{x}' A_j \mathbf{x}$. If (i) $q_k \geq 0$ for all $\mathbf{x}$ and (ii) $q, q_1, \ldots, q_{k-1}$ are distributed as noncentral chi squares, then $q_1, q_2, \ldots, q_k$ are independently distributed as noncentral chi squares.*

An important consequence of Theorem 2.11.4 is Cochran's theorem:

**Corollary 2.11.5** *Let $\mathbf{x} \sim N_p(\boldsymbol{\theta}, \Sigma)$, $\Sigma > 0$, and let $\mathbf{x}'\Sigma^{-1}\mathbf{x} = \sum_{j=1}^{k} q_j$ with $q_j = \mathbf{x}' A_j \mathbf{x} (j = 1, 2, \ldots, k)$. Then $q_1, q_2, \ldots, q_k$ are independently distributed as noncentral chi squares iff $\sum_{j=1}^{k} \rho(A_j) = p$.*

# Problems

**2.1** Show that

$$\phi_{x,y}(t_1, t_2) = \exp\left[-\tfrac{1}{2}(t_1^2 + t_2^2)\right] + \exp\left[-\delta - \tfrac{1}{2}c(t_1^2 + t_2^2)\right] \prod_{k=1}^{N} (b_k^2 t_1^2 - a_k^2 t_2^2)$$

is a cf of a bivariate distribution for some positive constants $\delta$ and $c$ and for any $(a_k, b_k) \in R^2$, $k = 1, 2, \ldots, N$ ($N$ being finite), but it is not the cf of a bivariate normal distribution. Show that the cf of $a_k x + b_k y$ is

$$\phi_{x,y}(a_k t, b_k t) = \exp\left[-\tfrac{1}{2}(a_k^2 - b_k^2)t^2\right],$$

which is the cf of a normal distribution.

**2.2** Let $x$ and $y$ be two random variables such that their joint density function is given by

$$f(x,y) = \begin{cases} 2\phi(x)\phi(y) & \text{for } \; 0 \leq y \leq x < \infty, \\ & \quad\;\; -\infty \leq x \leq y \leq 0, \\ & \quad\;\; \infty \geq y \geq -x \geq 0, \text{ and} \\ & \quad\;\; \infty \geq -y \geq x \geq 0, \\ 0 & \text{otherwise,} \end{cases}$$

where $\phi(x) = (2\pi)^{-1/2} \exp(-\tfrac{1}{2}x^2)$. Then show that $x$, $y$, $x+y$, and $x-y$ are normally distributed, while the joint distribution of $x,y$ is not bivariate normal. Show that $\operatorname{cor}(x,y) = 0$ but $x$ and $y$ are not independent. Modify the density function of $x$ and $y$ so that $x$, $y$, $a_1 x + b_1 y$, and $a_2 x + b_2 y$ are normally distributed.

**2.3** Given $\{(a_k, b_k) : k = 1, 2, \ldots\}$, a countable "distinct" sequence in $R^2$ such that for each $k$, $a_k x + b_k y$ is a normal random variable, then show that $(x,y)$ is a bivariate normal variable. [Hint: Use the following lemma: Given the functions $g_1, g_2$ in $C^{\infty}(R)$ (the class of continuous and infinitely differentiable functions) and a sequence $\{x_n\}$ such that $x_n \to x_0$ as $n \to \infty$, if $g_1(x_n) = g_2(x_n)$ for all $n = 1, 2, \ldots$ and $g_i(x) = \sum_{k=0}^{\infty} g_i^{(k)}(x_0)(x - x_0)^k / k!$ for $-\infty < x < \infty$ and $i = 1, 2$, then $g_1 = g_2$ on $R$.]

**2.4** Let the joint density of $x$ and $y$ be given by

$$\frac{1}{2}\left[\left[2\pi(1-\rho^2)^{1/2}\right]^{-1}\exp\left(-\frac{1}{2(1-\rho^2)}(x^2-2\rho xy+y^2)\right)\right]$$
$$+\frac{1}{2}\left[\left[2\pi(1-\tau^2)^{1/2}\right]^{-1}\exp\left(-\frac{1}{2(1-\tau^2)}(x^2-2\tau xy+y^2)\right)\right].$$

Show that the marginal densities of $x$ and $y$ are both $N(0,1)$, while the joint distribution of $x$ and $y$ is not bivariate normal. Further, show that $\text{cor}(x,y)=0$ if $\tau=-\rho$.

**2.5** Let the joint density of $x$ and $y$ be given by $f(x,y)=c$ for $x^2+y^2\leqslant k$ and 0 elsewhere. Show that $E(xy)=0$. Are $x$ and $y$ independent?

**2.6** Let $\mathbf{x}_1,\mathbf{x}_2,\dots,\mathbf{x}_n$ be independently and identically distributed (iid) as $N_p(\boldsymbol{\theta},\Sigma)$. Let $\bar{\mathbf{x}}=n^{-1}\Sigma_{i=1}^n\mathbf{x}_i$. Show that $\bar{\mathbf{x}}$ is independent of $(\mathbf{x}_1-\bar{\mathbf{x}},\dots,\mathbf{x}_{n-1}-\bar{\mathbf{x}})$. Hence show that $\bar{\mathbf{x}}$ and $\Sigma_{i=1}^n(\mathbf{x}_i-\bar{\mathbf{x}})(\mathbf{x}_i-\bar{\mathbf{x}})'$ are independently distributed.

**2.7** Let $\mathbf{x}_1,\mathbf{x}_2,\dots,\mathbf{x}_n$ be iid $N_p(\mathbf{0},I)$. Let $\mathbf{s}_r=\mathbf{x}_1+\cdots+\mathbf{x}_r$, $r\leqslant n$. Find the joint density of $(\mathbf{s}_m,\mathbf{s}_n)$ for $m<n$, and the conditional density of $\mathbf{s}_m$ given $\mathbf{s}_n=\mathbf{u}$.

**2.8** The random vector $\mathbf{x}=(x_1,\dots,x_k)'$ is said to have a multinomial distribution if its pdf is given by

$$f(\mathbf{x})=\begin{cases}\dfrac{x!}{\prod_{i=1}^k x_i!}\prod_{i=1}^k p_i^{x_i} & \text{if } \sum_{i=1}^k x_i=x,\quad x_i=0,1,\dots,x,\\ 0 & \text{otherwise,}\end{cases}$$

where $x_i$'s are integer-valued. Find its cf and $\text{Cov}(\mathbf{x})$. Show that the determinant of $\text{Cov}(\mathbf{x})$ is zero and hence the distribution is singular. However, the distribution of $(x_1,\dots,x_{k-1})$ obtained by replacing $x_k$ with $x-\Sigma_{i=1}^{k-1}x_i$ is nonsingular. What is the characteristic function of this nonsingular distribution? Obtain the marginal distribution of $(x_1,\dots,x_m)$, $m<k$, and its conditional distribution given the other variables.

**2.9** Let the pdf of the $p$-component $\mathbf{x}$ be given by

$$f(\mathbf{x})=\begin{cases}|A|^{1/2}\dfrac{[(p+2)\pi]^{\frac{1}{2}p}}{\Gamma\left(\frac{1}{2}p+1\right)} & \text{for}\quad (\mathbf{x}-\boldsymbol{\mu})'A(\mathbf{x}-\boldsymbol{\mu})\leqslant p+2,\\ 0 & \text{elsewhere.}\end{cases}$$

Show that $E(\mathbf{x})=\boldsymbol{\mu}$ and $\text{Cov}(\mathbf{x})=A^{-1}$.

**2.10** Let $\mathbf{x}=(x_1,\dots,x_p)'\sim N_p(\boldsymbol{\mu},\Sigma)$, where $\boldsymbol{\mu}=(\mu_1,\dots,\mu_p)'$ and $\Sigma=(\sigma_{ij})$. Let $f(u_1|u_2)$ denote the conditional pdf of $u_1$ given $u_2$, and $p(\mathbf{x})$ denote the pdf of $\mathbf{x}$. Write down explicitly

$$p(\mathbf{x})=f(x_1)f(x_2|x_1)f(x_3|x_1,x_2)\cdots f(x_p|x_1,\dots,x_{p-1}).$$

Choose suitable notation to write the means and variances.

**2.11** Let $\mathbf{x}_1, \mathbf{x}_2, \ldots, \mathbf{x}_n$ be iid $N_p(\mathbf{0}, \Sigma)$, and $S = \Sigma_{i=1}^n \mathbf{x}_i \mathbf{x}_i' = (s_{ij})$. Show that

$$P(s_{ii} \leqslant c_i,\ i=1,2,\ldots,p) \geqslant \prod_{i=1}^{p} P(s_{ii} \leqslant c_i),$$

and if $\sigma_{ij} = l_i l_j (\sigma_{ii}\sigma_{jj})^{1/2}$ and $0 \leqslant l_i^2 \leqslant 1$ for all $i=1,2,\ldots,p$, then

$$P(s_{ii} \geqslant c_i,\ i=1,2,\ldots,p) \geqslant \prod_{i=1}^{p} P(s_{ii} \geqslant c_i).$$

[Hint: Establish the result conditionally.]

**2.12** Let $\mathbf{x}_1, \mathbf{x}_2, \ldots, \mathbf{x}_N$ be iid $N_p(\boldsymbol{\theta}, \Sigma)$ where $\Sigma = (\sigma_{ij})$, $\sigma_{ii} = \sigma^2$, and $\sigma_{ij} = \rho\sigma^2$. Let $\bar{\mathbf{x}} = \Sigma_{i=1}^N \mathbf{x}_i / N$ and $S = \Sigma_{i=1}^N (\mathbf{x}_i - \bar{\mathbf{x}})(\mathbf{x}_i - \bar{\mathbf{x}})'$. Then show that if $\rho \geqslant 0$,

$$P\left( \frac{|\bar{x}_i - \theta_i|}{\sqrt{s_{ii}}} \leqslant c_i,\ i=1,2,\ldots,p \right) \geqslant \prod_{i=1}^{p} P\left( \frac{|\bar{x}_i - \theta_i|}{\sqrt{s_{ii}}} \leqslant c_i \right).$$

[Hint: Use Problem 2.11 after showing that $S = \Sigma_{i=1}^{N-1} \mathbf{y}_i \mathbf{y}_i'$, where $\mathbf{y}_1, \mathbf{y}_2, \ldots, \mathbf{y}_{N-1}$, $N^{\frac{1}{2}}(\bar{\mathbf{x}} - \boldsymbol{\theta})$ are independently distributed as $N_p(\mathbf{0}, \Sigma)$.]

**2.13** Show that $P(|x_i| \geqslant c_i,\ i=1,2,\ldots,p | \boldsymbol{\lambda})$ is a monotonically coordinatewise increasing function of $\lambda_1, \ldots, \lambda_p$ if $\mathbf{x} \sim N_p(\mathbf{0}, \Sigma)$, $\Sigma = (\sigma_{ij})$, $\sigma_{ii} = 1$, and $\sigma_{ij} = \lambda_i \lambda_j (\sigma_{ii}\sigma_{jj})^{1/2}$ for $0 \leqslant \lambda_i \leqslant 1$. [Hint: First prove the result when $\Sigma > 0$. Show that $\partial P / \partial \lambda_i \geqslant 0$ for all $i=1,2,\ldots,p$.]

**2.14** (i) Let $\mathbf{x}' = (x_1, \ldots, x_p)$ be a random vector such that $V(\mathbf{x})$ is p.d. If $\rho_{1.}^2$ is the multiple correlation between $x_1$ and $(x_2, \ldots, x_p)$ and $\rho_{12.3\ldots k}^2$ is the partial correlation between $x_1$ and $x_2$ when the best linear effects of $x_3, \ldots, x_k$ are eliminated, show that

$$\left(1 - \rho_{1.}^2\right) = \left(1 - \rho_{12}^2\right)\left(1 - \rho_{12.3}^2\right) \cdots \left(1 - \rho_{12.34\ldots p}^2\right)$$

$$= \left(1 - \rho_{13}^2\right)\left(1 - \rho_{13.2}^2\right) \prod_{k=4}^{p} \left(1 - \rho_{13.24\ldots k}^2\right).$$

(ii) If $V(\mathbf{x}) = (\sigma_{ij})$, $\sigma_{ii} = \sigma^2$, and $\sigma_{ij} = \sigma^2\rho$ for $i \neq j = 1,2,\ldots,p$, find the expressions for $\rho_{1.}^2$ and $\rho_{12.34\ldots k}$ for $k = 3,4,\ldots,p$.

(iii) Let the best linear estimate of $x_1$ on the basis of $(x_2, \ldots, x_p)$ be $\Sigma_{i=2}^p \beta_{1i} x_i$, and that of $x_2$ on the basis of $(x_1, x_3, \ldots, x_p)$ be $\Sigma_{i=3}^p \beta_{2i} x_i + \beta_{21} x_1$. Then show that

$$\rho_{12.34\ldots p}^2 = \beta_{12}\beta_{21}.$$

(iv) Show that

$$\rho_{12.34\ldots p} = \frac{\rho_{12.45\ldots p} - \rho_{13.4\ldots p}\rho_{23.4\ldots p}}{\sqrt{\left(1 - \rho_{13.4\ldots p}^2\right)\left(1 - \rho_{23.4\ldots p}^2\right)}}.$$

**2.15** Let $P\{x=1\}=p$, $P\{x=0\}=q=1-p$, $\mathbf{y}'=(y_1,\ldots,y_p)$, $(\mathbf{y}|x=1)\sim N_p(\boldsymbol{\mu}^{(1)},\Sigma)$, and $(\mathbf{y}|x=0)\sim N_p(\boldsymbol{\mu}^{(0)},\Sigma)$. Find $E(\mathbf{y})$, $E(y_iy_j)$, and $\rho_{x\mathbf{y}}$ (the multiple correlation). Show that $\boldsymbol{\mu}^{(1)}=\boldsymbol{\mu}^{(0)}$ iff $\rho_{x\mathbf{y}}=0$.

**2.16** Show that if $P\{x\geqslant 0, y\geqslant 0\}=\alpha$, where

$$\begin{pmatrix} x \\ y \end{pmatrix}\sim N_2\left(\begin{pmatrix} 0 \\ 0 \end{pmatrix},\begin{pmatrix} \sigma_1^2 & \rho\sigma_1\sigma_2 \\ \rho\sigma_1\sigma_2 & \sigma_2^2 \end{pmatrix}\right),$$

then $\rho=\cos(1-2\alpha)\pi$.

**2.17** Let $\mathbf{x}\sim N_p(\boldsymbol{\mu},\Sigma)$, $\Sigma>0$, where

$$\mathbf{x}'=(\overset{r}{\dot{\mathbf{x}}'},\overset{s}{\ddot{\mathbf{x}}'}) \quad\text{and}\quad \boldsymbol{\mu}'=(\overset{r}{\dot{\boldsymbol{\mu}}'},\overset{s}{\ddot{\boldsymbol{\mu}}'}), \qquad r+s=p.$$

Suppose $\ddot{\boldsymbol{\mu}}=\mathbf{0}$. Find the maximum-likelihood estimate of $\dot{\boldsymbol{\mu}}$, when $\Sigma$ is known, on the basis of a sample $\mathbf{x}_1,\ldots,\mathbf{x}_N$ of size $N$.

**2.18** Let $\mathbf{z}\sim CN_p(\boldsymbol{\theta},Q)$. Suppose a random sample of size $N$ is taken from this distribution. Find the maximum-likelihood estimates of $\boldsymbol{\theta}$ and $Q$. Show that they are sufficient statistics.

**2.19** Let $\mathbf{z}'=(\mathbf{z}_1',\mathbf{z}_2')$, where $\mathbf{z}_1$ and $\mathbf{z}_2$ are $r$ and $s$ complex vectors, respectively, with $r+s=p$. Suppose $\mathbf{z}\sim CN_p(\boldsymbol{\theta},Q)$. Find the marginal distribution of $\mathbf{z}_1$ and the conditional distribution of $\mathbf{z}_1$ given $\mathbf{z}_2$.

**2.20** Let $\mathbf{x}_1,\mathbf{x}_2,\ldots$ be independent and identically distributed random $p$-vectors with means $\boldsymbol{\theta}$ and covariance $I_p$. Find the limiting distribution of $n^{1/2}(\bar{\mathbf{x}}'\bar{\mathbf{x}}-\boldsymbol{\theta}'\boldsymbol{\theta})$, where $\bar{\mathbf{x}}=n^{-1}\Sigma_{i=1}^n\mathbf{x}_i$.

**2.21** A random $m$ vector $\mathbf{x}$ is said to have a multivariate Student's $t$ distribution (MSt) if its pdf is given by

$$\frac{\nu^{\nu/2}\Gamma\left[\frac{1}{2}(\nu+m)\right]|V|^{1/2}}{\pi^{\frac{1}{2}m}\Gamma\left(\frac{1}{2}\nu\right)}\left[\nu+(\mathbf{x}-\boldsymbol{\theta})'V(\mathbf{x}-\boldsymbol{\theta})\right]^{-\frac{1}{2}(m+\nu)},$$

$$-\infty<x_i<\infty, \quad i=1,2,\ldots,m,$$

where $\nu>0$, $V$ is an $m\times m$ positive definite and symmetric matrix, and $\boldsymbol{\theta}'=(\theta_1,\ldots,\theta_m)$. We shall write

$$\mathbf{x}\sim \mathrm{MSt}_m(\boldsymbol{\theta},V,\nu).$$

(i) Let $\mathbf{z}$ and $s^2$ be independently distributed, $\mathbf{z}\sim N_m(\mathbf{0},I_m)$, and $s^2\sim\chi^2_\nu$. Then show that the distribution of $\mathbf{x}=\boldsymbol{\theta}+\sqrt{\nu}\;V^{-1/2}\mathbf{z}/s$, where $V^{1/2}$ is a symmetric square root of a p.d. matrix $V$, is $\mathrm{MSt}_m(\boldsymbol{\theta},V,\nu)$. Hence or otherwise, show that the pdf of $y=(\mathbf{x}-\boldsymbol{\theta})'V(\mathbf{x}-\boldsymbol{\theta})/m$ has an $F$ distribution with $m$ and $\nu$ degrees of freedom. See Problem 2.26 for the definition of $F$-distribution.

(ii) Let $\mathbf{x}'=(\mathbf{x}_1',\mathbf{x}_2')$, where $\mathbf{x}_i$ is an $m_i$ vector and $m_1+m_2=m$. Find the marginal pdf of $\mathbf{x}_1$ and the conditional pdf of $\mathbf{x}_1$ given $\mathbf{x}_2$.

(iii) Find the marginal mean and covariance of $\mathbf{x}_1$ and the condition for their existence.

(iv) Find the conditional mean and variance of $\mathbf{x}_1$ given $\mathbf{x}_2$, and the condition for their existence.

**2.22** Let $\mathbf{x} \sim N_p(\boldsymbol{\theta}, \Sigma)$. Show that the cf of $\mathbf{x}'A\mathbf{x}$ is $|I-2it\Sigma A|^{-1/2}\exp[it\boldsymbol{\theta}'A(I-2it\Sigma A)^{-1}\boldsymbol{\theta}]$. Obtain the distribution of $\mathbf{x}'A\mathbf{x}$, where $A$ is p.s.d.

**2.23** Let $\mathbf{y} \sim N_n(\boldsymbol{\mu}, I)$. Let $K$ be an $r \times n$ matrix of constants with $\rho(K) = k < n$. Let $A = (a_{ij})$ be any $n \times n$ matrix with $a_{ij} = f_{ij}(K\mathbf{y})$, where $f_{ij}(\cdot)$ is a Borel function of the random vector $K\mathbf{y}$. Then the random variable $\omega = \mathbf{y}'A\mathbf{y}$ is distributed as a noncentral chi square if the following conditions hold with probability one:

(i) $A = L'AL$, where $L: n \times n$ is such that $LK' = 0$;

(ii) $A$ is idempotent;

(iii) $\operatorname{tr} A = m$, $m$ a constant positive integer;

(iv) $\boldsymbol{\mu}'A\boldsymbol{\mu} = \lambda$, $\lambda$ a constant.

**2.24** Let $\mathbf{y}$, $K$, and $L$ be defined as in Problem 2.23. Let the elements of the $n \times n$ matrices $A$ and $B$ be Borel functions of the vector $K\mathbf{y}$. Then $\mathbf{y}'A\mathbf{y}$ and $\mathbf{y}'B\mathbf{y}$ are independently distributed as noncentral chi squares if the following conditions hold with probability one:

(i) $L'AL = A$, $L'BL = B$;

(ii) $A = A^2$, $B = B^2$;

(iii) $\operatorname{tr} A = m_1$, $\operatorname{tr} B = m_2$, $m_1$ and $m_2$ constants;

(iv) $\boldsymbol{\mu}'A\boldsymbol{\mu} = \lambda_1$, $\boldsymbol{\mu}'B\boldsymbol{\mu} = \lambda_2$, $\lambda_1$ and $\lambda_2$ constants;

(v) $AB = 0$.

**2.25** Modify Theorems 2.11.3 and 2.11.5 when $\Sigma$ is singular.

**2.26** Let $v \sim \chi_p^2(\lambda^2)$ and $w \sim \chi_m^2$ be independently distributed. Then the random variable

$$F = \frac{v/p}{w/m}$$

is called noncentral $F$ with $p$ and $m$ degrees of freedom and noncentrality parameter $\lambda^2$, and is denoted by $F_{p,m}(\lambda^2)$. Show that its pdf is given by

$$\frac{pe^{-\frac{1}{2}\lambda^2}}{m\Gamma\left(\frac{1}{2}m\right)} \sum_{j=0}^{\infty} \frac{(\lambda^2/2)^j (pF/m)^{\frac{1}{2}p+j-1}\Gamma\left[\frac{1}{2}(p+m)+j\right]}{\Gamma\left(\frac{1}{2}p+j\right)(1+pF/m)^{\frac{1}{2}(p+m)+j} j!}.$$

Note that when $\lambda = 0$, it is called central $F$ or simply $F$ distribution with $p$ and $m$ d.f.

**2.27** Let $\mathbf{x} \sim N_p(\alpha\boldsymbol{\mu} + A\boldsymbol{\beta}, \Sigma)$ with $\alpha + \mathbf{b}'\boldsymbol{\beta} = 1$; $A - \boldsymbol{\mu}\mathbf{b}' = A_1$, a $p \times r$ matrix of rank $r$; and $\Sigma > 0$. Show that

$$(\mathbf{x}-\boldsymbol{\mu})'\Sigma^{-1}(\mathbf{x}-\boldsymbol{\mu}) - (\mathbf{x}-\boldsymbol{\mu})'\Sigma^{-1}A_1\left(A_1'\Sigma^{-1}A_1\right)^{-1}A_1'\Sigma^{-1}(\mathbf{x}-\boldsymbol{\mu})$$

is distributed as chi square with $p-r$ d.f. In particular, when $r = 1$, $b = 1$, $A = \mathbf{a}$, $\mathbf{a} - \boldsymbol{\mu} = \boldsymbol{\delta}$, and $\Delta^2 = \boldsymbol{\delta}'\Sigma^{-1}\boldsymbol{\delta}$, then $(\mathbf{x}-\boldsymbol{\mu})'\Sigma^{-1}(\mathbf{x}-\boldsymbol{\mu}) - \Delta^{-2}[(\mathbf{x}-\boldsymbol{\mu})'\Sigma^{-1}\boldsymbol{\delta}]^2$ is distributed as chi square with $p-1$ d.f.

# 3

# Wishart Distribution

## 3.1 Introduction

In this chapter, we consider a multivariate generalization of the chi-square distribution. Let $\mathbf{y}_1,\ldots,\mathbf{y}_n$ be iid $N_p(\mathbf{0},\Sigma)$. Then the distribution of $V=\Sigma_{i=1}^n \mathbf{y}_i\mathbf{y}_i'$ is called a Wishart distribution: a nonsingular (singular) Wishart distribution if $\Sigma>0$ ($\geqslant 0$) and a pseudo (nonsingular or singular) Wishart distribution if $n<\rho(\Sigma)$. If $E(\mathbf{y}_i)=\boldsymbol{\mu}_i\neq\mathbf{0}$ for at least one $i$, $i=1,2,\ldots,n$, then the distribution of $V$ is called the *noncentral* Wishart distribution. We shall confine our attention to the case when $E(\mathbf{y}_i)=\mathbf{0}$ for all $i=1,2,\ldots,n$. Note that if $p=1$, then $\Sigma^{-1}V$ has the chi-square distribution with $n$ d.f. Thus $V$ is indeed a multivariate generalization of the chi-square distribution.

## 3.2 Distribution of $V=YY'$ Where $Y\sim N_{p,n}(0,\Sigma,I_n)$

Let

$$Y=(\mathbf{y}_1,\mathbf{y}_2,\ldots,\mathbf{y}_n)$$

be a $p\times n$ ($p\leqslant n$) matrix whose columns are iid $N_p(\mathbf{0},\Sigma)$ with $\Sigma:p\times p$ positive definite. Then

$$V=YY'=\sum_{i=1}^n \mathbf{y}_i\mathbf{y}_i'$$

has a Wishart distribution with mean $n\Sigma$ and $n$ d.f. We shall write $V\sim W_p(\Sigma,n)$. Before we discuss and derive the density function of $V$ when $\Sigma>0$ and $n\geqslant p$, we need to show that $V$ is positive definite with probability one. This will be established when the observations $\mathbf{y}_i$, $i=1,2,\ldots,n$ $(\geqslant p)$, are independent and $V(\mathbf{y}_i)=\Sigma_i>0$ (or if the $\mathbf{y}_i$'s have nondegenerate continuous distributions). First of all, consider any $p$ observations from $\mathbf{y}_1,\ldots,\mathbf{y}_n$. Then, following Stein (1969) [see also Dykstra (1970)], we get

$$P(\rho(\mathbf{y}_1,\ldots,\mathbf{y}_p)=p)=1-P\left[\begin{array}{l}\text{at least one of}\\ \mathbf{y}_1,\ldots,\mathbf{y}_p\\ \text{is a linear combination of the others}\end{array}\right]$$

$$\geqslant 1-\sum_{i=1}^{p} P\left(\sum_{j=1}^{p} a_{ij}\mathbf{y}_j=\mathbf{0} \text{ with } a_{ii}=1 \text{ and the } a_{ij}\text{'s } \textit{are} \text{ constants}\right).$$

Since $\mathbf{y}_1,\ldots,\mathbf{y}_p$ are independent and $\Sigma_i>0$ for all $i$ (or nondegenerate continuous distributions),

$$P\left\{\mathbf{y}_i=-\sum_{\substack{j=1\\(j\neq i)}}^{p} a_{ij}\mathbf{y}_j \,\middle|\, \mathbf{y}_1,\ldots,\mathbf{y}_p \text{ except } \mathbf{y}_i\right\}=0$$

for all $i=1,2,\ldots,p$. Hence,

$$P\left[\rho(\mathbf{y}_1,\ldots,\mathbf{y}_p)=p\right]=1,$$

(3.2.1) $$P\left[\rho(V)=p\right]=P\left[\rho(YY')=p\right]=1,$$

when $n\geqslant p$ and $V(\mathbf{y}_i)=\Sigma_i>0$ for all $i$. However, when all the $\Sigma_i$'s are singular, nothing can be said except when $\Sigma_i\equiv\Sigma$ for all $i$ and $\rho(\Sigma)=r\leqslant n$. In this case

(3.2.2) $$P(\rho(V)=r)=P(\rho(Y)=r)=1.$$

This can be proved by using the characteristic property that $\mathbf{y}=B\mathbf{z}$ where $BB'=\Sigma$, $\rho(\Sigma)=\rho(B:p\times r)=r$, and $V(\mathbf{z})=I_r$.

Now we turn to the distribution of $V$ when $n\geqslant p$ and $\Sigma_i=\Sigma>0$. Since $V$ is symmetric, it has only $\frac{1}{2}p(p+1)$ random variables. Thus $V$ is a reduction from $p\times n$ random variables of $Y$ to only $\frac{1}{2}p(p+1)$ random variables. Obviously there are several ways of doing it, and so there are several ways of deriving the distribution of $V$. For various methods for the derivation of Wishart distribution, refer to Fisher (1915) for $p=2$, Wishart (1928), Ingham (1933), Madow (1938), Hsu (1939), Elfving (1947), Sverdrup (1947), Rasch (1948), Mahalanobis, Bose, and Roy (1937), Narian (1948), Ogawa (1953), Olkin and Roy (1954), James (1954), Mauldon (1955), Khatri and Ramachandran (1958), Kshirsagar (1959), and Khatri (1963b). For example, from Problem 1.21, we know that we can write (since $Y$ is of rank $p$ with probability one)

$$Y=\underset{\sim}{T}L,$$

where $\underset{\sim}{T}:p\times p$ is a lower triangular matrix with positive diagonal elements and $L$ is a $p\times n$ semiorthogonal matrix, $LL'=I_p$. Hence

$$V=\underset{\sim}{T}\underset{\sim}{T}'.$$

Thus, if we know the distribution of $\underset{\sim}{T}$, we can find the distribution of $V$. In order to find the distribution of $\underset{\sim}{T}$, we recall the following results from Problem 1.33.

Let $Y:p\times n$ be of rank $p$, and let $Y=\underset{\sim}{T}L$, where $\underset{\sim}{T}:p\times p$ is a lower triangular matrix with diagonal elements $t_{ii}>0$ and $L$ is a $p\times n$ semiorthogonal matrix, $LL'=I_p$. Then $J(Y\to\underset{\sim}{T},L)=g_{n,p}(L)\prod_{i=1}^{p}t_{ii}^{n-i}$, where $g_{n,p}(L)$ is a function of $L$ only.

**Lemma 3.2.1** *Let $Y\sim N_{p,n}(0,I_p,I_n)$, $p\leqslant n$, and let $Y=\underset{\sim}{T}L$. Then the* pdf *of $\underset{\sim}{T}$ is given by*

$$\left[2^{\frac{1}{2}np-p}\Gamma_p\left(\tfrac{1}{2}n\right)\right]^{-1}\left(\prod_{i=1}^{p}t_{ii}^{n-i}\right)\operatorname{etr}\left(-\tfrac{1}{2}\underset{\sim}{T}\underset{\sim}{T}'\right),$$

$$t_{ii}>0,\qquad -\infty<t_{ij}<\infty,\quad i>j,$$

*that is, $t_{ij}$, $1\leqslant j\leqslant i\leqslant p$, are independently distributed, with $t_{ij}\sim N(0,1)$, $i>j$, $1\leqslant j<i\leqslant p$, and $t_{ii}^2\sim\chi^2_{n-i+1}$, $i=1,2,\ldots,p$, where*

$$\Gamma_p\left(\tfrac{1}{2}n\right)=\pi^{\frac{1}{4}p(p-1)}\prod_{i=1}^{p}\Gamma\left(\frac{n-i+1}{2}\right). \tag{3.2.3}$$

PROOF. Since the pdf of $Y$ is given by

$$(2\pi)^{-\frac{1}{2}pn}\operatorname{etr}\left(-\tfrac{1}{2}YY'\right),$$

using the Jacobian of the transformation $Y=\underset{\sim}{T}L$ given by Problem 1.33, we find that the joint pdf of $\underset{\sim}{T}$ and $L$ is given by

$$(2\pi)^{-\frac{1}{2}pn}g_{n,p}(L)\left(\prod_{i=1}^{p}t_{ii}^{n-i}\right)\operatorname{etr}\left(-\tfrac{1}{2}\underset{\sim}{T}\underset{\sim}{T}'\right).$$

Integrating out $L$, we get the marginal density of $\underset{\sim}{T}$ as

$$c(2\pi)^{-\frac{1}{2}pn}\prod_{i=1}^{p}t_{ii}^{n-i}\operatorname{etr}\left(-\tfrac{1}{2}\underset{\sim}{T}\underset{\sim}{T}'\right), \tag{3.2.4}$$

where $c=\int g_{n,p}(L)\,dL$, integrated over the space $LL'=I_p$, is evaluated indirectly. Since Eq. (3.2.4) is a pdf, we have

$$(2\pi)^{-\frac{1}{2}pn} c \int_{\substack{t_{ii}>0 \\ -\infty<t_{ij}<\infty,\, i>j}} \left(\Pi t_{ii}^{n-i}\right) \operatorname{etr}\left(-\tfrac{1}{2} \underset{\sim}{T}\underset{\sim}{T}'\right) d\underset{\sim}{T} = 1.$$

Hence,

$$c^{-1} = (2\pi)^{-\frac{1}{2}pn} \int \Pi t_{ii}^{n-i} \exp\left(-\frac{1}{2} \sum_{i>j} t_{ij}^2\right) d\underset{\sim}{T}$$

$$= (2\pi)^{-\frac{1}{2}pn + \frac{1}{4}p(p-1)} \prod_{i=1}^{p} \int_0^\infty t_{ii}^{n-i} \exp\left(-\tfrac{1}{2} t_{ii}^2\right) dt_{ii}$$

$$= \pi^{\frac{1}{4}p(p-1) - \frac{1}{2}pn} 2^{-p} \prod_{i=1}^{p} \int_0^\infty \left(\tfrac{1}{2} t_{ii}^2\right)^{\frac{1}{2}(n-i+1)-1} \exp\left(-\tfrac{1}{2} t_{ii}^2\right) d\left(\tfrac{1}{2} t_{ii}^2\right)$$

$$= 2^{-p} \pi^{\frac{1}{4}p(p-1) - \frac{1}{2}pn} \prod_{i=1}^{p} \Gamma\left(\frac{n-i+1}{2}\right)$$

$$= 2^{-p} \pi^{-\frac{1}{2}pn} \Gamma_p\left(\tfrac{1}{2}n\right),$$

from Eq. (3.2.3). □

As a by-product, we get the following interesting result.

**Corollary 3.2.1**

$$\int_{LL' = I_p} g_{n,p}(L)\, dL = \frac{2^p \pi^{\frac{1}{2}pn}}{\Gamma_p\left(\frac{1}{2}n\right)}.$$

This result can be established directly using the explicit expression for $g_{n,p}(L)$ given in Problem 1.33.

Note that if we require that one element of each row of $L$ or diagonal elements be positive and denote this $L$ by $L_+$, then we get

**Corollary 3.2.2**

$$\int_{L_+ L_+' = I_p} g_{n,p}(L_+)\, dL_+ = \frac{\pi^{\frac{1}{2}pn}}{\Gamma_p\left(\frac{1}{2}n\right)}.$$

Notice that when $n = p$, $g_{p,p}(L_+) = f_p(L)$ defined in Theorem 1.11.5.

We now give a more general result than Lemma 3.2.1, which can be established by using Corollary 3.2.1.

**Lemma 3.2.2** *Let the* pdf *of* $Y: p\times n$, $p\leqslant n$, $\rho(Y)=p$, *be* $f(YY')$ *and* $Y=\underset{\sim}{T}L$ *as above. Then the* pdf *of* $\underset{\sim}{T}$ *is given by*

$$c\prod_{i=1}^{p} t_{ii}^{n-i} f(\underset{\sim}{T}\underset{\sim}{T}')=\frac{2^{p}\pi^{\frac{1}{2}pn}}{\Gamma_p\left(\frac{1}{2}n\right)}\left(\prod_{i=1}^{p} t_{ii}^{n-i}\right)f(\underset{\sim}{T}\underset{\sim}{T}').$$

*Also,* $\underset{\sim}{T}$ *and* $L$ *are independently distributed.*

Now, if we let

$$V=\underset{\sim}{T}\underset{\sim}{T}',$$

then $J(\underset{\sim}{T}\to V)=2^{-p}\prod_{i=1}^{p} t_{ii}^{-p+i-1}$ (see Problem 1.28). Hence we get

**Lemma 3.2.3** *Let the* pdf *of* $Y: p\times n$, $p\leqslant n$, $\rho(Y)=p$, *be given by* $f(YY')$ *and* $V=\underset{\sim}{T}\underset{\sim}{T}'=YY'$. *Then the* pdf *of* $V$ *is given by*

$$\frac{\pi^{\frac{1}{2}pn}}{\Gamma_p\left(\frac{1}{2}n\right)}|V|^{\frac{1}{2}(n-p-1)}f(V), \qquad V>0.$$

*Thus, we obtain the following interesting corollary, which can be established directly as in Khatri* (1963b).

**Corollary 3.2.3** *Let* $Y: p\times n$, $p\leqslant n$, $\rho(Y)=p$. *Then*

$$\int_{YY'=V} dY=\frac{\pi^{\frac{1}{2}pn}}{\Gamma_p\left(\frac{1}{2}n\right)}|V|^{\frac{1}{2}(n-p-1)}.$$

Letting $f$ be the normal density in Lemma 3.2.3, we obtain the Wishart distribution:

**Theorem 3.2.1** *Let* $Y\sim N_{p,n}(0,\Sigma,I_n)$, $p\leqslant n$, $\Sigma>0$, *and let* $V=YY'$. *Then the* pdf *of* $V$ *is given by*

$$\frac{|V|^{\frac{1}{2}(n-p-1)}\operatorname{etr}\left(-\frac{1}{2}\Sigma^{-1}V\right)}{2^{\frac{1}{2}pn}|\Sigma|^{\frac{1}{2}n}\Gamma_p\left(\frac{1}{2}n\right)}, \qquad V>0, \quad \Sigma>0,$$

*which is known as the Wishart distribution* $W_p(\Sigma,n)$. *In the sequel we shall write*

(3.2.5) $$C(p,n) \equiv \left[2^{\frac{1}{2}pn}\Gamma_p\left(\tfrac{1}{2}n\right)\right]^{-1}.$$

REMARK 3.2.1. When $\Sigma$ is singular, then by Definition 2.1.1 we can write $Y=BZ$, where $Z \sim N_{r,n}(0,I_r,I_n)$ and $B$ is a $p \times r$ matrix of rank $r$ with $\Sigma = BB'$. Then $V=BWB'$, and the distribution of $W=ZZ'$ is $W_r(I_r,n)$. In this sense, we shall say that the distribution of $V$ is $W_p(\Sigma,n)$ when $\Sigma$ is singular.

**Corollary 3.2.4** *Let $V \sim W_p(I,n)$, $n \geqslant p$ and let $V=\underset{\sim}{T}\underset{\sim}{T}'$. Then the elements of $\underset{\sim}{T}$ are independently distributed where $t_{ij} \sim N(0,1)$, $i>j$, and $t_{ii}^2 \sim \chi^2_{n-i+1}$, $i,j=1,2,\ldots,p$.*

The proof follows by using the Jacobian of the transformation $V=\underset{\sim}{T}\underset{\sim}{T}'$ in the density of $V$, and is similar to that of Lemma 3.2.1.

**Corollary 3.2.5** *Let $W \sim W_p(\Sigma,n)$ with $\Sigma>0$. Then there exists a random matrix $Y:p \times n$ such that $Y \sim N_{p,n}(0,\Sigma,I_n)$ and $W=YY'$.*

REMARK 3.2.2. This corollary is true even when $\Sigma$ is singular.

PROOF. Take $V=AWA'$ where $\Sigma^{-1}=A'A$. Then, using Theorem 1.11.4, it is easy to see that $V \sim W_p(I,n)$. Let us define a matrix $L:p \times n$ such that (i) $LL'=I_p$, (ii) the density function of $L$ is $cg_{n,p}(L)$ with $c^{-1}=2^p\pi^{\frac{1}{2}pn}/\Gamma_p(\frac{1}{2}n)$, and (iii) $L$ and $V$ are independently distributed. Let $V=\underset{\sim}{T}\underset{\sim}{T}'$, where $\underset{\sim}{T}$ is lower triangular with positive diagonal elements. Then, using Corollary 3.2.4 and the definition of $L$, the joint density of $\underset{\sim}{T}$ and $L$ is

$$(2\pi)^{-\frac{1}{2}pn}g_{n,p}(L)\left(\prod_{i=1}^{p} t_{ii}^{n-i}\right)\operatorname{etr}\left(-\tfrac{1}{2}\underset{\sim}{T}\underset{\sim}{T}'\right).$$

Let us define $X=\underset{\sim}{T}L$. Then, by Problem 1.33, the Jacobian of the transformation from $\underset{\sim}{T}$ and $L$ to $X$ is $J(\underset{\sim}{T},L \rightarrow X)=(g_{n,p}(L)\prod_{i=1}^{p}t_{ii}^{n-i})^{-1}$, and hence we have

$$X \sim N_{p,n}(0,I_p,I_n) \quad \text{and} \quad V=XX'.$$

Taking $Y=A^{-1}X$, we get the required Corollary 3.2.5. □

## 3.3 Some Properties of Wishart Distribution

### 3.3.1 Distribution of Sample Covariance

**Theorem 3.3.1** *Let $X \sim N_{p,N}(\boldsymbol{\theta}\mathbf{e}', \Sigma, I_N)$, $X = (\mathbf{x}_1, \ldots, \mathbf{x}_N)$ and let*

$$(3.3.1) \qquad V = \sum_{i=1}^{N} (\mathbf{x}_i - \bar{\mathbf{x}})(\mathbf{x}_i - \bar{\mathbf{x}})' = \sum_{i=1}^{N} \mathbf{x}_i \mathbf{x}_i' - N\bar{\mathbf{x}}\bar{\mathbf{x}}'$$

$$= XX' - N\bar{\mathbf{x}}\bar{\mathbf{x}}', \qquad \textit{where} \quad \bar{\mathbf{x}} = N^{-1} \sum_{i=1}^{N} \mathbf{x}_i,$$

$$= X\left[I_N - N^{-1}\mathbf{e}\mathbf{e}'\right]X', \qquad \mathbf{e}' = (1, 1, \ldots, 1) : 1 \times N.$$

*Then $V \sim W_p(\Sigma, n)$, where $n = N - 1$.*

PROOF. Let $\Gamma : N \times N$ be an orthogonal matrix with first column $(\mathbf{e}/N^{1/2})$, where $\mathbf{e}' = (1, 1, \ldots, 1)$. Let

$$(N^{1/2}\bar{\mathbf{x}}, Y) = X\Gamma, \qquad \text{where} \quad Y : p \times n, \quad n = N - 1.$$

Then $J(X \to N^{1/2}\bar{\mathbf{x}}, Y) = 1$. From the factorization of the pdf, it follows that $\bar{\mathbf{x}}$ and $Y$ are independently distributed. Also $Y \sim N_{p,n}(0, \Sigma, I_n)$. Thus from Eq. (3.3.1)

$$V = XX' - N\bar{\mathbf{x}}\bar{\mathbf{x}}' = YY',$$

where $Y \sim N_{p,n}(0, \Sigma, I_n)$. Hence, $V \sim W_p(\Sigma, n)$. □

REMARK 3.3.1. The positive definiteness of the sample covariance matrix $XX' - N\bar{\mathbf{x}}\bar{\mathbf{x}}' = X(I - N^{-1}\mathbf{e}\mathbf{e}')X'$ is assured because of normality, since in this case it is equal to $YY'$, where $n$ columns of $Y$ are independently distributed with variance matrix $\Sigma > 0$ and $n \geq p$, and the result follows from Section 3.1. For a proof that does not require normality, see Eaton and Perlman (1973), Okamoto (1973), and Das Gupta (1971).

### 3.3.2 Marginal Distribution

In this subsection we obtain marginal distributions of square submatrices on the main diagonal of a Wishart matrix $V \sim W_p(\Sigma, n)$.

**Theorem 3.3.2** *Let*

$$V = \begin{pmatrix} V_{11} & V_{12} \\ V_{12}' & V_{22} \end{pmatrix} \quad \textit{and} \quad \Sigma = \begin{pmatrix} \Sigma_{11} & \Sigma_{12} \\ \Sigma_{12}' & \Sigma_{22} \end{pmatrix},$$

*where $V_{11} : q \times q$ and $\Sigma_{11} : q \times q$, $1 \leq q \leq p$. Then $V_{11} \sim W_q(\Sigma_{11}, n)$.*

PROOF. Since

$$V=YY'=\begin{pmatrix}Y_1\\Y_2\end{pmatrix}(Y_1',Y_2')=\begin{pmatrix}Y_1Y_1' & Y_1Y_2'\\ Y_2Y_1' & Y_2Y_2'\end{pmatrix},$$

where $Y \sim N_{p,n}(0,\Sigma,I_n)$ and $Y_1 \sim N_{q,n}(0,\Sigma_{11},I_n)$, we find that $V_{11}=Y_1Y_1' \sim W_q(\Sigma_{11},n)$. □

**Theorem 3.3.3** *Let $V=(V_{ij})$ and $\Sigma=(\Sigma_{ij})$, where $V_{ij}:q_i\times q_j$ and $\Sigma_{ij}:q_i\times q_j$, $i,j=1,2,\ldots,k$, and $\sum_{i=1}^k q_i=p$. Then $V_{ii}\sim W_{q_i}(\Sigma_{ii},n)$, $q_i\geqslant 1$.*

PROOF. As above, let $Y_i$ be a $q_i\times n$ matrix, $i=1,2,\ldots,k$, and let

$$Y=\begin{bmatrix}Y_1\\ \vdots\\ Y_k\end{bmatrix}.$$

Then, since $V=YY'$, we get $V_{ii}=Y_iY_i'$, where $Y_i\sim N_{q_i}(0,\Sigma_{ii},I_n)$. Hence, $V_{ii}\sim W_{q_i}(\Sigma_{ii},n)$. In particular, if $q_i=1$, then $V_{ii}\sim\Sigma_{ii}\chi_n^2$. □

**Theorem 3.3.4** *Let $W\sim W_p(I,n)$, $n\geqslant p$, and let $W=(w_{ij})$, where $w_{ij}$ $(1\leqslant i,j\leqslant p)$ is a scalar. Then $w_{11},w_{22},\ldots,w_{pp}$ are independently distributed, each having the distribution of a $\chi_n^2$ random variable.*

PROOF. Since $W=YY'$, where $Y\sim N_{p,n}(0,I_p,I_n)$, we find that all the elements of $Y$ are independently normally distributed with mean zero and variance one. Hence the $w_{ii}$'s are independently distributed as $\chi_n^2$ random variables. □

An alternative proof is given in the proof of Theorem 3.5.1.

**Theorem 3.3.5** *Let $V\sim W_p(\Sigma,n)$, $n\geqslant p$, $\Sigma>0$, and let*

$$V=\begin{matrix} & \begin{matrix} r & s\end{matrix}\\ \begin{matrix} r\\ s\end{matrix} & \begin{pmatrix}V_{11} & V_{12}\\ V_{12}' & V_{22}\end{pmatrix}\end{matrix},\qquad \Sigma=\begin{matrix} & \begin{matrix} r & s\end{matrix}\\ \begin{matrix} r\\ s\end{matrix} & \begin{pmatrix}\Sigma_{11} & \Sigma_{12}\\ \Sigma_{12}' & \Sigma_{22}\end{pmatrix}\end{matrix},$$

*$V_{1.2}=V_{11}-V_{12}V_{22}^{-1}V_{12}'$, $\Sigma_{1.2}=\Sigma_{11}-\Sigma_{12}\Sigma_{22}^{-1}\Sigma_{12}'$, and $\beta=\Sigma_{12}\Sigma_{22}^{-1}$, where $r+s=p$. Then $V_{1.2}$ and $(V_{12},V_{22})$ are independently distributed, where $V_{1.2}\sim W_r(\Sigma_{1.2},n-s)$, $V_{12}$ given $V_{22}$ is $N_{r,s}(\beta V_{22},\Sigma_{1.2},V_{22})$, and $V_{22}\sim W_s(\Sigma_{22},n)$.*

PROOF. Let $V=YY'$, where $Y\sim N_{p,n}(0,\Sigma,I_n)$. Let $Y'=(Y_1',Y_2')$, with $Y_1$ and $Y_2$ being $r\times n$ and $s\times n$ matrices. Since $\rho(Y_2)=s$, by Corollary 1.5.8 we can find a matrix $G:(n-s)\times n$ of rank $n-s$ such that $Y_2G'=0$, $GG'=I_{n-s}$, and $(Y_2',G')=Q'$ is nonsingular. Now

$$I=Q'(QQ')^{-1}Q=Y_2'(Y_2Y_2')^{-1}Y_2+G'G.$$

We observe the following relationships:

$$V_{22}=Y_2Y_2', \qquad V_{12}=Y_1Y_2',$$

and

$$V_{1.2}=Y_1\big(I-Y_2'(Y_2Y_2')^{-1}Y_2\big)Y_1'=Y_1G'GY_1'.$$

Since $Y\sim N_{p,n}(0,\Sigma,I)$, by Theorem 2.2.7 the conditional distribution of $Y_1$ given $Y_2$ is

$$N_{r,n}(\beta Y_2,\Sigma_{1.2},I),$$

where

$$\beta=\Sigma_{12}\Sigma_{22}^{-1} \quad\text{and}\quad \Sigma_{1.2}=\Sigma_{11}-\beta\Sigma_{12}'.$$

Let us use the transformation $Y_1Q'=(V_{12},X)$ given $Y_2$. Then $V_{1.2}=XX'$ and $V_{12}=Y_1Y_2'$. The Jacobian of the transformation is

$$J(Y_1\to V_{12},X)=|Q'^{-1}|^r=|QQ'|^{-r/2}=|V_{22}|^{-r/2},$$

and

$$(Y_1-\beta Y_2)(Y_1-\beta Y_2)'=[(V_{12},X)-\beta Y_2Q'](QQ')^{-1}[(V_{12},X)-\beta Y_2Q']'$$

$$=(V_{12}-\beta V_{22})V_{22}^{-1}(V_{12}-\beta V_{22})'+XX'.$$

Hence, the conditional density of $V_{12}$ and $X$ given $Y_2$ is

$$\text{const. etr}\Big[-\tfrac{1}{2}\Sigma_{1.2}^{-1}XX'-\tfrac{1}{2}\Sigma_{1.2}^{-1}(V_{12}-\beta V_{22})V_{22}^{-1}(V_{12}-\beta V_{22})'\Big],$$

and this shows that

$$X\sim N_{r,n-s}(0,\Sigma_{1.2},I_{n-s}),$$

$$(V_{12}\text{ given }Y_2)\sim N_{r,s}(\beta V_{22},\Sigma_{1.2},V_{22}),$$

and $X$ and $(V_{12}',Y_2)$ are independently distributed. This means that $V_{1.2}=XX'$ and $(V_{12}',V_{22})$ are independently distributed. $V_{1.2}\sim W_r(\Sigma_{1.2},n-s)$, $(V_{12}\text{ given }V_{22})\sim N_{r,s}(\beta V_{22},\Sigma_{1.2},V_{22})$, and $V_{22}\sim W_s(\Sigma_{22},n)$. This proves the theorem. □

REMARK 3.3.2. Theorem 3.3.5 is true when $\Sigma$ is p.s.d.

### 3.3.3 *Additive Property*

**Theorem 3.3.6** *Let $V \sim W_p(\Sigma, n)$ and $W \sim W_p(\Sigma, m)$ be independently distributed. Then $V+W \sim W_p(\Sigma, n+m)$.*

PROOF. Since we can write $V = XX'$ and $W = YY'$, where $X \sim N_{p,n}(0, \Sigma, I_n)$ and $Y \sim N_{p,m}(0, \Sigma, I_m)$, and since $V$ and $W$ are independently distributed, so will be $X$ and $Y$. Hence the matrix $(X, Y) \sim N_{p,n+m}(0, \Sigma, I_{n+m})$. Hence $V + W = (X, Y)(X, Y)' \sim W_p(\Sigma, n+m)$. □

### 3.3.4 *Distribution of $AVA'$*

**Theorem 3.3.7** *Let $V \sim W_p(\Sigma, n)$, $n \geqslant p$, $\Sigma$ p.d., and let $A: k \times p$, $k \leqslant p$, be a matrix of constants. Then $AVA' \sim W_k(A\Sigma A', n)$.*

PROOF. Since $V = YY'$, where $Y \sim N_{p,n}(0, \Sigma, I_n)$, we find that $AVA' = (AY)(AY)'$, where $AY \sim N_k(0, A\Sigma A', I_n)$. Hence $AVA' \sim W_k(A\Sigma A', n)$. □

Thus as a special case of the above result, we find that for a $p \times 1$ nonnull vector $\mathbf{a}$,

$$\mathbf{a}' V \mathbf{a} \sim W_1(\mathbf{a}'\Sigma\mathbf{a}, n),$$

that is, $(\mathbf{a}'V\mathbf{a})/(\mathbf{a}'\Sigma\mathbf{a})$ has a chi-square distribution with $n$ d.f. if $\mathbf{a}'\Sigma\mathbf{a} \neq 0$.

The converse of this theorem is not true in general; for a counterexample, see Mitra (1969). However, if we have further information, then the converse is true. This is contained in

**Theorem 3.3.7a** *Let $\mathbf{a}$ be a nonnull $p$ vector such that $\mathbf{a}'\Sigma\mathbf{a} \neq 0$. Let $V$ be a random* p.d. *matrix such that $V = XBX'$, where the column vectors of $X$ are independently and normally distributed and $B$ is a symmetric matrix with nonrandom elements. If, for all nonnull vectors $\mathbf{a}$, $\mathbf{a}'V\mathbf{a}/\mathbf{a}'\Sigma\mathbf{a}$ is distributed as chi square with $n$* d.f., *then $V \sim W_p(\Sigma, n)$.*

The proof follows easily from Corollary 2.11.2.

### 3.3.5 *Generalized Variance and Its Distribution*

Let $V \sim W_p(\Sigma, n)$, $n \geqslant p$, $\Sigma$ positive definite. Then the determinant of $n^{-1}V$ is called the generalized variance [introduced by Wilks (1932); see also Frisch (1929)]. From the previous result it follows that

$$W \equiv (\Sigma^{-1/2}) V (\Sigma^{-1/2})' \sim W_p(I, n),$$

where $\Sigma^{1/2}$ is any nonsingular factorization of $\Sigma$, $\Sigma=(\Sigma^{1/2})(\Sigma^{1/2})'$. Writing $W=\underset{\sim}{T}\underset{\sim}{T}'$, we find from Corollary 3.2.4 that the elements of $\underset{\sim}{T}$ are independently distributed with $t_{ij}\sim N(0,1)$, $1\leqslant j<i\leqslant p$ and $t_{ii}^2\sim\chi_{n-i+1}^2$, $i=1,2,\ldots,p$. Hence

$$|W|=|\Sigma^{-1/2}V\Sigma^{-1/2}|=|\underset{\sim}{T}\underset{\sim}{T}'|=\Pi t_{ii}^2.$$

Hence,

**Theorem 3.3.8** *Let $V\sim W_p(\Sigma,n)$. Then*

$$|V|=|\Sigma|\prod_{i=1}^{p} u_i,$$

*where $u_i$'s are independently distributed as $\chi_{n-i+1}^2$.*

In the special case of $p=2$, we get

$$|W|\equiv[|V|/|\Sigma|]=(u_1u_2).$$

In the following lemma, it will be shown that $2(u_1u_2)^{1/2}\sim\chi_{2n-2}^2$.

**Lemma 3.3.1** *Let $x\sim\chi_n^2$ and $y\sim\chi_{n-1}^2$ be independently distributed. Let $z^2=4(xy)$. Then $z\sim\chi_{2n-2}^2$.*

The joint pdf of $x$ and $y$ is given by

$$\text{const}\, x^{\frac{1}{2}n-1}y^{\frac{1}{2}(n-1)-1}e^{-\frac{1}{2}(x+y)}.$$

Letting

$$z^2=4xy,$$

we get $J(y\to z)=\frac{1}{2}(z/x)$. Hence, the joint pdf of $z$ and $x$ is given by

$$\text{const}\,(z^2)^{\frac{1}{2}n-1}x^{-1/2}e^{-\frac{1}{2}[x+z^2(4x)^{-1}]}.$$

Hence, the pdf of $z$ is given by

$$\text{const}\, z^{n-2}h(z),$$

where

$$h(z)=\int_0^\infty x^{-1/2}e^{-\frac{1}{2}[x+z^2(4x)^{-1}]}dx.$$

Hence

$$h'(z)=-\tfrac{1}{4}z\int_0^\infty x^{-3/2}\exp\{-\tfrac{1}{2}[x+z^2(4x)^{-1}]\}\,dx$$
$$=-\tfrac{1}{2}\int_0^\infty u^{-1/2}\exp\{-\tfrac{1}{2}[u+z^2(4u)^{-1}]\}\,du$$

[by putting $(4x)^{-1}z^2=u$],

$$=-\tfrac{1}{2}h(z),$$

which gives

$$h(z)=\exp\left(-\tfrac{1}{2}z+c\right).$$

Thus, the pdf of $z$ is

$$\operatorname{const} z^{\frac{1}{2}(2n-2)-1}e^{-\frac{1}{2}z},$$

which is $\chi^2_{2n-2}$.

For general $p$, a closed-form density is not available. However, we can easily write the $h$th moment as follows.

**Lemma 3.3.2**

$$E(|W|^h)=\int_{W>0}\frac{|W|^{\frac{1}{2}(n-p-1)+h}\operatorname{etr}\left(-\frac{1}{2}W\right)}{2^{\frac{1}{2}pn}\Gamma_p\left(\frac{1}{2}n\right)}dW$$

$$=2^{ph}\frac{\Gamma_p\left(\frac{1}{2}n+h\right)}{\Gamma_p\left(\frac{1}{2}n\right)}.$$

[The result is valid for complex $h$ such that $\operatorname{Re}h>-\frac{1}{2}(n-p+1)$.] From this, an approximate density can be obtained on the lines given in Chapter 7. For another approximation, see Problem 3.31.

### *3.3.6 Characteristic Function*

Let $V\sim W_p(\Sigma,n)$, $\Sigma>0$, $n\geqslant p$, and let $\theta$ be a symmetric $p\times p$ matrix. Since $\operatorname{tr}\theta V=\Sigma_{i,j}\theta_{ij}v_{ji}$, where $\theta=(\theta_{ij})$ and $V=(v_{ij})$, the characteristic function of $V$ is

$$\phi_V(\theta)=E(e^{i\operatorname{tr}\theta V}),\qquad V\sim W_p(\Sigma,n).$$

Since $\Sigma>0$ and $\theta$ is a symmetric matrix, there exists a nonsingular matrix $\Gamma$ such that $\theta=\Gamma D_\alpha\Gamma'$, $\Sigma^{-1}=\Gamma\Gamma'$, where $D_\alpha=\operatorname{diag}(\alpha_1,\ldots,\alpha_p)$ and $\alpha_1,\ldots,\alpha_p$ are the characteristic roots of $\Sigma\theta$ (see Theorem 1.9.2). Hence

$$\phi_V(\theta)=E(e^{i\operatorname{tr}D_\alpha\Gamma'V\Gamma})$$

$$\equiv E(e^{i\operatorname{tr}D_\alpha U}),$$

where from Theorem 3.3.7, $U\equiv\Gamma'V\Gamma\sim W_p(I,n)$. Thus

$$\phi_V(\theta)=E(e^{i\Sigma\alpha_j u_{jj}}),$$

where $U=((u_{ij}))$ and $u_{11},u_{22},\ldots,u_{pp}$ are independently distributed as chi-square random variables with $n$ d.f. Hence

$$\begin{aligned}\phi_V(\theta) &= \prod_{j=1}^{p} E(e^{i\alpha_j u_{jj}}) \\ &= \prod_{j=1}^{p} (1-2i\alpha_j)^{-\frac{1}{2}n} \\ &= |I-2iD_\alpha|^{-\frac{1}{2}n} = |I-2i\Sigma\theta|^{-\frac{1}{2}n}.\end{aligned}$$

Thus, we get

**Theorem 3.3.9** *If $V \sim W_p(\Sigma, n)$, $\Sigma > 0$, and $n \geqslant p$, then the characteristic function of $V$ is given by $|I-2i\Sigma\theta|^{-\frac{1}{2}n}$, where $\theta$ is any symmetric matrix.*

We point out that Theorem 3.3.9 is valid even when $n < p$ and $\Sigma \geqslant 0$. Let $V \sim W_p(\Sigma, n)$. Then $V = YY'$, where $Y = (\mathbf{y}_1, \ldots, \mathbf{y}_n)$ and the $\mathbf{y}_j$'s are independent $N_p(\mathbf{0}, \Sigma)$. By Definition 2.2.1, we can write

$$\mathbf{y}_j = B\mathbf{z}_j, \qquad \mathbf{z}_j \sim N_r(\mathbf{0}, I_r), \quad \text{and} \quad \rho(B : p \times r) = r = \rho(\Sigma).$$

Hence, the characteristic function of $V$ is

$$\begin{aligned}\phi_V(\theta) &= E\,\mathrm{etr}\left(i\theta \sum_{j=1}^{n} B\mathbf{z}_j\mathbf{z}_j' B'\right) \\ &= \prod_{j=1}^{n} E\exp(i\mathbf{z}_j' B'\theta B\mathbf{z}_j).\end{aligned} \tag{3.3.2}$$

From Section 2.2.11,

$$\begin{aligned}E\exp(i\mathbf{z}_j' B'\theta B\mathbf{z}_j) &= |I-2iB'\theta B|^{-1/2} \\ &= |I-2i\Sigma\theta|^{-1/2},\end{aligned}$$

which gives the required result of Theorem 3.3.9 without assuming that $\Sigma > 0$ and $n \geqslant p$.

Next, we obtain the characteristic function of the noncentral Wishart matrix $V = YY'$, where the columns $\mathbf{y}_j$ of $Y$ are independently distributed as $N_p(\boldsymbol{\mu}_j, \Sigma)$. Then, as before, by Definition 2.2.1, we have

$$\mathbf{y}_j = \boldsymbol{\mu}_j + B\mathbf{z}_j, \qquad \mathbf{z}_j \sim N_r(\mathbf{0}, I_r), \qquad BB' = \Sigma.$$

Then

$$(3.3.3)\quad \phi_V(\theta) = E\,\mathrm{etr}(i\theta V)$$

$$= \prod_{j=1}^{n} E\,\mathrm{etr}(i\theta \mathbf{y}_j \mathbf{y}_j')$$

$$= \prod_{j=1}^{n} E\exp(i\mathbf{z}_j' B'\theta B\mathbf{z}_j + 2i\boldsymbol{\mu}_j'\theta B\mathbf{z}_j + i\boldsymbol{\mu}_j'\theta\boldsymbol{\mu}_j)$$

$$= \prod_{j=1}^{n} \left\{ |I - 2iB'\theta B|^{-1/2} \exp\left[ i\boldsymbol{\mu}_j'\theta\boldsymbol{\mu}_j - 2\boldsymbol{\mu}_j'\theta B(I - 2iB'\theta B)^{-1} B'\theta\boldsymbol{\mu}_j \right] \right\}$$

$$= |I - 2i\Sigma\theta|^{-\frac{1}{2}n}\,\mathrm{etr}\left[ i\theta(I - 2i\Sigma\theta)^{-1}\mu\mu' \right],$$

where the cf of a quadratic form from Section 2.2.11 is used and $\mu = (\boldsymbol{\mu}_1, \boldsymbol{\mu}_2, \ldots, \boldsymbol{\mu}_n)$. This gives

**Theorem 3.3.10** *Let $V$ have a noncentral Wishart distribution denoted as $V \sim W_p(\Sigma, n, \Omega)$. Then the characteristic function of $V$ is given by*

$$\phi_V(\theta) = |I - 2i\Sigma\theta|^{-\frac{1}{2}n}\,\mathrm{etr}\left[ i\theta(I - 2i\Sigma\theta)^{-1}\Omega \right],$$

*where $\theta$ is symmetric and $\Omega$ stands for the noncentral parameter matrix, which is* p.s.d.

From the above result, the following theorem is obvious, and hence its proof is omitted.

**Theorem 3.3.11** *Let $V_j \sim W_p(\Sigma, n_j, \Omega_j)$, $j = 1, 2, \ldots, k$, and let them be independently distributed. Then*

$$V = \sum_{j=1}^{k} V_j \sim W_p(\Sigma, n, \Omega),$$

*where*

$$n = \sum_{j=1}^{k} n_j \quad \text{and} \quad \Omega = \sum_{j=1}^{k} \Omega_j.$$

**Theorem 3.3.12** *Let $V \sim W_p(D_\sigma, n, \Omega)$, where $V = (v_{ij})$, $\Omega = (w_{ij})$, and $D_\sigma =$* $\mathrm{diag}(\sigma_1, \ldots, \sigma_p)$. *Then $v_{ii}/\sigma_i$ $(i = 1, 2, \ldots, p)$ are independently distributed as noncentral chi square with $n$* d.f. *and noncentrality parameter $w_{ii}/\sigma_i$.*

## 3.4 Distribution of the Characteristic Roots and Vectors when $\Sigma = I$

### *3.4.1 Characteristic Roots*

Let $W \sim W_p(I,n)$, $n \geqslant p$. Then there exists a unique orthogonal matrix $Q$ with positive diagonal elements such that

$$W = QD_wQ', \tag{3.4.1}$$

where $D_w = \text{diag}(w_1, w_2, \ldots, w_p)$, and $w_1 > w_2 > \cdots > w_p > 0$ are the characteristic roots of $W$. Note that for the sake of convenience of notation, we are writing $Q$ rather than the $Q_+$ used in Chapter 1 and in the beginning of this chapter. Since the Jacobian of the transformation (3.4.1) is, by Theorem 1.11.5,

$$J(W \to Q, D_w) = \prod_{i<j} (w_i - w_j) f_p(Q), \tag{3.4.2}$$

where $f_p(Q)$ is defined in that theorem, we find that the joint pdf of $Q$ and $\mathbf{w} = (w_1, w_2, \ldots, w_p)'$ is given by

$$C(p,n)|D_w|^{\frac{1}{2}(n-p-1)}\left[\exp\left(-\frac{1}{2}\sum_{i=1}^{p} w_i\right)\right]\left[\prod_{i<j}(w_i - w_j)\right]f_p(Q), \tag{3.4.3}$$

where $C(p,n)$ is defined by Eqs. (3.2.3) and (3.2.5). Thus $Q$ and $\mathbf{w}$ are independently distributed. Integrating out $Q$ over the region $QQ' = I$, with the help of Corollary 3.2.2, we get the pdf of $\mathbf{w}$ as

$$\frac{\pi^{\frac{1}{2}p^2}}{2^{\frac{1}{2}pn}\Gamma_p\left(\frac{1}{2}p\right)\Gamma_p\left(\frac{1}{2}n\right)}\left(\prod_{i=1}^{p} w_i^{\frac{1}{2}(n-p-1)} e^{-\frac{1}{2}w_i}\right)\prod_{i<j}(w_i - w_j). \tag{3.4.4}$$

Thus, we get

**Theorem 3.4.1** *Let $W \sim W_p(I,n)$, $p \leqslant n$. Then the* pdf *of the characteristic roots $w_1 > w_2 > \cdots > w_p$ of $W$ is given by Eq.* (3.4.4).

REMARK 3.4.1. That the roots of $W$ are distinct with probability one follows from the results of Okamoto (1973).

### *3.4.2 Characteristic Vectors*

From (3.4.3) and (3.4.4) we get

**Theorem 3.4.2** *The density of $Q$ is given by*

$$\pi^{-\frac{1}{2}p^2}\Gamma_p\left(\tfrac{1}{2}p\right)f_p(Q)\,dQ, \qquad QQ' = I.$$

## 3.5 Distribution of Sample Correlations

### *3.5.1 Matrix of Correlations and Correlation in the Bivariate Case*

**Theorem 3.5.1** *Let $W \sim W_p(I,n)$, $n \geqslant p$. Let for $i \neq j$, $1 \leqslant i < j \leqslant p$*

$$r_{ij} = \frac{w_{ij}}{(w_{ii}w_{jj})^{1/2}} \quad \text{and} \quad r_{ii} = 1, \qquad i = 1,2,\ldots,p.$$

*Then the* pdf *of $R = (r_{ij})$, the matrix of sample correlations, is given by*

$$\frac{\left[\Gamma\left(\frac{1}{2}n\right)\right]^p}{\Gamma_p\left(\frac{1}{2}n\right)} |R|^{\frac{1}{2}(n-p-1)}.$$

PROOF. Letting

$$W = \operatorname{diag}\left(w_{11}^{1/2},\ldots,w_{pp}^{1/2}\right) R \operatorname{diag}\left(w_{11}^{1/2},\ldots,w_{pp}^{1/2}\right),$$

the Jacobian of the transformation $J(W \to R, w_{11},\ldots,w_{pp}) = \prod_{i<j}(w_{ii}w_{jj})^{1/2} = \prod_{r=1}^{p}(w_{rr})^{\frac{1}{2}(p-1)}$. Hence, the joint pdf of $R$ and $w_{11},\ldots,w_{pp}$ is given by

$$p(R, w_{ii}, i=1,\ldots,p) = C(p,n)\left(\prod_{i=1}^{p} w_{ii}^{\frac{1}{2}(n-2)}\right)|R|^{\frac{1}{2}(n-p-1)}\exp\left(-\tfrac{1}{2}\Sigma w_{ii}\right).$$

Integrating out $w_{ii}$, we get the pdf of $R$ as given in Theorem 3.5.1. □

From this it also follows that $R$ and $w_{11}, w_{22},\ldots,w_{pp}$ are independently distributed—the distributions of the $w_{ii}$'s being chi squares with $n$ d.f., which has already been established.

If $W \sim W_p(\Sigma, n)$, the distribution of $R$ is difficult to obtain. An explicit expression is given by Fisher (1915, p. 62) for $p=2$ and $p=3$, and a multiple series expansion is given by Ali, Fraser, and Lee (1970). We consider only the case when $p=2$. This will be the distribution of the sample correlation from a sample of size $N=n+1$, when the population correlation is different from zero. For simplicity we write

$$R = \begin{pmatrix} 1 & r \\ r & 1 \end{pmatrix}.$$

Also, since the sample correlation $r$ is unchanged by a scale transformation, we may assume without any loss of generality that $\Sigma$ is of the form

$$\Sigma = \begin{pmatrix} 1 & \rho \\ \rho & 1 \end{pmatrix}.$$

Thus the joint pdf of $r$, $w_{11}$, and $w_{22}$ can be written as

$$(3.5.1)\qquad (1-\rho^2)^{-\frac{1}{2}n}C(2,n)(w_{11}w_{22})^{\frac{1}{2}n-1}(1-r^2)^{\frac{1}{2}(n-3)}$$
$$\times\exp\left(-\frac{1}{2(1-\rho^2)}(w_{11}+w_{22})\right)\exp\left(\frac{r\rho}{(1-\rho^2)}w_{11}^{1/2}w_{22}^{1/2}\right).$$

Let $u_{ii}=w_{ii}/(1-\rho^2)$, $i=1,2$; we find that the joint pdf of $u_{11}$, $u_{22}$, and $r$ is given by

$$f_0(r)(u_{11}u_{22})^{\frac{1}{2}n-1}\exp\left[-\tfrac{1}{2}(u_{11}+u_{22})\right]\exp\left(r\rho u_{11}^{1/2}u_{22}^{1/2}\right)$$
$$=f_0(r)(u_{11}u_{22})^{\frac{1}{2}n-1}\exp\left[-\tfrac{1}{2}(u_{11}+u_{22})\right]\sum_{j=0}^{\infty}\frac{\left(\rho r u_{11}^{1/2}u_{22}^{1/2}\right)^j}{j!},$$

where

$$f_0(r)=C(2,n)(1-\rho^2)^{\frac{1}{2}n}(1-r^2)^{\frac{1}{2}(n-3)}.$$

Integrating out $u_{11}$ and $u_{22}$, we get the pdf of $r$ as

$$(3.5.2)\qquad C(2,n)(1-\rho^2)^{\frac{1}{2}n}(1-r^2)^{\frac{1}{2}(n-3)}\sum_{j=0}^{\infty}\frac{(r\rho)^j}{j!}\Gamma^2\left(\tfrac{1}{2}n+\tfrac{1}{2}j\right)2^{n+j}$$
$$=\left[\frac{\pi\Gamma(n-1)}{2^{n-2}}\right]^{-1}(1-\rho^2)^{\frac{1}{2}n}(1-r^2)^{\frac{1}{2}(n-3)}\sum_{j=0}^{\infty}\frac{(2\rho r)^j}{j!}\Gamma^2\left(\tfrac{1}{2}n+\tfrac{1}{2}j\right),$$

by using the duplication formula $\Gamma(r)\Gamma(r+\frac{1}{2})=\pi^{1/2}\Gamma(2r)/2^{2r-1}$.

This distribution of $r$ was given by Fisher (1915), who also gave another form of the density:

$$(3.5.3)\qquad \frac{(1-\rho^2)^{\frac{1}{2}n}(1-r^2)^{\frac{1}{2}(n-3)}}{\pi(n-2)!}\left[\frac{d^{n-1}}{dx^{n-1}}\left\{\frac{\cos^{-1}(-x)}{(1-x^2)^{1/2}}\right\}\Bigg|_{x=r\rho}\right].$$

For establishing Eq. (3.5.3), use the transformations $w_{11}=(1-\rho^2)u/y$ and $w_{22}=(1-\rho^2)uy$ in Eq. (3.5.1), and integrate over $u$. This will lead to the joint density of $y$ and $r$ as

$$(3.5.4)\qquad 2^2C(2,n)(n-1)!(1-\rho^2)^{\frac{1}{2}n}(1-r^2)^{\frac{1}{2}(n-3)}(2y)^{n-1}(y^2-2\rho ry+1)^{-n}.$$

Integration over $y$ using $\int_0^{\infty}dy/(y^2-2\rho ry+1)=\cos^{-1}(-\rho r)/(1-\rho^2r^2)^{1/2}$ gives Eq. (3.5.3). Since the series in Eq. (3.5.2) converges very slowly, Hotelling (1953) gave the following form for the pdf after integrating over

$y$ from Eq. (3.5.4) in a different way:

$$\frac{n-1}{(2\pi)^{1/2}}\frac{\Gamma(n)}{\Gamma\left(n+\frac{1}{2}\right)}(1-\rho^2)^{\frac{1}{2}n}(1-r^2)^{\frac{1}{2}(n-3)} \times (1-\rho r)^{-n+\frac{1}{2}}{}_2F_1\left(\tfrac{1}{2},\tfrac{1}{2};n+\tfrac{1}{2};\tfrac{1}{2}(1+\rho r)\right), \tag{3.5.5}$$

where ${}_2F_1$ is a hypergeometric function given by

$${}_2F_1(a,b;c;x)=\sum_{j=0}^{\infty}\frac{\Gamma(a+j)\Gamma(b+j)}{\Gamma(a)\Gamma(b)}\frac{\Gamma(c)}{\Gamma(c+j)}\frac{x^j}{j!}. \tag{3.5.6}$$

Hence we get

**Theorem 3.5.2** *The* pdf *of the sample correlation when the population correlation* $\rho$ *is different from zero is given by Eq.* (3.5.1) *or* (3.5.3) *or* (3.5.5).

The cumulative distribution of $r$ has been tabulated by David (1938) for $\rho=0(.1).9$, $n+1\equiv N=3(1)25, 50, 100, 200, 400$, and $r=-1(.05)1$.

### *3.5.2 Multiple Correlation*

In this section we derive the distribution of the sample multiple correlation based on $N$ observations from a *normal* distribution. Let

$$\mathbf{z}=\begin{matrix}1\\p-1\end{matrix}\begin{pmatrix}y\\\mathbf{x}\end{pmatrix}\sim N_p(\boldsymbol{\mu},\Sigma),\qquad\text{where}\qquad \Sigma=\begin{matrix}\\1\\p-1\end{matrix}\begin{matrix}\begin{matrix}1 & p-1\end{matrix}\\\begin{pmatrix}\sigma_{11} & \boldsymbol{\sigma}_{12}'\\\boldsymbol{\sigma}_{12} & \Sigma_{22}\end{pmatrix}\end{matrix}.$$

Then, by definition, the multiple correlation $\rho_{y\mathbf{x}}\equiv\rho$ is given by

$$\rho^2=1-\frac{1}{\sigma_{11}\sigma^{11}}=\frac{\boldsymbol{\sigma}_{12}'\Sigma_{22}^{-1}\boldsymbol{\sigma}_{12}}{\sigma_{11}}. \tag{3.5.7}$$

Suppose a sample of size $N$ is taken from $N(\boldsymbol{\mu},\Sigma)$, and let $S$ denote the sample covariance. Then if

$$S=\begin{pmatrix}s_{11} & \mathbf{s}_{12}'\\\mathbf{s}_{12} & S_{22}\end{pmatrix}\equiv n^{-1}V=n^{-1}\begin{pmatrix}v_{11} & \mathbf{v}_{12}'\\\mathbf{v}_{12} & V_{22}\end{pmatrix}\qquad\text{with}\quad n=N-1,$$

the sample multiple correlation is defined by

$$r^2\equiv r_{y\mathbf{x}}^2=1-\frac{1}{s_{11}s^{11}}=\frac{\mathbf{s}_{12}'S_{22}^{-1}\mathbf{s}_{12}}{s_{11}}=\frac{\mathbf{v}_{12}'V_{22}^{-1}\mathbf{v}_{12}}{v_{11}},$$

and $V\sim W_p(\Sigma,n)$. By Theorem 3.3.5, $v_{1.2}=v_{11}-\mathbf{v}_{12}'V_{22}^{-1}\mathbf{v}_{12}$ and $(\mathbf{v}_{12},V_{22})$ are independently distributed, $v_{1.2}/\sigma_{1.2}\sim\chi^2_{n-p+1}$, $V_{22}\sim W_{p-1}(\Sigma_{22},n)$, and given

$V_{22}$, $\mathbf{v}_{12} \sim N_{p-1}(V_{22}\boldsymbol{\beta}, \sigma_{1.2}V_{22})$, where $\sigma_{1.2} = \sigma_{11} - \boldsymbol{\sigma}'_{12}\Sigma_{22}^{-1}\boldsymbol{\sigma}_{12} = \sigma_{11}(1-\rho^2)$ and $\boldsymbol{\beta} = \Sigma_{22}^{-1}\boldsymbol{\sigma}_{12}$. From Section 2.2.11, it is easy to see that given $V_{22}$, $\mathbf{v}'_{12}V_{22}^{-1}\mathbf{v}_{12}/\sigma_{1.2} \sim \chi^2_{p-1}(\delta)$, where $\delta^2 = \boldsymbol{\beta}'V_{22}\boldsymbol{\beta}/\sigma_{1.2}$. Hence, given $V_{22}$, the conditional distribution of $(n-p+1)r^2/(p-1)(1-r^2)$ $[=(n-p+1)\mathbf{v}'_{12}V_{22}^{-1}\mathbf{v}_{12}/(p-1)v_{1.2}]$ is noncentral $F$ with $p-1$ and $n-p+1$ d.f. and the noncentrality parameter $\delta^2$. This gives the density of $r^2$ given $V_{22}$ as

(3.5.8) $$p(r^2|\delta^2) = \text{const } e^{-\frac{1}{2}\delta^2}(r^2)^{\frac{1}{2}(p-3)}(1-r^2)^{\frac{1}{2}(n-p-1)} \times {}_1F_1\left(\frac{n}{2}; \frac{p-1}{2}; \frac{\delta^2 r^2}{2}\right),$$

where

$$\delta^2 = \frac{\boldsymbol{\beta}'V_{22}\boldsymbol{\beta}}{\sigma_{1.2}} \quad \text{and} \quad {}_1F_1(a;b;x) = \frac{\Gamma(b)}{\Gamma(a)}\sum_{\alpha=0}^{\infty}\frac{\Gamma(a+\alpha)}{\Gamma(b+\alpha)}\frac{x^\alpha}{\alpha!}.$$

Let

$$\theta^2 = \boldsymbol{\beta}'\Sigma_{22}\boldsymbol{\beta} \quad \text{and} \quad \sigma^2 = \sigma_{1.2} = \sigma_{11} - \theta^2.$$

Then

$$\frac{\boldsymbol{\beta}'V_{22}\boldsymbol{\beta}}{\boldsymbol{\beta}'\Sigma_{22}\boldsymbol{\beta}} = \frac{\sigma^2\delta^2}{\theta^2} \sim \chi^2_n.$$

Hence,

(3.5.9) $$p(r^2) = \text{const}\int_0^\infty e^{-\frac{1}{2}\delta^2}(r^2)^{\frac{1}{2}(p-3)}(1-r^2)^{\frac{1}{2}(n-p-1)} \times {}_1F_1\left(\frac{n}{2}; \frac{p-1}{2}; \frac{\delta^2 r^2}{2}\right)(\delta^2)^{\frac{1}{2}n-1}\exp\left(-\frac{1}{2}\frac{\sigma^2\delta^2}{\theta^2}\right)d\delta^2$$
$$= \text{const}(r^2)^{\frac{1}{2}(p-3)}(1-r^2)^{\frac{1}{2}(n-p-1)}{}_2F_1\left(\frac{n}{2}, \frac{n}{2}; \frac{p-1}{2}; \rho^2 r^2\right),$$

where

$$\rho^2 \equiv \rho^2_{y\mathbf{x}} = \frac{\boldsymbol{\sigma}'_{12}\Sigma_{22}^{-1}\boldsymbol{\sigma}_{12}}{\sigma_{11}} = \frac{\boldsymbol{\beta}'\Sigma_{22}\boldsymbol{\beta}}{\sigma_{11}} = \frac{\theta^2}{\sigma^2+\theta^2}.$$

The constant can easily be evaluated and checked to be equal to

(3.5.10) $$\frac{(1-\rho^2)^{\frac{1}{2}n}}{\beta\left(\frac{p-1}{2}, \frac{n-p+1}{2}\right)}.$$

Hence, we get

**Theorem 3.5.3** *The* pdf *of the multiple correlation $r_{y,\mathbf{x}}$ between $y$ and $\mathbf{x}$ is given by Eq.* (3.5.9).

For the large-sample study of multiple $r$, one can refer to Fisher (1928), Khatri (1966b), Gurland and Milton (1970), Gajjar (1967), and Lee (1971). Tables of upper percentage points are given by Lee (1972) and Kramer (1963).

### *3.5.3 Partial Correlation*

In this section we derive the distribution of the sample partial correlation based on a sample of size $N$ from a normal distribution. Let

$$\mathbf{z}=\begin{matrix}1\\1\\p-2\end{matrix}\begin{bmatrix}x_1\\x_2\\\mathbf{x}_3\end{bmatrix}\sim N_p(\boldsymbol{\mu},\Sigma),\qquad \text{where}\quad \Sigma=\begin{bmatrix}\sigma_{11}&\sigma_{12}&\boldsymbol{\sigma}_{13}'\\\sigma_{12}&\sigma_{22}&\boldsymbol{\sigma}_{23}'\\\boldsymbol{\sigma}_{13}&\boldsymbol{\sigma}_{23}&\Sigma_{33}\end{bmatrix}.$$

Then, by definition, the partial correlation between $x_1$ and $x_2$ (in the normal case) is given by (see Section 2.2.6)

$$\rho_{12\cdot3\ldots p}=\frac{\operatorname{Cov}(x_1,x_2|\mathbf{x}_3)}{\left[\operatorname{Var}(x_1|\mathbf{x}_3)\operatorname{Var}(x_2|\mathbf{x}_3)\right]^{1/2}}.$$

Suppose a sample of size $N$ is taken from $N_p(\boldsymbol{\mu},\Sigma)$, and let $S$ denote the sample covariance. Then, if

$$S=\begin{bmatrix}s_{11}&s_{12}&\mathbf{s}_{13}'\\s_{12}&s_{22}&\mathbf{s}_{23}'\\\mathbf{s}_{13}&\mathbf{s}_{23}&S_{33}\end{bmatrix},$$

the sample partial correlation $r_{12.3\ldots p}$ is defined as

$$\begin{aligned}r_{12.34\ldots p}&=-s^{12}/s^{11}s^{22}\\&=\frac{s_{12}-\mathbf{s}_{13}'S_{33}^{-1}\mathbf{s}_{23}}{\left(s_{11}-\mathbf{s}_{13}'S_{33}^{-1}\mathbf{s}_{13}\right)^{1/2}\left(s_{22}-\mathbf{s}_{23}'S_{33}^{-1}\mathbf{s}_{23}\right)^{1/2}}.\end{aligned}$$

Since

$$n\begin{pmatrix}s_{11}-\mathbf{s}_{13}'S_{33}^{-1}\mathbf{s}_{13}&s_{12}-\mathbf{s}_{13}'S_{33}^{-1}\mathbf{s}_{23}\\s_{12}-\mathbf{s}_{13}'S_{33}^{-1}\mathbf{s}_{23}&s_{22}-\mathbf{s}_{23}'S_{33}^{-1}\mathbf{s}_{23}\end{pmatrix}\sim W_2(\Delta,n-p+2),$$

where

$$\Delta=\begin{pmatrix}\sigma_{11}-\boldsymbol{\sigma}_{13}'\Sigma_{33}^{-1}\boldsymbol{\sigma}_{13}&\sigma_{12}-\boldsymbol{\sigma}_{13}'\Sigma_{33}^{-1}\boldsymbol{\sigma}_{23}\\\sigma_{12}-\boldsymbol{\sigma}_{13}'\Sigma_{33}^{-1}\boldsymbol{\sigma}_{23}&\sigma_{22}-\boldsymbol{\sigma}_{23}'\Sigma_{33}^{-1}\boldsymbol{\sigma}_{23}\end{pmatrix},$$

we get

**Theorem 3.5.4** *The distribution of the sample partial correlation* $r_{12.3\ldots p}$ *is just that of ordinary correlation (in the bivariate case) based on* $n-p+2$ d.f.

## 3.6 Multivariate Beta Distributions

In this section, the univariate $F$ distribution is generalized to the multivariate $F$ distribution or the multivariate Beta distributions of first and second types. The following theorem is in this direction:

**Theorem 3.6.1** *Let* $W_1$ *and* $W_2$ *be independently distributed as Wishart* $W_p(I,n_i)$, $i=1,2$, $p\leqslant n_i$. *Let*

$$Z=W_1^{-1/2}W_2W_1^{-1/2\prime},$$

*where* $W_1^{1/2}$ *is any nonsingular factorization of* $W_1$, $W_1=W_1^{1/2}W_1^{1/2\prime}$. *Then the* pdf *of* $Z$, *a multivariate Beta matrix of type II, is*

$$\frac{C(p,n_1)C(p,n_2)}{C(p,n_1+n_2)}\,\frac{|Z|^{\frac{1}{2}(n_2-p-1)}}{|I+Z|^{\frac{1}{2}(n_1+n_2)}},\qquad Z>0. \tag{3.6.1}$$

This will be written as $Z\sim M\beta_{\mathrm{II}}(p,n_2,n_1)$.

PROOF. The joint pdf of $W_1$ and $W_2$ is given by

$$\prod_{i=1}^{2}\left[C(p,n_i)|W_i|^{\frac{1}{2}(n_i-p-1)}\operatorname{etr}\left(-\tfrac{1}{2}W_i\right)\right],$$

where

$$C(p,n_i)=\left[2^{\frac{1}{2}pn_i}\Gamma_p\left(\tfrac{1}{2}n_i\right)\right]^{-1},\qquad i=1,2.$$

Hence, the joint pdf of $Z$ and $W_1$ (since the Jacobian of the transformation is $|W_1|^{\frac{1}{2}(p+1)}$) is

$$\left[\prod_{i=1}^{2}C(p,n_i)\right]|Z|^{\frac{1}{2}(n_2-p-1)}|W_1|^{\frac{1}{2}(n_1+n_2-p-1)}\operatorname{etr}\left[-\tfrac{1}{2}W_1(I+Z)\right].$$

Integrating out $W_1$ we get the pdf of $Z$ as Eq. (3.6.1). □

Hence, the pdf of the characteristic roots $l_1 > l_2 > \cdots > l_p > 0$ of $Z = W_1^{-1/2} W_2 W_1^{-1/2'}$ is given by

$$\text{(3.6.2)} \quad \frac{\pi^{\frac{1}{2}p} C(p,n_1) C(p,n_2)}{\Gamma_p\left(\frac{1}{2}p\right) C(p,n_1+n_2)} \left[ \prod_{i=1}^{p} l_i^{-(n_2-p-1)} (1+l_i)^{-\frac{1}{2}(n_1+n_2)} \right] \prod_{i<j} (l_i - l_j).$$

Note that the ch roots $l_1 > l_2 > \cdots > l_p > 0$ of $Z = W_1^{-1/2} W_2 W_1^{-1/2'}$ are the ch roots of $W_1^{-1} W_2$ or $W_2 W_1^{-1}$. Since $\text{ch}[(CW_1C')^{-1} C' W_2 C] = \text{ch}\,(W_1^{-1} W_2)$, the roots are invariant under nonsingular transformations. Hence, we get

**Theorem 3.6.2** *Let $V_1$ and $V_2$ be independently distributed as $W_p(\Sigma, n_i)$, $\Sigma > 0$, $i = 1,2$; $p \leq (n_1, n_2)$. Then the* pdf *of the* ch *roots $l_1 > l_2 > \cdots > l_p > 0$ of $V_1^{-1} V_2$ is given by Eq.* (3.6.2).

**Corollary 3.6.1** *The distribution of the roots of $V_1^{-1} V_2$ is invariant under the nonsingular group of transformations $V_1 \to C V_1 C'$ and $V_2 \to C V_2 C'$.*

**Theorem 3.6.3** *Let $W_1$ and $W_2$ be independently distributed $W_p(I, n_i)$, $i = 1,2$, $p \leq n_i$. Let*

$$\text{(3.6.3)} \qquad F = (W_1 + W_2)^{-1/2} W_2 (W_1 + W_2)^{-1/2'},$$

*where $(W_1 + W_2)^{1/2}$ is any nonsingular factorization of $W_1 + W_2$ in the sense $W_1 + W_2 = (W_1 + W_2)^{1/2} [(W_1 + W_2)^{1/2}]'$. Then the* pdf *of $F$, a multivariate Beta matrix of type I, is given by*

$$\text{(3.6.4)} \qquad \frac{C(p,n_1) C(p,n_2)}{C(p,n_1+n_2)} |F|^{\frac{1}{2}(n_2-p-1)} |I - F|^{\frac{1}{2}(n_1-p-1)}, \qquad 0 < F < I.$$

This will be denoted as $F \sim M\beta_{\text{I}}(p, n_2, n_1)$.

PROOF. The joint pdf of $W_1$ and $W_2$ is given by

$$\prod_{i=1}^{2} \left[ C(p,n_i) |W_i|^{\frac{1}{2}(n_i-p-1)} \text{etr}\left(-\tfrac{1}{2} W_i\right) \right].$$

Let $W = W_1 + W_2$ and $W_2 = W_2$. Then the joint pdf of $W$ and $W_2$ is given by

$$\left[ \prod_{i=1}^{2} C(p,n_i) \right] |W - W_2|^{\frac{1}{2}(n_1-p-1)} |W_2|^{\frac{1}{2}(n_2-p-1)} \text{etr}\left(-\tfrac{1}{2} W\right).$$

Let $F = W^{-1/2}W_2W^{-1/2'}$, where $W = W^{1/2}W^{1/2'}$. Then the joint pdf of $F$ and $W$ [since $J(W_2 \to F) = |W|^{\frac{1}{2}(p+1)}$] is

(3.6.5) $$C(p,n_1)C(p,n_2)|W|^{\frac{1}{2}(n_1-p-1)}|I-F|^{\frac{1}{2}(n_1-p-1)}|F|^{\frac{1}{2}(n_2-p-1)} \times |W|^{\frac{1}{2}n_2}\operatorname{etr}\left(-\tfrac{1}{2}W\right).$$

Integrating out $W$ from Eq. (3.6.5), we get the pdf of $F$ as Eq. (3.6.4). □

Hence, the pdf of the ch roots $f_1 > f_2 > \cdots > f_p > 0$ of $F$ is given by

(3.6.6) $$\frac{\pi^{\frac{1}{2}p^2}C(p,n_1)C(p,n_2)}{\Gamma_p\left(\frac{1}{2}p\right)C(p,n_1+n_2)} \times \left[\prod_{i=1}^{p} f_i^{\frac{1}{2}(n_2-p-1)}(1-f_i)^{\frac{1}{2}(n_1-p-1)}\right]\prod_{i<j}(f_i-f_j).$$

Clearly, from Eq. (3.6.5), $W$ and $F$ are independently distributed. Hence, we get

**Theorem 3.6.4** *Let $W_i \sim W_p(I,n_i)$, $i=1,2$, be independently distributed. Let $F$ be given by Eq.* (3.6.3) *and $W = W_1 + W_2$. Then $F$ and $W$ are independently distributed.*

REMARK 3.6.1. The converse of Theorem 3.6.4 has been established by Olkin and Rubin (1962) and for the complex case by Carter (1975).

Since the ch roots of $F = (W_1+W_2)^{-1/2}W_2(W_1+W_2)^{-1/2'}$ are the roots of $(W_1+W_2)^{-1/2'}(W_1+W_2)^{-1/2}W_2$ or $(W_1+W_2)^{-1}W_2$ or $[\Sigma^{1/2}(W_1+W_2)(\Sigma^{1/2})']^{-1}[\Sigma^{1/2}W_2(\Sigma^{1/2})']$ with $\Sigma = (\Sigma^{1/2})(\Sigma^{1/2})' > 0$, the roots $f_i$ of $F$ are related to the roots $\lambda_i$ of $W_1^{-1}W_2$ by the following relationship:

$$\lambda_i = \frac{f_i}{1-f_i},$$

and the distribution of the roots $f_i$ is invariant under the transformation $W_1 \to CW_1C'$ and $W_2 = CW_2C'$, $C$ (n.s.). Hence, we get

**Theorem 3.6.5** *Let $V_1 \sim W_p(\Sigma,n_1)$ and $V_2 \sim W_p(\Sigma,n_2)$, $\Sigma > 0$, $p \leqslant (n_1,n_2)$, be independently distributed. Then the* pdf *of the* ch *roots $f_1 > \ldots > f_p > 0$ of $(V_1+V_2)^{-1/2}V_2(V_1+V_2)^{-1/2'}$ is given by Eq.* (3.6.6).

**Theorem 3.6.6** *Let $W_1 \sim W_p(I,n_1)$, $n_1 \geqslant p$, and $Y \sim N_{p,n_2}(0,I_{n_2})$, $n_2 < p$, be independently distributed. Then the* pdf *of*

(3.6.7) $$F = Y'(W_1+YY')^{-1}Y$$

*is given by*

$$\text{(3.6.8)} \qquad \frac{C(p,n_1)C(n_2,p)}{C(p,n_1+n_2)}|F|^{\frac{1}{2}(p-n_2-1)}|I-F|^{\frac{1}{2}(n_1-p-1)}, \qquad 0<F<I.$$

PROOF. The joint pdf of $W_1$ and $Y$ is given by

$$C(p,n_1)(2\pi)^{-\frac{1}{2}pn_2}\left[|W_1|^{\frac{1}{2}(n_1-p-1)}\operatorname{etr}\left(-\tfrac{1}{2}W_1\right)\right]\operatorname{etr}\left(-\tfrac{1}{2}YY'\right).$$

Letting $W=W_1+YY'$, we find that the joint pdf of $W$ and $Y$ is given by

$$C(p,n_1)(2\pi)^{-\frac{1}{2}pn_2}|W-YY'|^{\frac{1}{2}(n_1-p-1)}\operatorname{etr}\left(-\tfrac{1}{2}W\right).$$

Letting $E=W^{-1/2}Y$, we find that $J(Y\to E)=|W|^{\frac{1}{2}n_2}$. Hence, the joint pdf of $E$ and $W$ is given by

$$C(p,n_1)(2\pi)^{-\frac{1}{2}pn_2}|W|^{\frac{1}{2}(n_1+n_2-p-1)}|I_p-EE'|^{\frac{1}{2}(n_1-p-1)}\operatorname{etr}\left(-\tfrac{1}{2}W\right)$$

$$=C(p,n_1)(2\pi)^{-\frac{1}{2}pn_2}|W|^{\frac{1}{2}(n_1+n_2-p-1)}|I_{n_2}-E'E|^{\frac{1}{2}(n_1-p-1)}\operatorname{etr}\left(-\tfrac{1}{2}W\right).$$

Hence the joint pdf of $W$ and $F$ is given by (see Lemma 3.2.3)

$$\text{(3.6.9)} \qquad =C(p,n_1)C(n_2,p)\left[|W|^{\frac{1}{2}(n_1+n_2-p-1)}\operatorname{etr}\left(-\tfrac{1}{2}W\right)\right]$$

$$\times|F|^{\frac{1}{2}(p-n_2-1)}|I-F|^{\frac{1}{2}(n_1-p-1)}. \qquad \square$$

Clearly $W$ and $F$ are independently distributed. Hence, we get

**Theorem 3.6.7** *Let $W_1\sim W_p(I,n_1)$ and $Y\sim N_{p,n_2}(0,I_{n_2})$, $n_1\geqslant p$, $n_2<p$, be independently distributed. Then $W=W_1+YY'$ and $F=W^{-1/2}YY'W^{-1/2'}$ are independently distributed.*

Integrating out $W$, we get the pdf of $F$ as Eq. (3.6.8). Hence the pdf of the ch roots $f_1>f_2>\cdots>f_{n_2}>0$ of $F$ is given by

$$\text{(3.6.10)} \qquad \frac{\pi^{\frac{1}{2}n_2}}{\Gamma_{n_2}\left(\frac{1}{2}n_2\right)}\frac{C(p,n_1)C(n_2,p)}{C(p,n_1+n_2)}$$

$$\times\left[\prod_{i=1}^{n_2}f_i^{\frac{1}{2}(p-n_2-1)}(1-f_i)^{\frac{1}{2}(n_1-p-1)}\right]\prod_{i<j}(f_i-f_j).$$

Since the distributions of the roots $f_i$ are invariant under the transformation $W_1\to CW_1C'$ and $Y\to CY\Gamma$, where $C$ is n.s. and $\Gamma$ orthogonal, we get

**Theorem 3.6.8** *Let $V_1 \sim W_p(\Sigma, n_1)$, $\Sigma > 0$, $p \leqslant n_1$, and $Y \sim N_{p,n_2}(0, \Sigma)$, $p > n_2$, be independently distributed. Then the* pdf *of the* ch *roots $f_1 > f_2 > \cdots > f_{n_2} > 0$ of $F = Y'(V_1 + YY')^{-1}Y$ is given by Eq.* (3.6.10).

**Theorem 3.6.9** *Let*

$$V = \begin{matrix} & r & s \\ r & V_{11} & V_{12} \\ s & V_{12}' & V_{22} \end{matrix} \sim W_p(\Sigma, n), \qquad \text{where } \Sigma = \begin{matrix} & r & s \\ r & \Sigma_1 & 0 \\ s & 0' & \Sigma_2 \end{matrix} > 0,$$

*with $r + s = p$ and $r \leqslant s$. Then from Theorem* 3.3.5, *$V_{1.2} \equiv V_{11} - V_{12}V_{22}^{-1}V_{12}' \sim W_r(\Sigma_1, n-s)$, and $V_{12}$ given $V_{22}$ is $N_{r,s}(0, \Sigma_1, V_{22})$. Hence given $V_{22}$, $V_{12}V_{22}^{-1/2} \sim N_{r,s}(0, \Sigma_1, I)$. Since it does not depend on $V_{22}$ and since $r \leqslant s$, $P \equiv V_{12}V_{22}^{-1}V_{12}' \sim W_r(\Sigma_1, s)$ and is independent of $V_{1.2}$ (and $V_{22}$).*

Let $f_1 > f_2 > \ldots > f_r$ denote the nonzero ch roots of

$$V_{11}^{-1}P \equiv V_{11}^{-1}V_{12}V_{22}^{-1}V_{12}' = \left(V_{11} - V_{12}V_{22}^{-1}V_{12}' + V_{12}V_{22}^{-1}V_{12}'\right)^{-1}V_{12}V_{22}^{-1}V_{12}'$$
$$= (V_{1.2} + P)^{-1}P.$$

The distribution of these roots is obtainable from Eq. (3.6.6) by changing $p \to r, n_1 \to n - s, n_2 \to s$. Hence it is given by

$$\text{(3.6.11)} \qquad \frac{\pi^{\frac{1}{2}r^2} C(r, n-s) C(r, s)}{\Gamma_r\left(\frac{1}{2}r\right) C(r, n)} \prod_{i=1}^{r} f_i^{\frac{1}{2}(s-r-1)} (1 - f_i)^{\frac{1}{2}(n-r-s-1)} \prod_{i<j} (f_i - f_j).$$

Thus we get

**Theorem 3.6.10** *Let $V \sim W_p(\Sigma, n)$ with*

$$\Sigma = \begin{matrix} & r & s \\ r & \Sigma_1 & 0 \\ s & 0' & \Sigma_2 \end{matrix} > 0.$$

*Suppose $V$ is partitioned as in Theorem* 3.6.9 *and $r \leqslant s$. Then the distributions of the nonzero* ch *roots $f_1 > f_2 > \cdots > f_r$ of $V_{11}^{-1}V_{12}V_{22}^{-1}V_{12}'$ are given by Eq.* (3.6.11); *they are known as sample canonical correlations between two vectors of dimensions $r$ and $s$, respectively.*

REMARK 3.6.2. The above theorems remain valid even when $n_1 < p$, $n_2 < p$, and $n_1 + n_2 \geqslant p$. For this, one may refer to Mitra (1969) and Khatri (1970). In these cases, the density may not exist. When $n_1 \geqslant p$, theorems like 3.6.1, 3.6.3, 3.6.6, etc. were established by Khatri (1959a).

## 3.7 Complex Wishart Distribution

Let the columns of a $p\times n$ random matrix $Y$ be independently distributed as complex multivariate normal with mean zero and positive definite Hermitian covariance matrix $\Sigma$. Then the distribution of the matrix $A=YY^*$ is said to have a complex Wishart distribution, denoted by $CW_p(\Sigma,n)$. The pdf is given by [see Goodman (1963), Khatri (1965a), and Srivastava (1965a)]

$$\tilde{C}(n,p)|A|^{n-p}|\Sigma|^{-n}\operatorname{etr}(-\Sigma^{-1}A),$$

where

$$\tilde{C}(n,p)=\left[\pi^{\frac{1}{2}p(p-1)}\Gamma(n)\cdots\Gamma(n-p+1)\right]^{-1}.$$

All the results corresponding to the real Wishart case can easily be obtained, and are left as exercises for the reader. In the remainder of the book, not much attention will be paid to the complex case, for which the reader is referred to Goodman (1963), Khatri (1965a, b), Srivastava (1965a), Giri (1965), Saxena (1966), and Krishnaiah (1976).

## Problems

**3.1** Let $W\sim W_p(I,n)$, where $W$ is partitioned as follows:

$$W=\begin{matrix} \\ r \\ s\end{matrix}\begin{pmatrix} \overset{r}{W_{11}} & \overset{s}{W_{12}} \\ W_{12}' & W_{22}\end{pmatrix},\qquad r+s=p.$$

Find the joint distribution of $W_{11}$, $W_{22}$, and $R=W_{11}^{-1/2}W_{12}W_{22}^{-1}W_{12}'W_{11}^{-1/2'}$, where $W_{11}=(W_{11}^{1/2})(W_{11}^{1/2})'$. Are $W_{11}$, $W_{22}$, and $R$ independently distributed? Find their distributions.

**3.2** Let $W\sim W_p(I,n)$, $n\geqslant p$. Show that:

(i) $EW=nI$.

(ii) $E(WAW)=n(n+1)A+n(\operatorname{tr}A)I_p$ when $A$ is symmetric.

(iii) $E(W^{-1})=(n-p-1)^{-1}I_p$.

(iv) $E(W^{-1}AW^{-1})=[(n-p)(n-p-3)]^{-1}A+[(n-p)(n-p-1)(n-p-3)]^{-1}(\operatorname{tr}A)I_p$.

(v) If $W\sim W_p(\Sigma,n)$, the expressions on the right sides of (i)–(iv) modify respectively to $n\Sigma$, $n(n+1)\Sigma A\Sigma+n(\operatorname{tr}A\Sigma)\Sigma$, $(n-p-1)^{-1}\Sigma^{-1}$, and $[(n-p)(n-p-3)]^{-1}\Sigma^{-1}A\Sigma^{-1}+[(n-p)(n-p-1)(n-p-3)]^{-1}(\operatorname{tr}A\Sigma^{-1})\Sigma^{-1}$.

(vi) Find $E[(\mathbf{a}'W\mathbf{a})/(\mathbf{a}'\mathbf{a})-(\mathbf{a}'W^{-1}\mathbf{a})/(\mathbf{a}'W^{-2}\mathbf{a})]$ where $\mathbf{a}\neq 0$.

[Hint: Use lower-triangular factorization for (ii) and upper-triangular factorization for (iii) and (iv). Due to symmetry, it is sufficient to establish the results for a diagonal element and an off-diagonal element only. Further, first establish the results when $A$ is diagonal and then generalize to any symmetric matrix $A$. If $P = WD_a W$, then $p_{ij} = \Sigma_\alpha a_\alpha w_{i\alpha} w_{j\alpha}$, etc. For (vi), let $B = (\mathbf{a}'W\mathbf{a}/\mathbf{a}'\mathbf{a}) - (\mathbf{a}'W^{-1}\mathbf{a})/(\mathbf{a}'W^{-2}\mathbf{a})$, and $V = \Gamma W \Gamma'$, where $\Gamma' = ((\mathbf{a}'\mathbf{a})^{-1/2}\mathbf{a}, C)$ and is orthogonal. Then $B = v_{11} - [v^{11}/\Sigma_{j=1}^{p}(v^{1i})^2]$. Let $V = \underset{\sim}{T}\underset{\sim}{T}'$, where $\underset{\sim}{T}$ is a lower triangular matrix with positive diagonal elements.]

**3.3** Let $\mathbf{y}_1,\ldots,\mathbf{y}_m$, $m>p$, be iid $N_p(\mathbf{0},\Sigma)$, $\Sigma>0$. Let $Y=(\mathbf{y}_1,\ldots,\mathbf{y}_m)$.

(i) Show that the pdf of

$$\mathbf{v} = (v_1,\ldots,v_p)' = (YY')^{-1/2}\mathbf{y}_1$$

is given by

$$f(\mathbf{v}) = \frac{\Gamma(\frac{1}{2}m)}{\pi^{\frac{1}{2}p}\Gamma\left[\frac{1}{2}(m-p)\right]}(1-\mathbf{v}'\mathbf{v})^{\frac{1}{2}(m-p)-1}, \qquad \mathbf{v}'\mathbf{v}<1, \quad m>p.$$

[Hint: The joint pdf of $A = \Sigma_{i=2}^{m}\mathbf{y}_i\mathbf{y}_i'$ and $\mathbf{y}_1$ can easily be written. Also, it can easily be shown that

$$|A| = |A+\mathbf{z}_1\mathbf{z}_1'|\,|I-(A+\mathbf{z}_1\mathbf{z}_1')^{-1/2}\mathbf{z}_1\mathbf{z}_1'(A+\mathbf{z}_1\mathbf{z}_1')^{-1/2}|.]$$

(ii) For any fixed nonnull vector $\mathbf{l}$ (or a random vector $\mathbf{l}$ distributed independently of $\mathbf{y}$) such that $\mathbf{l}'\mathbf{l}=1$, show that the distribution of $z=\mathbf{l}'\mathbf{v}$, where $\mathbf{v}$ is as defined in (i), is given by

$$h(z) = \left[\,\beta\left(\tfrac{1}{2},\tfrac{1}{2}(m-1)\right)\right]^{-1}(1-z^2)^{\frac{1}{2}(m-1)-1}, \qquad -1<z<1.$$

[Hint: The distribution of $\mathbf{v}$ is invariant under the transformation $\mathbf{v}\to L\mathbf{v}$ where $L$ is orthogonal.]

(iii) Let $p \geqslant 2$, and $\mathbf{v}$ be as above. Show that the joint pdf of $v_1$ and $q=\mathbf{v}'\mathbf{v}$ is given by

$$g(v_1,q) = \frac{(1-q)^{\frac{1}{2}(m-p)-1}(q-v_1^2)^{\frac{1}{2}(p-1)-1}}{\beta\left(\frac{1}{2},\frac{1}{2}(p-1)\right)\beta\left(\frac{1}{2}p,\frac{1}{2}(m-p)\right)}.$$

What is the joint pdf of $z$ and $q$?

**3.4** Let $nS \sim W_p(I,n)$, and let

$$S = I + n^{-1/2}V, \qquad \text{where} \quad V=(v_{ij}).$$

Let $\boldsymbol{\delta}$ be a fixed vector. Show that

(i) $E(v_{11}^2)=2$,

(ii) $E(v_{12}^2)=1$,

(iii) $E(\boldsymbol{\delta}'V\boldsymbol{\delta})^2 = 2(\boldsymbol{\delta}'\boldsymbol{\delta})^2$,

(iv) $E(\boldsymbol{\delta}'V^2\boldsymbol{\delta}) = (p+1)(\boldsymbol{\delta}'\boldsymbol{\delta})$.

**3.5** Let $V \sim W_p(\Sigma, n)$, $n \geq p$. Show that

(i) $E(\operatorname{tr}_j V) = n(n-1)\cdots(n-j+1)\operatorname{tr}_j \Sigma$, $j = 1, 2, \ldots, p$,

(ii) $E(V^2) = n^2\Sigma^2 + n(\Sigma \operatorname{tr} \Sigma + \Sigma^2)$.

**3.6** Let $V \sim W_p(\Sigma, n)$, $\Sigma > 0$. Find the distribution of $(\mathbf{a}'\Sigma^{-1}\mathbf{a})/(\mathbf{a}'V^{-1}\mathbf{a})$ for any nonnull vector $\mathbf{a}$.

**3.7** Let $V \sim W_p(\Sigma, n)$, where

$$V = \overset{\begin{matrix} p-1 & 1 \end{matrix}}{\begin{pmatrix} V_1 & \mathbf{v}_{1p} \\ \mathbf{v}'_{1p} & v_{pp} \end{pmatrix}} \begin{matrix} p-1, \\ 1 \end{matrix} \qquad \Sigma > 0.$$

$V^{-1} = (v^{ij})$ and $\Sigma^{-1} = (\sigma^{ij})$. Show that $\sigma^{pp}/v^{pp} \sim \chi^2_{n-p+1}$ and is independent of $V_1$.

**3.8** Let $V \sim W_p(\Sigma, n)$, where $V = (v_{ij})$ and $\Sigma = (\sigma_{ij})$. For notational convenience, let $A_r$ denote the top left-hand corner $r \times r$ matrix of the $p \times p$ matrix $A$, $r = 1, 2, \ldots, p$. Thus $A_1$ is a $1 \times 1$ matrix, the element in the first row and the first column of $A$. Show that if $|\Sigma_0| \equiv 1$ and $|V_0| \equiv 1$, then

$$\frac{|V_r|}{|V_{r-1}|} \frac{|\Sigma_{r-1}|}{|\Sigma_r|}, \qquad r = 1, 2, \ldots, p$$

are independently distributed as $\chi^2_{n-r+1}$, $r = 1, 2, \ldots, p$.

**3.9** Let $V \sim W_p(\Sigma_I, n)$, where $\Sigma_I = \sigma^2[(1-\rho)I + \rho\mathbf{e}\mathbf{e}'] > 0$, and $\mathbf{e}' = (1, 1, \ldots, 1) : 1 \times p$. Find the maximum-likelihood estimates of $\sigma^2$ and $\rho$.

**3.10** Let $\mathbf{y} \sim N_p(\boldsymbol{\theta}, \Sigma)$ and $V \sim W_p(\Sigma, n)$ be independently distributed, where

$$\boldsymbol{\theta}' = (\overset{r}{\boldsymbol{\theta}'_1}, \overset{s}{\boldsymbol{\theta}'_2}) \quad \text{and} \quad \mathbf{y} = (\overset{r}{\mathbf{y}'_1}, \overset{s}{\mathbf{y}'_2}), \qquad r + s = p.$$

Suppose it is known that $\boldsymbol{\theta}_2 = \mathbf{0}$. Find the maximum-likelihood estimate of $\boldsymbol{\theta}_1$ and $\Sigma$.

**3.11** Generalize Problem 2.24 to Wishartness when

$$Y \sim N_{p,n}(\mu, \Sigma, I_n) \quad \text{and} \quad W = YAY'.$$

**3.12** Let $X \sim N_{p,n}(0, \Sigma, I_n)$. Then in order that the random matrix $Q = XAX' + M_1X' + XM_2' + C$, $A$ symmetric of rank $r$, shall be distributed as the sum of two independent random matrices $V$ and $Y$, where $V \sim W_p(\Sigma, r)$ and $Y$ is normal, it is necessary and sufficient that all the following three conditions be satisfied:

$$\text{(i) } A^2 = A, \qquad \text{(ii) } (M_1 - M_2)A = 0, \qquad \text{(iii) } M_1A = 0.$$

**3.13** Let $X \sim N_{p,n}(0, \Sigma, I_n)$. Let $Q_1 = XAX' + L_1X' + XL_2' + C$ and $Q_2 = XBX' + M_1X' + XM_2' + C_2$, where $A$ and $B$ are symmetric matrices. Then in order that

$Q_1$ and $Q_2$ shall be independently distributed, it is necessary and sufficient that all the following conditions be satisfied:

$$\text{(i) } AB=0, \quad \text{(ii) } L_1B=0, \quad \text{(iii) } (M_1-M_2)A=0,$$
$$\text{(iv) } (L_1-L_2)B=0, \quad \text{(v) } M_1A=0,$$
$$\text{(vi) } \begin{pmatrix} L_1M_1' & L_1(M_1-M_2)' \\ (L_1-L_2)M_1' & (L_1-L_2)(M_1-M_2)' \end{pmatrix}=0.$$

**3.14** Let $A$ be an idempotent matrix of rank $r$, and let $X \sim N_{p,n}(0,\Sigma,I_n)$. Let

$$XAX' = \sum_{i=1}^{k} XA_iX'.$$

Then in order that the matrix quadratic forms $XA_iX'$, $i=1,2,\ldots,k$, be mutually independently distributed as Wishart $W_p(\Sigma,\rho(A_i))$, it is necessary and sufficient that one of the following three equivalent conditions be satisfied:

(i) $A_i^2 = A_i$, $i=1,2,\ldots,k$.

(ii) $A_iA_j=0$, $i \neq j$, $i,j=1,2,\ldots,k$.

(iii) $\Sigma_{i=1}^k \rho(A_i)=\rho(A)=r$.

**3.15** Let $X \sim N_{p,N}(\mu,\Sigma,I_N)$, where the $p \times N$ matrix

$$\mu = \begin{pmatrix} \mu_1,\ldots,\mu_N \\ 0 \end{pmatrix},$$

$\mu_1,\ldots,\mu_N$ being scalars. Find the distribution of $W=XX'$.

**3.16** If the distribution of $r$ is given by Eq. (3.4.2), show that

$$E(\sin^{-1} r) = \sin^{-1} \rho.$$

For exact moments of $\sin^{-1} r$, see Khatri (1968b).

**3.17** In Eq. (3.5.9), if $(n-p+1)/2=k$ is a positive integer, show that the cumulative distribution function is

$$\sum_{j=0}^{k} b_j F_{p-1+2j,2k}\left[x(1-\rho^2)\right],$$

where

$$b_k = \binom{k}{j}(\rho^2)^j(1-\rho^2)^{k-j}$$

and $F_{r,s}(y)$ is the cdf of the ratio $\chi_r^2/\chi_s^2$ of independent chi-square random variables.

**3.18** Let

$$\tilde{\rho} = \frac{\rho}{(1-\rho^2)^{1/2}} \quad \text{and} \quad \tilde{r} = \frac{r}{(1-r^2)^{1/2}}.$$

Show that

$$\tilde{r}=\frac{b_{21}+\tilde{\rho}b_{11}}{b_{22}},$$

where $b_{11}$, $b_{22}$, and $b_{21}$ are independently distributed, $b_{11}\sim\chi^2_n$, $b_{22}\sim\chi^2_{n-1}$, and $b_{21}\sim N(0,1)$. Extend to multiple correlation.

**3.19** Let

$$\mathbf{u}\sim N_n(0,I),$$

$$\bar{u}=n^{-1}\sum_{i=1}^{n}u_i,$$

$$S^2=\sum_{i=1}^{n}(u_i-\bar{u})^2.$$

Show that $(u_1-\bar{u})/S$ is independently distributed of $\bar{u}$ and $S$. Find the distribution of

$$\left(\frac{n}{n-1}\right)^{1/2}\frac{u_1-\bar{u}}{S}.$$

**3.20** Let $V\sim W_p(\Sigma,n)$, $n\geqslant p$, and $X\sim N_{p,q}(0,\Sigma,I)$ be independently distributed. Let $V_1=V+XX'$ and $Z=\underset{\sim}{T}^{-1}X$, where $V_1=\underset{\sim}{T}\underset{\sim}{T}'$ and $\underset{\sim}{T}$ is lower triangular. Show that $V_1$ and $Z$ are independently distributed.

**3.21** If $V\sim W_p(\Sigma,n)$, $n\geqslant p$, and $X_i\sim N_{p,q_i}(0,\Sigma,I)$, $i=1,2,\ldots,k$, are mutually independently distributed, then $V_k=V+\sum_{i=1}^{k}X_iX_i'$ and $Z_i=\underset{\sim}{T}_i^{-1}X_i$, $i=1,2,\ldots,k$, where $V_i=V+\sum_{j=1}^{i}X_jX_j'=\underset{\sim}{T}_i\underset{\sim}{T}_i'$ and the $\underset{\sim}{T}_i$'s are lower triangular and mutually independently distributed. Suppose that $p\geqslant q_i$. Show that the distribution of $W_i=Z_i'Z_i=X_i'(V+\sum_{j=1}^{i}X_jX_j')^{-1}X_i$ is given by

$$\text{const}\,|W_i|^{\frac{1}{2}(p-q_i-1)}|I-W_i|^{\frac{1}{2}(n+e_{i-1}-p-1)},\qquad e_i=\sum_{j=1}^{i}q_j.$$

What is the distribution of $Z_iZ_i'$ if $p\leqslant q_i$?

**3.22** In Problem 3.21, let

$$U_i=W_i(I-W_i)^{-1}=(I-W_i)^{-1}-I=X_i'\left(V+\sum_{j=1}^{i-1}X_iX_i'\right)^{-1}X_i,$$

where $p\geqslant q_i$. Then show that the pdf of $U_i$ is given by

$$\text{const}\,|U_i|^{\frac{1}{2}(p-q_i-1)}|I+U_i|^{-\frac{1}{2}(n+e_i)}.$$

**3.23** Let $U=U'$ have the density

$$\pi^{-\frac{1}{4}p(p+1)}2^{-\frac{1}{2}p}\operatorname{etr}\left(-\tfrac{1}{2}U^2\right),$$

that is, the elements $u_{ij}$ $(i\leqslant j)$ of $U$ are independently normally distributed with mean 0, and

$$E(u_{ii}^2)=1 \quad \text{and} \quad Eu_{ij}^2=\tfrac{1}{2}, \qquad i<j.$$

Then the pdf of the ch roots $\lambda_1>\cdots>\lambda_p$ of $U$ is given by

$$2^{-\frac{1}{2}p}\left\{\prod_{i=1}^{p}\Gamma\left[\tfrac{1}{2}(p-i+1)\right]\right\}^{-1} e^{-\frac{1}{2}\Sigma\lambda_i^2}\prod_{i<j}(\lambda_i-\lambda_j).$$

**3.24** (i) Let $Z\sim M\beta_{\text{II}}(p,n_2,n_1)$. Then show that the distribution of $F=I_p-(I_p+Z)^{-1}\sim M\beta_{\text{I}}(p,n_2,n_1)$. Conversely, if $F\sim M\beta_{\text{I}}(p,n_2,n_1)$, then $Z=(I-F)^{-1}-I\sim M\beta_{\text{II}}(p,n_2,n_1)$.

(ii) Let $Z\sim M\beta_{\text{II}}(p,n_2,n_1)$. Find the first two moments of the elements of $Z$, and $I_p-(I_p+Z)^{-1}=F$. Establish the conditions necessary for their existence.

**3.25** Let $Z\sim M\beta_{\text{II}}(p,n_2,n_1)$, and $L$ be an $r\times p$ matrix such that $LL'=I_r$. Then show that $LZL'\sim M\beta_{\text{II}}(r,n_2,n_1-p+r)$. If

$$Z=\begin{pmatrix} Z_{11} & Z_{12} \\ Z_{12}' & Z_{22}\end{pmatrix},$$

then show that $Z_{1.2}=Z_{11}-Z_{12}Z_{22}^{-1}Z_{12}'$, $Z_{12}[Z_{22}^{-1}-(I+Z_{22})^{-1}]Z_{12}'$, and $Z_{22}$ are independently distributed. If $Z_{12}$ is an $r\times s$ matrix with $r\leqslant s$ and $r+s=p$, then show that if $I+Z_{1.2}=(I+Z_{1.2})^{1/2}[(I+Z_{1.2})^{1/2}]'$, then

$$Z_{1.2}\sim M\beta_{\text{II}}(r,n_2-p+r,n_1),$$
$$Z_{22}\sim M\beta_{\text{II}}(s,n_2,n_1-p+s),$$

and

$$(I_r+Z_{1.2})^{-1/2}Z_{12}\left[Z_{22}^{-1}-(I_s+Z_{22})^{-1}\right]Z_{12}'\left[(I_r+Z_{1.2})^{-1/2}\right]'$$
$$\sim M\beta_{\text{II}}(r,s,n_1+n_2-s),$$

and they are independently distributed.

**3.26** Let $U\sim W_2(I,n)$ and $V\sim W_2(I,m)$ be independently distributed. Let $\lambda_1>\lambda_2$ be the ch roots of $V^{-1}U$. Find the distribution of

(i) $\lambda_1+\lambda_2$,

(ii) $\lambda_1$,

(iii) $\lambda_2$.

**3.27** Let $U\sim W_2(D_\alpha,n)$, where $D_\alpha=\text{diag}(\alpha_1,\alpha_2)$. Let

$$U=H\begin{pmatrix}\lambda_1 & 0\\ 0 & \lambda_2\end{pmatrix}H, \qquad \text{where} \quad H=\begin{pmatrix}\cos\phi & \sin\phi\\ -\sin\phi & \cos\phi\end{pmatrix}.$$

(i) Find the joint distribution of $\lambda_1$, $\lambda_2$, and $\phi$.

(ii) Find the distribution of the largest root of $U$.

**3.28** Let $U \sim W_p(I,m)$ and $V \sim W_p(I,n)$ be independently distributed. Let $\lambda_1 > \cdots > \lambda_p$ be the roots of $V^{-1}U$. Find the limiting distribution of the $\lambda_i$'s when $n \to \infty$.

**3.29** Obtain results corresponding to Subsections 3.3.1–3.3.3 for the complex Wishart distribution.

**3.30** Let

$$z_N = \tfrac{1}{2}\ln\frac{1+r_N}{1-r_N} \quad \text{and} \quad \xi = \tfrac{1}{2}\ln\frac{1+\rho}{1-\rho},$$

where $r_N$ is the sample correlation coefficient. Show that as $n \to \infty$

$$\mathcal{L}\left[N^{1/2}(z_N - \xi)\right] \to N(0,1).$$

**3.31** Let $nS \sim W_p(\Sigma, n)$. Using the central limit theorem (Theorem 2.10.1 or 2.10.2), show that as $n \to \infty$,

$$n^{1/2}\left(\frac{|S|}{|\Sigma|} - 1\right) \to N(0, 2p).$$

**3.32** Let $A_N = \Sigma_{\alpha=1}^{N}(\mathbf{x}_\alpha - \bar{\mathbf{x}}_N)(\mathbf{x}_\alpha - \bar{\mathbf{x}}_N)'$, where $\mathbf{x}_1, \mathbf{x}_2, \ldots$ are iid $N(\boldsymbol{\mu}, \Sigma)$. Show that the asymptotic distribution of $N^{-1/2}[A_N - (N-1)\Sigma]$ is normal with mean 0 and covariance

$$E(b_{ijN} b_{klN}) = \sigma_{ik}\sigma_{jl} + \sigma_{il}\sigma_{jk}.$$

**3.33** Let $r_N$ be the sample correlation coefficient of a sample of $N$ from a bivariate normal distribution with correlation $\rho$. Show that as $N \to \infty$

$$\mathcal{L}\left(N^{1/2}\frac{r_N - \rho}{1-\rho^2}\right) \to N(0,1).$$

Note that

$$r_N = \sum_{i=1}^{N} \frac{(x_i - \bar{x})(y_i - \bar{y})}{\left[\Sigma(x_i - \bar{x})^2 \Sigma(y_i - \bar{y})^2\right]^{1/2}},$$

where $(x_i, y_i)$, $i = 1, 2, \ldots, N$, are independent samples from

$$N_2\left(\mathbf{0}, \begin{pmatrix} \sigma_{11} & \sigma_{12} \\ \sigma_{12} & \sigma_{22} \end{pmatrix}\right).$$

Here $\sigma_{12} = \sigma_1 \sigma_2 \rho$.

# 4

# Inference on Location; Hotelling's $T^2$

## 4.1 Introduction

In this chapter several testing and confidence-interval problems on the location (or mean) of the multivariate normal distribution are considered, and tests based on the likelihood ratio and Roy's union–intersection principle are obtained; these problems are special cases of Chapter 6. The distribution and properties of these tests are also discussed. Since the tests based on sufficient statistics are as powerful as those based on original observations, we shall consider only tests based on sufficient statistics. For example, if $\mathbf{x}_1, \mathbf{x}_2, \ldots, \mathbf{x}_N$ are independently distributed as $N_p(\boldsymbol{\mu}, \Sigma)$, we need only consider tests based on $\bar{\mathbf{x}}$ and $V \equiv \sum_{i=1}^{N}(\mathbf{x}_i - \bar{\mathbf{x}})(\mathbf{x}_i - \bar{\mathbf{x}})'$, where $\bar{\mathbf{x}} = N^{-1}\sum_{i=1}^{N}\mathbf{x}_i$. Also, the constant in any probability density function will be denoted by a generic symbol $c$, since it is irrelevant in constructing tests.

## 4.2 Union–Intersection Principle

In this section we describe Roy's union–intersection principle; for the likelihood principle, see Fraser (1976) or Lehmann (1959). Let $\phi_H$ denote the likelihood function when the simple hypothesis $H$ is true. By the Neymann–Pearson lemma, the uniformly most powerful test of size $\alpha$ for testing a simple hypothesis $H$ against a simple alternative $A$ is based on the critical region

$$w(H,A,\alpha) : \phi_A \geqslant \lambda \phi_H,$$

where $\lambda$ is determined from $P[\mathbf{x} \in w(H,A,\alpha) | H] = \alpha$. We observe that $\lambda$

depends on $H$, $A$, and $\alpha$, and $\alpha$ can be chosen in advance depending on $A$ and $H$. We denote these $\alpha$ and $\lambda$ by $\alpha_{A,H}$ and $\lambda(H,A,\alpha_{A,H})$ when we want to emphasize their dependence on hypotheses and the size of the test.

Let us extend the above procedure for testing a simple hypothesis $H$ against a composite alternative $A=\bigcup_{A_1}(A_1\in\Omega)$, where $A_1$ denotes a simple alternative hypothesis. Now, by the above procedure, the critical region for testing $H$ against $A_1$ is

$$w(H,A_1,\alpha_{A_1,H}):\phi_{A_1}\geqslant\lambda(H,A_1,\alpha_{A_1,H})\phi_H.$$

A test for this situation can be obtained in the following two ways:

1. Suppose $\alpha_{A_1,H}=\alpha$ for every $A_1\in\Omega$. Then, the rejection and acceptance regions for testing $H$ against $A$ can be written as

$$w_1(H,\alpha)=\bigcup_{A_1\in\Omega} w(H,A_1,\alpha)$$

and

$$\bar{w}_1(H,\alpha)=\bigcap_{A_1\in\Omega}\bar{w}(H,A_1,\alpha),$$

respectively, where $\bar{w}$ is the complementary region of $w$. This is known as Roy's (1953) *type-I test* for $H$ against $A$. The level of significance is given by

$$P[\mathbf{x}\in w_1(H,\alpha)|H]=\delta(H,\alpha).$$

Thus, given $\delta$ and $H$, we should be able to choose $\alpha$. To carry out the test procedure, we assume that this can be done.

2. Instead of taking $\alpha_{A_1,H}=\alpha$ as in (1), let us assume that $\lambda(H,A_1,\alpha_{A_1,H})=\lambda$, a preassigned constant for all $A_1\in\Omega$. Then, set up the rejection and acceptance regions as

$$w_2(H,\lambda)=\bigcup_{A_1\in\Omega} w(H,A_1,\alpha_{A_1,H})$$

and

$$\bar{w}_2(H,\lambda)=\bigcap_{A_1\in\Omega}\bar{w}(H,A_1,\alpha_{A_1,H}).$$

This is known as Roy's (1953) *type-II test* for $H$ against $A$. The level of significance is given by

$$P[\mathbf{x}\in w_2(H,\lambda)|H]=\delta_2(H,\lambda).$$

As before, it is assumed that one can find $\lambda$, given $\delta_2$ and $H$, from

$\delta_2(H,\lambda)=\delta_2$. Since

$$w_2(H,\lambda):\bigcup_{A_1\in\Omega}\{\phi_{A_1}\geqslant\lambda\phi_H\}=\operatorname*{Sup}_{A_1\in\Omega}\phi_{A_1}\geqslant\lambda\phi_H,$$

it can be seen that this test is equivalent to the *likelihood-ratio test procedure* for testing the simple hypothesis $H$ against the composite alternative $A$.

We now consider the case when the hypothesis $H$ and the alternative $A$ are composite and disjoint hypotheses. Suppose $H$ can be written as an intersection of simple hypotheses $H_i$ ($i\in$continuum), i.e., $H=\bigcap_i H_i$, and for each $i$, there are alternatives in $A$, say $A_{ji}$ ($j\in$continuum) such that $A=\bigcup_i\bigcup_j A_{ji}$. Then, an *extension of the type-I test* is obtained by taking $\alpha_{A_{ji},H_i}=\alpha$ for all $i$ and $j$, and setting the rejection region as the union of all the rejection regions obtained for simple hypothesis $H_i$ against simple alternative $A_{ji}$. This extended type-I test will have rejection and acceptance regions given by

$$w_1(\alpha):\bigcup_i\bigcup_j w(H_i,A_{ji},\alpha)$$

and

$$\bar{w}_1(\alpha):\bigcap_i\bigcap_j \bar{w}(H_i,A_{ji},\alpha),$$

respectively. The size of the critical region is given by

$$P[\mathbf{x}\in w_1(\alpha)|H]=\delta(\alpha,H).$$

We assume that given $\delta$, it will be possible for us to find $\alpha$ from $\delta=\delta(\alpha,H)$, i.e., $\alpha=\alpha(\delta,H)$.

The extension of the type-II test may not lead to the usual likelihood-ratio test. In fact, it is not at all clear how to extend it in a meaningful way. For this reason, no consideration will be given to this case.

The heuristic test procedure given above as an extension of the type-I test is known as Roy's union–intersection principle. In place of an extended type-II test, we consider the likelihood-ratio test procedure. In this book we shall consider as far as possible tests based on the likelihood-ratio principle and the union–intersection principle. As a special case of the union–intersection principle we get a procedure known as the *step-down procedure*. This will be illustrated in Example 4.2.2 below. Let us consider some simple illustrations.

EXAMPLE 4.2.1. Let $x_{ij}$ ($j=1,2,\dots,n_i$ and $i=1,2,\dots,k$) be independent $N(\mu_i,\sigma^2)$. The problem is to test $H:(\mu_1=\mu_2=\cdots=\mu_k,\ \sigma^2>0)$ against $A(\mu_i\neq\mu_{i'}$ for at least one pair $(i,i')$, $\sigma^2>0)$. Such types of alternatives will be denoted as $A\neq H$. We shall see in this situation that there is no unique way of writing $H$ and $A$ as the intersection and union of simple hypotheses. This leads to different kinds of test procedures of type I.

First of all we observe that the given data will be reduced to the class of complete sufficient statistics $(\bar{x}_1,\dots,\bar{x}_k,v)$ for $(\mu_1,\dots,\mu_k,\sigma^2)$, where

$$\bar{x}_i=\sum_{j=1}^{n_i}\frac{x_{ij}}{n_i},\quad i=1,2,\dots,k,\quad \text{and}\quad v=\sum_{i=1}^{k}\sum_{j=1}^{n_i}(x_{ij}-\bar{x}_i)^2,$$

are independently distributed as $N(\mu_i,\sigma^2/n_i)$ $(i=1,2,\dots,k)$ and $\sigma^2\chi^2_{n-k}$ with $n-k$ $[=\sum_{i=1}^{k}(n_i-1)]$ d.f. respectively.

As mentioned above, there are several ways of expressing $H$ and $A$ as intersection and union of simple hypotheses. For example, we can write

$$H=\bigcap_{\mathbf{a}} H_{\mathbf{a}}\left(\sum_{i=1}^{k}a_i\mu_i=0\right)$$

and

$$A=\bigcup_{\mathbf{a}} A_{\mathbf{a}},\qquad A_{\mathbf{a}}\neq H_{\mathbf{a}},$$

where $\sum_{i=1}^{k}a_i=0$ and the union and the intersections are taken for all nonnull $\mathbf{a}$ satisfying this condition. Given $\mathbf{a}$, we observe that

$$\sum_{i=1}^{k}a_i\bar{x}_i\sim N\left(\sum_{i=1}^{k}a_i\mu_i,\sigma^2\sum_{i=1}^{k}\frac{a_i^2}{n_i}\right)$$

and the critical region for testing $H_{\mathbf{a}}$ against $A_{\mathbf{a}}$ is

$$w(H_{\mathbf{a}},A_{\mathbf{a}},\alpha):(n-k)\left(\sum_{i=1}^{k}a_i\bar{x}_i\right)^2\Big/ v\left(\sum_{i=1}^{k}\frac{a_i^2}{n_i}\right)\geqslant c,$$

where $c$ is a constant which depends only on $\alpha$, because under $H_{\mathbf{a}}$,

$$(n-k)^{1/2}\left(\sum_{i=1}^{k}a_i\bar{x}_i\right)\Big/\left[v\left(\sum_{i=1}^{k}\frac{a_i^2}{n_i}\right)\right]^{1/2}$$

is distributed as Student's $t$ with $n-k$ d.f. Hence, the critical region for testing $H$ against $A\neq H$ is

$$w:\bigcup_{\mathbf{a}}\left\{(n-k)\left(\sum_{i=1}^{k}a_i\bar{x}_i\right)^2\Big/ v\left(\sum_{i=1}^{k}\frac{a_i^2}{n_i}\right)\geqslant c\right\}$$

or

$$w: \operatorname*{Sup}_{\mathbf{a}}\left[(n-k)\left(\sum_{i=1}^{k} a_i\bar{x}_i\right)^2 \Big/ v\left(\sum_{i=1}^{k}\frac{a_i^2}{n_i}\right)\right] \geqslant c,$$

where the union and the supremum are taken over all nonnull $\mathbf{a}$ satisfying $\Sigma_{i=1}^{k} a_i = 0$. Taking $D = \operatorname{diag}(n_1, \ldots, n_k)$, $\bar{\mathbf{x}}' = (\bar{x}_1, \bar{x}_2, \ldots, \bar{x}_k)$, and $\mathbf{a}' = (a_1, \ldots, a_k)$, we see from Problem 1.38 that

$$\frac{(n-k)\left(\sum_{i=1}^{k} a_i\bar{x}_i\right)^2}{v\left(\sum_{i=1}^{k}\frac{a_i^2}{n_i}\right)} = \frac{(n-k)\mathbf{a}'(\bar{\mathbf{x}}\bar{\mathbf{x}}')\mathbf{a}}{v\mathbf{a}'D^{-1}\mathbf{a}} \leqslant \frac{(n-k)\sum_{i=1}^{k} n_i(\bar{x}_i-\bar{x})^2}{v}$$

with $\bar{x} = \Sigma_{i=1}^{k} n_i\bar{x}_i/n$, and hence the critical region for $H$ against $A$ is

$$w: F_{k-1,n-k} \geqslant c_1,\ F_{k-1,n-k} = \frac{(n-k)\sum_{i=1}^{k} n_i(\bar{x}_i-\bar{x})^2}{(k-1)v},$$

which is the usual likelihood-ratio test procedure.

Now, we consider a different decomposition of $H$ and $A$ as

$$H = \bigcap_{\substack{i,j\\ i\neq j}} H_{ij}(\mu_i = \mu_j), \qquad A = \bigcup_{\substack{i,j\\ i\neq j}} A_{ij}(\mu_i \neq \mu_j).$$

From the univariate theory, the critical region for testing $H_{ij}(\mu_i = \mu_j)$ against $A_{ij}$ is

$$w_{ij}(H_{ij}, A_{ij}, \alpha): \sqrt{n-k}\left(\frac{n_i n_j}{n_i + n_j}\right)^{1/2} \frac{|\bar{x}_i - \bar{x}_j|}{\sqrt{v}} \geqslant c,$$

where $c$ is a constant depending on $\alpha$ only. Hence, the critical region for testing $H$ against $A$ under the above decomposition is

$$w_1: \sqrt{\frac{n-k}{v}} \operatorname*{Sup}_{i,j}\left[\left(\frac{n_i n_j}{n_i + n_j}\right)^{1/2} |\bar{x}_i - \bar{x}_j|\right] \geqslant c,$$

where $c$ is to be determined from $P(\mathbf{x} \in w_1 | H) = \delta$. When $n_1 = n_2 = \cdots = n_k = n_0$ (say), then the above critical region is

$$w_1: \frac{\bar{x}_{\max} - \bar{x}_{\min}}{v^{\frac{1}{2}}} \geqslant c,$$

where $c$ is a constant to be determined from

$$P\left( \frac{\bar{x}_{\max} - \bar{x}_{\min}}{v^{\frac{1}{2}}} \geqslant c \,\middle|\, H \right) = \delta.$$

Thus, we have obtained two test procedures from the union–intersection principle by considering two different decompositions of the hypothesis and the alternative. Possibly many more tests can be obtained by considering different decompositions. For example, if $n_1 = n_2 = \cdots = n_k$, then $(\bar{x}_{\max} - \bar{x})/v^{\frac{1}{2}}$ is also obtained as a union–intersection test procedure by decomposing $H = \bigcap_{i=1}^{k} \{ \mu_i = \bar{\mu},\ \bar{\mu} = k^{-1} \Sigma \mu_i \}$ and $A = \bigcup_{i=1}^{k} \{ \mu_i \neq \bar{\mu} \}$. This fact should be kept in mind while obtaining tests from this principle.

EXAMPLE 4.2.2. Let $(x_i, y_i)$ $(i = 1, 2, \ldots, n)$ be independently distributed as bivariate normal with means $\mu_1, \mu_2$, variances $\sigma_1^2, \sigma_2^2$, and correlation $\rho$. Suppose a test procedure is required for testing

$$H = H\{ \mu_1 = \mu_2 = 0,\ \sigma_1 > 0, \sigma_2 > 0,\ |\rho| < 1 \}$$

against the alternatives

$$A : A\{ \mu_1 \neq 0 \text{ or } \mu_2 \neq 0,\ \sigma_1 > 0, \sigma_2 > 0,\ |\rho| < 1 \}.$$

We notice that

$$H = H_1(\mu_1 = 0,\ \sigma_1 > 0) \cap H_2\left( \mu_2 = \frac{\rho \sigma_2 \mu_1}{\sigma_1},\ \sigma_2^2(1 - \rho^2) > 0,\ \frac{\rho \sigma_2}{\sigma_1} \in R \right)$$

and

$$A = A_1(\mu_1 \neq 0,\ \sigma_1 > 0) \cup A_2\left( \mu_2 \neq \frac{\rho \sigma_2 \mu_1}{\sigma_1},\ \sigma_2^2(1 - \rho^2) > 0,\ \frac{\rho \sigma_2}{\sigma_1} \in R \right).$$

In the step-down procedure, we first test $H_1$ against $A_1$ at level of significance $\alpha_1$, say. If it is rejected, we reject $H$ and do not proceed further. But if $H_1$ is not rejected, we proceed for testing $H_2$ against $A_2$ at level of significance $\alpha_2$, say. If it is rejected, we reject $H$; otherwise we accept $H$. The level of significance will depend on the levels of the previous steps. Clearly in the *step-down* procedure, the rejection can take place at any step, while the acceptance will take place only at the final step. Observe that in *union–intersection* procedure, we combine all the above steps in a single step, while in the step-down, we consider the testing of hypotheses in various steps with various sizes. However, there is no unique way of ordering these steps. For this reason, we shall not consider this procedure in detail.

For the above problem, let us observe that the parameters occurring in $H_1$ and $H_2$ are

$$\mu_1, \quad \nu=\mu_2-\beta\mu_1, \quad \beta=\frac{\rho\sigma_2}{\sigma_1}, \quad \sigma_1^2, \text{ and } \sigma_{2.1}^2=\sigma_2^2(1-\rho^2),$$

and the joint density of $(x,y)$ can be written as

$$\text{const}(\sigma_1\sigma_{2.1})^{-1}\exp\left[-(x-\mu_1)^2(2\sigma_1^2)^{-1}-(y-\nu-\beta x)^2(2\sigma_{2.1}^2)^{-1}\right].$$

The complete sufficient statistics for these parameters are

$$\bar{x}, \quad \bar{y}, \quad s_1^2, \quad s_{2.1}^2, \quad b,$$

where $\bar{x}=\Sigma_{i=1}^n x_i/n$, $\bar{y}=\Sigma y_i/n$, $s_1^2=\Sigma(x_i-\bar{x})^2/(n-1)$, $s_{2.1}^2=\Sigma[y_i-\bar{y}-b(x_i-\bar{x})]^2/(n-2)$, and $b=\Sigma(y_i-\bar{y})(x_i-\bar{x})/\Sigma(x_i-\bar{x})^2$. Now we can use the univariate results and get the following test procedures:

$$H_1 \text{ is rejected if } n\bar{x}^2/s_1^2=F_{1,n-1}\geqslant c_1$$

and

$$H_2 \text{ is rejected if } ns_1^2\left(s_1^2+\frac{n\bar{x}^2}{n-1}\right)^{-1}\frac{(\bar{y}-b\bar{x})^2}{s_{2.1}^2}=F_{1,n-2}\geqslant c_2,$$

where under $H$, $F_{1,n-1}$ and $F_{1,n-2}$ are independently distributed and their respective distributions are $F$ with $(1,n-1)$ and $(1,n-2)$ d.f. and $P(F_{1,n-1}\geqslant c_1|H_1)=\alpha_1$, $P(F_{1,n-2}\geqslant c_2|H_2)=\alpha_2$. Notice that $H_2$ can be tested whether $H_1$ is true or not. The total size of the test is the product of the sizes of the above two tests. This gives the step-down procedure as

$$\text{reject } H \text{ if } F_{1,n-1}\geqslant c_1 \text{ or } F_{1,n-2}\geqslant c_2,$$

where $P(F_{1,n-1}<c_1 \text{ and } F_{1,n-2}<c_2|H)=(1-\alpha_1)(1-\alpha_2)$ and the size of the test is $\alpha=\alpha_1+\alpha_2-\alpha_1\alpha_2$.

## 4.3 Test for $\boldsymbol{\mu}=\mathbf{0}$, Where $\mathbf{x}\sim N_p(\boldsymbol{\mu},\Sigma)$

Let $\mathbf{x}_1,\ldots,\mathbf{x}_N$ be iid $N_p(\boldsymbol{\mu},\Sigma)$, where $(\boldsymbol{\mu},\Sigma)\in\Omega\equiv\{(\boldsymbol{\mu},\Sigma): -\infty<\mu_i<\infty,\ i=1,2,\ldots,p \text{ and } \Sigma>0\}$. Let $\Omega_0$ be the region defined by $\Omega_0=\{(\boldsymbol{\mu},\Sigma):\boldsymbol{\mu}=\mathbf{0} \text{ and } \Sigma>0\}$. Suppose we wish to test the hypothesis $H:(\boldsymbol{\mu},\Sigma)\in\Omega_0$ against the alternative $A:(\boldsymbol{\mu},\Sigma)\in\Omega$.

### *4.3.1 Likelihood-Ratio Test*

In this subsection we derive the likelihood-ratio test for this problem on the basis of a sample $\mathbf{x}_1,\ldots,\mathbf{x}_N$ of size $N$ from $N_p(\boldsymbol{\mu},\Sigma)$. Let

$$\bar{\mathbf{x}}=N^{-1}\sum_{i=1}^N \mathbf{x}_i \quad\text{and}\quad V\equiv\sum_{i=1}^N(\mathbf{x}_i-\bar{\mathbf{x}})(\mathbf{x}_i-\bar{\mathbf{x}})'.$$

Then $\bar{\mathbf{x}}$ and $V$ are sufficient statistics for $\boldsymbol{\mu}$ and $\Sigma$, and are independently distributed:

$$N^{1/2}\bar{\mathbf{x}}\equiv\mathbf{y}\sim N_p(\boldsymbol{\theta},\Sigma) \quad \text{and} \quad V\sim W_p(\Sigma,n),$$

where

$$\boldsymbol{\theta}=N^{1/2}\boldsymbol{\mu} \quad \text{and} \quad n=N-1.$$

Hence the likelihood function of $\boldsymbol{\mu}$ and $\Sigma$ is given by

$$L(\boldsymbol{\mu},\Sigma)=c|\Sigma|^{-\frac{1}{2}N}|V|^{\frac{1}{2}(n-p-1)}\operatorname{etr}\left\{-\tfrac{1}{2}\Sigma^{-1}\left[V+N(\bar{\mathbf{x}}-\boldsymbol{\mu})(\bar{\mathbf{x}}-\boldsymbol{\mu})'\right]\right\},$$

where $c$ is a constant. Under $A$, the likelihood estimate of $\boldsymbol{\mu}$ is $\bar{\mathbf{x}}$, and that of $\Sigma$ is $N^{-1}V$. Under $H$, the likelihood estimate of $\Sigma$ is $N^{-1}[V+N\bar{\mathbf{x}}\bar{\mathbf{x}}']\equiv N^{-1}[V+\mathbf{y}\mathbf{y}']$. Hence

$$\sup_{\Omega} L(\boldsymbol{\theta},\Sigma)=c|N^{-1}V|^{-N/2}|V|^{\frac{1}{2}(n-p-1)}\exp\left(-\tfrac{1}{2}Np\right)$$

and

$$\sup_{\Omega_0} L(\boldsymbol{\theta},\Sigma)=c|N^{-1}(V+\mathbf{y}\mathbf{y}')|^{-\frac{1}{2}N}|V|^{\frac{1}{2}(n-p-1)}\exp\left(-\tfrac{1}{2}Np\right).$$

Thus the likelihood ratio $\lambda$ is given by

$$\begin{aligned}\lambda&=\frac{\sup_{\Omega_0} L(\boldsymbol{\theta},\Sigma)}{\sup_{\Omega} L(\boldsymbol{\theta},\Sigma)}\\&=\left[\frac{|V|}{|V+\mathbf{y}\mathbf{y}'|}\right]^{\frac{1}{2}N}\\&=\left[\frac{1}{1+\mathbf{y}'V^{-1}\mathbf{y}}\right]^{\frac{1}{2}N}\\&=\left[\frac{1}{1+T^2/n}\right]^{\frac{1}{2}N}, \qquad n=N-1,\end{aligned}$$

where

$$T^2=n\mathbf{y}'V^{-1}\mathbf{y}=Nn\bar{\mathbf{x}}'V^{-1}\bar{\mathbf{x}}=N(N-1)\bar{\mathbf{x}}'V^{-1}\bar{\mathbf{x}}$$

is known as Hotelling's $T^2$ statistic. Since $\lambda$ is a monotone decreasing function of $T^2$, the hypothesis $H$ is rejected for large values of $T^2$.

Thus, the hypothesis $H$ is rejected if

$$T^2\geqslant T_0^2,$$

where $P(T^2\geqslant T_0^2|H)=\alpha$ and $\alpha$ is the level of significance.

### 4.3.2 Union–Intersection Test Procedure

We shall show that the above test procedure can be obtained on the basis of the union–intersection procedure. Let $\mathbf{a}$ be a $p\times 1$ nonnull vector. Then

$$\Omega_0 = \bigcap_{\mathbf{a}} \{\mathbf{a}'\boldsymbol{\mu}=0, \mathbf{a}'\Sigma\mathbf{a}>0\} = \bigcap_{\mathbf{a}} H_{\mathbf{a}} \quad \text{(say)}.$$

Given $\mathbf{a}$, the complete sufficient statistics are $\mathbf{a}'\bar{\mathbf{x}}$ and $\mathbf{a}'V\mathbf{a}$; they are independently distributed, $\mathbf{a}'\bar{\mathbf{x}}\sim N(\mathbf{a}'\boldsymbol{\mu},\mathbf{a}'\Sigma\mathbf{a}/N)$, and $\mathbf{a}'V\mathbf{a}/\mathbf{a}'\Sigma\mathbf{a}\sim\chi^2_n$. Hence, by the univariate results, the test procedure for testing $H_{\mathbf{a}}$ against $A_{\mathbf{a}}\neq H_{\mathbf{a}}$ is

$$\text{reject } H_{\mathbf{a}} \quad \text{if} \quad \frac{N(N-1)(\mathbf{a}'\bar{\mathbf{x}})^2}{\mathbf{a}'V\mathbf{a}} \geqslant c.$$

Then, taking the union of the above critical regions over all $\mathbf{a}$, we get

$$\text{reject } H \quad \text{if} \quad N(N-1)\sup_{\mathbf{a}} \frac{\mathbf{a}'\bar{\mathbf{x}}\bar{\mathbf{x}}'\mathbf{a}}{\mathbf{a}'V\mathbf{a}} \geqslant c,$$

or by using Corollary 1.10.1,

$$\text{reject } H \quad \text{if} \quad N(N-1)\bar{\mathbf{x}}'V^{-1}\bar{\mathbf{x}} \geqslant c,$$

which is the same as derived in Subsection 4.3.1.

### 4.3.3 Distribution of $T^2$

Let $\mathbf{z}=\Sigma^{-1/2}\mathbf{y}$ and $W=\Sigma^{-1/2}V\Sigma^{-1/2}$, where $\Sigma=(\Sigma^{1/2})^2$. Then $\mathbf{z}\sim N_p(N^{1/2}\Sigma^{-1/2}\boldsymbol{\mu},I)$ and $W\sim W_p(I,n)$ are independently distributed. Let $\Gamma$ be a random orthogonal matrix with last row as $\mathbf{z}'/(\mathbf{z}'\mathbf{z})^{1/2}$. Let $\mathbf{u}'=\mathbf{z}'\Gamma'$ and $W^*=\Gamma W\Gamma'$. Conditionally, given $\Gamma$, we have $W^*\sim W_p(I,n)$, and hence $W^*$ is independently distributed of $\Gamma$ (and $\mathbf{z}$).

Thus, since $\mathbf{u}'=\mathbf{z}'\Gamma'=(0,0,\ldots,0,\sqrt{\mathbf{z}'\mathbf{z}}\,)$, we get

$$\begin{aligned} T^2 &= n\mathbf{y}'V^{-1}\mathbf{y} = n\mathbf{z}'W^{-1}\mathbf{z} \\ &= n\mathbf{u}'W^{*-1}\mathbf{u} \\ &= n(\mathbf{z}'\mathbf{z})/t^2_{pp}, \end{aligned}$$

where

$$W^* = \underset{\sim}{T}\underset{\sim}{T}'$$

and $\underset{\sim}{T}$ is a *lower* triangular matrix. Since $\mathbf{z}'\mathbf{z}$ has a noncentral chi-square distribution with $p$ d.f. and noncentrality parameter $\delta^2=N\boldsymbol{\mu}'\Sigma^{-1}\boldsymbol{\mu}$, and is independently distributed of $t^2_{pp}$, which has a central chi-square distribution with $n-p+1$ d.f., it follows that

$$\frac{n-p+1}{p}\frac{T^2}{n}\equiv\frac{N-p}{p}\frac{T^2}{n}$$

has a noncentral $F$ distribution with $p$ and $N-p$ d.f. and noncentrality parameter $\delta^2$. Alternatively the distribution can be derived with the help of Problem 3.6. The above proof is attributed to A. S. Bowker.

Tang (1938) has given tables for $\beta\equiv$(the error of the second kind)$\equiv 1-$(power of the test), for $\alpha=0.01$ and $\alpha=0.05$, where $\alpha$ is the level of significance.

### *4.3.4 Invariance*

The problem of testing $\Omega_0$ vs $\Omega$ (i.e., testing $\boldsymbol{\mu}=\mathbf{0}$ vs $\boldsymbol{\mu}\neq\mathbf{0}$, when $\Sigma>0$) remains invariant under the group of nonsingular linear transformations [see Lehmann (1959) for details on invariance], but is not invariant under the additive group of transformations. The sufficient statistics for this problem are $(\mathbf{y}, V)$, which are independently distributed:

$$\mathbf{y}\equiv N^{1/2}\overline{\mathbf{x}}\sim N_p(\boldsymbol{\theta},\Sigma) \text{ and } V\sim W_p(\Sigma,n); \qquad \boldsymbol{\theta}=N^{1/2}\boldsymbol{\mu};$$

and if $\mathbf{y}\to A\mathbf{y}+\mathbf{b}$, then $\boldsymbol{\theta}\to A\boldsymbol{\theta}+\mathbf{b}$, and the hypothesis $\Omega_0$ is changed unless $\mathbf{b}=\mathbf{0}$. Next, consider the transformation

$$\mathbf{y}\to C\mathbf{y} \quad\text{and}\quad V\to CVC',$$

where $C$ is nonsingular. Then it can easily be seen that the hypothesis and the alternative both remain unchanged. Under this transformation Hotelling's $T^2$ statistic, or equivalently

$$\mathbf{y}'V^{-1}\mathbf{y},$$

remains invariant. For if $\mathbf{y}^*=C\mathbf{y}$ and $V^*=CVC'$, then

$$\mathbf{y}^{*\prime}V^{*-1}y^*=\mathbf{y}'V^{-1}\mathbf{y}.$$

In order to show that it is also maximal invariant, we need to show that if

$$\mathbf{y}'V^{-1}\mathbf{y}=\mathbf{y}^{*\prime}V^{*-1}\mathbf{y}^*,$$

then there exists a nonsingular matrix $C$ such that

$$\mathbf{y}^*=C\mathbf{y} \quad\text{and}\quad V^*=CVC'.$$

Since $\mathbf{y}'V^{-1}\mathbf{y}=\mathbf{y}^{*\prime}V^{*-1}\mathbf{y}^*$, we get from Theorem 1.9.3

$$y^{*\prime}V^{*-1/2}=\mathbf{y}'V^{-1/2}\Gamma,$$

where $\Gamma$ is an orthogonal matrix. Hence

$$y^{*\prime}=\mathbf{y}'V^{-1/2}\Gamma V^{*1/2},$$

and choosing $C = V^{*1/2}\Gamma' V^{-1/2}$, we have

$$y^* = C\mathbf{y} \quad \text{and} \quad CVC' = V^*.$$

Thus $\mathbf{y}'V^{-1}\mathbf{y}$ is a maximal invariant. The maximal invariant on the parameter space is $\delta^2 = N\boldsymbol{\mu}'\Sigma^{-1}\boldsymbol{\mu}$.

### *4.3.5 Optimality*

The $T^2$ test is admissible and minimax. These properties, however, will not be discussed in this book; see Stein (1956), Kiefer and Schwartz (1965), Giri, Kiefer, and Stein (1963), and Salaevski (1968). In this section, we establish another interesting property:

**Theorem 4.3.1** *Given the observations* $\mathbf{x}_1, \ldots, \mathbf{x}_N$ *from* $N_p(\boldsymbol{\mu}, \Sigma)$, *of all tests of* $\boldsymbol{\mu} = \mathbf{0}$ *that are invariant under the group of nonsingular linear transformations, the* $T^2$ *test is uniformly most powerful.*

PROOF. First, we note that a test based on sufficient statistics $\bar{\mathbf{x}}$ and $V$ is as powerful as any other test based on all the observations. Thus we may confine our attention to tests based only on $\bar{\mathbf{x}}$ and $V$. Second, under the transformation $\bar{\mathbf{x}} \to C\bar{\mathbf{x}}$ and $V \to CVC'$ where $C$ is a nonsingular matrix $T^2$ is a maximal invariant (Subsection 4.3.4). The maximal invariant in the parameter space is $\delta^2 \equiv N\boldsymbol{\mu}'\Sigma^{-1}\boldsymbol{\mu}$.

Hence, applying the Neyman–Pearson fundamental lemma to find the most powerful test for testing $\delta^2 = 0$ vs $\delta^2 > 0$, we have the test

$$\frac{f(T^2, \delta^2)}{f(T^2, 0)} \geqslant c,$$

where $f(T^2, \delta^2)$ denotes the density of $T^2$. The ratio of the densities is given by

$$\frac{\Gamma\left(\frac{1}{2}p\right)}{\Gamma\left(\frac{1}{2}N\right)} e^{-\frac{1}{2}\delta^2} \sum_{\alpha=0}^{\infty} \frac{\left(\frac{1}{2}\delta^2\right)^{\alpha} \Gamma\left(\frac{1}{2}N + \alpha\right)}{\alpha! \Gamma\left(\frac{1}{2}p + \alpha\right)} \left(\frac{T^2/n}{1 + T^2/n}\right)^{\alpha}.$$

This is a strictly increasing function of $T^2$. Hence the test is to reject the hypothesis for larger values of $T^2$. Since this test does not depend upon the alternative $\delta^2$, the test is uniformly most powerful. □

This proves the theorem. In fact, it can be shown that among all tests whose power depends on $\delta^2$, the $T^2$ test is most powerful. This result is due to Simaika (1941).

### 4.3.6 Confidence Interval for $\boldsymbol{\mu}$

Let $\mathbf{x}_1,\ldots,\mathbf{x}_N$ be a random sample of size $N$ from $N_p(\boldsymbol{\mu},\Sigma)$. Based on this sample, we wish to find a $(1-\alpha)$ confidence interval for $\boldsymbol{\mu}$. Let $R_n$ be a region in the $p$-dimensional Euclidean space such that

$$R_n=\left\{\mathbf{z}:(\bar{\mathbf{x}}-\mathbf{z})'V^{-1}(\bar{\mathbf{x}}-\mathbf{z})\leqslant T_\alpha^2\right\},$$

where $T_\alpha^2$ is defined by

$$P\left(T^2\leqslant nNT_\alpha^2\right)=1-\alpha$$

and $T^2$ is Hotelling's $T^2$ based on $p$ and $N-p$ d.f. Then

$$\begin{aligned}P\left[\boldsymbol{\mu}\in R_n\right]&=P\left[(\bar{\mathbf{x}}-\boldsymbol{\mu})'V^{-1}(\bar{\mathbf{x}}-\boldsymbol{\mu})\leqslant T_\alpha^2\right]\\&=P\left[T^2\leqslant nNT_\alpha^2\right]=1-\alpha.\end{aligned}$$

Hence, $R_n$ is a confidence region with confidence coefficient $1-\alpha$. For confidence region of fixed radius, the reader is referred to Srivastava (1967a) and (1971).

We observe from Subsection 4.3.2 that $(\bar{\mathbf{x}}-\boldsymbol{\mu})'V^{-1}(\bar{\mathbf{x}}-\boldsymbol{\mu})\leqslant T_\alpha^2$ is equivalent to

$$\mathbf{a}'\bar{\mathbf{x}}-T_\alpha(\mathbf{a}'V\mathbf{a})^{1/2}\leqslant\mathbf{a}'\boldsymbol{\mu}\leqslant\mathbf{a}'\bar{\mathbf{x}}+T_\alpha(\mathbf{a}'V\mathbf{a})^{1/2}\qquad\text{for all}\quad\mathbf{a}\neq\mathbf{0}.$$

Hence, this gives simultaneous confidence bounds on all linear functions of $\boldsymbol{\mu}$. This result can now be used to get a confidence bound on $\boldsymbol{\mu}'\boldsymbol{\mu}$, since

$$\operatorname*{Sup}_{\mathbf{a}}\left[\mathbf{a}'\bar{\mathbf{x}}-T_\alpha(\mathbf{a}'V\mathbf{a})^{1/2}\right]\leqslant\operatorname*{Sup}_{\mathbf{a}}(\mathbf{a}'\boldsymbol{\mu})\leqslant\operatorname*{Sup}_{\mathbf{a}}\left[\mathbf{a}'\bar{\mathbf{x}}+T_\alpha(\mathbf{a}'V\mathbf{a})^{1/2}\right]$$

with confidence greater than or equal to $1-\alpha$, and the above inequality implies

$$\left[\mathrm{ch}_p(V)\right]^{1/2}(T-T_\alpha)\leqslant(\boldsymbol{\mu}'\boldsymbol{\mu})^{1/2}\leqslant\left[\mathrm{ch}_1(V)\right]^{1/2}(T+T_\alpha),$$

or

$$(\bar{\mathbf{x}}'\bar{\mathbf{x}})^{1/2}-\left[\mathrm{ch}_1(V)\right]^{1/2}T_\alpha\leqslant(\boldsymbol{\mu}'\boldsymbol{\mu})^{1/2}\leqslant(\bar{\mathbf{x}}'\bar{\mathbf{x}})^{1/2}+T_\alpha\left[\mathrm{ch}_1(V)\right]^{1/2},$$

where $\mathrm{ch}_1(V)$ and $\mathrm{ch}_p(V)$ are respectively the maximum and the minimum roots of $V$ and $T=(\bar{\mathbf{x}}'V^{-1}\bar{\mathbf{x}})^{1/2}$.

## 4.4 Test for a Subvector

Let $\mathbf{y}\sim N_p(\boldsymbol{\theta},\Sigma)$ and $V\sim W_p(\Sigma,n)$ be independently distributed. Thus in terms of the previous notation $\mathbf{y}=N^{1/2}\bar{\mathbf{x}}$. Let

$$\boldsymbol{\theta}'=(\overset{r}{\boldsymbol{\theta}_1'},\overset{s}{\boldsymbol{\theta}_2'})\quad\text{and}\quad\mathbf{y}'=(\overset{r}{\mathbf{y}_1'},\overset{s}{\mathbf{y}_2'}),$$

where $r+s=p$. Suppose it is given that $\boldsymbol{\theta}_2=\mathbf{0}$, that is, the parameter space is $\Omega\equiv\{(\boldsymbol{\theta},\Sigma):\boldsymbol{\theta}_2=\mathbf{0},\ \Sigma>0\}$. Let $\Omega_0\equiv\{(\boldsymbol{\theta},\Sigma):\boldsymbol{\theta}=\mathbf{0},\ \Sigma>0\}$. In this section we wish to derive the likelihood-ratio test and its distribution for testing $H:(\boldsymbol{\theta},\Sigma)\in\Omega_0$ vs $A:(\boldsymbol{\theta},\Sigma)\in\Omega$. The likelihood function is given by

$$L(\boldsymbol{\theta}_1,\Sigma)=c|\Sigma|^{-\frac{1}{2}N}|V|^{\frac{1}{2}(n-p-1)}\operatorname{etr}\left\{-\tfrac{1}{2}\Sigma^{-1}\left[V+(\mathbf{y}-B\boldsymbol{\theta}_1)(\mathbf{y}-B\boldsymbol{\theta}_1)'\right]\right\},$$

where $B'=(\mathbf{I}_r,\mathbf{0})$ and $c$ is a constant. Hence,

$$\operatorname*{Sup}_{\Omega_0}L(\boldsymbol{\theta}_1,\Sigma)=c|V|^{\frac{1}{2}(n-p-1)}e^{-\frac{1}{2}Np}N^{-\frac{1}{2}Np}|V+\mathbf{y}\mathbf{y}'|^{-\frac{1}{2}N},$$

and

$$\begin{aligned}L(\boldsymbol{\theta}_1,\Sigma)&\leqslant L\left(\boldsymbol{\theta}_1,N^{-1}\left[V+(\mathbf{y}-B\boldsymbol{\theta}_1)(\mathbf{y}-B\boldsymbol{\theta}_1)'\right]\right)\\&\leqslant L\left(\hat{\boldsymbol{\theta}}_1,\hat{\Sigma}\right),\end{aligned}$$

where the first inequality follows by Theorem 1.10.4, the second inequality follows by Theorem 1.10.3,

$$\hat{\boldsymbol{\theta}}_1=(B'V^{-1}B)^{-1}B'V^{-1}\mathbf{y},$$

and

$$N\hat{\Sigma}=V+\left[I-B(B'V^{-1}B)^{-1}B'V^{-1}\right]\mathbf{y}\mathbf{y}'\left[I-V^{-1}B(B'V^{-1}B)^{-1}B'\right].$$

Since

$$\begin{aligned}|N\hat{\Sigma}|&=|V|\cdot\left\{1+\mathbf{y}'\left[V^{-1}-V^{-1}B(B'V^{-1}B)^{-1}B'V^{-1}\right]\mathbf{y}\right\}\\&=|V|\left(1+\mathbf{y}_2'V_{22}^{-1}\mathbf{y}_2\right),\end{aligned}$$

where

$$V=\begin{pmatrix}V_{11}&V_{12}\\V_{12}'&V_{22}\end{pmatrix},$$

we get

$$\operatorname*{Sup}_{\Omega}L(\boldsymbol{\theta},\Sigma)=c|V|^{\frac{1}{2}(n-p-1)}e^{-\frac{1}{2}Np}N^{-\frac{1}{2}Np}|V|^{-\frac{1}{2}N}\left(1+\mathbf{y}_2'V_{22}^{-1}\mathbf{y}_2\right)^{-\frac{1}{2}N}.$$

Hence, the likelihood ratio $\lambda$ is given by

$$\lambda=\frac{\operatorname*{Sup}_{\Omega_0}L(\boldsymbol{\theta}_1,\Sigma)}{\operatorname*{Sup}_{\Omega}L(\boldsymbol{\theta}_1,\Sigma)}=\left(\frac{1+\mathbf{y}_2'V_{22}^{-1}\mathbf{y}_2}{1+\mathbf{y}'V^{-1}\mathbf{y}}\right)^{\frac{1}{2}N}.$$

Let

$$L=\frac{1+\mathbf{y}'V^{-1}\mathbf{y}}{1+\mathbf{y}_2'V_{22}^{-1}\mathbf{y}_2}.$$

Then the hypothesis $H$ is rejected for larger values of $L$. In the next subsection we derive its distribution.

### 4.4.1 Distribution of L

As in the previous section, we observe that

$$\mathbf{y}_2'V_{22}^{-1}\mathbf{y}_2 = \mathbf{y}'\left(V^{-1} - V^{-1}B(B'V^{-1}B)^{-1}B'V^{-1}\right)\mathbf{y}$$
$$= \mathbf{y}'V^{-1}\mathbf{y} - \left[\mathbf{y}_1 + (V^{11})^{-1}V^{12}\mathbf{y}_2\right]'V^{11}\left[\mathbf{y}_1 + (V^{11})^{-1}V^{12}\mathbf{y}_2\right],$$

where

$$V^{-1} = \begin{pmatrix} V^{11} & V^{12} \\ (V^{12})' & V^{22} \end{pmatrix},$$

$$V^{11} = \left(V_{11} - V_{12}V_{22}^{-1}V_{12}'\right)^{-1} = V_{1.2}^{-1} \quad \text{and} \quad V^{12} = -V_{1.2}^{-1}V_{12}V_{22}^{-1}.$$

Hence

$$L = 1 + \frac{(\mathbf{y}_1 - V_{12}V_{22}^{-1}\mathbf{y}_2)'V_{1.2}^{-1}(\mathbf{y}_1 - V_{12}V_{22}^{-1}\mathbf{y}_2)}{1 + \mathbf{y}_2'V_{22}^{-1}\mathbf{y}_2}.$$

Recalling the results of Theorem 2.3.5 and Theorem 3.3.5, we know that

(i) given $\mathbf{y}_2$, $\mathbf{y}_1 \sim N_r(\boldsymbol{\theta}_1 + \beta\mathbf{y}_2, \Sigma_{1.2})$, $\beta = \Sigma_{12}\Sigma_{22}^{-1}$, and $\Sigma_{1.2} = \Sigma_{11} - \Sigma_{12}\Sigma_{22}^{-1}\Sigma_{12}'$;

(ii) given $V_{22}$, $V_{12} \sim N_{r,s}(\beta V_{22}, \Sigma_{1.2}, V_{22})$;

(iii) given $\mathbf{y}_2$ and $V_{22}$, $V_{12}V_{22}^{-1}\mathbf{y}_2 \sim N_r(\beta\mathbf{y}_2, \mathbf{y}_2'V_{22}^{-1}\mathbf{y}_2\Sigma_{1.2})$;

(iv) $V_{1.2}$ is distributed independently of $(V_{12}, V_{22})$;

(v) $V_{1.2} \sim W_r(\Sigma_{1.2}, n-s)$.

Thus, given $\mathbf{y}_2$ and $V_{22}$, $(\mathbf{y}_1 - V_{12}V_{22}^{-1}\mathbf{y}_2) \sim N_r\left(\theta_1, \left(1 + \mathbf{y}_2'V_{22}^{-1}\mathbf{y}_2\right)\Sigma_{1.2}\right)$. Hence, given $\mathbf{y}_2$ and $V_{22}$

$$\frac{(n-p+1)(\mathbf{y}_1 - V_{12}V_{22}^{-1}\mathbf{y}_2)'V_{1.2}^{-1}(\mathbf{y}_1 - V_{12}V_{22}^{-1}\mathbf{y}_2)}{\left(1 + \mathbf{y}_2'V_{22}^{-1}\mathbf{y}_2\right)r}$$

has a noncentral $F$ distribution with $r$ and $n-p+1$ d.f. and noncentrality parameter

$$\frac{\boldsymbol{\theta}_1'\Sigma_{1.2}^{-1}\boldsymbol{\theta}_1}{1 + \mathbf{y}_2'V_{22}^{-1}\mathbf{y}_2}.$$

Thus, under $H$, the distribution of $(n-p+1)(L-1)/r$ is a central $F$ distribution with $r$ and $n-p+1$ d.f.

### 4.4.2 *Invariance*

Let $\mathcal{G}=\{g\}$ be the group of nonsingular linear transformations of the form

$$g=\begin{pmatrix} g_{11} & g_{12} \\ 0 & g_{22} \end{pmatrix},$$

where $g_{11}: r\times r$ and $g_{22}: s\times s$. A set of maximal invariants under this group is given by

$$\{\mathbf{y}'V^{-1}\mathbf{y}, \mathbf{y}_2'V_{22}^{-1}\mathbf{y}_2\}.$$

It is, however, more convenient to work with an equivalent set of maximal invariants

$$\{R_1+R_2\equiv 1+\mathbf{y}'V^{-1}\mathbf{y},\ R_2\equiv 1+\mathbf{y}_2'V_{22}^{-1}\mathbf{y}_2\}.$$

Hence

$$R_1=(\mathbf{y}_1-V_{12}V_{22}^{-1}\mathbf{y}_2)'V_{1.2}^{-1}(\mathbf{y}_1-V_{12}V_{22}^{-1}\mathbf{y}_2).$$

The corresponding set of maximal invariants in the parameter space is given by

$$\delta_1+\delta_2=\boldsymbol{\theta}'\Sigma^{-1}\boldsymbol{\theta} \quad \text{and} \quad \delta_2=\boldsymbol{\theta}_2'\Sigma^{-1}\boldsymbol{\theta}_2.$$

The problem of testing $H$ vs $A$ is equivalent to testing $H:\delta_1=0,\ \delta_2=0$ against $A:\delta_1>0,\ \delta_2=0$. Since $\delta_2=0$ under $H$ as well as $A$ (i.e., $\boldsymbol{\theta}_2=0$), we get

$$\delta_1=\boldsymbol{\theta}_1'\Sigma_{1.2}^{-1}\boldsymbol{\theta}_1.$$

Hence, the likelihood-ratio test is an invariant test under the group of triangular transformations.

### 4.4.3 *Optimum Property*

From the distribution result in Subsection 4.4.1, we easily get the following:

**Theorem 4.4.1** *Conditionally given* $\mathbf{y}_2$ *and* $V_{22}$, *the likelihood-ratio test of* $H:\boldsymbol{\theta}_1=\mathbf{0}\ \boldsymbol{\theta}_2=\mathbf{0}$ vs $A:\boldsymbol{\theta}_1\neq\mathbf{0},\ \boldsymbol{\theta}_2=\mathbf{0}$ *is uniformly the most powerful invariant test.*

This problem has been considered by Rao (1949) and Giri (1964) among others.

It may be noted that the test given by $L$ is also the likelihood-ratio test procedure for testing the hypothesis $H(\boldsymbol{\mu}_1=\Sigma_{12}\Sigma_{22}^{-1}\boldsymbol{\mu}_2,\ \Sigma>0)$ against the alternative $A(\boldsymbol{\mu}_1\neq\Sigma_{12}\Sigma_{22}^{-1}\boldsymbol{\mu}_2,\ \Sigma>0)$ (see Problem 4.10).

### *4.4.4 Other Invariant Tests*

For the above problem, one may propose Hotelling's $T^2$ test based on all $r+s$ variates and a test based on $R_1=\mathbf{y}'V^{-1}\mathbf{y}-\mathbf{y}_{22}'V_{22}^{-1}\mathbf{y}_2$ proposed by Rao (1949, 1966). Both these tests are invariant under the group of transformations given in Subsection 4.4.2. Some power comparisons of these two tests with the likelihood-ratio test have been given by Subramaniam and Subramaniam (1973) and Marden and Perlman (1977). These comparisons reveal that Hotelling's $T^2$ has lower power in most practical situations.

## 4.5 Comparing Means of Correlated Variables

Let $\mathbf{x}\sim N_p(\boldsymbol{\mu},\Sigma)$ and suppose a random sample $\mathbf{x}_1,\mathbf{x}_2,\ldots,\mathbf{x}_N$ of size $N$ is given. Then the sufficient statistics are $\bar{\mathbf{x}}=N^{-1}\Sigma_{i=1}^N\mathbf{x}_i$ and $V=\Sigma_{\alpha=1}^N(\mathbf{x}_\alpha-\bar{\mathbf{x}})(\mathbf{x}_\alpha-\bar{\mathbf{x}})'$, which are independently distributed; $\bar{\mathbf{x}}\sim N_p(\boldsymbol{\mu},N^{-1}\Sigma)$ and $V\sim W_p(\Sigma,n)$, where $n=N-1$. In this section several problems on comparing the means $\mu_1,\mu_2,\ldots,\mu_p$ are considered. In particular, we consider the hypothesis of testing the equality of the $\mu_i$'s when the variables are interchangeable with respect to variances and covariances—the intraclass correlation model, that is, when $\Sigma$ is of the form

$$\Sigma_I=\sigma^2\left[(1-\rho)I_p+\rho\mathbf{e}\mathbf{e}'\right],\qquad \mathbf{e}'=(1,\ldots,1):1\times p.$$

For some further results on this model and related models, which include the problem of testing covariances, see Srivastava (1965b), Krishnaiah and Pathak (1967), Olkin and Press (1969), and Selliah (1964).

### *4.5.1 Relation Between Means when $\Sigma>0$*

Consider a general case when the hypothesis $H:(\boldsymbol{\mu},\Sigma)\in\Omega_0\equiv\{(\boldsymbol{\mu},\Sigma):\boldsymbol{\mu}=B\boldsymbol{\xi},\ \Sigma>0,\ \boldsymbol{\xi}\in R^r,\ r\leqslant p,\ B:p\times r$ is known and of rank $r\}$ against the alternative $A:(\boldsymbol{\mu},\Sigma)\in\Omega\equiv\{(\boldsymbol{\mu},\Sigma):\boldsymbol{\mu}\in R^p$ and $\Sigma>0\}$. A special case of this hypothesis is the equality of means when $B'=(1,1,\ldots,1)=\mathbf{e}'$ (say). Another special case is the hypothesis of testing a subvector $\boldsymbol{\mu}_2=\mathbf{0}$, that is, the hypothesis when $B'=(I_r,0)$. First, we consider the likelihood-ratio test procedure. For this, the likelihood based on the sufficient statistics $\bar{\mathbf{x}}$ and $V$ is

$$\text{(4.5.1)}\quad L(\boldsymbol{\mu},\Sigma)=c|V|^{\frac{1}{2}(n-p-1)}|\Sigma|^{-\frac{1}{2}N}\operatorname{etr}\left\{-\tfrac{1}{2}\Sigma^{-1}\left[V+N(\bar{\mathbf{x}}-\boldsymbol{\mu})(\bar{\mathbf{x}}-\boldsymbol{\mu})'\right]\right\},$$

where $c$ is a constant. Then, as in Subsection 4.3.1,

$$\text{(4.5.2)} \qquad \sup_{\Omega} L(\boldsymbol{\mu},\Sigma)=c(N)^{\frac{1}{2}N}|V|^{\frac{1}{2}(n-p-1)}|V|^{-\frac{1}{2}N}e^{-\frac{1}{2}Np}.$$

Under $H\in\Omega_0$, the likelihood is

$$\begin{aligned}\text{(4.5.3)} \quad L(\boldsymbol{\xi},\Sigma) &= c|V|^{\frac{1}{2}(n-p-1)}|\Sigma|^{-\frac{1}{2}N}\operatorname{etr}\left\{-\tfrac{1}{2}\Sigma^{-1}\left[V+N(\overline{\mathbf{x}}-B\boldsymbol{\xi})(\overline{\mathbf{x}}-B\boldsymbol{\xi})'\right]\right\}\\ &\leqslant L\left(\xi,N^{-1}\left[V+N(\overline{\mathbf{x}}-B\boldsymbol{\xi})(\overline{\mathbf{x}}-B\boldsymbol{\xi})'\right]\right)\\ &\leqslant L(\hat{\xi},\hat{\Sigma}),\end{aligned}$$

where the two inequalities follow respectively from Theorem 1.10.4 and Theorem 1.10.3,

$$\text{(4.5.4)} \qquad \hat{\xi}=(B'V^{-1}B)^{-1}B'V^{-1}\overline{\mathbf{x}},$$

and

$$\begin{aligned}\text{(4.5.5)} \quad N\hat{\Sigma} &= V+N\left[I-B(B'V^{-1}B)^{-1}B'V^{-1}\right]\overline{\mathbf{x}}\overline{\mathbf{x}}'\left[I-V^{-1}B(B'V^{-1}B)^{-1}B'\right]\\ &\equiv V+N(I-TV^{-1})\overline{\mathbf{x}}\overline{\mathbf{x}}'(I-V^{-1}T),\end{aligned}$$

where $T=B(B'V^{-1}B)^{-1}B'$.

Hence

$$\text{(4.5.6)} \qquad \sup_{\Omega_0} L(\boldsymbol{\mu},\Sigma)=cN^{\frac{1}{2}N}|V|^{\frac{1}{2}(n-p-1)}|N\hat{\Sigma}|^{-\frac{1}{2}N}e^{-\frac{1}{2}Np}.$$

Using the above results, the likelihood ratio is

$$\begin{aligned}\text{(4.5.7)} \qquad \lambda &= \frac{\sup_{\Omega_0} L(\boldsymbol{\mu},\Sigma)}{\sup_{\Omega} L(\boldsymbol{\mu},\Sigma)}\\ &=\left(\frac{|V|}{|V+N(I-TV^{-1})\overline{\mathbf{x}}\overline{\mathbf{x}}'(I-V^{-1}T)|}\right)^{\frac{1}{2}N}\\ &=\left[1+N\overline{\mathbf{x}}'(V^{-1}-V^{-1}TV^{-1})\overline{\mathbf{x}}\right]^{-\frac{1}{2}N}.\end{aligned}$$

We may observe from Corollary 1.9.2, that

$$V^{-1}=V^{-1}B(B'V^{-1}B)^{-1}B'V^{-1}+B_1(B_1'VB_1)^{-1}B_1',$$

where $B_1'B=0$, and the rank of $B_1$ is $p-r$. Thus, the likelihood-ratio test procedure is reject $H\in\Omega_0$ if

$$\text{(4.5.8)} \qquad N\overline{\mathbf{x}}'(V^{-1}-V^{-1}TV^{-1})\overline{\mathbf{x}}=N\overline{\mathbf{x}}'B_1(B_1'VB_1)^{-1}B_1'\overline{\mathbf{x}}\geqslant c,$$

where $N^{1/2}B_1'\bar{\mathbf{x}}$ and $B_1'VB_1$ are independently distributed, $N^{1/2}B_1'\bar{\mathbf{x}} \sim N(N^{1/2}B_1'\boldsymbol{\mu}, B_1'\Sigma B_1)$ under $A$, and $B_1'VB_1 \sim W_p(B_1'\Sigma B_1, n)$. Hence, from Subsection 4.3.3, the distribution of

$$N\bar{\mathbf{x}}'(V^{-1} - V^{-1}TV^{-1})\bar{\mathbf{x}}\frac{n-p+r+1}{p-r}$$

is noncentral $F$ with $p-r$ and $n-p+r+1$ d.f. and noncentrality parameter $\delta = N\boldsymbol{\mu}'B_1(B_1'\Sigma B_1)^{-1}B_1'\boldsymbol{\mu} = N(\boldsymbol{\mu}'\Sigma^{-1}\boldsymbol{\mu} - \boldsymbol{\mu}'\Sigma^{-1}B(B'\Sigma^{-1}B)^{-1}B'\Sigma^{-1}\boldsymbol{\mu})$.

The above test procedure can also be obtained as a union–intersection test procedure. Note that for any vector $\mathbf{a}$, $\mathbf{a}'B = \mathbf{0} \Rightarrow \mathbf{a}'\boldsymbol{\mu} = 0$. Hence, the hypothesis $H$ can be written as

$$H = \bigcap_{\mathbf{a}} H_{\mathbf{a}}(\mathbf{a}'\boldsymbol{\mu} = 0, \mathbf{a}'\Sigma\mathbf{a} > 0),$$

where $\bigcap_{\mathbf{a}}$ denotes the intersection over all nonnull vectors $\mathbf{a}$ for which $\mathbf{a}'B = \mathbf{0}$. We may observe from the univariate theory (as in Section 3.2), that

$$H_{\mathbf{a}} \text{ is rejected} \quad \text{if} \quad N(\mathbf{a}'\bar{\mathbf{x}})^2/\mathbf{a}'V\mathbf{a} \geqslant c_1,$$

where $c_1$ is a constant depending on $\alpha$ only. By the union–intersection test procedure,

$$\text{reject } H \quad \text{if} \qquad \underset{\substack{\mathbf{a} \\ \text{subject to } \mathbf{a}'B = \mathbf{0}}}{\text{Sup}} \quad \frac{N(\mathbf{a}'\bar{\mathbf{x}})^2}{\mathbf{a}'V\mathbf{a}} \geqslant c,$$

and the use of Problem 1.38 gives a union–intersection test procedure that is the same as derived by the likelihood-ratio principle. This consideration gives simultaneous confidence bounds on $\mathbf{a}'\boldsymbol{\mu}$ (as mentioned in Subsection 4.3.6) for all nonnull vector $\mathbf{a}$ subject to $\mathbf{a}'B = \mathbf{0}$.

### *4.5.2 Intraclass Correlation Model*

Here, we assume that the covariance matrix has the intraclass correlation model. That is, $\Sigma = \Sigma_I = \sigma^2(1-\rho)I + \sigma^2\rho\mathbf{e}\mathbf{e}'$ with $\mathbf{e}' = (1, 1, \ldots, 1) : 1 \times p$. From the sufficient statistics $\bar{\mathbf{x}}$ and $V$, we shall obtain complete sufficient statistics for the parameters $\boldsymbol{\mu}$, $\sigma$, and $\rho$ or equivalently for $\boldsymbol{\mu}$, $\sigma_1^2 \equiv \sigma^2(1-\rho)$, and $\sigma_2^2 \equiv \sigma^2(1-\rho+p\rho)$. We use the orthogonal transformation

$$\mathbf{y} = \Gamma\bar{\mathbf{x}} \quad \text{and} \quad W = \Gamma V \Gamma',$$

where $\Gamma$ is an orthogonal matrix whose last row is $\mathbf{e}'/\sqrt{p}$. Then

$$\mathbf{y} \sim N_p(\Gamma\boldsymbol{\mu}, D/N) \quad \text{and} \quad W \sim W_p(D, n),$$

where $D = \text{diag}(\sigma_1^2 I_{p-1}, \sigma_2^2)$ with $\sigma_1^2 = \sigma^2(1-\rho)$ and $\sigma_2^2 = \sigma^2(1-\rho+p\rho)$. Then, the sufficient statistics for $(\Gamma\boldsymbol{\mu}, \sigma_1^2, \sigma_2^2)$ are $(\mathbf{y}, \text{tr}\, W - w_{pp}, w_{pp})$. We observe from Theorem 3.3.4 that the diagonal elements of $W$ are independently distributed, and hence $\operatorname{tr} W - w_{pp}$ and $w_{pp}$ are independently distributed, $w_{pp}/\sigma_2^2 \sim \chi_n^2$ and $(\operatorname{tr} W - w_{pp})/\sigma_1^2 \sim \chi_{n(p-1)}^2$. Thus, in terms of the original variables, the sufficient statistics for $(\boldsymbol{\mu}, \sigma_1^2, \sigma_2^2)$ are $(\bar{\mathbf{x}}, \operatorname{tr} V - \mathbf{e}' V \mathbf{e}/p, \mathbf{e}' V \mathbf{e}/p)$; they are independently distributed: $(\operatorname{tr} V - \mathbf{e}' V \mathbf{e}/p)/\sigma_1^2 \sim \chi_{n(p-1)}^2$, $\mathbf{e}' V \mathbf{e}/p\sigma_2^2 \sim \chi_n^2$, and $\bar{\mathbf{x}} \sim N_p(\boldsymbol{\mu}, \Sigma_I/N)$, where $N = n+1$. It can be shown as in Section 2.8 that these statistics are complete. Hence, we consider test procedures based on the above complete sufficient statistics only.

Let us define the parametric subregions as follows:

$$\Omega_1 = \{(\boldsymbol{\mu}, \Sigma_I) : \mu_1 = \cdots = \mu_p, \sigma_1^2 > 0, \sigma_2^2 > 0\},$$

$$\Omega_2 = \{(\boldsymbol{\mu}, \Sigma_I) : \mu_1 = \cdots = \mu_k, \mu_{k+1} = \cdots = \mu_p, \\ p-1 \geqslant k \geqslant 2, \sigma_1^2 > 0, \sigma_2^2 > 0\},$$

$$\Omega_3 = \{(\boldsymbol{\mu}, \Sigma_I) : \mu_1 = \cdots = \mu_k, \sigma_1^2 > 0, \sigma_2^2 > 0, \\ p \geqslant k \geqslant 2 \text{ and } -\infty < \mu_j < \infty \text{ for } j = k+1, \ldots, p\},$$

and

$$\Omega_4 = \{(\boldsymbol{\mu}, \Sigma_I) : -\infty < \mu_j < \infty, j = 1, 2, \ldots, p, \sigma_1^2 > 0, \sigma_2^2 > 0\}.$$

Consider the following testing problems:

(a) $\Omega_1$ vs $\Omega_4$, (b) $\Omega_1$ vs $\Omega_2$, (c) $\Omega_1$ vs $\Omega_3$,

(d) $\Omega_2$ vs $\Omega_3$, (e) $\Omega_2$ vs $\Omega_4$, (f) $\Omega_3$ vs $\Omega_4$.

Problem (a) has been considered by Wilks (1946) and is a special case of Problem (f). Problems (b) to (e) have been considered by Olkin and Shrikhande (1970). Notice that the above subregions $\Omega_1$, $\Omega_2$, and $\Omega_3$ have certain structure on the mean vector $\boldsymbol{\mu}$. For if $\boldsymbol{\mu} = B\boldsymbol{\xi}$, where $B$ is a $p \times r$ matrix of rank $r$, then

$$\text{for } \Omega_1, \quad B' = (1, \ldots, 1) \equiv \mathbf{e}',$$

$$\text{for } \Omega_2, \quad B' = \begin{pmatrix} 1 & 1 & \cdots & 1 & 0 & 0 & \cdots & 0 \\ 0 & 0 & \cdots & 0 & 1 & 1 & \cdots & 1 \end{pmatrix},$$

and

$$\text{for } \Omega_3, \quad B' = \left(\begin{array}{cccc|c} 1 & 1 & \cdots & 1 & \mathbf{0}' \\ \hline & & 0 & & I_{p-k} \end{array}\right).$$

In all these regions, there is one property—if for any vector $\mathbf{a}$, $\mathbf{a}'B = \mathbf{0}$, then $\mathbf{a}'\mathbf{e} = 0$, and so $\mathbf{a}'\Sigma_I\mathbf{a} = \sigma_1^2(\mathbf{a}'\mathbf{a})$. Thus, let us replace the parametric subregions $\Omega_1$, $\Omega_2$, and $\Omega_3$ by the following two subregions:

$$\Omega_{(1)}: \{(\boldsymbol{\mu}, \Sigma_I): \boldsymbol{\mu} = B\boldsymbol{\xi},\ -\infty < \xi_i < \infty\ (i=1,2,\ldots,r), \sigma_1^2 > 0,\ \sigma_2^2 > 0\},$$

where $\mathbf{a}'B = \mathbf{0} \Rightarrow \mathbf{a}'\mathbf{e} = 0$ for any $\mathbf{a} \neq \mathbf{0}$, $B: p \times r$ and $\rho(B) = r$, and

$$\Omega_{(2)}: \{(\boldsymbol{\mu}, \Sigma_I): \boldsymbol{\mu} = B_1\boldsymbol{\eta},\ -\infty < \eta_i < \infty\ (i=1,2,\ldots,q > r), \sigma_1^2 > 0,\ \sigma_2^2 > 0\}.$$

Then the above six problems reduce to the following two problems for testing of hypotheses:

(i) $\Omega_{(1)}$ vs $\Omega_4$ and

(ii) $\Omega_{(1)}$ vs $\Omega_{(2)}$

when $\rho(B, B_1) = \rho(B_1) = q > r$. Observe (a), (e), and (f) are particular cases of (i), and (b), (c), and (d) are particular cases of (ii). We shall develop the test procedures for (i) and (ii).

### *4.5.3 Test Procedure for Problem* (i)

Using complete sufficient statistics under $\Omega_4$, the likelihood function for $(\boldsymbol{\mu}, \sigma_1^2, \sigma_2^2)$ is given by

$$(4.5.9)\quad L(\boldsymbol{\mu}, \sigma_1^2, \sigma_2^2) = c_1 \sigma_1^{-N(p-1)} \sigma_2^{-N} v_1^{\frac{1}{2}n(p-1)-1} v_2^{\frac{1}{2}n-1} \times \exp\left(-\frac{1}{2}\frac{v_1}{\sigma_1^2} - \frac{1}{2}\frac{v_2}{\sigma_2^2} - \tfrac{1}{2}N(\bar{\mathbf{x}} - \boldsymbol{\mu})'\Sigma_I^{-1}(\bar{\mathbf{x}} - \boldsymbol{\mu})\right),$$

where

$$N = n+1, \qquad v_1 + v_2 = \operatorname{tr} V,$$
$$v_2 = \mathbf{e}'V\mathbf{e}/p, \qquad \Sigma_I = \sigma_1^2 I + (\sigma_2^2 - \sigma_1^2)\mathbf{e}\mathbf{e}'/p,$$

and $c_1$ is a constant depending on $n$ and $p$ only. It is easy to see that

$$(4.5.10)\quad \sup_{\Omega_4} L(\boldsymbol{\mu}, \sigma_1^2, \sigma_2^2) = c_2 \left(\frac{v_1}{N(p-1)}\right)^{-\frac{1}{2}N(p-1)} \left(\frac{v_2}{N}\right)^{-\frac{1}{2}N} v_1^{\frac{1}{2}n(p-1)-1} v_2^{\frac{1}{2}n-1},$$

where $c_2 = c_1 e^{-\frac{1}{2}Np}$. To obtain supremum over $\Omega_{(1)}$, we note from Theorem 4.4.1(iv) that

$$\Sigma_I^{-1} = \sigma_1^{-2} I - \frac{(\sigma_2^2 - \sigma_1^2)\mathbf{e}\mathbf{e}'}{p\sigma_1^2\sigma_2^2}.$$

Also, we have the following identities:

$$I \equiv \left[I - B(B'B)^{-1}B'\right] + B(B'B)^{-1}B',$$
$$0 \equiv \left[I - B(B'B)^{-1}B'\right]B.$$

Since $\mathbf{a}'B = \mathbf{0} \Rightarrow \mathbf{a}'\mathbf{e} = 0$ for all nonnull vectors $\mathbf{a}$, it follows from the above that

$$\mathbf{0} = \left[I - B(B'B)^{-1}B'\right]\mathbf{e}.$$

Then

$$\mathbf{e} = B(B'B)^{-1}B'\mathbf{e}.$$

Hence

$$(\bar{\mathbf{x}} - B\boldsymbol{\xi})'\Sigma_1^{-1}(\bar{\mathbf{x}} - B\boldsymbol{\xi})$$
$$= \sigma_1^{-2}(\bar{\mathbf{x}} - B\boldsymbol{\xi})'(\bar{\mathbf{x}} - B\boldsymbol{\xi}) - \frac{(\sigma_2^2 - \sigma_1^2)(\bar{\mathbf{x}} - B\boldsymbol{\xi})'\mathbf{e}\mathbf{e}'(\bar{\mathbf{x}} - B\boldsymbol{\xi})}{p\sigma_1^2\sigma_2^2}$$
$$= \frac{1}{\sigma_1^2}\bar{\mathbf{x}}'\left[I - B(B'B)^{-1}B'\right]\bar{\mathbf{x}}$$
$$+ (\bar{\mathbf{x}} - B\boldsymbol{\xi})'\left[\frac{1}{\sigma_1^2}B(B'B)^{-1}B' - \frac{\sigma_2^2 - \sigma_1^2}{p\sigma_1^2\sigma_2^2}\right.$$
$$\left. \times B(B'B)^{-1}B'\mathbf{e}\mathbf{e}'B(B'B)^{-1}B'\right](\bar{\mathbf{x}} - B\boldsymbol{\xi})$$
$$= \frac{1}{\sigma_1^2}\bar{\mathbf{x}}'\left[I - B(B'B)^{-1}B'\right]\bar{\mathbf{x}} + (\mathbf{y} - \boldsymbol{\xi})'\left[\frac{1}{\sigma_1^2}B'B - \frac{\sigma_2^2 - \sigma_1^2}{p\sigma_1^2\sigma_2^2}B'\mathbf{e}\mathbf{e}'B\right](\mathbf{y} - \boldsymbol{\xi}),$$

where $\mathbf{y} = (B'B)^{-1}B'\bar{\mathbf{x}}$ and

$$B'\left(\frac{1}{\sigma_1^2}I - \frac{\sigma_2^2 - \sigma_1^2}{p\sigma_1^2\sigma_2^2}\mathbf{e}\mathbf{e}'\right)B$$

is positive definite. Hence, under $\Omega_{(1)}$, the likelihood $L(\boldsymbol{\mu}, \sigma_1^2, \sigma_2^2)$ is

(4.5.11) $L(B\boldsymbol{\xi}, \sigma_1^2, \sigma_2^2)$

$$= c_1\sigma_1^{-N(p-1)}\sigma_2^{-N}v_1^{\frac{1}{2}n(p-1)-1}v_2^{\frac{1}{2}n-1}\exp\left(-\frac{v_2}{2\sigma_2^2}\right)$$
$$\times \exp\left(-\frac{N}{2}(\mathbf{y} - \boldsymbol{\xi})'B'\Sigma_I^{-1}B(\mathbf{y} - \boldsymbol{\xi}) - \frac{1}{2\sigma_1^2}\left\{v_1 + N\bar{\mathbf{x}}'\left[I - B(B'B)^{-1}B'\right]\bar{\mathbf{x}}\right\}\right).$$

From the above, it is easy to see that

$$(4.5.12)\quad \operatorname*{Sup}_{\Omega_{(1)}} L(\boldsymbol{\mu},\sigma_1^2,\sigma_2^2)=c_2\left(\frac{v_1+N\bar{\mathbf{x}}'\left[I-B(B'B)^{-1}B'\right]\bar{\mathbf{x}}}{N(p-1)}\right)^{-\frac{1}{2}N(p-1)}$$

$$\times v_1^{\frac{1}{2}n(p-1)-1}(v_2/N)^{-\frac{1}{2}N}v_2^{\frac{1}{2}n-1}.$$

Using the above results, the likelihood ratio is

$$(4.5.13)\qquad \lambda=\frac{\operatorname*{Sup}_{\Omega_{(1)}} L(\boldsymbol{\mu},\sigma_1^2,\sigma_2^2)}{\operatorname*{Sup}_{\Omega_4} L(\boldsymbol{\mu},\sigma_1^2,\sigma_2^2)}$$

$$=\left[\frac{v_1}{v_1+N\bar{\mathbf{x}}'\left[I-B(B'B)^{-1}B'\right]\bar{\mathbf{x}}}\right]^{\frac{1}{2}N(p-1)},$$

and the likelihood-ratio test procedure is

$$(4.5.14)\qquad \text{reject } \Omega_{(1)} \text{ vs } \Omega_4 \quad \text{if} \quad N\bar{\mathbf{x}}'\left[I-B(B'B)^{-1}B'\right]\bar{\mathbf{x}}/v_1 \geqslant c,$$

where $c$ is a constant to be determined from $P[F_{p-r,n(p-1)}>n(p-1)c/(p-r)]=\alpha$, because $n(p-1)N\bar{\mathbf{x}}'[I-B(B'B)^{-1}B']\bar{\mathbf{x}}/(p-r)v_1$ is distributed as noncentral $F_{p-r,n(p-1)}(\delta)$ with $(p-r,n(p-1))$ d.f. and noncentrality parameter $\delta=N\boldsymbol{\mu}'[I-B(B'B)^{-1}B']\boldsymbol{\mu}/\sigma_1^2$ and $\delta=0$ under $\Omega_{(1)}$.

The above test procedure can be viewed as a union–intersection test procedure from the following considerations: If $H\in\Omega_{(1)}$ and $A\in\Omega_4$, then

$$H=\bigcap_{\mathbf{a}} H_{\mathbf{a}}\left(\mathbf{a}'\boldsymbol{\mu}=0,\ \sigma_1^2>0,\sigma_2^2>0\right) \quad\text{and}\quad A=\bigcup_{\mathbf{a}} A_{\mathbf{a}}\left(\mathbf{a}'\boldsymbol{\mu}\neq 0,\ \sigma_1^2>0,\sigma_2^2>0\right),$$

where $\bigcap_{\mathbf{a}}$ and $\bigcup_{\mathbf{a}}$ denote the intersection and the union over all nonnull vectors $\mathbf{a}$ for which $\mathbf{a}'B=\mathbf{0}$. We may observe that $\mathbf{a}'\bar{\mathbf{x}}$ is distributed as $N(\mathbf{a}'\boldsymbol{\mu},\mathbf{a}'\mathbf{a}\sigma_1^2/N)$ because $\mathbf{a}'B=\mathbf{0} \Rightarrow \mathbf{a}'\mathbf{e}=0$. Hence, using univariate theory,

$$H_{\mathbf{a}} \text{ is rejected against } A_{\mathbf{a}} \quad \text{if} \quad N(\mathbf{a}'\bar{\mathbf{x}})^2/\mathbf{a}'\mathbf{a}v_1 \geqslant c_0,$$

where $c_0$ is a constant depending on $\alpha$, $n$, and $p$ only. By the union–intersection test procedure,

$$\text{reject } H \text{ vs } A \quad \text{if} \quad \operatorname*{Sup}_{\substack{\mathbf{a}\\ \text{subject to } \mathbf{a}'B=\mathbf{0}}} \frac{N(\mathbf{a}'\bar{\mathbf{x}})^2}{\mathbf{a}'\mathbf{a}v_1} \geqslant c_0,$$

and the use of Problem 1.38 gives a union–intersection test procedure that is the same as derived by the likelihood-ratio principle. However, by the latter method the simultaneous confidence bounds of $\mathbf{a}'\boldsymbol{\mu}$ for all nonnull vectors $\mathbf{a}$ subject to $\mathbf{a}'B=\mathbf{0}$ can easily be obtained as in Subsection 1.3.6.

These are given by

$$\mathbf{a}'\bar{\mathbf{x}}-(c_0\mathbf{a}'\mathbf{a}v_1/N)^{1/2}\leqslant\mathbf{a}'\boldsymbol{\mu}\leqslant\mathbf{a}'\bar{\mathbf{x}}+(c_0\mathbf{a}'\mathbf{a}v_1/N)^{1/2}$$

for all nonnull vectors $\mathbf{a}$ subject to $\mathbf{a}'B=\mathbf{0}$. Such Scheffe-type confidence intervals are given by Bhargava and Srivastava (1973). For Tukey-type confidence intervals, we have to consider the test procedure

$$\text{reject } H \text{ vs } A \text{ if} \qquad \sup_{\substack{\mathbf{a}\\ \text{subject to } \mathbf{a}'B=\mathbf{0}}}\left\{\frac{\sqrt{N}\,|\mathbf{a}'\bar{\mathbf{x}}|}{\sqrt{v_1}\,\sum_{j=1}^{p}|a_j|}\right\}\geqslant c_1$$

and obtain the confidence interval as

$$\mathbf{a}'\bar{\mathbf{x}}-\left(c_1\sum_{j=1}^{p}|a_j|\frac{\sqrt{v_1}}{\sqrt{N}}\right)\leqslant\mathbf{a}'\boldsymbol{\mu}\leqslant\mathbf{a}'\bar{\mathbf{x}}+\left(c_1\sum_{j=1}^{p}|a_j|\frac{\sqrt{v_1}}{\sqrt{N}}\right).$$

For the comparison of these intervals, see Bhargava and Srivastava (1973).

### *4.5.4 Likelihood-Ratio Procedure for Problem (ii)*

It may be observed that we have already obtained $\sup_{\Omega_{(1)}}L(\boldsymbol{\mu},\sigma_1^2,\sigma_2^2)$ in Subsection 4.4.3 and it is given by Eq. (4.5.12). In the same way, we can establish

$$\text{(4.5.15)}\quad \sup_{\Omega_{(2)}}L(\boldsymbol{\mu},\sigma_1^2,\sigma_2^2)=c_2\left(\frac{v_1+N\bar{\mathbf{x}}'\left[I-B_1(B_1'B_1)^{-1}B_1'\right]\bar{\mathbf{x}}}{N(p-1)}\right)^{\frac{1}{2}N(p-1)}$$
$$\times v_1^{\frac{1}{2}n(p-1)-1}(v_2/N)^{-\frac{1}{2}N}v_2^{\frac{1}{2}n-1}.$$

Hence, using Eqs. (4.5.12) and (4.5.15), the likelihood ratio is

$$\text{(4.5.16)}\qquad \lambda=\frac{\sup_{\Omega_{(1)}}L(\boldsymbol{\mu},\sigma_1^2,\sigma_2^2)}{\sup_{\Omega_{(2)}}L(\boldsymbol{\mu},\sigma_1^2,\sigma_2^2)}$$
$$=\left[\frac{v_1+N\bar{\mathbf{x}}'\left[I-B_1(B_1'B_1)^{-1}B_1'\right]\bar{\mathbf{x}}}{v_1+N\bar{\mathbf{x}}'\left[I-B(B'B)^{-1}B'\right]\bar{\mathbf{x}}}\right]^{\frac{1}{2}N(p-1)},$$

and the likelihood-ratio test procedure is

$$\text{(4.5.17)}\qquad \text{reject } \Omega_{(1)} \text{ vs } \Omega_{(2)} \quad \text{if} \quad N\bar{\mathbf{x}}'P\bar{\mathbf{x}}/v\geqslant c,$$

where $v=v_1+N\bar{\mathbf{x}}'[I-B_1(B_1'B_1)^{-1}B_1']\bar{\mathbf{x}}$, $P=B_1(B_1'B_1)^{-1}B_1'-B(B'B)^{-1}B'$, and $c$ is to be determined from $P(N\bar{\mathbf{x}}'P\bar{\mathbf{x}}/v\geqslant c|\Omega_{(1)})=\alpha$.

Since $\rho(B_1,B)=\rho(B_1)=q$, there exists a matrix $Z:q\times r$ such that $B=B_1Z$. This gives a solution for $Z$ as $Z=(B_1'B_1)^{-1}B_1'B$. Hence

$$\left[I-B_1(B_1'B_1)^{-1}B_1'\right]B=0 \quad \text{or} \quad B=B_1(B_1'B_1)^{-1}B_1'B.$$

Since $\mathbf{e}=B(B'B)^{-1}B'\mathbf{e}$, we have

$$\mathbf{e}=B_1(B_1'B_1)^{-1}B_1'B(B'B)^{-1}B'\mathbf{e}=B_1(B_1'B_1)^{-1}B_1'\mathbf{e}.$$

If $P=B_1(B_1'B_1)^{-1}B_1'-B(B'B)^{-1}B'$ and $Q=I-B_1(B_1'B_1)^{-1}B_1'$, then from the above expressions it is easy to verify that

$$Q\mathbf{e}=P\mathbf{e}=\mathbf{0},\quad P^2=P,\quad Q^2=Q,\quad PQ=0,$$
$$\operatorname{tr}Q=p-q,\quad \operatorname{tr}P=q-r.$$

This means that

$$P\Sigma_I Q=0,\qquad \frac{P}{\sigma_1^2}\Sigma_I\frac{P}{\sigma_1^2}=\frac{P}{\sigma_1^2},\qquad \frac{Q}{\sigma_1^2}\Sigma_I\frac{Q}{\sigma_1^2}=\frac{Q}{\sigma_1^2}.$$

Since $\bar{\mathbf{x}}\sim N_p(\boldsymbol{\mu},N^{-1}\Sigma_I)$, it is easy to show, using Corollaries 2.11.2 and 2.11.4, that $N\bar{\mathbf{x}}'P\bar{\mathbf{x}}/\sigma_1^2$ and $N\bar{\mathbf{x}}'Q\bar{\mathbf{x}}/\sigma_1^2$ are independently distributed as noncentral chi squares with respective d.f. $q-r$ and $p-q$ and with respective noncentral parameters $\delta_1=N\boldsymbol{\mu}'P\boldsymbol{\mu}/\sigma_1^2$ and $\delta_2=N\boldsymbol{\mu}'Q\boldsymbol{\mu}/\sigma_1^2$. Since $v_1/\sigma_1^2\sim\chi^2_{n(p-1)}$ and is independently distributed of $\bar{\mathbf{x}}$, we get from the above the following result: $[n(p-1)+p-q]N\bar{\mathbf{x}}P\bar{\mathbf{x}}/(v_1+N\bar{\mathbf{x}}'Q\bar{\mathbf{x}})(q-r)$ is distributed as noncentral $F_{q-r,n(p-1)+p-q}(\delta_1,\delta_2)$, where $\delta_1=N\boldsymbol{\mu}'P\boldsymbol{\mu}/\sigma_1^2$ and $\delta_2=N\boldsymbol{\mu}'Q\boldsymbol{\mu}/\sigma_1^2$. Under $\Omega_{(2)}$, $\delta_2=0$ and $\delta_1=N\boldsymbol{\mu}'(I-B(B'B)^{-1}B')\boldsymbol{\mu}/\sigma_1^2$; and under $\Omega_{(1)}$, $\delta_1=\delta_2=0$. Hence, we can calculate $c$ of Eq. (4.5.17) from

$$P\left(F_{q-r,n(p-1)+p-q}\geqslant c\frac{n(p-1)+p-q}{q-r}\right)=\alpha.$$

When $p=q$, we get the result of Subsection 4.5.3.

It can be shown that the test procedure (4.5.17) can also be obtained by the union–intersection principle, and this is left as Problem 4.3.

### *4.5.5 Solutions of Problems* (a) *to* (f)

In this subsection, we write down the solutions of the six problems (a) to (f) raised in Subsection 4.5.2.

(a) Reject $\Omega_1$ vs $\Omega_4$ if $F_1=N[\bar{\mathbf{x}}'\bar{\mathbf{x}}-p^{-1}(\bar{\mathbf{x}}'\mathbf{e})^2]/v_1\geqslant c_1$, where $P(F_1\geqslant c_1|\Omega_1)=\alpha$, and $n(p-1)F_1/(p-1)$ is distributed as noncentral $F$ with $p-1$ and $n(p-1)$ d.f. and noncentral parameter $\delta_1=N[\boldsymbol{\mu}'\boldsymbol{\mu}-p^{-1}(\boldsymbol{\mu}'\mathbf{e})^2]/\sigma_1^2$.

(b) Before giving the test procedure, let us write $\mathbf{e}'\bar{\mathbf{x}}/p=z$, $z_1=(\bar{x}_1+\cdots+\bar{x}_k)/k$, and $z_2=(\bar{x}_{k+1}+\cdots+\bar{x}_p)/(p-k)$; then $pz=kz_1+(p-k)z_2$, $\bar{\mathbf{x}}'P\bar{\mathbf{x}}=kz_1^2+(p-k)z_2^2-pz^2=k(p-k)(z_1-z_2)^2/p$, and $v_1+N\bar{\mathbf{x}}'(I-B_1(B_1'B_1)^{-1}B_1')\bar{\mathbf{x}}=\operatorname{tr}V-p^{-1}(\mathbf{e}'V\mathbf{e})+N\bar{\mathbf{x}}'\bar{\mathbf{x}}-k_1z_1^2-(p-k)z_2^2=v$ (say). Then

$$\text{reject } \Omega_1 \text{ vs } \Omega_2 \quad \text{if} \quad F_2=\frac{Nk(p-k)(z_1-z_2)^2}{pv}\geqslant c_2,$$

where $(np-n+p-2)F_2$ is distributed as noncentral $F$ with $(1,np-n+p-2)$ d.f. and noncentral parameter $Nk(p-k)(\bar{\mu}_1-\bar{\mu}_2)^2/p\sigma_1^2$, $\bar{\mu}_1=\Sigma_{i=1}^k\mu_i/k$, $\bar{\mu}_2=\Sigma_{i=k+1}^p\mu_i/(p-k)$, and $\bar{\mu}=\Sigma_{i=1}^p\mu_i/p$.

(c) Let $\mathbf{y}_2'=(\bar{x}_{k+1},\ldots,\bar{x}_p)$ and $\mathbf{y}_1'=(\bar{x}_1,\bar{x}_2,\ldots,\bar{x}_k)$. Then

$$\text{reject } \Omega_1 \text{ vs } \Omega_3 \quad \text{if} \quad F_3=N\big(kz_1^2+\mathbf{y}_2'\mathbf{y}_2-pz^2\big)/v_c\geqslant c_3,$$

where $v_c=v_1+N(\mathbf{y}_1'\mathbf{y}_1-kz_1^2)$, and $(np-n+k-1)F_3/(p-k)$ is distributed as noncentral $F$ with $(p-k,np-n+k-1)$ d.f. and noncentrality parameter $\delta=N(k\bar{\mu}_1^2+\Sigma_{i=k+1}^p\mu_i^2-p\bar{\mu}^2)/\sigma_1^2$.

(d) Reject $\Omega_2$ vs $\Omega_3$ if $F_4=N[\mathbf{y}_2'\mathbf{y}_2-(p-k)z_2^2]/v_c\geqslant c_4$, where $(np-n+k-1)F_4/(p-k-1)$ is distributed as noncentral $F$ with $(p-k-1,np-n+k-1)$ d.f. and noncentral parameter $N(\Sigma_{i=k+1}^p\mu_i^2-(p-k)\bar{\mu}_2^2)\sigma_1^2$.

(e) Reject $\Omega_2$ vs $\Omega_4$ if $F_5=N[\mathbf{y}_1'\mathbf{y}_1+\mathbf{y}_2'\mathbf{y}_2-kz_1^2-(p-k)z_2^2]/v_1\geqslant c_5$, where $n(p-1)F_5/(p-2)$ is distributed as noncentral $F$ with $(p-2,n(p-1))$ d.f. and noncentral parameter $N[\boldsymbol{\mu}'\boldsymbol{\mu}-k\bar{\mu}_1^2-(p-k)\bar{\mu}_2^2]/\sigma_1^2$.

(f) Reject $\Omega_3$ vs $\Omega_4$ if $F_6=N(\mathbf{y}_1'\mathbf{y}_1-kz_1^2)/v_1\geqslant c_6$, where $n(p-1)F_6/(k-1)$ is distributed as noncentral $F$ with $(k-1,n(p-1))$ d.f. and noncentral parameter $N(\Sigma_{i=1}^k\mu_i^2-k\bar{\mu}_1^2)/\sigma_1^2$.

*Note*: The confidence bounds and other properties of test procedures given in this section can be given in the same way as those developed in Subsection 4.5.3, and they are left as exercises to the readers.

## 4.6 Two-Sample Problem

Let $\mathbf{x}_1,\ldots,\mathbf{x}_m$, $\mathbf{x}_{m+1},\ldots,\mathbf{x}_N$ be independently distributed, where $\mathbf{x}_i\sim N_p(\mu_1,\Sigma)$, $i=1,2,\ldots,m$, and $\mathbf{x}_i\sim N_p(\boldsymbol{\mu}_2,\Sigma)$, $i=m+1,\ldots,N$. Then the problem of testing $\boldsymbol{\mu}_1=\boldsymbol{\mu}_2$ can easily be reduced to the one-sample problem of Section 4.3, and turns out to be Hotelling's $T^2$ test based on $N-2$ d.f. For tests when the point of change $m$ is not known, see Sen and Srivastava (1973, 1975a–c). The $k$-sample problem is left as Problem 4.9 and shown to be a special case of the regression model in Chapter 6.

## 4.7 Behrens–Fisher Problem

In this section, we consider the problem of testing the equality of mean vectors from two normal populations with unequal covariance matrices. This problem is known as the Behrens–Fisher problem. Several exact and approximate solutions are available [see references in Yao (1965)]. We, however, present here an exact solution at the sacrifice of degrees of freedom, which is a multivariate extension of the univariate solution given by Scheffé (1943); the multivariate extension was given by Bennett (1951). While little is known about any optimal properties of multivariate tests, in the univariate case this technique gives the shortest confidence intervals obtained by using the $t$ distribution. It should be mentioned that one should resort to the Behrens–Fisher situation only if the variances appear quite far apart; the usual $t$ (or $T^2$) tests should be applied if there is even a slight departure from the equality of variances. Some results in this direction have been obtained by Carter, Khatri, and Srivastava (1979).

Let $\{\mathbf{x}_\alpha^{(i)}\}$ $(\alpha=1,2,\ldots,N_i;\ i=1,2)$ be samples from $N(\boldsymbol{\mu}^{(i)},\Sigma_i)$, where $(\boldsymbol{\mu}^{(i)},\Sigma_i)\in\Omega\equiv\{(\boldsymbol{\mu}^{(i)},\Sigma_i):-\infty<\mu_\alpha^{(i)}<\infty,\ \alpha=1,2,\ldots,p,\ \Sigma_i>0,\ i=1,2\}$. Let $\Omega_0\equiv\{(\boldsymbol{\mu}^{(i)},\Sigma_i,i=1,2):\boldsymbol{\mu}^{(1)}=\boldsymbol{\mu}^{(2)},\ \Sigma_i>0,\ i=1,2\}$. We wish to test the hypothesis $H:(\boldsymbol{\mu}^{(i)},\ \Sigma_i,\ i=1,2)\in\Omega_0$ against the alternative $A:(\boldsymbol{\mu}^{(i)},\ \Sigma_i,\ i=1,2)\in\Omega$. If $N_1=N_2\equiv N$, then a test for $H$ can be taken as

$$T^2=N\bar{\mathbf{y}}'S^{-1}\bar{\mathbf{y}},\qquad \bar{\mathbf{y}}=N^{-1}\sum_{\alpha=1}^{N}\mathbf{y}_\alpha,$$

where

$$\mathbf{y}_\alpha=\mathbf{x}_\alpha^{(1)}-\mathbf{x}_\alpha^{(2)},\quad \alpha=1,2,\ldots,N,\qquad \bar{\mathbf{y}}=\bar{\mathbf{x}}^{(1)}-\bar{\mathbf{x}}^{(2)},$$

and

$$S=(N-1)^{-1}\sum_{\alpha=1}^{N}\left(\mathbf{x}_\alpha^{(1)}-\mathbf{x}_\alpha^{(2)}-\bar{\mathbf{x}}^{(1)}+\bar{\mathbf{x}}^{(2)}\right)\left(\mathbf{x}_\alpha^{(1)}-\mathbf{x}_\alpha^{(2)}-\bar{\mathbf{x}}^{(1)}+\bar{\mathbf{x}}^{(2)}\right)',$$

and where $T^2$ has Hotelling's $T^2$ distribution with $N-1$ d.f. Thus, we have lost $N-1$ d.f. from the situation where $\Sigma_1=\Sigma_2$. To extend the result when $N_1\neq N_2$, we proceed as follows:

Without loss of generality, we may assume $N_1\leqslant N_2$. Define

$$\mathbf{y}_\alpha=\mathbf{x}_\alpha^{(1)}-\left(N_1N_2^{-1}\right)^{1/2}\mathbf{x}_\alpha^{(2)}+\left(N_1N_2\right)^{-1/2}\sum_{\beta=1}^{N_1}\mathbf{x}_\beta^{(2)}-N_2^{-1}\sum_{\beta=1}^{N_2}\mathbf{x}_\beta^{(2)},$$
$$\alpha=1,2,\ldots,N_1.$$

Then

$$E(\mathbf{y}_\alpha)=\boldsymbol{\mu}^{(1)}-\boldsymbol{\mu}^{(2)}, \qquad \alpha=1,2,\dots,N_1,$$

and

$$\text{Cov}(\mathbf{y}_\alpha)=\Sigma_1+(N_1N_2^{-1})\Sigma_2, \qquad \alpha=1,2,\dots,N_1,$$

and the $\mathbf{y}_\alpha$'s are independently normally distributed random $p$-vectors. Hence, the problem reduces to that of testing that the mean of $\mathbf{y}$ is zero on the basis of a random sample of size $N_1$. The test statistic can now be taken as Hotelling's $T^2$ based on $N_1-1$ d.f. Thus we have lost $N_2-1$ d.f. as compared to the situation when it is given that $\Sigma_1=\Sigma_2$.

## Problems

**4.1** Develop a step-down procedure for testing the hypothesis $H(\boldsymbol{\mu}=\mathbf{0},\ \Sigma>0)$ vs $A(\boldsymbol{\mu}\neq\mathbf{0},\ \Sigma>0)$ when a random sample of size $N$ is taken from a normal population with mean $\boldsymbol{\mu}$ and covariance matrix $\Sigma>0$.

**4.2** Let $\mathbf{y}\sim N_p(\boldsymbol{\mu},I)$. Obtain the likelihood-ratio test (LRT) and its distribution for testing $H:\boldsymbol{\mu}=\mathbf{0}$ vs $A:\boldsymbol{\mu}\neq\mathbf{0}$. Show that the LRT is uniformly the most powerful invariant under the group of orthogonal transformations. Show that it is a union–intersection test procedure.

**4.3** Show that the test procedure given in Eq. (4.5.4) can also be obtained by the union–intersection principle.

**4.4** Let $\bar{\mathbf{x}}_i$, $V_i$, $i=1,2,\dots,q$, be all independently distributed, where $\bar{\mathbf{x}}_i\sim N_p(\boldsymbol{\mu}_i,N_i^{-1}\Sigma)$ and $V_i\sim W_p(\Sigma,n_i)$, $i=1,2,\dots,q$. Devise a test for testing the hypothesis $H:\Sigma_{i=1}^{q}a_i\boldsymbol{\mu}_i=\mathbf{0}$ vs $A\neq H$, where the $a_i$'s are given numbers, and obtain the simultaneous confidence bounds on $\Sigma_{i=1}^{p}a_i\boldsymbol{\mu}_i$.

**4.5** Let $\mathbf{y}\sim N_p(\boldsymbol{\mu},\Sigma)$ and $V\sim W_p(\Sigma,n)$ be independently distributed, where

$$\mathbf{y}'=(\overset{r}{\mathbf{y}_1'},\overset{s}{\mathbf{y}_2'}) \quad\text{and}\quad \boldsymbol{\mu}'=(\overset{r}{\boldsymbol{\mu}_1'},\overset{s}{\boldsymbol{\mu}_2'}), \qquad r+s=p.$$

Find the likelihood-ratio and union–intersection test procedures for testing $\boldsymbol{\mu}_2=\mathbf{0}$ vs $\boldsymbol{\mu}_2\neq\mathbf{0}$. Show that the likelihood-ratio test is uniformly most powerful invariant similar test.

**4.6** Let $\mathbf{y}\sim N_p(\boldsymbol{\mu},\Sigma)$ and $V\sim W_p(\Sigma,n)$ be independently distributed, where

$$\mathbf{y}'=(\overset{r}{\mathbf{y}_1'},\overset{s}{\mathbf{y}_2'},\overset{t}{\mathbf{y}_3'}) \quad\text{and}\quad \boldsymbol{\mu}'=(\overset{r}{\boldsymbol{\mu}_1'},\overset{s}{\boldsymbol{\mu}_2'},\overset{t}{\boldsymbol{\mu}_3'})$$

with $r+s+t=p$. Suppose it is given that $\boldsymbol{\mu}_3=\mathbf{0}$. Find the likelihood-ratio test and its distribution for testing $H:\boldsymbol{\mu}_2=\mathbf{0}$ ($\boldsymbol{\mu}_1$ anything, $\Sigma>0$) vs $A\neq H$. Show that the likelihood-ratio test is uniformly most powerful invariant similar test under the group of certain transformations.

**4.7** Let $\mathbf{y}_i \sim N_p(\boldsymbol{\mu}_i, \Sigma)$ and $V_i \sim W_p(\Sigma, n_i)$, $i=1,2$, be all independently distributed, where

$$\mathbf{y}_i' = (\overset{r}{\dot{\mathbf{y}}_i'}, \overset{s}{\ddot{\mathbf{y}}_i'}) \quad \text{and} \quad \boldsymbol{\mu}_i' = (\overset{r}{\dot{\boldsymbol{\mu}}_i}, \overset{s}{\ddot{\boldsymbol{\mu}}_i}), \qquad r+s=p.$$

Suppose it is known that $\dot{\boldsymbol{\mu}}_1 = \dot{\boldsymbol{\mu}}_2$. Derive the likelihood-ratio test and its distribution for this problem. Show that this test can be obtained by the union–intersection principle.

**4.8** Let $\mathbf{x}_\alpha^{(i)}$ $(\alpha=1,2,\ldots,N_i,\ i=1,2,\ldots,q)$ be independent samples from $N(\boldsymbol{\mu}^{(i)}, \Sigma_i)$ $(i=1,2,\ldots,q)$ respectively. Devise a test for testing the hypothesis $H: \Sigma_{i=1}^q a_i \boldsymbol{\mu}^{(i)} = \mathbf{0}$ against $A \neq H$, where the $a_i$'s are given numbers. [Hint: Without loss of generality, let $N_1$ be the smallest of $N_i$. Define

$$\mathbf{y}_\alpha = a\mathbf{x}_\alpha^{(1)} + \sum_{i=2}^{q} a_i \left(\frac{N_1}{N_i}\right)^{1/4} \left[\mathbf{x}_\alpha^{(i)} - N_1^{-1} \sum_{\beta=1}^{N_1} \mathbf{x}_\beta^{(i)} + (N_1 N_i)^{-1/2} \sum_{\gamma=1}^{N_i} x_\gamma^{(i)}\right],$$

$\alpha = 1,2,\ldots,N_1$.]

**4.9** Let $\mathbf{y}_i \sim N_p(\boldsymbol{\mu}_i, \Sigma)$ and $V_i \sim W_p(\Sigma, n_i)$, $i=1,2,\ldots,q$, be all independently distributed. Obtain the likelihood-ratio test for $H: \boldsymbol{\mu}^{(1)} = \cdots = \boldsymbol{\mu}^{(q)}$ against $A \neq H$.

**4.10** Let $\mathbf{y} \sim N_p(\boldsymbol{\mu}, \Sigma)$ and $V \sim W_p(\Sigma, n)$ be independently distributed, and let

$$\boldsymbol{\mu}' = (\boldsymbol{\mu}_1', \boldsymbol{\mu}_2') \quad \text{and} \quad \Sigma = \begin{pmatrix} \Sigma_{11} & \Sigma_{12} \\ \Sigma_{12}' & \Sigma_{22} \end{pmatrix},$$

where $\mu_i: 1 \times p_i$, and $\Sigma_{ij}: p_i \times p_j$ $(i,j=1,2)$ are submatrices. Derive the likelihood-ratio and union–intersection test procedures for testing $H: (\boldsymbol{\mu}_1 = \Sigma_{12}\Sigma_{22}^{-1}\boldsymbol{\mu}_2,\ \Sigma > 0)$ against an alternative $A: (\boldsymbol{\mu}_1 \neq \Sigma_{12}\Sigma_{22}^{-1}\boldsymbol{\mu}_2,\ \Sigma > 0)$. Obtain the simultaneous confidence bounds on $\boldsymbol{\mu}_1 - \Sigma_{12}\Sigma_{22}^{-1}\boldsymbol{\mu}_2$.

# 5

# Linear Regression Models Estimation

## 5.1 Introduction

Let $\mathbf{x}$ be a vector of $N$ random variables. If, for a given $m \times N$ matrix $Z$ and for a fixed vector $\boldsymbol{\beta}$,

$$\mathbf{x} = Z'\boldsymbol{\beta} + \boldsymbol{\varepsilon},$$

where $\boldsymbol{\varepsilon}$ is a random vector such that $E\boldsymbol{\varepsilon} = \mathbf{0}$, then $\boldsymbol{\beta}$ is called a vector of *regression parameters* and the model is called the *regression model*. The matrix $Z$ is sometimes called a *design matrix*. For design of experiments, the elements of $Z$ are either 0 or 1. For experiments that involve concomitant variables or regression, the matrix $Z$ consists of observations on these elements, and the remaining ones are merely constants, either 0 or 1, depending on the structure of the design. The random vector $\boldsymbol{\varepsilon}$ is called a vector of errors. In this chapter, *no* assumption on the distribution of $\boldsymbol{\varepsilon}$ is made except for the existence of the first two moments. Thus our model is

$$(5.1.1) \qquad \mathbf{x} = Z'\boldsymbol{\beta} + \boldsymbol{\varepsilon}, \qquad E\boldsymbol{\varepsilon} = \mathbf{0}, \qquad E\boldsymbol{\varepsilon}\boldsymbol{\varepsilon}' = V,$$

where $Z$ is known, $\boldsymbol{\beta}$ is unknown, and $V$ may be partially known. If $V = \Sigma_{i=1}^{k} \sigma_i^2 V_i$, where $\sigma_1^2, \ldots, \sigma_k^2$ are unknown and distinct, and $V_1, \ldots, V_k$ are known p.s.d. matrices, then we can find a matrix $U = (U_1, U_2, \ldots, U_k)$ such that $U_i U_i' = V_i$ $(i = 1, 2, \ldots, k)$, and hence we can redefine the model (5.1.1) as

$$(5.1.2) \qquad \mathbf{x} = Z'\boldsymbol{\beta} + U\boldsymbol{\eta}, \qquad E\boldsymbol{\eta} = \mathbf{0}, \qquad E\boldsymbol{\eta}\boldsymbol{\eta}' = D_\sigma,$$

where $D_\sigma = \operatorname{diag}(\sigma_1^2 I_{c_1}, \sigma_2^2 I_{c_2}, \ldots, \sigma_k^2 I_{c_k})$, $\Sigma_{i=1}^{k} c_i \geqslant N$, and $c_i$ is the rank of $V_i$ $(i = 1, 2, \ldots, k)$. If $k = 1$, the model is known as the *fixed-effect model* or the *ordinary linear regression model*. When $k>1$, then we get the mixed-effect model as a special case of Eq. (5.1.2). For example, in a one-way mixed-effect model, we have

$$x_{ij}=\mu+\alpha_j+\varepsilon_{ij}, \qquad i=1,2,\ldots,n_j, \quad j=1,2,\ldots,k,$$

where $\mu$ is fixed, the random effects $\alpha_j$ and the errors $\varepsilon_{ij}$ are independently distributed with zero means, and $E\alpha_j^2=\sigma_1^2$ and $E\varepsilon_{ij}^2=\sigma^2$ for all $i$ and $j$. Let

$$\mathbf{x}'=(x_{11},\ldots,x_{1n_1},x_{21},\ldots,x_{2n_2},\ldots,x_{k1},\ldots,x_{kn_k}),$$

$$e'_N=(1,1,\ldots,1), \qquad N=\sum_{i=1}^{k} n_i,$$

$$\boldsymbol{\alpha}'=(\alpha_1,\alpha_2,\ldots,\alpha_k),$$

$$\boldsymbol{\varepsilon}'=(\varepsilon_{11},\ldots,\varepsilon_{1n_1},\varepsilon_{21},\ldots,\varepsilon_{2n_2},\ldots,\varepsilon_{k1},\ldots,\varepsilon_{kn_k}).$$

Then the above model can be written as

$$\mathbf{x}=\mathbf{e}_N\mu+U\boldsymbol{\eta}, \qquad \boldsymbol{\eta}'=(\boldsymbol{\alpha}',\boldsymbol{\varepsilon}')$$

and

$$U=\left[\begin{array}{cccc|cccc} \mathbf{e}_{n_1} & \mathbf{0} & \cdots & \mathbf{0} & I_{n_1} & 0 & \cdots & 0 \\ \mathbf{0} & \mathbf{e}_{n_2} & \cdots & \mathbf{0} & 0 & I_{n_2} & \cdots & 0 \\ \vdots & \vdots & \ddots & \vdots & \vdots & \vdots & \ddots & \vdots \\ \mathbf{0} & \mathbf{0} & \cdots & \mathbf{e}_{n_k} & 0 & 0 & \cdots & I_{n_k} \end{array}\right].$$

In the model (5.1.2), if we take $U=I_N$, we get a linear regression model with heteroscedastic variances. The model (5.1.2) is known in the literature as the *variance-component* model. The next section describes another related model, known as the *growth-curve model*, which has been introduced by Potthoff and Roy (1964). In this chapter we shall be concerned with the general problem of estimating $\boldsymbol{\beta}$ and $\sigma_i^2$ $(i=1,2,\ldots,k)$ or their estimable linear functions. In particular, the least-squares theory and MINQUE theory will be discussed; the estimability criterion will be discussed in Section 5.3.

For nonparametric testing and estimation of $\boldsymbol{\beta}$, the reader is referred to Srivastava and Saleh (1970) and Srivastava (1970a, b, 1972, 1975).

## 5.2 Growth-Curve Model

In this section, we describe another linear model, known as growth-curve model [see, for example, Rao (1959) and Potthoff and Roy (1964)]. The model (5.1.1) can be rewritten in an alternative way provided $N=N_1p$,

$m=m_1q$, and $Z'=B\otimes A'=(b_{ij}A')$, $A'$ and $B'$ being $N_1\times m_1$ and $q\times p$ matrices, respectively. For this, we can partition $\mathbf{x}$ and $\boldsymbol{\varepsilon}$ into $p$ subvectors, each of order $N_1$, and $\boldsymbol{\beta}$ into $q$ subvectors, each of order $m_1$. That is,

$$\mathbf{x}'=(\mathbf{x}_1',\dots,\mathbf{x}_p'),\qquad \boldsymbol{\varepsilon}'=(\boldsymbol{\varepsilon}_1',\dots,\boldsymbol{\varepsilon}_p'),\quad\text{and}\quad \boldsymbol{\beta}'=(\boldsymbol{\beta}_1',\dots,\boldsymbol{\beta}_q'),$$

where $\mathbf{x}_i$ and $\boldsymbol{\varepsilon}_i$ $(i=1,2,\dots,p)$ are $N_1$-vectors and $\boldsymbol{\beta}_j$ $(j=1,2,\dots,q)$ are $m_1$-vectors. If

$$X'=(\mathbf{x}_1,\dots,\mathbf{x}_p),\qquad \varepsilon'=(\boldsymbol{\varepsilon}_1,\dots,\boldsymbol{\varepsilon}_p),\quad\text{and}\quad \xi'=(\boldsymbol{\beta}_1,\dots,\boldsymbol{\beta}_q),$$

then the model (5.1.1) can be rewritten as

$$X=B\xi A+\varepsilon, \tag{5.2.1}$$

where we shall assume that

$$E(\varepsilon)=0,\quad\text{and}\quad E\boldsymbol{\varepsilon}_i\boldsymbol{\varepsilon}_j'=\sigma_{ij}W,\quad i,j=1,2,\dots,p. \tag{5.2.2}$$

In this model, $\xi$ and $\Sigma=(\sigma_{ij})$ are unknown parameters, and $B$, $A$, and $W$ are known matrices. Notice that the growth-curve model differs from the linear model (5.1.1) or (5.1.2) only in the case of second moments of the random variables, and thus gives us a new model.

The following are some of the illustrations where the above growth-curve model can arise:

EXAMPLE 5.2.1. Let us suppose that the $N$ individuals can be divided into $k$ homogeneous groups, having $N_1,N_2,\dots,N_k$ individuals, according to their growth pattern. For example, the growth patterns of boys and girls are different, and hence they form two different groups. We shall assume that the growth of an individual is a polynomial in time $t$ measured at $t=1,2,\dots,T$. If $y_{ijt}$ denotes the observed growth of the $i$th individual in the $j$th group at time $t$, then

$$y_{ijt}=\beta_{0j}+\beta_{1j}t+\cdots+\beta_{qj}t^q+\varepsilon_{ijt} \tag{5.2.3}$$

for $i=1,2,\dots,N_j$, $j=1,2,\dots,k$, and $t=1,2,\dots,T$, where

$$E\varepsilon_{ijt}=0,\qquad E\varepsilon_{ijt}\varepsilon_{i'j't'}=\begin{cases}\sigma_{tt'} & \text{if } i=i' \text{ and } j=j',\\ 0 & \text{otherwise.}\end{cases} \tag{5.2.4}$$

Let us denote

$$\mathbf{y}_t'=(y_{11t},\dots,y_{N_11t},y_{12t},\dots,y_{N_22t},\dots,y_{N_kkt}),$$
$$\boldsymbol{\varepsilon}_t'=(\varepsilon_{11t},\dots,\varepsilon_{N_11t},\varepsilon_{12t},\dots,\varepsilon_{N_22t},\dots,\varepsilon_{N_kkt}),$$
$$\xi=(\beta_{ij},\ i=0,1,\dots,q,\ j=1,2,\dots,k).$$

Then

$$\mathbf{y}_t'=(1,t,\dots,t^q)\xi A+\boldsymbol{\varepsilon}_t',$$

where $A=\operatorname{diag}(\mathbf{e}'_{N_1},\mathbf{e}'_{N_2},\ldots,\mathbf{e}'_{N_k})$ with $\mathbf{e}'_j=(1,1,\ldots,1):1\times j$, and $E\boldsymbol{\varepsilon}_t\boldsymbol{\varepsilon}'_{t'}=(I_N)\sigma_{tt'}$, $E\boldsymbol{\varepsilon}_t=0$. Hence, if

$$\varepsilon'=(\boldsymbol{\varepsilon}_1,\ldots,\boldsymbol{\varepsilon}_T),\qquad Y'=(\mathbf{y}_1,\mathbf{y}_2,\ldots,\mathbf{y}_T),$$

and

$$B=\begin{bmatrix}1 & 1 & \cdots & 1\\ 1 & 2 & \cdots & 2^q\\ \vdots & \vdots & & \vdots\\ 1 & T & \cdots & T^q\end{bmatrix},$$

then the above model is the same as Eq. (5.2.1); that is,

$$Y=B\xi A+\varepsilon \tag{5.2.5}$$

with $E(\varepsilon)=0$ and $E\boldsymbol{\varepsilon}_t\boldsymbol{\varepsilon}'_{t'}=\sigma_{tt'}I_N$.

EXAMPLE 5.2.2. Suppose that we want to find out the effect of $v$ treatments on the growth pattern of certain plants. We decide to use incomplete block design. Let $y_{ijt}$ be the growth of the plant grown in the $i$th block having the $j$th treatment at time $t$. Then, for $i=1,2,\ldots,b$, $j=1,2,\ldots,v$, and $t=1,2,\ldots,T$,

$$y_{ijt}=\sum_{\alpha=0}^{q}(\mu_\alpha+\beta_{\alpha i}+\tau_{\alpha j})t^\alpha+\varepsilon_{ijt},$$

where $\sum_{\alpha=0}^{q}\beta_{\alpha i}t^\alpha$ is the polynomial growth effect due to the $i$th block, $\sum_{\alpha=0}^{q}\tau_{\alpha j}t^\alpha$ is the polynomial growth effect due to the $j$th treatment, and $\varepsilon_{ijt}$ is the random error such that

$$E(\varepsilon_{ijt})=0,\qquad E\varepsilon_{ijt}\varepsilon_{i'j't'}=\begin{cases}\sigma_{tt'} & \text{for } i=i' \text{ and } j=j',\\ 0 & \text{otherwise.}\end{cases}$$

Here, $(i,j)=1$ if the plant in the $i$th block received the $j$th treatment; otherwise $(i,j)=0$. Let us denote by $\mathbf{y}_t$ a column vector arranged according to blockwise observations at time $t$, and by $\boldsymbol{\varepsilon}_t$ a random vector arranged in the same way as $\mathbf{y}_t$. Then

$$\mathbf{y}'_t=(1,t,\ldots,t^q)\xi A+\epsilon'_t,$$

where

$$\xi=\begin{bmatrix}\mu_0 & \beta_{01} & \cdots & \beta_{0b} & \tau_{01} & \cdots & \tau_{0v}\\ \mu_1 & \beta_{11} & \cdots & \beta_{1b} & \tau_{11} & \cdots & \tau_{1v}\\ \vdots & \vdots & & \vdots & \vdots & & \vdots\\ \mu_q & \beta_{q1} & \cdots & \beta_{qb} & \tau_{q1} & \cdots & \tau_{qv}\end{bmatrix}$$

and $A'$ is the design matrix due to incomplete block design. Thus, we get the growth-curve model given in Eqs. (5.2.1) and (5.2.2).

For further details, one can refer to Potthoff and Roy (1964).

## 5.3 Estimability Criterion

We shall consider the linear model (5.1.1) for the estimability of $C\boldsymbol{\beta}$, a linear function of parameter $\boldsymbol{\beta}$. $C\boldsymbol{\beta}$ will be said to be estimable iff there exists a linear function $A\mathbf{x}-\mathbf{d}$ such that $EA\mathbf{x}-\mathbf{d}=C\boldsymbol{\beta}$ for all permissible values of $\boldsymbol{\beta}$. A necessary and sufficient condition for the estimability of $C\boldsymbol{\beta}$ is derived in this section. Let $E\boldsymbol{\varepsilon}\boldsymbol{\varepsilon}'=V$. There are two main situations, namely, (i) $V$ is nonsingular and $\boldsymbol{\beta}$ has no restrictions, and (ii) $V$ is singular. If $\boldsymbol{\beta}$ has some linear restrictions—namely, $D\boldsymbol{\beta}=\boldsymbol{\delta}$, where $D$ and $\boldsymbol{\delta}$ are given—then redefining the model (5.1.1) as

$$\begin{pmatrix}\mathbf{y}\\ \boldsymbol{\delta}\end{pmatrix}=\begin{pmatrix}Z'\\ D\end{pmatrix}\boldsymbol{\beta}+\begin{pmatrix}\boldsymbol{\varepsilon}\\ \mathbf{0}\end{pmatrix},$$

we have

$$E\begin{pmatrix}\boldsymbol{\varepsilon}\\ \mathbf{0}\end{pmatrix}(\boldsymbol{\varepsilon}',\mathbf{0}')=\begin{pmatrix}V & 0\\ 0 & 0\end{pmatrix}, \qquad \text{a singular matrix.}$$

Thus, the considerations of the above two situations will exhaust other possibilities in which there are linear restrictions on the parameters $\boldsymbol{\beta}$, irrespective of $V$ being nonsingular or singular. The following two subsections consider the two situations mentioned above as (i) and (ii).

### *5.3.1 Necessary and Sufficient Condition with V Nonsingular and No Restrictions on* $\boldsymbol{\beta}$

The condition of estimability of $C\boldsymbol{\beta}$ is

$$E(A\mathbf{x})=C\boldsymbol{\beta}+\mathbf{d} \qquad \text{for all } \boldsymbol{\beta},$$

which gives

$$(AZ'-C)\boldsymbol{\beta}=\mathbf{d} \qquad \text{for all } \boldsymbol{\beta}.$$

Hence,

$$\mathbf{d}=\mathbf{0} \quad \text{and} \quad AZ'=C.$$

Thus, a necessary and sufficient condition for estimability of $C\boldsymbol{\beta}$ is

$$\rho(Z)=\rho(Z,C'). \tag{5.3.1}$$

Since

$$\rho(I_m-C^-C,C^-C)=m \quad \text{and} \quad \rho\begin{pmatrix}I_N & -Z'C^-\\ 0 & I_c\end{pmatrix}=N+c,$$

we get from Corollary 1.5.3

$$\rho\begin{pmatrix} Z' \\ C \end{pmatrix} = \rho\begin{pmatrix} I_N & -Z'C^- \\ 0 & I_c \end{pmatrix}\begin{pmatrix} Z' \\ C \end{pmatrix}(I_m - C^-C, C^-C)$$

$$= \rho\begin{pmatrix} Z'(I_m - C^-C) & 0 \\ 0 & C \end{pmatrix} = \rho(C) + \rho[Z'(I_m - C^-C)].$$

Similarly, it can be shown that

$$\rho\begin{pmatrix} Z' \\ C \end{pmatrix} = \rho(Z) + \rho[(I - ZZ^-)C'].$$

Hence, Eq. (5.3.1) can be rewritten as

(5.3.2a) $$\rho[Z'(I_m - C^-C)] = \rho(Z) - \rho(C)$$

or

(5.3.2b) $$(I - ZZ^-)C' = 0,$$

which gives a necessary and sufficient condition for the estimability of $C\boldsymbol{\beta}$. The condition (5.3.2a) was given by Roy and Roy (1959) and later reconsidered by Milliken (1971).

### *5.3.2 Necessary and Sufficient Condition with V Singular*

In this situation, $N - \rho(V)$ variables of $\mathbf{x}$ take constant values with probability one. Since

$$\text{Cov}[(I - VV^-)\mathbf{x}] = 0,$$

we have

(5.3.3) $$(I - VV^-)\mathbf{x} = (I - VV^-)Z'\boldsymbol{\beta}$$

with probability one, and this equation must be consistent. Letting

$$\mathbf{x}_1 = (I - VV^-)\mathbf{x} \quad \text{and} \quad H = (I - VV^-)Z',$$

we get from Eq. (5.3.3) a general solution of $\boldsymbol{\beta}$ as

(5.3.4) $$\boldsymbol{\beta} = H^-\mathbf{x}_1 + (I - H^-H)\boldsymbol{\omega}$$

where $\boldsymbol{\omega}$ is an arbitrary vector. Thus, for the estimability of $C\boldsymbol{\beta}$, $\boldsymbol{\beta}$ must satisfy Eq. (5.3.4) and

$$(AZ' - C)\boldsymbol{\beta} = \mathbf{d} \qquad \text{for some } A \text{ and } \mathbf{d}.$$

Using Eq. (5.3.4), we get

$$(AZ' - C)[H^-\mathbf{x}_1 + (I - H^-H)\boldsymbol{\omega}] = \mathbf{d},$$

or

$$(AZ' - C)(I - H^-H)\boldsymbol{\omega} = \mathbf{d} - (AZ' - C)H^-\mathbf{x}_1,$$

for all vectors $\boldsymbol{\omega}$. This gives a necessary and sufficient condition and a solution for $A$ and $\mathbf{d}$ from

(5.3.5) $AZ'(I-H^-H)=C(I-H^-H)$ and $\mathbf{d}=(AZ'-C)H^-\mathbf{x}_1$.

Notice that if we determine $A$ from $AZ'(I-H^-H)=C(I-H^-H)$, then $\mathbf{d}$ can be determined easily. Thus, a necessary and sufficient condition for the solvability of $A$ is

$$\rho[Z'(I-H^-H)]=\rho\begin{pmatrix} Z'(I-H^-H) \\ C(I-H^-H) \end{pmatrix},$$

which is equivalent to

(5.3.6) $$\rho\begin{pmatrix} Z' \\ H \end{pmatrix}=\rho\begin{bmatrix} Z' \\ H \\ C \end{bmatrix}.$$

Since $H=(I-VV^-)Z'$, a necessary and sufficient condition for the estimability of $C\boldsymbol{\beta}$, when $V$ is singular, is also given by Eq. (5.3.1) or (5.3.2), the same as in the nonsingular case. From Eq. (5.3.5), it is obvious that

$$\mathbf{d}=\mathbf{0} \quad \text{when} \quad V \text{ is nonsingular (i.e., } \mathbf{x}_1=\mathbf{0}),$$

while, in general

$$\mathbf{d}\neq\mathbf{0} \quad \text{when} \quad V \text{ is singular,}$$

since $\mathbf{d}=\mathbf{0}$ iff $(I-VV^-)\mathbf{x}=\mathbf{x}_1=\mathbf{0}$ (that is, $\mathbf{x}$ lies in the space generated by the column vectors of $V$).

## 5.4 Estimation of $\boldsymbol{\beta}$ in the Ordinary Linear Regression Model

The ordinary linear regression model is given by

(5.4.1) $\mathbf{x}=Z'\boldsymbol{\beta}+\boldsymbol{\varepsilon}$, $E\boldsymbol{\varepsilon}=\mathbf{0}$, and $E\boldsymbol{\varepsilon}\boldsymbol{\varepsilon}'=\sigma^2W$,

where $\sigma^2$ and $\boldsymbol{\beta}$ are unknown parameters, and $W$ and $Z$ are given matrices. We shall be concerned with the problem of estimating $\boldsymbol{\beta}$ when $V\equiv\sigma^2W$ is nonsingular as well as when $V$ is singular. Of all the estimable functions of $\boldsymbol{\beta}$ given in Section 5.3, we shall choose the one that has the least total variance. Such an estimate will be called BLUE (*best linear unbiased estimate*). First of all, we shall try to rewrite the model (5.4.1); for this purpose let us assume that the rank of $Z$ is $r$ ($\leqslant m<N$). Then, by Theorem 1.5.3, we can write

(5.4.2) $$Z=Z_2Z_1,$$

where $Z_2$ and $Z_1$ are $m\times r$ and $r\times N$ matrices of rank $r$. Then, by

Corollary 1.5.8, we can find a matrix $Z_3 : m \times (m-r)$ such that

$$Z_3' Z_2 = 0 \quad \text{and} \quad |(Z_2, Z_3)| \neq 0. \tag{5.4.3}$$

Let us denote

$$Z_2' \boldsymbol{\beta} = \boldsymbol{\beta}_1 \quad \text{and} \quad Z_3' \boldsymbol{\beta} = \boldsymbol{\beta}_2. \tag{5.4.4}$$

Then, using Eq. (5.4.3), Eq. (5.4.4) is equal to

$$\left[ \begin{pmatrix} Z_2' \\ Z_3' \end{pmatrix} (Z_2, Z_3) \right] (Z_2, Z_3)^{-1} \boldsymbol{\beta} = \begin{pmatrix} \boldsymbol{\beta}_1 \\ \boldsymbol{\beta}_2 \end{pmatrix},$$

and this gives

$$\boldsymbol{\beta} = Z_2 (Z_2' Z_2)^{-1} \boldsymbol{\beta}_1 + Z_3 (Z_3' Z_3)^{-1} \boldsymbol{\beta}_2. \tag{5.4.5}$$

Thus, given $\boldsymbol{\beta}$, we can obtain $\boldsymbol{\beta}_1$ and $\boldsymbol{\beta}_2$ from Eq. (5.4.4), whereas given $\boldsymbol{\beta}_1$ and $\boldsymbol{\beta}_2$, we can determine $\boldsymbol{\beta}$ from Eq. (5.4.5). This shows that the transformation is one-to-one and onto. Since

$$Z' \boldsymbol{\beta} = Z_1' Z_2' \boldsymbol{\beta} = Z_1' \boldsymbol{\beta}_1,$$

the model can be rewritten as

$$\mathbf{x} = Z_1' \boldsymbol{\beta}_1 + \boldsymbol{\varepsilon}, \qquad E(\boldsymbol{\varepsilon}) = \mathbf{0}, \quad \text{and} \quad E(\boldsymbol{\varepsilon}\boldsymbol{\varepsilon}') = \sigma^2 W. \tag{5.4.6}$$

From the above, it follows that $\boldsymbol{\beta}_2$ is not estimable, and hence $C\boldsymbol{\beta}$ will be estimable iff

$$CZ_3 = 0, \quad \text{or equivalently,} \quad (I - ZZ^-)C' = 0, \tag{5.4.7}$$

which is the same as Eq. (5.3.2b). To estimate $\boldsymbol{\beta}_1$ for the model (5.4.6), let us choose an $r \times N$ matrix $G$ such that

$$GZ_1' = I_r. \tag{5.4.8}$$

Notice that such a $G$ always exists, because $G$ is a $g$ inverse of $Z_1'$. For obtaining BLUE, we have to choose those $G$'s that make the total variance of $G\mathbf{x}$ a minimum subject to Eq. (5.4.8). Thus, $G$ is to be chosen so as to

$$\text{minimize} \quad \operatorname{tr} GWG' \qquad \text{subject to the condition} \quad GZ_1' = I_r. \tag{5.4.9}$$

Using Lagrange multipliers $\Lambda$, a symmetric matrix, we wish to minimize

$$\operatorname{tr} GWG' - 2 \operatorname{tr} (GZ_1' - I_r)\Lambda. \tag{5.4.10}$$

Equating the coefficients of the differentials of $G$ and $\Lambda$ to zero, we get

$$GW = \Lambda Z_1 \quad \text{and} \quad GZ_1' = I_r. \tag{5.4.11}$$

Since $W$ is p.s.d., a solution of Eq. (5.4.11) will give the required solution for Eq. (5.4.9).

In the next two subsections, we obtain the solution of Eq. (5.4.11), respectively, when $W$ is nonsingular and when $W$ is singular. Then the BLUE for $\boldsymbol{\beta}_1$ will be given by $G\mathbf{x}$, where $G$ is the solution of Eq. (5.4.11). It can be shown that this is also obtainable by the generalized least-squares method, that is, by minimizing [see Rao (1973b)]

$$(\mathbf{x}-Z_1'\boldsymbol{\beta}_1)'W_1^-(\mathbf{x}-Z_1'\boldsymbol{\beta}_1), \tag{5.4.12}$$

where $W_1=W+Z'UZ$, $U$ is a symmetric matrix, and $\rho(W_1)=\rho(W,Z')$.

### *5.4.1 Solution of Eq. (5.4.11) when W Is Nonsingular*

From Eq. (5.4.11), we get

$$G=\Lambda Z_1W^{-1} \quad \text{and} \quad GZ_1'=\Lambda Z_1W^{-1}Z_1'=I.$$

Hence, the unique solution $\hat{G}$ for $G$ is

$$\hat{G}=(Z_1W^{-1}Z_1')^{-1}Z_1W^{-1},$$

and the unique BLUE for $\boldsymbol{\beta}_1$ is

$$\hat{\boldsymbol{\beta}}_1=\hat{G}\mathbf{x}=(Z_1W^{-1}Z_1')^{-1}Z_1W^{-1}\mathbf{x} \tag{5.4.13}$$

with (covariance matrix of $\hat{\boldsymbol{\beta}}_1$)

$$\operatorname{Cov}(\hat{\boldsymbol{\beta}}_1)=(Z_1W^{-1}Z_1')^{-1}\sigma^2. \tag{5.4.14}$$

Hence, using Eqs. (5.4.5) and (5.4.7), the BLUE for $C\boldsymbol{\beta}$ is

$$C\hat{\boldsymbol{\beta}}=CZ_2(Z_2'Z_2)^{-1}\hat{\boldsymbol{\beta}}_1=C(ZW^{-1}Z')^-ZW^{-1}\mathbf{x}, \tag{5.4.15}$$

because $Z_2(Z_2'Z_2)^{-1}(Z_1W^{-1}Z_1')^{-1}(Z_2'Z_2)^{-1}Z_2'$ is a $g$ inverse of $ZW^{-1}Z'$. Further, Eq. (5.4.15) is also the least-squares estimate (LSE) of $C\boldsymbol{\beta}$, as can be shown by minimizing

$$(\mathbf{x}-Z'\boldsymbol{\beta})'W^{-1}(\mathbf{x}-Z'\boldsymbol{\beta}).$$

The covariance matrix of $C\hat{\boldsymbol{\beta}}$ is

$$\sigma^2C(ZW^{-1}Z')^-C'. \tag{5.4.16}$$

### *5.4.2 Solution of Eq. (5.4.11) when W Is Singular*

Let $R$ be an arbitrary symmetric matrix such that if $W_1=W+Z_1'RZ_1$,

$$\rho(W_1)=\rho(W,Z')=\rho(W,Z_1')\leqslant N.$$

Observing that

$$W_1=(W,Z_1')\begin{pmatrix} W^- & 0 \\ 0 & R\end{pmatrix}\begin{pmatrix} W \\ Z_1\end{pmatrix},$$

and using the Frobenius theorem on ranks (namely, Corollary 1.5.5), it is easy to see that for any $T$

$$\rho(TW_1) \geq \rho[T(W, Z_1')] + \rho(W_1) - \rho(W, Z_1'),$$

while by Theorem 1.5.1,

$$\rho(TW_1) \leq \rho[T(W, Z_1')].$$

Thus, for any $T$,

$$\rho(TW_1) = \rho[T(W, Z_1')] = \rho(TW_1T'). \tag{5.4.17}$$

Thus, if $T = I - W_1W_1^-$, then $TW_1 = 0$, and from Eq. (5.4.17), $\rho[T(W, Z_1')] = \rho(TW_1) = 0$. This gives $T(W, Z_1') = 0$. Similarly, if $T = Z_1W_1^-$, then $\rho(Z_1W_1^-Z_1') = r$, since $Z_1' = W_1W_1^-Z_1' = W_1T'$ and $TW_1T' = Z_1W_1^-Z_1'$ because $W_1^-$ is a symmetric $g$ inverse of $W_1$. Hence,

$$(W_1, Z_1') = W_1W_1^-(W_1Z_1') \quad \text{and} \quad \rho(Z_1W_1^-Z_1') = r, \tag{5.4.18}$$

$W_1^-$ being a symmetric $g$ inverse of $W_1$. Now, Eq. (5.4.11) is equivalent to

$$GW_1 = (\Lambda + R)Z_1 \quad \text{and} \quad GZ_1' = I_r. \tag{5.4.19}$$

Since $Z_1 = Z_1W_1^-W_1$, $GW_1 = (\Lambda + R)Z_1$ can be solved for $G$, and its general solution is given by

$$G = (\Lambda + R)Z_1W_1^- + G_1(I - W_1W_1^-), \tag{5.4.20}$$

where $G_1$ is an arbitrary matrix. Using Eqs. (5.4.20) and (5.4.18) in $GZ_1' = I_r$, we get

$$I_r = (\Lambda + R)Z_1W_1^-Z_1', \quad \text{or} \quad \Lambda + R = (Z_1W_1^-Z_1')^{-1}. \tag{5.4.21}$$

Using this in Eq. (5.4.20), the general solution of $G$ satisfying Eq. (5.4.11) or (5.4.19) is

$$G = (Z_1W_1^-Z_1')^{-1}Z_1W_1^- + G_1(I - W_1W_1^-). \tag{5.4.22}$$

Since $E\boldsymbol{\varepsilon} = \mathbf{0}$ and $E\boldsymbol{\varepsilon}\boldsymbol{\varepsilon}' = \sigma^2 W$, on account of Eq. (5.4.18),

$$(I - W_1W_1^-)\boldsymbol{\varepsilon} = \mathbf{0} \qquad \text{with probability one,}$$

and using $(I - W_1W_1^-)Z_1' = 0$, we get

$$(I - W_1W_1^-)\mathbf{x} = (I - W_1W_1^-)\boldsymbol{\varepsilon} = \mathbf{0}$$

with probability one. Hence, the best linear estimate of $\boldsymbol{\beta}_1$ is

$$\hat{\boldsymbol{\beta}}_1 = (Z_1W_1^-Z_1')^{-1}Z_1W_1^-\mathbf{x}, \tag{5.4.23}$$

and its covariance matrix is

$$\sigma^2\left[(Z_1W_1^-Z_1')^{-1} - R\right],$$

where $W_1$ and $R$ are defined at the beginning of the subsection. The BLUE for the estimable linear function $C\boldsymbol{\beta}$ is

$$C\hat{\boldsymbol{\beta}} = CZ_2(Z_2'Z_2)^{-1}\hat{\boldsymbol{\beta}}_1 = C(ZW_1^{-}Z')^{-}ZW_1^{-}\mathbf{x}, \tag{5.4.24}$$

and its covariance matrix is given by

$$\sigma^2[C(ZW_1^{-}Z')^{-}C' - CR_1C'],$$

where $R_1 = Z_2(Z_2'Z_2)^{-1}R(Z_2'Z_2)^{-1}Z_2'$ and $W_1 = W + Z'R_1Z$, with $\rho(W_1) = \rho(W, Z')$. Notice that $\hat{\boldsymbol{\beta}}_1$ is a generalized least-squares estimate of $\boldsymbol{\beta}_1$, and it is obtained by minimizing the sum of squares

$$(\mathbf{x} - Z_1'\boldsymbol{\beta}_1)'W_1^{-}(\mathbf{x} - Z_1'\boldsymbol{\beta}_1) = \boldsymbol{\varepsilon}'W_1^{-}\boldsymbol{\varepsilon}. \tag{5.4.25}$$

These and similar results were established by Rao and Mitra (1971), Mitra (1973), Mitra and Moore (1973), Rao (1971, 1972, 1973b, 1974), Zyskind and Martin (1969), and others referred to by them.

## 5.5 BLUE Under Two Different Models

Let us consider the following two models:

$$\mathbf{x} = Z_0'\boldsymbol{\beta} + \boldsymbol{\varepsilon}, \qquad E\boldsymbol{\varepsilon} = \mathbf{0}, \quad \text{and} \quad E\boldsymbol{\varepsilon}\boldsymbol{\varepsilon}' = \sigma^2 I \tag{5.5.1}$$

and

$$\mathbf{x} = Z'\boldsymbol{\beta} + \boldsymbol{\eta}, \qquad E\boldsymbol{\eta} = \mathbf{0}, \quad \text{and} \quad E\boldsymbol{\eta}\boldsymbol{\eta}' = \Sigma. \tag{5.5.2}$$

Assuming that every BLUE under the model (5.5.1) is also a BLUE under the model (5.5.2) with the same variances, we would like to determine the structure on $Z$ and $\Sigma$.

The BLUE for $Z_0\boldsymbol{\beta}$ under the model (5.5.1) is

$$Z_0'(Z_0Z_0')^{-}Z_0\mathbf{x} \equiv P_{Z_0}\mathbf{x} = \mathbf{y} \quad \text{(say)}, \tag{5.5.3}$$

while under the model (5.5.2), the first two moments of $\mathbf{y}$ are

$$E(\mathbf{y}|(5.5.2)) = P_{Z_0}Z'\boldsymbol{\beta} \quad \text{and} \quad \operatorname{Cov}(\mathbf{y}|(5.5.2)) = P_{Z_0}\Sigma P_{Z_0}. \tag{5.5.4}$$

$\mathbf{y}$ is unbiased for $Z_0\boldsymbol{\beta}$ under the model (5.5.2) iff

$$P_{Z_0}Z' = Z_0' \quad \text{or} \quad Z - Z_0 = G_1(I - P_{Z_0}), \tag{5.5.5}$$

where $G_1$ is an arbitrary matrix. In order that $\mathbf{y}$ may also be BLUE under the model (5.5.2), it must be uncorrelated with the estimates of zero [see, for example, Rao (1973a, p. 317)]. The estimates of zero under the model (5.5.2) are $(I - Z^{-}Z)'\mathbf{x}$, and hence for $\mathbf{y}$ to be BLUE, we must have

$$\operatorname{Cov}(\mathbf{y}, (I - Z^{-}Z)'\mathbf{x}|(5.5.2)) = P_{Z_0}\Sigma(I - Z^{-}Z) = 0,$$

which gives

$$P_{Z_0}\Sigma = Z_0'G_2Z, \tag{5.5.6}$$

where $G_2$ is an arbitrary symmetric matrix such that $Z_0'G_2Z_0$ is p.s.d. and $\rho(Z_0'G_2Z_0)=\rho(Z_0'G_2Z)$. We are interested in obtaining a p.s.d. solution of $\Sigma$ from Eq. (5.5.6) in which $Z$ satisfies Eq. (5.5.5). For this, we can use Theorem 2.2 of Khatri and Mitra (1976, p. 580) and for completeness it is given as follows (for the real case):

**Lemma 5.5.1** *The equation $AX=C$ has a nonnegative definite* (*or* p.s.d.) *solution iff $CA'$ is nonnegative definite and $\rho(CA')=\rho(C)$, in which case a general nonnegative definite solution is*

$$C'(CA')^{-}C+(I-A^{-}A)U(I-A^{-}A)',$$

*where $(CA')^{-}$ and $A^{-}$ are arbitrary g inverses of $CA'$ and $A$, respectively, and $U$ is an arbitrary nonnegative definite matrix.*

From Eq. (5.5.5), $P_{Z_0}Z'=Z_0'$, and hence $Z_0'G_2ZP_{Z_0}'=Z_0'G_2Z_0$ is p.s.d. and $\rho(Z_0'G_2Z_0)=\rho(Z_0'G_2Z)$; thus the above lemma is applicable, and a general solution of (5.5.6) is given by

$$\Sigma=Z'G_2'Z_0(Z_0'G_2Z_0)^{-}Z_0'G_2Z+(I-P_{Z_0})G_3(I-P_{Z_0}), \tag{5.5.7}$$

where $G_3$ is an arbitrary p.s.d. matrix and $G_2$ is an arbitrary symmetric matrix such that

$$Z_0'G_2Z_0 \text{ is p.s.d. and } \rho(Z_0'G_2Z_0)=\rho(Z_0'G_2Z).$$

Using Eq. (5.5.5), Eq. (5.5.7) can be rewritten as

$$\begin{aligned}\Sigma=&Z_0'G_2Z_0+Z_0'G_2G_1(I-P_{Z_0})+(I-P_{Z_0})G_1'G_2'Z_0\\&+(I-P_{Z_0})G_4(I-P_{Z_0}),\end{aligned} \tag{5.5.8}$$

where $G_4$ is an arbitrary p.s.d. matrix, $G_2$ is an arbitrary matrix such that $Z_0'G_2Z_0$ is a p.s.d. matrix, and $G_1$ is an arbitrary matrix defined in Eq. (5.5.5). Under Eq. (5.5.8), Eq. (5.5.4) gives

$$\text{Cov}(\mathbf{y}|(5.5.2))=Z_0'G_2Z_0. \tag{5.5.9}$$

In order that $\text{Cov}(\mathbf{y}|(5.5.2))=\text{Cov}(\mathbf{y}|(5.5.1))$, we must have

$$Z_0'G_2Z_0=\sigma^2P_{Z_0} \quad\text{or}\quad G_2=\sigma^2(Z_0Z_0')^{-}+G_5-P_{Z_0}G_5P_{Z_0}, \tag{5.5.10}$$

where $G_5$ is an arbitrary matrix and $(Z_0Z_0')^{-}$ is a symmetric $g$ inverse of $Z_0Z_0'$. It may be observed that with $G_2$ given by Eq. (5.5.10), the rank condition $\rho(Z_0'G_2Z)=\rho(Z_0)$ will be satisfied. Hence, using Eq. (5.5.10) in Eq. (5.5.8), we get

$$\Sigma=\sigma^2P_{Z_0}+(I-P_{Z_0})G_6'Z_0+Z_0'G_6(I-P_{Z_0})+(I-P_{Z_0})G_4(I-P_{Z_0}), \tag{5.5.11}$$

where $G_4$ is an arbitrary p.s.d. matrix and $G_6=[G_5+\sigma^2(Z_0Z_0')^-]G_1$ is an arbitrary matrix. Thus, the structures of $Z$ and $\Sigma$ of Eq. (5.5.2) are given by Eqs. (5.5.5) and (5.5.11), respectively, and under this structure the BLUE of $Z_0\boldsymbol{\beta}$ in the model (5.5.1) is the same as in the model (5.5.2) with the same covariance matrix. This problem was solved by Rao (1967) and Mitra and Rao (1968).

## 5.6 Estimation of Location Parameters for the Growth-Curve Model

Let us rewrite the growth-curve model given by Eqs. (5.2.1) and (5.2.2) as

(5.6.1) $\quad X=B\xi A+\varepsilon, \qquad E(\operatorname{vec}\varepsilon)=\mathbf{0} \quad \text{and} \quad \operatorname{Cov}(\operatorname{vec}\varepsilon)=\Sigma\otimes W,$

where $X$, $B$, $A$, and $W$ are known matrices of respective order $p\times n$, $p\times q$, $m\times n$, and $n\times n$; $\varepsilon$ is a $p\times n$ random error matrix; and $\xi$ and $\Sigma$ are unknown matrices of respective orders $q\times m$ and $p\times p$. Here, $(\operatorname{vec}\varepsilon)'$ is a row vector obtained from $\varepsilon$ by arranging its rows in succession; for example, if $\boldsymbol{\varepsilon}'=(\boldsymbol{\varepsilon}_1,\boldsymbol{\varepsilon}_2,\ldots,\boldsymbol{\varepsilon}_p)$, then

$$(\operatorname{vec}\varepsilon)'=(\boldsymbol{\varepsilon}_1',\boldsymbol{\varepsilon}_2',\ldots,\boldsymbol{\varepsilon}_p').$$

$\xi$ is known as the location-parameter matrix, and $\Sigma$ is known as the scale-parameter matrix. The model (5.6.1) can be rewritten as

$$(\operatorname{vec}X)=(B\otimes A')\operatorname{vec}\xi+\operatorname{vec}\varepsilon, \qquad (5.6.2)$$
$$E\operatorname{vec}\varepsilon=\mathbf{0}$$
$$\operatorname{Cov}(\operatorname{vec}\varepsilon)=\Sigma\otimes W.$$

Let us consider the situation when $\Sigma\otimes W$ is known and p.d. Then, by Eq. (5.4.15), the BLUE for the estimable function $(C\otimes D')\operatorname{vec}\xi$ is

$$(C\otimes D')\big[(B'\otimes A)(\Sigma\otimes W)^{-1}(B\otimes A')\big]^- \qquad (5.6.3)$$
$$\times(B'\otimes A)(\Sigma\otimes W)^{-1}(\operatorname{vec}X).$$

This is equivalent to the BLUE for the estimable linear function $C\xi D$ and is given by

$$C(B'\Sigma^{-1}B)^-B'\Sigma^{-1}XW^{-1}A'(AW^{-1}A')^-D. \qquad (5.6.4)$$

However, $\Sigma$ is usually unknown and should be replaced by an estimate. An unbiased estimate of $\Sigma$ is given by

$$S=\frac{XW^{-1}X'-XW^{-1}A'(AW^{-1}A')^-AW^{-1}X'}{f},$$

where $f=n-\rho(A)>p$. Hence, an estimate for the estimable linear function $C\xi D$ is given by

$$C(B'S^{-1}B)^{-}B'S^{-1}XW^{-1}A'(AW^{-1}A')^{-}D. \tag{5.6.5}$$

This estimate is no longer BLUE, and will be called an *empirical estimate*. However, from Theorem 1.10.3, it can be shown that the estimate (5.6.5) can be written as $C\hat{\xi}D$, where $\hat{\xi}$ is such that

$$|(X-B\xi A)W^{-1}(X-B\xi A)'| \quad \text{is minimum at} \quad \xi=\hat{\xi}. \tag{5.6.6}$$

If the distribution of $\varepsilon$ is symmetric, i.e., the distributions of $\varepsilon$ and $-\varepsilon$ are the same, then the estimate given by Eq. (5.6.5) is an unbiased estimate of $C\xi D$. This is left to the reader to verify.

When $W$ is singular and $\Sigma$ is nonsingular, we can use the results of Subsection 5.4.2 and obtain an estimate of $C\xi D$ as

$$C(B'S^{-1}B)^{-}B'S^{-1}XW_1^{-}A'(AW_1^{-}A')^{-}D, \tag{5.6.7}$$

where $W_1=W+A'RA$, $R$ is a symmetric matrix such that $\rho(W_1)=\rho(W,A')$, and

$$S=X[W_1^{-}-W_1^{-}A'(AW_1^{-}A')^{-}AW_1^{-}]X'/f,$$

with $f=\rho(W,A')-\rho(A)$ $(>p)$ and $\rho(W_1^{-})=\rho(W_1)$. The condition $\rho(W_1^{-})=\rho(W_1)$ is imposed so that $S$ can be written as

$$S=\varepsilon[W_1^{-}-W_1^{-}A'(AW_1^{-}A')^{-}AW_1^{-}]\varepsilon'/f.$$

The estimate (5.6.7) can also be written as $C\hat{\xi}D$, where $\hat{\xi}$ is such that

$$|(X-B\xi A)W_1^{-}(X-B\xi A)'| \quad \text{is minimum at} \quad \xi=\hat{\xi}. \tag{5.6.8}$$

The details for singular $W$ are omitted and are left to the reader.

## 5.7 Best Quadratic Unbiased Estimator of the Variance

Let $x_1,x_2,\ldots,x_N$ be independently distributed so that

$$\mathbf{x}=Z'\boldsymbol{\beta}+\boldsymbol{\varepsilon}, \qquad E(\boldsymbol{\varepsilon})=\mathbf{0}, \qquad \mathrm{Cov}(\boldsymbol{\varepsilon})=\sigma^2 I_N, \tag{5.7.1}$$

where $\mathbf{x}=(x_1,\ldots,x_N)'$, $\boldsymbol{\varepsilon}=(\varepsilon_1,\ldots,\varepsilon_N)'$, $Z$ is an $s\times N$ known matrix, and $\boldsymbol{\beta}$ is the $s$ vector of unknown regression parameters. Let

$$M=I-Z'(ZZ')^{-}Z \quad \text{and} \quad f=\mathrm{tr}\,M. \tag{5.7.2}$$

Then it is well known that $\mathbf{x}'M\mathbf{x}/f$ is the best (uniformly minimum-variance quadratic unbiased) estimator of $\sigma^2$ if $x_1,\ldots,x_N$ are independently normally distributed (Gauss–Markoff model). In this section we derive the minimum variance quadratic unbiased estimator (MVQUE) of $\sigma^2$ *without* the assumption of normality. However, we shall assume that

(5.7.3a) $$\gamma_i = E(\varepsilon_i^4)/\sigma^4 < \infty, \qquad i=1,2,\ldots,N,$$

and

(5.7.3b) $$\gamma_{1i} = E(\varepsilon_i^3)/\sigma^3 < \infty, \qquad i=1,2,\ldots,N.$$

Under these assumptions we shall obtain a set of necessary and sufficient conditions that the elements of the (symmetric) matrix $P$ must satisfy in order that the quadratic estimator

(5.7.4) $$q(\mathbf{x}) \equiv \mathbf{x}'P\mathbf{x}$$

may be the UMVQUE. We shall consider two situations, namely (i) estimate with invariance and (ii) estimate without invariance. The first case has been considered by Hsu (1938) and Rao (1952), and the second case by Khatri (1978a). For multivariate generalizations, see Kleffe (1978).

### 5.7.1 *Uniformly Minimum-Variance Quadratic Unbiased Estimate with Invariance (*UMVIQUE*)*

Let $q(\mathbf{x}) = \mathbf{x}'P\mathbf{x}$ be a quadratic estimator of $\sigma^2$ such that

(i) $Eq = \sigma^2$ for all $\boldsymbol{\beta}$ (unbiased);

(ii) $q(\mathbf{x}) = q(\mathbf{x} - Z'\boldsymbol{\beta})$ for all $\boldsymbol{\beta}$ (invariance);

(iii) if $q_1$ is another quadratic estimator of $\sigma^2$ satisfying (i) and (ii), then $\sigma_q^2 \leqslant \sigma_{q_1}^2$.

We shall now obtain a set of necessary and sufficient conditions on $P$ such that $q = \mathbf{x}'P\mathbf{x}$ satisfies (i), (ii), and (iii). The invariance and unbiasedness conditions require that

(5.7.5) $$ZP = 0 \quad \text{and} \quad \operatorname{tr} P = 1.$$

A general symmetric matrix $B$ satisfying $ZB = 0$ is given by

(5.7.6) $$B = MTM,$$

where $T$ is an arbitrary symmetric matrix, and $M$ is given by Eq. (5.7.2). We now consider a quadratic unbiased estimator of zero with the invariance property, say $\mathbf{x}'R\mathbf{x}$. Then $ZR = 0$ and $\operatorname{tr} R = 0$. Hence, from Eq. (5.7.6),

(5.7.7) $$R = MTM \quad \text{and} \quad \operatorname{tr} MT = 0,$$

since $M^2 = M$, where $T$ is an arbitrary symmetric matrix. Hence, $\mathbf{x}'P\mathbf{x}$ is UMVIQUE iff

(5.7.8) $$\operatorname{Cov}(\mathbf{x}'P\mathbf{x}, \mathbf{x}'R\mathbf{x}) = 0$$

for all matrices $P$ and $R$ satisfying Eqs. (5.7.5) and (5.7.7). We observe that

$$\operatorname{Cov}(\mathbf{x}'P\mathbf{x},\mathbf{x}'R\mathbf{x})=\operatorname{Cov}(\boldsymbol{\varepsilon}'P\boldsymbol{\varepsilon},\boldsymbol{\varepsilon}'R\boldsymbol{\varepsilon})$$
$$=\operatorname{Cov}\left(\sum_{i,j}p_{ij}\varepsilon_i\varepsilon_j,\ \sum_{k,l}r_{kl}\varepsilon_k\varepsilon_l\right).$$

Since $\sum p_{ii}r_{ii}=-\sum_{i\neq j}p_{ii}r_{jj}$, we get from Eq. (5.7.3)

$$\text{(5.7.9)}\qquad \frac{\operatorname{Cov}(\mathbf{x}'P\mathbf{x},\mathbf{x}'R\mathbf{x})}{\sigma^4}=\sum_{i=1}^{N}(\gamma_i-3)p_{ii}r_{ii}+2\sum_{i,j}p_{ij}r_{ij}.$$

Hence, from Eqs. (5.7.5) and (5.7.7), $\mathbf{x}'P\mathbf{x}$ is UMVIQUE of $\sigma^2$ iff

$$\sum_{i=1}^{N}(\gamma_i-3)p_{ii}\sum_{j,k}m_{ij}m_{ik}t_{jk}+2\sum_{i,j}p_{ij}t_{ji}=0 \quad\text{and}\quad \sum p_{ii}=1$$

for all $t_{jk}=t_{kj}$ satisfying $\sum_{i,j}m_{ij}t_{ji}=0$. Thus, for some constant $\lambda$,

$$\sum_{j,k}t_{jk}\left(\sum_{i=1}^{N}(\gamma_i-3)p_{ii}m_{ik}m_{ij}+2p_{kj}-\lambda m_{kj}\right)=0 \quad\text{and}\quad \sum p_{ii}=1.$$

Since this is true for all $t_{jk}$ such that $T=(t_{ij})$ is symmetric, the necessary and sufficient condition for $\mathbf{x}'P\mathbf{x}$ to be UMVIQUE is

$$\text{(5.7.10)}\qquad \sum_{i=1}^{N}(\gamma_i-3)p_{ii}m_{ik}m_{ij}+2p_{kj}=\lambda m_{kj}\qquad \text{for all } j,k,$$
$$\sum_{i=1}^{N}p_{ii}=1.$$

In the matrix notation, this condition can be written as

$$\text{(5.7.11)}\qquad P=MD_\tau M \quad\text{and}\quad \operatorname{tr}P=1,$$

where $D_\tau=\operatorname{diag}(\tau_1,\ldots,\tau_N)$ and $2\tau_j=\lambda-(\gamma_j-3)p_{jj}$. It should be noted, however, that once the $p_{jj}$'s and $\lambda$'s are determined, $P$ is determined completely. Hence, we need to solve only the following system of $N+1$ equations:

$$\text{(5.7.12)}\qquad \sum_{i=1}^{N}(\gamma_i-3)p_{ii}m_{ij}^2+2p_{jj}=\lambda m_{jj},\qquad j=1,2,\ldots,N,$$
$$\sum_{j=1}^{N}p_{jj}=1,$$

which is necessary but *not* sufficient—a subset of the system of equations (5.7.10) or (5.7.11).

Next, we calculate the variance of the estimator $\mathbf{x}'P\mathbf{x}$ where $P$ satisfies Eq. (5.7.10) or (5.7.11). From Eq. (5.7.9),

$$\text{(5.7.13)} \qquad \operatorname{Var}(\mathbf{x}'P\mathbf{x}) = \left( \sum_{i=1}^{N} (\gamma_i - 3) p_{ii}^2 + 2 \sum_{i,j} p_{ij}^2 \right) \sigma^4.$$

From Eq. (5.7.10) and the fact that $M^2 = M$ and $\operatorname{tr} M = f$,

$$4 \sum_{j,k} p_{jk}^2 = \sum_{j,k} \left( \lambda m_{jk} - \sum_{i=1}^{N} (\gamma_i - 3) p_{ii} m_{ik} m_{ij} \right)^2$$

$$= \lambda^2 f - 2\lambda \sum_{i=1}^{N} (\gamma_i - 3) p_{ii} m_{ii} + \sum_{i,j} (\gamma_i - 3)(\gamma_j - 3) p_{ii} p_{jj} m_{ij}^2.$$

From Eq. (5.7.12),

$$\sum_{j=1}^{N} 2 p_{jj}^2 (\gamma_j - 3) = \lambda \sum_{j=1}^{N} (\gamma_j - 3) p_{jj} m_{jj} - \sum_{j=1}^{N} (\gamma_j - 3) p_{jj} \sum_{i=1}^{N} (\gamma_i - 3) p_{ii} m_{ij}^2$$

and

$$\sum_{j=1}^{N} \sum_{i=1}^{N} (\gamma_i - 3) p_{ii} m_{ij}^2 + 2 = \lambda f \quad \text{or} \quad \sum_{i=1}^{N} (\gamma_i - 3) p_{ii} m_{ii} = \lambda f - 2.$$

These results when substituted in Eq. (5.7.13) give the variance of $\mathbf{x}'P\mathbf{x}$ under the condition (5.7.10) as

$$\text{(5.7.14)} \qquad \operatorname{Var}(\mathbf{x}'P\mathbf{x}) = \lambda \sigma^4.$$

Hsu (1938) obtained these results by minimizing the variance (5.7.13). He, however, wrote Eq. (5.7.12) in an alternative way to obtain solutions for $\tau_1, \ldots, \tau_N$ and $\lambda$, rather than for $p_{11}, \ldots, p_{NN}$ and $\lambda$. This can be done as follows. From Eq. (5.7.11),

$$p_{jj} = \sum_{i=1}^{N} m_{ij}^2 \tau_i, \qquad j = 1, 2, \ldots, N.$$

Substituting these values in Eq. (5.7.12), we get

$$\text{(5.7.15)} \qquad \sum_{k=1}^{N} \left( \sum_{i=1}^{N} (\gamma_i - 3) m_{ij}^2 m_{ik}^2 + 2 m_{kj}^2 \right) \tau_k = \lambda m_{jj}, \qquad j = 1, \ldots, N,$$

$$\sum_{i=1}^{N} \sum_{j=1}^{N} m_{ij}^2 \tau_i = \sum_{j=1}^{N} m_{jj} \tau_j = 1.$$

At the beginning of this section, it was mentioned that under the assumption of normality $f^{-1}\mathbf{x}'M\mathbf{x}$, where $f=\operatorname{tr} M$, is the best estimator of $\sigma^2$. It is natural now to ask under what conditions $f^{-1}\mathbf{x}'M\mathbf{x}$ will be UMVIQUE without the assumption of normality. Clearly from Eq. (5.7.11), we must have

(5.7.16) $$MD_\delta M=(\lambda f-2)M\equiv\lambda_1 M,$$

where $D_\delta=\operatorname{diag}(\delta_1,\ldots,\delta_N)$ and $\delta_j=(\gamma_j-3)m_{jj}$. These necessary and sufficient conditions were given by Rao (1952). However, there is some redundancy in Eq. (5.7.16); we need only the diagonal conditions, namely,

(5.7.17) $$\sum_{i=1}^{N}\delta_i m_{ij}^2=\lambda_1 m_{jj},\qquad j=1,2,\ldots,N.$$

These conditions were given by Hsu (1938), who obtained them by minimizing the variance of the estimator. The equivalence of the two conditions follows from

**Lemma 5.7.1** *Let $M$ be a symmetric idempotent matrix and $D_\omega$ be any diagonal matrix. Suppose the diagonal elements of $MD_\omega M$ are zero. Then $MD_\omega M=0$.*

PROOF. Let $C=MD_\omega M$. Then, since $M^2=M$,

$$\operatorname{tr} CC'=\operatorname{tr} MD_\omega M^2D_\omega M=\operatorname{tr} MD_\omega MD_\omega=\operatorname{tr} CD_\omega=\sum_{i=1}^{N}c_{ii}\omega_i.$$

By hypothesis $c_{ii}=0$. Hence, $\operatorname{tr} CC'=0$. Consequently $C=0$. □

### *5.7.2* UMVQUE *(Without Invariance)*

In this subsection we give conditions under which a quadratic estimator $q=\mathbf{x}'P\mathbf{x}$ is UMVQUE without satisfying the invariance property, that is, for $q$ satisfying only (i) and (iii) of Subsection 5.7.1. In order that $\mathbf{x}'P\mathbf{x}$ shall be unbiased, we must have

(5.7.18) $$\operatorname{tr} P=1\quad\text{and}\quad ZPZ'=0.$$

Let $B$ be a symmetric matrix such that $ZBZ'=0$. Then a general symmetric solution of $B$ is

(5.7.19) $$B=(MT+T'M),\qquad M=I-Z'(ZZ')^{-}Z,$$

where $T$ is any arbitrary matrix. We now consider a quadratic unbiased estimator of zero, say, $\mathbf{x}'R\mathbf{x}$. Then we must have

(5.7.20) $$\begin{aligned}&\operatorname{tr} R=0,\\ &ZRZ'=0,\qquad\text{i.e.,}\quad R=(MT+T'M),\\ &\operatorname{tr} MT=0.\end{aligned}$$

Since

$$\mathbf{x}'P\mathbf{x}=\boldsymbol{\varepsilon}'P\boldsymbol{\varepsilon}+2\boldsymbol{\beta}'ZP\boldsymbol{\varepsilon} \quad \text{and} \quad \mathbf{x}'R\mathbf{x}=\boldsymbol{\varepsilon}'R\boldsymbol{\varepsilon}+2\boldsymbol{\beta}'ZR\boldsymbol{\varepsilon},$$

we get

$$\begin{aligned}\operatorname{Cov}(\mathbf{x}'P\mathbf{x},\mathbf{x}'R\mathbf{x})=&\,\sigma^4\Big(\sum_i(\gamma_i-3)p_{ii}r_{ii}+2\sum_{i,j}p_{ij}r_{ij}\Big)\\ &+2\sigma^3\left[\sum_{i=1}^N(\beta'ZP)_{ii}r_{ii}\gamma_{1i}+\sum_{i=1}^N(\beta'ZR)_{ii}p_{ii}\gamma_{1i}\right]\\ &+4\sigma^2\boldsymbol{\beta}'ZPRZ'\boldsymbol{\beta},\end{aligned}$$

where $\gamma_{1i}=E(\varepsilon_i^3)/\sigma^3$, and $(A)_{ij}$ denotes the $(i,j)$th element of $A$.

Hence $\mathbf{x}'P\mathbf{x}$ is UMVQUE iff the above covariance is equal to zero for all $R$ satisfying $\operatorname{tr}R=0$ and $ZRZ'=0$. However, the solution of $P$ obtained from above is not free from $\sigma^2$ and $\boldsymbol{\beta}$ even if the $\gamma_i$'s and $\gamma_{1i}$'s are known. In order that the estimates may be free from $\sigma^2$ and $\boldsymbol{\beta}$, we must have, for all $R$ satisfying $\operatorname{tr}R=0$ and $ZRZ'=0$,

(5.7.21) $$\sum_{i=1}^N(\gamma_i-3)p_{ii}r_{ii}+2\sum_{i,j}^N p_{ij}r_{ij}=0,$$

(5.7.22) $$ZPRZ'=0,$$

(5.7.23) $$Z\sum_{i=1}^N(\mathbf{p}_i r_{ii}\gamma_{1i}+\mathbf{r}_i p_{ii}\gamma_{1i})=0,$$

where $P=(\mathbf{p}_1,\mathbf{p}_2,\ldots,\mathbf{p}_N)$ and $R=(\mathbf{r}_1,\ldots,\mathbf{r}_N)$. Since $R=(MT+T'M)$, Eq. (5.7.21) can be rewritten as

$$2\sum_{i,j}^N(\gamma_i-3)p_{ii}m_{ij}t_{ji}+2\sum_{i,j}^N t_{ji}\Big(\sum_k p_{ki}m_{kj}+\sum_k p_{ik}m_{jk}\Big)=0,$$

where $\sum_{i,j}m_{ij}t_{ji}=0$.

Hence, for some $\lambda$,

(5.7.24) $$(\gamma_i-3)p_{ii}m_{ij}+2\sum_k p_{ki}m_{kj}=\lambda m_{ij} \qquad \text{for all} \quad i,j=1,\ldots,N.$$

In matrix notation, this can be rewritten as

(5.7.25) $$D_\tau M=PM,$$

where $D_\tau=\operatorname{diag}(\tau_1,\ldots,\tau_N)$ and $2\tau_i=\lambda-(\gamma_i-3)p_{ii}$. Since $ZPZ'=0$, $P$ is also of the form (5.7.19), that is, $P=MT_0+T_0'M$ for some arbitrary matrix $T_0$. Hence

$$MT_0M+T_0'M=D_\tau M, \qquad MT_0+MT_0'M=MD_\tau,$$
$$\text{and} \quad MPM=MT_0M+MT_0'M.$$

This gives

$$P = D_\tau M + MD_\tau - MD_\tau M = D_\tau - (I-M)D_\tau(I-M). \tag{5.7.26}$$

Let us now consider Eq. (5.7.22). Since $R = MT + T'M$ and $MZ' = 0$, we get

$$ZPMTZ' = 0$$

for all arbitrary matrices $T$. Since $Z$ is of rank $\geqslant 1$ for $N \geqslant 2$, we get

$$ZPM = 0 = ZD_\tau M. \tag{5.7.27}$$

Hence, from Eqs. (5.7.26) and (5.7.27) and the fact that $ZM = 0$, we get

$$ZP = 0, \qquad \text{or} \quad PZ' = 0. \tag{5.7.28}$$

Since $M = I - Z'(ZZ')^- Z$ and $I - M = Z'(ZZ')^- Z$, it follows from Eq. (5.7.28) that

$$P(I-M) = 0, \qquad \text{that is,} \quad P = PM. \tag{5.7.29}$$

Hence, from Eq. (5.7.25),

$$P = D_\tau M = MD_\tau. \tag{5.7.30}$$

Finally, we consider the last condition (5.7.23). Since $ZP = 0$, this gives

$$ZR\begin{bmatrix} p_{11}\gamma_{11} \\ p_{22}\gamma_{12} \\ \vdots \\ p_{NN}\gamma_{1N} \end{bmatrix} = 0, \qquad \text{or} \quad ZT'M\begin{bmatrix} p_{11}\gamma_{11} \\ \vdots \\ p_{NN}\gamma_{1N} \end{bmatrix} = 0$$

for all arbitrary matrices $T$. Hence, as in Eq. (5.7.27),

$$M\begin{bmatrix} p_{11}\gamma_{11} \\ \vdots \\ p_{NN}\gamma_{1N} \end{bmatrix} = 0.$$

That is,

$$\sum_{j=1}^{N} \gamma_{1j} p_{jj} m_{ij} = 0 \qquad \text{for} \quad i = 1, 2, \ldots, N,$$

or from Eq. (5.7.30),

$$\sum_{j=1}^{N} \gamma_{1j} \tau_j m_{jj} m_{ij} = 0 \qquad \text{for} \quad i = 1, 2, \ldots, N. \tag{5.7.31}$$

From Eq. (5.7.30), we can rewrite Eq. (5.7.24) as

$$(\gamma_i - 3) m_{ii} \tau_i m_{ij} + 2\sum_k m_{ki} m_{kj} \tau_i = \lambda m_{ij},$$

or

$$[(\gamma_i-3)m_{ii}\tau_i+2\tau_i-\lambda]m_{ij}=0, \qquad i,j=1,2,\dots,N.$$

Thus, Eqs. (5.7.21)–(5.7.23) are equivalent to

(5.7.32)
$$P=MD_\tau=D_\tau M,$$

$$\sum_{i=1}^{N}\tau_i m_{ii}=1,$$

$$\{\tau_i[(\gamma_i-3)m_{ii}+2]-\lambda\}m_{ij}=0, \qquad i,j=1,2,\dots,N,$$

$$\sum_{j=1}^{N}\gamma_{1j}\tau_j m_{jj}m_{ij}=0, \qquad i=1,2,\dots,N.$$

Let us consider the case when $\gamma_{1i}\equiv\alpha$, $i=1,2,\dots,N$, and $\gamma_i\equiv\gamma$, $i=1,2,\dots,N$. Then, in order for the estimator to be free of $\gamma$ and $\alpha$, we must have

$$\tau_i=\frac{1}{f}, \quad m_{ii}=\frac{f}{N}, \text{ and } \sum_{j=1}^{N}m_{ij}=0, \qquad i=1,2,\dots,N.$$

Thus, the usual estimate of $\sigma^2$, $f^{-1}\mathbf{x}'M\mathbf{x}$, will be UMVQUE iff

$$m_{ii}=N^{-1}f \text{ and } \sum_{j=1}^{N}m_{ij}=0, \qquad i=1,\dots,N.$$

## 5.8 Estimation of Variance Components

In this section, we consider the variance-component model (5.1.2), which is given by

(5.8.1)
$$\mathbf{x}=Z'\boldsymbol{\beta}+U\boldsymbol{\eta}, \qquad E\boldsymbol{\eta}=\mathbf{0}, \text{ and } E\boldsymbol{\eta}\boldsymbol{\eta}'=D_\sigma,$$

where $\boldsymbol{\eta}'=(\boldsymbol{\eta}_1',\boldsymbol{\eta}_2',\dots,\boldsymbol{\eta}_k')$, with $\boldsymbol{\eta}_i$ a $c_i$ vector; $D_\sigma=\operatorname{diag}(\sigma_1^2 I_{c_1},\dots,\sigma_k^2 I_{c_k})$, with $c=\sum_{i=1}^{k}c_i\geqslant N$; $\boldsymbol{\beta}:m\times 1$; $\sigma_1^2,\dots,\sigma_k^2$ $(\sigma_1^2\neq\sigma_2^2\neq\cdots\neq\sigma_k^2)$ are unknown parameters; and $Z:m\times N$ and $U:N\times c$ are known matrices. As in Section 5.1, we shall write $U=(U_1,\dots,U_k)$ and $V_i=U_iU_i'$ with $\rho(V_i)=\rho(U_i)=c_i$ $(i=1,2,\dots,k)$. In this section, we shall be concerned primarily with the problem of estimating $\sigma_1^2,\dots,\sigma_k^2$ or linear functions of them. These estimates can be used to obtain an estimate of $\boldsymbol{\beta}$ (see, for example, Problem 5.8), but we shall not go into the details for such estimates. Let a linear function of $\sigma_1^2,\dots,\sigma_k^2$ be denoted by

(5.8.2)
$$f=\sum_{i=1}^{k}f_i\sigma_i^2,$$

where $f_1, f_2, \ldots, f_k$ are known constants. The problem is to estimate $f$. For this, a quadratic-form estimator $q = \mathbf{x}'A\mathbf{x}$ will be considered. Since

$$Eq = E(\mathbf{x}'A\mathbf{x}) = E\,\mathrm{tr}\,A\mathbf{x}\mathbf{x}' = \mathrm{tr}\,AE(\mathbf{x}\mathbf{x}') = \mathrm{tr}\,AV + \boldsymbol{\beta}'ZAZ'\boldsymbol{\beta},$$

where

$$V = UD_\sigma U' = \sum_{i=1}^{k} \sigma_i^2 V_i$$

is nonsingular, the estimator $q$ will be unbiased iff

$$ZAZ' = 0 \quad \text{and} \quad \mathrm{tr}\,AV_i = f_i, \quad i = 1, 2, \ldots, k. \tag{5.8.3}$$

In addition to unbiasedness, we may sometimes desire that the estimate be invariant under the translation of the parameter $\boldsymbol{\beta}$. That is, if instead of $\boldsymbol{\beta}$, we consider $\mathbf{r} = \boldsymbol{\beta} - \boldsymbol{\beta}_0$ as the unknown parameter, the estimate $\mathbf{x}'A\mathbf{x}$ should remain unchanged. This can happen iff

$$\mathbf{x}'A\mathbf{x} = (\mathbf{x} - Z'\boldsymbol{\beta}_0)'A(\mathbf{x} - Z'\boldsymbol{\beta}_0) \text{ for all } \boldsymbol{\beta}_0,$$

that is, iff

$$AZ' = 0. \tag{5.8.4}$$

Thus, a quadratic estimator $\mathbf{x}'A\mathbf{x}$ will be unbiased iff Eq. (5.8.3) is satisfied, and it is unbiased as well as invariant iff Eqs. (5.8.3) and (5.8.4) are both satisfied. However, neither Eq. (5.8.3) nor Eqs. (5.8.3) and (5.8.4) together give a unique solution for $A$. In fact there will be several $A$'s satisfying these conditions. Thus, some suitable criterion is needed to obtain $A$. Several criteria have been proposed in the literature, but they all appear to be *ad hoc*. A unified theory known as MINQUE (minimum-norm quadratic unbiased estimator) has been developed by Rao (1970, 1971). Mitra (1971) gave an alternative derivation of this MINQUE theory on the lines of Gauss–Markov theorem. Several extensions of this theory have been carried out in the literature. For an extensive bibliography, the reader is referred to the excellent lecture notes of Kleffe (1977).

In the next two subsections, we describe the MINQUE theory *without* invariance and *with* invariance [that is, $A$ satisfying Eq. (5.8.3) only, and $A$ satisfying Eqs. (5.8.3) and (5.8.4)]. In further subsections, we consider other aspects of MINQUE theory.

### *5.8.1* MINQUE *Theory (Without Invariance)*

The main problem here is to obtain $A$ by minimizing a certain norm subject to Eq. (5.8.3). To define an appropriate norm, let us assume that the error vector $\boldsymbol{\eta}$ is *known*. Then a reasonable unbiased estimate of

$f=\sum_{i=1}^{k} f_i\sigma_i^2$ based on $\boldsymbol{\eta}$ is given by

$$(5.8.5) \qquad \boldsymbol{\eta}'\Delta\boldsymbol{\eta} \quad \text{with} \quad \Delta=\operatorname{diag}\left(\frac{f_1}{c_1}I_{c_1},\ldots,\frac{f_k}{c_k}I_{c_k}\right),$$

while any quadratic estimator $\mathbf{x}'A\mathbf{x}$ can be written as

$$(5.8.6) \qquad (\boldsymbol{\beta}'Z+\boldsymbol{\eta}'U')A(Z'\boldsymbol{\beta}+U\boldsymbol{\eta})$$

$$=(\boldsymbol{\eta}',\boldsymbol{\beta}')\left[\begin{pmatrix} U' & 0 \\ 0 & I\end{pmatrix}\begin{pmatrix} A & AZ' \\ ZA & 0\end{pmatrix}\begin{pmatrix} U & 0 \\ 0 & I\end{pmatrix}\right]\begin{pmatrix}\boldsymbol{\eta} \\ \boldsymbol{\beta}\end{pmatrix},$$

where $ZAZ'=0$ for unbiasedness. Hence, Rao (1970) proposes to determine $A$ by minimizing

$$(5.8.7) \qquad \|R\| \equiv \left\|\begin{pmatrix} U' & 0 \\ 0 & I\end{pmatrix}\begin{pmatrix} A & AZ' \\ ZA & 0\end{pmatrix}\begin{pmatrix} U & 0 \\ 0 & I\end{pmatrix}-\begin{pmatrix} \Delta & 0 \\ 0 & 0\end{pmatrix}\right\|$$

subject to the condition (5.8.3), where $\|G\|$ denotes the Euclidean norm $(\operatorname{tr} GG')^{1/2}$. Notice that Eq. (5.8.7) is equal to

$$\|R\|^2=\operatorname{tr}(U'AU-\Delta)^2+2\operatorname{tr} AZ'ZAUU'$$

$$=\operatorname{tr} AWAW+2\operatorname{tr} AWAZ'Z-\sum_{i=1}^{k}\frac{f_i^2}{c_i}$$

because, under Eq. (5.8.3),

$$\operatorname{tr} AU\Delta U'=\operatorname{tr} U'AU\Delta=\sum_{i=1}^{k}\frac{f_i}{c_i}\operatorname{tr} U_i'AU_i$$

$$=\sum_{i=1}^{k}\frac{f_i}{c_i}\operatorname{tr} AV_i=\sum_{i=1}^{k}\frac{f_i^2}{c_i}$$

and

$$W=UU'=\sum_{i=1}^{k}U_iU_i'=\sum_{i=1}^{k}V_i.$$

Hence, the problem of minimizing $\|R\|$ subject to Eq. (5.8.3) is equivalent to minimizing

$$(5.8.8) \qquad \operatorname{tr} AWA(W+2Z'Z)$$

subject to Eq. (5.8.3). The solution of this problem is given in Lemma 5.8.1, and the estimate $\mathbf{x}'\hat{A}\mathbf{x}$ is known as MINQUE without invariance, where $\hat{A}$ is a solution matrix of Eq. (5.8.8), given in Lemma 5.8.1. We require the following notation for this purpose. Let $W$ be positive definite, and denote

$$(5.8.9) \qquad P=Z'(ZW^{-1}Z')^{-}Z, \qquad P_W=PW^{-1},$$

and

(5.8.10) $$W_1 = W + Z'TZ,$$

$T$ being a symmetric matrix. Let $L$ be an $N \times N$ nonsingular matrix such that

(5.8.11) $$L'WL = I, \qquad L'W_1L = D_\delta = \text{diag}(\delta_1, \delta_2, \ldots, \delta_N),$$

and let $M_i$ be a matrix with its $(r,s)$th element equal to the $(r,s)$th element of $L'V_iL$ divided by $\delta_r + \delta_s$. Note that for any matrices $B$, $C$, and $R$,

(5.8.12) $$\operatorname{tr} B(D_\delta M_i + M_i D_\delta) = \operatorname{tr} B(L'V_iL)$$

and

(5.8.13) $$\operatorname{tr} P_W' C P_W WRW_1 = \operatorname{tr} P_W WRW_1 P_W' C = \operatorname{tr} PRZ'SZC,$$

where

$$W_1 P_W' = Z'(ZW^{-1}Z')^- Z + Z'T(ZW^{-1}Z')(ZW^{-1}Z')^- Z$$
$$= Z'\left[(ZW^{-1}Z')^- + T\right]Z = Z'SZ \quad \text{(say)}.$$

With this notation, the solution of Eq. (5.8.8) is now contained in the following.

**Lemma 5.8.1** *Let $A$ be a symmetric matrix such that $ZAZ' = 0$ and $\operatorname{tr} AV_j = f_j$, $j = 1, 2, \ldots, k$. If $W$ and $W_1 = W + Z'TZ$ are positive definite, then the* minimum *value of* $\operatorname{tr} AWAW_1$ *over all $A$ is attained at $A = \hat{A}$, given by*

$$\hat{A} = \sum_{i=1}^{k} \lambda_i (C_i - P_W' C_i P_W) \equiv \sum_{i=1}^{k} \lambda_i A_i \quad \textit{(say)},$$

*where the $\lambda_i$'s are to be determined from $\sum_{i=1}^{k} \lambda_i \operatorname{tr} A_i V_j = f_j$, $j = 1, 2, \ldots, k$, $A_i = C_i - P_W C_i P_W$, $C_i = LM_iL'$, $L'WL = I$, $P_W = Z'(ZW^{-1}Z')^- ZW^{-1}$, and $M_i$ is defined above.*

PROOF. Let

$$A = \hat{A} + R,$$

$R$ being a symmetric matrix. Since $ZAZ' = Z\hat{A}Z' = 0$ and $\operatorname{tr} AV_j = \operatorname{tr} \hat{A}V_j = f_j$, $j = 1, 2, \ldots, k$, we get

(5.8.14) $$ZRZ' = 0 \quad \text{and} \quad \operatorname{tr} RV_j = 0 \quad \text{for } j = 1, 2, \ldots, k.$$

Then, it is easy to verify that

$$P_W WRW_1 P_W' = 0 \;\Rightarrow\; \operatorname{tr} P_W' C_i P_W WRW_1 = 0.$$

Hence,

$$\begin{aligned}\operatorname{tr}\hat{A}WRW_1 &= \sum_{i=1}^{k} \lambda_i \operatorname{tr}(C_i - P'_W C_i P_W) WRW_1 = \sum_{i=1}^{k} (\operatorname{tr} C_i WRW_1)\lambda_i \\ &= \sum_{i=1}^{k} \lambda_i \operatorname{tr}\left(LM_i L' L'^{-1} L^{-1} R L'^{-1} D_\delta L^{-1}\right) \\ &= \sum_{i=1}^{k} \lambda_i \operatorname{tr} M_i L^{-1} R L^{-1\prime} D_\delta \\ &= \frac{1}{2} \sum_{i=1}^{k} \lambda_i \operatorname{tr}(M_i D_\delta + D_\delta M_i) L^{-1} R L'^{-1} \\ &= \frac{1}{2} \sum_{i=1}^{k} \lambda_i \operatorname{tr} V_i R = 0.\end{aligned}$$

Thus, it is easy to see that

$$\begin{aligned}\operatorname{tr} AWAT &= \operatorname{tr}\hat{A}W\hat{A}W_1 + \operatorname{tr} RWRW_1 \\ &= \operatorname{tr}\hat{A}W\hat{A}W_1 + \operatorname{tr}(A-\hat{A})W(A-\hat{A})W_1\end{aligned}$$

which proves Lemma 5.8.1. □

It may be noted that Lemma 5.8.1 is applicable even when some prior information about the order of magnitude of $\boldsymbol{\beta}$ compared to $\boldsymbol{\eta}$, and some prior information on $\sigma_1^2, \ldots, \sigma_k^2$ in terms of the magnitudes of their ratios, are available. For example, if the average value of the components of $\boldsymbol{\beta}$ is $t$ times the average standard deviation of the $\boldsymbol{\eta}$ variables and if the ratio of the magnitude of $\sigma_1^2, \ldots, \sigma_k^2$ are $\alpha_1^2, \ldots, \alpha_k^2$, then we need to minimize

$$\operatorname{tr} AW(\alpha)AW(\alpha) + 2t^2 \operatorname{tr} AZ'ZAW(\alpha)$$

subject to Eq. (5.8.3) and $W(\alpha) = \Sigma_{i=1}^k \alpha_i^2 V_i$. This is easily obtainable from the above Lemma 5.8.1.

### *5.8.2* MINQUE *with Invariance*

The MINQUE with invariance of $f = \Sigma_{i=1}^k f_i \sigma_i^2$ is obtained by a quadratic form $\mathbf{x}'A\mathbf{x}$ where $A$ is determined by minimizing $\|R\|$ or $\operatorname{tr} AWAW$ subject to Eqs. (5.8.3) and (5.8.4). The solution to this problem is given by

$$\hat{A} = \sum_{i=1}^{k} \lambda_i^{(1)} (I - P'_W) W^{-1} V_i W^{-1} (I - P_W), \tag{5.8.15}$$

where $\lambda_i^{(1)}, \lambda_i^{(2)}, \ldots, \lambda_i^{(k)}$ is a solution from

$$\sum_{i=1}^{k} \lambda_i^{(1)} \operatorname{tr}\left[(I - P_W)' W^{-1} V_i W^{-1} (I - P_W) V_j\right] = f_j \tag{5.8.16}$$

for $j=1,2,\ldots,k$. The proof of Eq. (5.8.15) is similar to that given for Lemma 5.8.1, and the details are left to the reader.

Since the MINQUE uses the condition of unbiasedness, it may be negative. In the following examples we obtain MINQUE with invariance as well as without invariance.

### 5.8.3 Examples

In the model (5.8.1), let us assume that

$$Z'\boldsymbol{\beta}=\mu\mathbf{e}$$

and $U$ is nonsingular, where $\mathbf{e}'=(1,1,\ldots,1):1\times N$.

EXAMPLE 5.8.1. If $k=1$, then $W=UU'$, $c_1=N$, and $\sigma_1^2=\sigma^2$ (say). Here MINQUE with invariance is

$$\mathbf{x}'\hat{A}\mathbf{x}=\frac{\mathbf{x}'[W^{-1}(I-P_W)]\mathbf{x}}{N-1}=\frac{\mathbf{x}'W^{-1}\mathbf{x}-(\mathbf{x}'W^{-1}\mathbf{e})^2/(\mathbf{e}'W^{-1}\mathbf{e})}{N-1},$$

while MINQUE without invariance is the same. Notice that it is nonnegative for all $\mathbf{x}$.

EXAMPLE 5.8.2. If $k\geqslant 2$, then $\sum_{j=1}^{k}c_j=N$. Using Eqs. (5.8.15) and (5.8.16), MINQUE with invariance for $f=\sum_{i=1}^{k}f_i\sigma_i^2$ is given by

$$\hat{f}_{mi}=\sum_{i=1}^{k}\mathbf{x}'\begin{bmatrix}A_{1i}\\ \vdots\\ A_{ki}\end{bmatrix}(A_{i1},A_{i2},\ldots,A_{ik})\lambda_i\mathbf{x},$$

where $\lambda_1,\lambda_2,\ldots,\lambda_k$ are determined from

$$\sum_{i=1}^{k}\lambda_i\operatorname{tr}A_{ji}A_{ij}=f_j,\qquad j=1,2,\ldots,k,$$

and $W=UU'$, so that

$$W^{-1}-\frac{W^{-1}\mathbf{e}\mathbf{e}'W^{-1}}{\mathbf{e}'W^{-1}\mathbf{e}}=(A_{ij}),$$

where $A_{ij}$ is a $c_i\times c_j$ matrix. When $W=I_N$ or $U=I_N$, then

$$A_{ii}=I_{c_i}-\frac{\mathbf{e}_{c_i}\mathbf{e}'_{c_i}}{N},\qquad A_{ij}=-\frac{\mathbf{e}_{c_i}\mathbf{e}'_{c_j}}{N},$$

$\mathbf{e}_r$ being an $r\times 1$ vector of unit elements. Then it is easy to verify that

$$\frac{N-2}{N}\lambda_jc_j+c_j\sum_{i=1}^{k}\frac{\lambda_ic_i}{N^2}=f_j\qquad\text{for}\quad j=1,2,\ldots,k$$

and these give the unique solution for $\lambda_1, \lambda_2, \ldots, \lambda_k$ as

$$\lambda_j = \left[ \frac{f_j}{c_j} - \sum_{i=1}^{k} \frac{f_i}{N(N-1)} \right] \frac{N}{N-2} \qquad \text{for } j=1,2,\ldots,k.$$

Using these in $\hat{f}_{mi}$, we get

(5.8.17)

$$\hat{f}_{mi} = \frac{N}{N-2} \sum_{i=1}^{k} \frac{f_i}{c_i} \sum_{j=c_1+\cdots+c_{i-1}+1}^{c_1+\cdots+c_i} (x_j - \bar{x})^2 - \frac{\sum_{i=1}^{k} f_i}{(N-1)(N-2)} \sum_{j=1}^{N} (x_j - \bar{x})^2.$$

Notice that since the above estimate is the difference of two sums of squares, it can be negative for some values of $\mathbf{x}$. The MINQUE with invariance for $\sigma_i^2$ is

$$\hat{\sigma}_i^2 = \frac{N}{N-2} \frac{1}{c_i} \sum_{j=c_1+\cdots+c_{i-1}+1}^{c_1+\cdots+c_i} (x_j - \bar{x})^2 - \frac{1}{(N-1)(N-2)} \sum_{j=1}^{N} (x_j - \bar{x})^2$$

which can be negative for some values of $\mathbf{x}$.

EXAMPLE 5.8.3. If $k \geqslant 2$ and $W = I_N$ (or $U = I$), then $\sum_{j=1}^{k} c_j = N$. Then using Lemma 5.8.1, MINQUE without invariance for $f = \sum_{i=1}^{k} f_i \sigma_i^2$ can be calculated as follows:

Here $W_1 = I_N + 2\mathbf{e}_N \mathbf{e}_N'$, $L = (\mathbf{e}_N/\sqrt{N}, L_1)$ is an orthogonal matrix, $L' W_1 L = \operatorname{diag}(1+2N, I_{N-1})$, $L_1' = (L_{1c_1}, \ldots, L_{1c_k})$,

$$L' V_i L = \begin{bmatrix} c_i/N & \mathbf{e}_{c_i}' L_{1c_i}/\sqrt{N} \\ L_{1c_i}' \mathbf{e}_{c_i}/\sqrt{N} & L_{1c_i}' L_{1c_i} \end{bmatrix},$$

and hence

$$C_i = \tfrac{1}{2} V_i - \frac{1}{2(1+N)} \left[ \mathbf{e}_N (\mathbf{0} \mid \mathbf{e}_{c_i}' \mid \mathbf{0}) + \begin{bmatrix} \mathbf{0} \\ \mathbf{e}_{c_i} \\ \mathbf{0} \end{bmatrix} \mathbf{e}_N \right] + \frac{c_i N}{(N+1)(1+2N)} P$$

with $P = \mathbf{e}_N \mathbf{e}_N'/N$ and $\mathbf{e}_N' C_i \mathbf{e}_N = c_i/2(1+2N)$. This gives

$$A_i = C_i - P C_i P = C_i - \frac{c_i}{2N(1+2N)} P$$

$$= \tfrac{1}{2} V_i - \frac{1}{2(1+N)} \left[ \mathbf{e}_N (\mathbf{0} \mid \mathbf{e}_{c_i}' \mid \mathbf{0}) + \begin{bmatrix} \mathbf{0} \\ \mathbf{e}_{c_i} \\ \mathbf{0} \end{bmatrix} \mathbf{e}_N' \right] + \frac{c_i(N-1)}{2N(N+1)} P,$$

$$\sum A_i = \tfrac{1}{2}(I_N - P),$$

and

$$\operatorname{tr} A_i V_j = \begin{cases} \dfrac{(N-1)c_i}{2(1+N)} + \dfrac{c_i^2(N-1)}{2N^2(N+1)} & \text{if } i=j, \\[2ex] \dfrac{c_i(N-1)}{2N(N+1)} \dfrac{c_j}{N} & \text{if } i \neq j. \end{cases}$$

Hence, the solution of the equations

$$\sum_{i=1}^{k} \lambda_i \operatorname{tr} A_i V_j = f_j \qquad \text{for } j=1,2,\ldots,k$$

is given by

$$\lambda_i = \left( \frac{f_i}{c_i} - \sum_{j=1}^{k} \frac{f_j}{N(N+1)} \right) \frac{2(N+1)}{N-1} \qquad \text{for } i=1,2,\ldots,k.$$

Using the above calculations in Lemma 5.8.1, we get MINQUE without invariance for $f=\sum_{i=1}^{k} f_i \sigma_i^2$ as

$$(5.8.18) \qquad \hat{f}_{m\omega} = \sum_{i=1}^{k} \lambda_i \mathbf{x}' A_i \mathbf{x}$$

$$= \frac{N+1}{N-1} \sum_{i=1}^{k} \frac{f_i}{c_i} \sum_{j=c_1+\cdots+c_{i-1}+1}^{c_1+c_2+\cdots+c_i} (x_j - \bar{x}_N)^2$$

$$- \frac{2\sum f_i}{N-1} \bar{x}_N^2 + \frac{2\bar{x}_N}{N-1} \sum_{i=1}^{k} \frac{f_i}{c_i} \sum_{j=c_1+\cdots+c_{i-1}+1}^{c_1+c_2+\cdots+c_i} x_j$$

$$- \frac{\sum f_i}{N(N-1)} \sum_{i=1}^{k} (x_i - \bar{x}_N)^2,$$

where $\bar{x}_N = \sum_{j=1}^{N} x_j / N$. In particular, when $c_1 = c_2 = \cdots = c_k = 1$, $N=k$, and $f_1 = 1$, $f_2 = \cdots = f_N = 0$, then

$$\hat{\sigma}_1^2 = x_1^2 - \sum_{j \neq j'=1}^{N} \frac{x_j x_{j'}}{N(N-1)},$$

which can be negative if $x_1$ is small compared to other observations.

Other examples can be given in the case of mixed designs.

### *5.8.4 Locally Uniformly Minimum-Variance Unbiased Estimate* (LUMVUE)

The following lemma gives the LMVUE (locally minimum variance unbiased estimate) of $f=\sum_{i=1}^{k} f_i \sigma_i^2$. Let us denote $\sigma' = (\sigma_1, \sigma_2, \ldots, \sigma_k)$.

**Lemma 5.8.2** *Let $\boldsymbol{\beta}=\boldsymbol{\beta}_0$ and $\boldsymbol{\sigma}=\boldsymbol{\sigma}_0$ be given points in the parametric space for the model* (5.8.1). *When* $\mathbf{x}$ *is normally distributed, then the* LMVUE *for $f=\Sigma_{i=1}^k f_i\sigma_i^2$ at $\boldsymbol{\beta}=\boldsymbol{\beta}_0$ and $\boldsymbol{\sigma}=\boldsymbol{\sigma}_0$ is*

$$\hat{f}=\sum_{i=1}^{k}\lambda_i^{(2)}(\mathbf{x}-Z'\boldsymbol{\beta}_0)'A_i(\mathbf{x}-Z'\boldsymbol{\beta}_0),$$

*where*

$$W(\boldsymbol{\sigma})=\sum_{i=1}^{k}\sigma_i^2V_i,$$

$$A_i=\left[W(\boldsymbol{\sigma}_0)\right]^{-1}(V_i-P_WV_iP_W')\left[W(\boldsymbol{\sigma}_0)\right]^{-1},$$

$$P_W=Z\left\{Z'\left[W(\boldsymbol{\sigma}_0)\right]^{-1}Z\right\}^{-}Z'\left[W(\boldsymbol{\sigma}_0)\right]^{-1},$$

*and $\lambda_1^{(1)},\lambda_2^{(2)},\ldots,\lambda_k^{(2)}$ are the solution of $\Sigma_{i=1}^k\lambda_i^{(2)}\operatorname{tr}A_iV_j=f_j$ for $j=1,2,\ldots,k$.*

PROOF. It is easy to see that the proposed estimate is unbiased for $f$ under any $\boldsymbol{\beta}$ and $W_\sigma=W(\boldsymbol{\sigma})$. To prove that its variance is minimum at $\boldsymbol{\beta}=\boldsymbol{\beta}_0$ and $\boldsymbol{\sigma}=\boldsymbol{\sigma}_0$, let us consider any arbitrary measurable function $g(\mathbf{x})$ of $\mathbf{x}$ with zero expectation under $\boldsymbol{\beta}$ and $\boldsymbol{\sigma}$, and has bounded variance at $\boldsymbol{\beta}=\boldsymbol{\beta}_0$ and $\boldsymbol{\sigma}=\boldsymbol{\sigma}_0$. The density function of $\mathbf{x}$ is proportional to

$$h(\mathbf{x})=h(\mathbf{x}|\boldsymbol{\beta},\boldsymbol{\sigma})=\exp\left[-\tfrac{1}{2}(\mathbf{x}-Z'\boldsymbol{\beta})'W_\sigma^{-1}(\mathbf{x}-Z'\boldsymbol{\beta})\right]. \tag{5.8.19}$$

Then it is easy to see that

$$\frac{\partial}{\partial\boldsymbol{\beta}}h(\mathbf{x})=ZW_\sigma^{-1}(\mathbf{x}-Z'\boldsymbol{\beta})h(\mathbf{x}), \tag{5.8.20}$$

$$\left(\frac{\partial}{\partial\boldsymbol{\beta}}\right)\left(\frac{\partial}{\partial\boldsymbol{\beta}}\right)'h(\mathbf{x})=\left[ZW_\sigma^{-1}(\mathbf{x}-Z'\boldsymbol{\beta})(\mathbf{x}-Z'\boldsymbol{\beta})'W_\sigma^{-1}z'-ZW_\sigma^{-1}Z'\right]h(\mathbf{x}),$$

and for $i=1,2,\ldots,k$,

$$2\frac{\partial}{\partial\sigma_i^2}h(\mathbf{x})=\left[(\mathbf{x}-Z'\boldsymbol{\beta})'W_\sigma^{-1}V_iW_\sigma^{-1}(\mathbf{x}-Z'\boldsymbol{\beta})\right]h(\mathbf{x}). \tag{5.8.21}$$

Since $g(\mathbf{x})$ has a zero expectation, we have

$$\int g(\mathbf{x})h(\mathbf{x})\,d\mathbf{x}=0,$$

and using Eqs. (5.8.21) and (5.8.20), we get

$$\operatorname{Cov}\left(g(\mathbf{x}),(\mathbf{x}-Z'\boldsymbol{\beta})'W_\sigma^{-1}V_iW_\sigma^{-1}(\mathbf{x}-Z'\boldsymbol{\beta})\right)=0, \tag{5.8.22}$$

and

$$\operatorname{Cov}\left(g(\mathbf{x}),ZW_\sigma^{-1}(\mathbf{x}-Z'\boldsymbol{\beta})(\mathbf{x}-Z'\boldsymbol{\beta})'W_\sigma^{-1}Z'\right)=0. \tag{5.8.23}$$

From Eqs. (5.8.22) and (5.8.23), it is easy to see that

$$\operatorname{Cov}(g(\mathbf{x}),\hat{f})=0$$

when the true density function of $\mathbf{x}$ is given at $\boldsymbol{\beta}=\boldsymbol{\beta}_0$ and $\boldsymbol{\sigma}=\boldsymbol{\sigma}_0$. This shows that $\hat{f}$ defined in Lemma 5.8.2 is the LMVUE of $f=\Sigma_{i=1}^k f_i\sigma_i^2$. □

*Note 5.8.1* It may be observed from Lemma 5.8.2 that MINQUE defined in Lemma 5.8.1 with $W_1=W$ is LMVUE at $\boldsymbol{\beta}=\mathbf{0}$ and $\boldsymbol{\sigma}=\mathbf{1}$ when $\mathbf{x}$ is normally distributed.

*Note 5.8.2* We can rewrite $\hat{f}$ as

$$\hat{f}=(\mathbf{x}-Z'\boldsymbol{\beta}_0)[W(\boldsymbol{\sigma}_0)]^{-1}(G-P_WGP'_W)[W(\boldsymbol{\sigma}_0)]^{-1}(\mathbf{x}-Z'\boldsymbol{\beta}_0),$$

where $G=\Sigma_{i=1}^k\lambda_i^{(2)}V_i$. Notice that $\hat{f}$ will be free from $\boldsymbol{\beta}_0$ iff

$$(G-P_WGP'_W)[W(\boldsymbol{\sigma}_0)]^{-1}Z'=0 \tag{5.8.24}$$

$$\Leftrightarrow\quad (I-P_W)G[W(\boldsymbol{\sigma}_0)]^{-1}Z'=0$$

$$\Leftrightarrow\quad [I-Z(Z'Z)^{-}Z']G[W(\boldsymbol{\sigma}_0)]^{-1}Z'=0.$$

Taking $M=I-Z(Z'Z)^{-}Z'$ and using the condition (5.8.24), $\hat{f}$ can be rewritten as

$$\hat{f}=\mathbf{x}'[MW(\boldsymbol{\sigma}_0)M]^{+}G[MW(\boldsymbol{\sigma}_0)M]^{+}\mathbf{x}, \tag{5.8.25}$$

where $(\cdot)^{+}$ is the Moore–Penrose inverse of $(\cdot)$.

The question of the independence of $\boldsymbol{\sigma}_0$ is not easy to answer from Eqs. (5.8.24) and (5.8.25). We state the following lemma and its proof is omitted. [For the proof, one may refer to Kleffe (1977).]

**Lemma 5.8.3** *There exists an UMVUE for each linear parametric function $f=\Sigma f_i\sigma_i^2$ iff*

(i) $MGW^{-}Z'=0$ *for all* $W\in\nu$ $\{W(\boldsymbol{\sigma}),\ \sigma_i^2\geqslant 0$ *with* $|W(\boldsymbol{\sigma})|\neq 0\}$ *and for all* $G\in$ span $\nu$, *and*

(ii) *for arbitrary* $B\in\nu_M$ $\{MVM,\ V\in\nu\}$ *and* $A\in$ span $\nu_M$, *we have* $AB^{+}A\in$ span $\nu_M$.

For some further results on MVUE, one can refer to Rao (1971), and Seely (1970).

## Problems

**5.1** Let the linear model be

$$\mathbf{x}=Z'\boldsymbol{\beta}+\sum_{i=1}^{k}U_i\boldsymbol{\xi}_i=Z'\boldsymbol{\beta}+U\boldsymbol{\xi},$$

where $Z$ and $U=(U_1,\ldots,U_k)$ are given matrices, $\boldsymbol{\beta}$ is unknown, and for $i\neq j$, $i,j=1,2,\ldots,k$,

$$E(\boldsymbol{\xi}_i)=\mathbf{0},\qquad \operatorname{Var}(\boldsymbol{\xi}_i)=\Sigma_i,\quad\text{and}\quad \operatorname{Cov}(\boldsymbol{\xi}_i,\boldsymbol{\xi}_j)=0.$$

Assume that $\Sigma_1,\Sigma_2,\ldots,\Sigma_k$ are completely unknown. If $\Sigma_1=\Sigma_2$, then the number $k$ will be reduced to $k-1$. Obtain MINQUE (with or without invariance) for $\sum_{i=1}^{k}\operatorname{tr}\Sigma_iF_i=f$, where $F_i$, $i=1,2,\ldots,k$, are known symmetric matrices. [Hint: Let $\mathbf{x}'A\mathbf{x}$ be the required estimate of $f$. Then, conditions for unbiasedness (without invariance) are

$$ZAZ'=0\quad\text{and}\quad U_i'AU_i=F_i\quad\text{for } i=1,2,\ldots,k, \tag{5.P.1}$$

while conditions for unbiasedness (with invariance) are

$$AZ'=0\quad\text{and}\quad U_i'AU_i=F_i\quad\text{for } i=1,2,\ldots,k. \tag{5.P.2}$$

Show that the norm to be minimized subject to (5.P.1) is

$$\operatorname{tr}AWAW+2t^2\operatorname{tr}AZ'ZAW, \tag{5.P.3}$$

while the norm to be minimized subject to (5.P.2) is

$$\operatorname{tr}AWAW, \tag{5.P.4}$$

where $W=UU'$ or $W=\sum_{i=1}^{k}U_i\Sigma_{i,0}U_i'$ if the prior knowledge of $\Sigma_{i,0}$ on $\Sigma_i$ is available. Show that

$$A_{\text{op}}=R\left(\sum_{i=1}^{k}U_i\Lambda_iU_i'\right)R,\qquad R=W^{-1}(I-P_W),$$

is optimum under (5.P.2), where $\Lambda_1,\Lambda_2,\ldots,\Lambda_k$ are to be determined from

$$F_i=\sum_{j=1}^{k}(U_i'RU_j)\Lambda_j(U_j'RU_i),\qquad i=1,2,\ldots,k,$$

while

$$A_{\text{opt}}=L\left(\sum_{i=1}^{k}U_i\Lambda_iU_i'\right)L'-P_W'L\left(\sum_{i=1}^{k}U_i\Lambda_iU_i'\right)L'P_W$$

is optimum under (5.P.1), where $\Lambda_1,\Lambda_2,\ldots,\Lambda_k$ are to be determined from

$$F_j=\sum_{i=1}^{k}(U_j'LU_i)\Lambda_i(U_j'LU_i)'-\sum_{i=1}^{k}(U_j'P_W'LU_i)\Lambda_i(U_j'P_W'LU_i)'$$

and the others are defined in Section 5.8.]

**5.2** Let the linear model be given by

$$X = Z'\beta + \sum_{i=1}^{k} U_i \xi_i,$$

where $X$, $Z$, and $U_i$ are known matrices of respective orders $n\times p, m\times n, m\times p, n\times r_i$ $(i=1,2,\ldots,k)$. Defining $\mathrm{vec}\,\xi_i$ as a column vector obtained by writing successively the first column, then the second column, then the third column, etc., of $\xi_i$, assume that $\mathrm{Var}(\mathrm{vec}\,\xi_i)=\Sigma_i \otimes I_{r_i}$, $i=1,2,\ldots,k$, and $E\xi_i=0$. Show that MINQUEs (with or without invariance) of $\sum_{i=1}^{k} f_i\Sigma_i = F$ are given by

$$X'A_{\mathrm{opt}}X \qquad \text{(without invariance)}$$

and

$$X'A_{\mathrm{op}}X \qquad \text{(with invariance)},$$

where $A_{\mathrm{opt}}$ and $A_{\mathrm{op}}$ are given respectively in Problem 5.1.

**5.3** Let $A$ be the class of symmetric matrices such that $AZ'=0$ and $\mathrm{tr}\,AV_i = p_i$, $i=1,2,\ldots,k$. Then the minimum of $\sum_{i=1}^{s}\mathrm{tr}(AW_iAW_i)$, where $W_i$ $(i=1,2,\ldots,k)$ are nonnegative matrices, is attained at $A_{\mathrm{op}}$ satisfying the equations

$$M\left(\sum_{i=1}^{s} W_i A_{\mathrm{op}} W_i\right)M = M\left(\sum_{i=1}^{k}\lambda_i V_i\right)M \quad \text{and} \quad \mathrm{tr}\,A_{\mathrm{op}}V_i = p_i$$

for $i=1,2,\ldots,k$ and $M = I - Z(Z'Z)^{-}Z'$.

**5.4** Let $A$ be the class of symmetric matrices such that $AZ'=0$ and the diagonal elements of $A$ are fixed, $a_{ii}=p_i$ $(i=1,2,\ldots,n)$. Then the minimum of $\mathrm{tr}\,AWAW$ is attained at

$$A_{\mathrm{op}} = W^{-1}(I-P_\omega)\Delta(I-P_\omega)'W^{-1},$$

where $\Delta$ is a diagonal matrix. Let $\boldsymbol{\delta}$ be the $n$ vector of the diagonal elements of $\Delta$, and $\mathbf{p}$ be the $n$ vector of constants $p_1,\ldots,p_n$. Then $\boldsymbol{\delta}$ satisfies the equation

$$M_2\boldsymbol{\delta} = \mathbf{p},$$

where $M_2$ is the matrix obtained from $W^{-1}(I-P_\omega)$ by replacing each of its elements with its square.

**5.5** Let the linear model be given by

$$\mathbf{x} = Z'\boldsymbol{\beta} + \boldsymbol{\varepsilon}, \qquad E(\boldsymbol{\varepsilon}) = \mathbf{0}, \qquad \mathrm{Var}(\boldsymbol{\varepsilon}) = \sum_{i=1}^{k}\lambda_i V_i = W_\lambda \quad \text{(say)},$$

where $Z, V_1,\ldots,V_k$ are known matrices and $\boldsymbol{\beta},\lambda_1,\ldots,\lambda_k$ are unknowns such that Var $(\boldsymbol{\varepsilon})$ is p.s.d. Let $G$ denote a $g$ inverse of $\sum_{i=1}^{k}V_i + Z'Z$. Here, $V_1, V_2,\ldots,V_k$ are p.s.d. Show that there exists a BLUE for $Z'\boldsymbol{\beta}$ iff

$$ZGV_j[I - Z'(ZZ')^{-}Z] = 0 \qquad \text{for } j=1,2,\ldots,k.$$

Under the above condition, $Z'(ZGZ')^{-}ZG\mathbf{x}$ is one choice of a BLUE for $X\boldsymbol{\beta}$. [Hint: Taking $Z=Z_2Z_1$ and $\boldsymbol{\beta}_1 = Z_2'\boldsymbol{\beta}$ as in Section 5.2, the necessary and sufficient condition for $R\mathbf{x}$ with $RZ_1' = I_r$ to be BLUE for $\boldsymbol{\beta}_1$ is $R(\Sigma_{i=1}^k \lambda_i V_i)[I - Z_1'(Z_1 Z_1')^{-1} Z_1] = 0$ for all $\lambda_1, \ldots, \lambda_k$, i.e., $RV_i[I - Z'(ZZ')^- Z] = 0$ for $i = 1, 2, \ldots, k$. Notice that $R = (Z_1 G Z_1')^{-1} Z_1 G$ satisfies the condition $RZ_1' = I_r$.]

**5.6** In Problem 5.5, $X\boldsymbol{\beta}$ will have a BLUE iff $W_\lambda$ admits a representation

$$W_\lambda = Z'\Sigma_0 Z + W'\Sigma W,$$

where $W$ is a matrix such that $\rho(Z', W') = \rho(Z) + \rho(W) = N$, and $\Sigma, \Sigma_0$ are p.s.d. matrices dependent on $\lambda$, while $W$ is independent of $\sigma$. [Hint: Some $\rho(Z', W') = \rho(Z) + \rho(W) = N$,

$$Z(Z'Z + W'W)^{-1} W' = 0, \qquad W(Z'Z + W'W)^{-1} W'W = W$$

and

$$Z(Z'Z + W'W)^{-1} Z'Z = Z.$$

Under $W_\sigma = Z'\Sigma_0 Z + W'\Sigma W$, prove that $Z'Z(Z'Z + W'W)^{-1}\mathbf{x}$ is BLUE for $Z'\boldsymbol{\beta}$. Notice that $EW(Z'Z + W'W)^{-1}\mathbf{x} = \mathbf{0}$, and hence the necessary and sufficient condition for $Z'Z(Z'Z + W'W)^{-1}\mathbf{x}$ to be BLUE for $Z'\boldsymbol{\beta}$ is

$$Z(Z'Z + W'W)^{-1} V_i (Z'Z + W'W)^{-1} W' = 0$$

$$\Leftrightarrow \quad V_i = Z'\Lambda_{1i} Z + W'\Lambda_{2i} W \quad \text{for some } \Lambda_{1i} \text{ and } \Lambda_{2i}.]$$

**5.7** Show that the representation of $W_\lambda$ in Problem 5.6 when it holds is unique.

**5.8** Consider the linear model as given in Problem 5.5. Let $\hat{W} = \Sigma_{i=1}^k \hat{\lambda}_i V_i$, in which $\hat{\lambda}_i$ $(i = 1, 2, \ldots, k)$ are MINQUEs (with invariance) for $\lambda_1, \lambda_2, \ldots, \lambda_k$. Assume that $\mathbf{x}$ is normally distributed. Then the proposed estimate

$$Z'\hat{\boldsymbol{\beta}} = Z'(Z\hat{W}^{-1}Z')^- Z\hat{W}^{-1}\mathbf{x}$$

is an unbiased estimate of $Z'\boldsymbol{\beta}$. [Hint: Show that $E(\mathbf{x}|\mathbf{u}) = E[Z'(ZZ')^- Z\mathbf{x} + \mathbf{u}|\mathbf{u}] = \mathbf{u} + E(Z'(ZZ')^- Z\mathbf{x}) = \mathbf{u} + Z'\boldsymbol{\beta}$, where $\mathbf{u} = [I - Z'(ZZ')^- Z]\mathbf{x}$ and $Z\mathbf{x}$ are independently distributed. Notice that $\hat{W} = f(\mathbf{u})$ is such that $f(-\mathbf{u}) = f(\mathbf{u})$. Then show that

$$EZ'(Z\hat{W}^{-1}Z')^- Z\hat{W}^{-1}\mathbf{u} = \mathbf{0}.]$$

**5.9** From the linear model of Problem 5.5, deduce the following model (known as derived linear model):

$$E(M\mathbf{x} \otimes M\mathbf{x}) = (M \otimes M) D\boldsymbol{\lambda},$$

where

$$\boldsymbol{\lambda}' = (\lambda_1, \lambda_2, \ldots, \lambda_k), \quad M = I - Z'(ZZ')^- Z, \quad \text{and} \quad D = (\operatorname{vec} V_1, \operatorname{vec} V_2, \ldots, \operatorname{vec} V_k),$$

$\operatorname{vec}\xi$ being defined in Problem 5.2. Use this model to obtain various types of estimates for $\boldsymbol{\lambda}$.

**5.10** Let $\boldsymbol{\varepsilon}$ be a random vector. Denote

$$E(\boldsymbol{\varepsilon} \otimes \boldsymbol{\varepsilon}\boldsymbol{\varepsilon}') = \zeta \quad \text{and} \quad E(\boldsymbol{\varepsilon}\boldsymbol{\varepsilon}' \otimes \boldsymbol{\varepsilon}\boldsymbol{\varepsilon}') = \psi,$$

where $\zeta$ and $\psi$ are matrices of order $N^2 \times N$ and $N^2 \times N^2$. Show that

$$E(\boldsymbol{\varepsilon}' A \boldsymbol{\varepsilon})(\boldsymbol{\varepsilon}' B \boldsymbol{\varepsilon}) = \operatorname{tr}[(A \otimes B)\psi]$$

and

$$E(\boldsymbol{\varepsilon}'\mathbf{a})(\boldsymbol{\varepsilon}' A \boldsymbol{\varepsilon}) = \operatorname{tr}[(a' \otimes A)\zeta].$$

Use the above results to obtain the variance of an unbiased estimate (5.6.4) for $f = \sum_{i=1}^{k} f_i \sigma_i^2$. Find the expression when $\boldsymbol{\varepsilon} \sim N(0, V_\sigma)$.

# 6

# Linear Models—Testing of Hypotheses for Regression Parameters

## 6.1 Introduction

In Chapter 5, we considered the problem of estimating the parameters involved in linear models. In this chapter, we shall be concerned with the testing of any estimable linear function of regression parameters of the linear models. For this purpose, we shall assume that the errors are normally distributed. In Chapter 5 we made no assumption on the distribution of the errors.

## 6.2 Tests on the Regression Parameter $\boldsymbol{\beta}$

Let us consider a simple linear model

(6.2.1) $$\mathbf{x}=Z'\boldsymbol{\beta}+\boldsymbol{\varepsilon},$$

where $\boldsymbol{\varepsilon}\sim N(\mathbf{0},\sigma^2V)$, $V$ is a known $N\times N$ p.s.d. matrix, $Z$ is a known $m\times N$ matrix, and $\boldsymbol{\beta}$ and $\sigma^2$ are unknown parameters. In this section we consider the problem of testing the hypothesis $H:C\boldsymbol{\beta}=\mathbf{0}$ vs $A:C\boldsymbol{\beta}\neq\mathbf{0}$. The case where $V$ is singular is treated separately from that where $V$ is nonsingular.

### *6.2.1 Likelihood-Ratio Test for $C\,\boldsymbol{\beta}=\mathbf{0}$ when $V$ Is P.D.*

As in Section 5.2, let us write

(6.2.2) $$Z=Z_2Z_1,\qquad \beta=Z_2(Z_2'Z_2)^{-1}\boldsymbol{\beta}_1+Z_3(Z_3'Z_3)^{-1}\boldsymbol{\beta}_2,$$

where $Z_2$ and $Z_1$ are $m\times r$ and $r\times N$ matrices of ranks $r$ $[=\rho(Z)]$ and $\rho(Z_2,Z_3)=m$ with $Z_3'Z_2=0$. With this notation, the likelihood function for

$\boldsymbol{\beta}_1$ and $\sigma^2$ is given by

$$(6.2.3)\quad L(\boldsymbol{\beta}_1,\sigma^2)=\text{const}\,\sigma^{-N}\exp\left(-\frac{(\mathbf{x}-Z_1'\boldsymbol{\beta}_1)'V^{-1}(\mathbf{x}-Z_1'\boldsymbol{\beta}_1)}{2\sigma^2}\right)$$

with $\text{const}=(2\pi)^{-N/2}|V|^{-1/2}$. Taking

$$(6.2.4)\quad \boldsymbol{\omega}=\left(Z_1V^{-1}Z_1'\right)^{-1}Z_1V^{-1}\mathbf{x}\quad\text{and}\quad s^2=\mathbf{x}'V^{-1}\mathbf{x}-\mathbf{x}'V^{-1}Z_1'\boldsymbol{\omega},$$

the likelihood (6.2.3) can be rewritten as

$$(6.2.5)\quad L(\boldsymbol{\beta}_1,\sigma^2)=\text{const}\,\sigma^{-N}\exp\left(-\frac{s^2}{2\sigma^2}-\frac{(\boldsymbol{\omega}-\boldsymbol{\beta}_1)'(Z_1V^{-1}Z_1')(\boldsymbol{\omega}-\boldsymbol{\beta}_1)}{2\sigma^2}\right).$$

This will be maximized at $\boldsymbol{\beta}_1=\boldsymbol{\omega}$ and $\sigma^2=s^2/N$. Hence

$$(6.2.6)\qquad \operatorname*{Max}_{\Omega} L(\boldsymbol{\beta}_1,\sigma^2)=\text{const}\left(\frac{s^2}{N}\right)^{-N/2}e^{-N/2}.$$

Since $C\boldsymbol{\beta}$ is estimable, we can write from Eq. (6.2.2)

$$(6.2.7)\quad C\boldsymbol{\beta}=C_1\boldsymbol{\beta}_1,\qquad\text{where}\quad CZ_3=0\text{ and }C_1=CZ_2(Z_2'Z_2)^{-1}.$$

Hence, if $\rho(C)=l$, where $C:l\times m$, $l\leqslant r$, then there exists a matrix $C_2$: $(r-l)\times r$ with $\rho(C_2)=r-l$ and $C_2C_1'=0$. [See the remark at the end of this subsection for the case when $\rho(C)<l$.] Thus

$$(6.2.8)\qquad C\boldsymbol{\beta}=\mathbf{0}\;\Leftrightarrow\;\boldsymbol{\beta}_1=C_2'\boldsymbol{\eta},\quad\boldsymbol{\eta}\text{ being unknown.}$$

Under $C\boldsymbol{\beta}=\mathbf{0}$, Eq. (6.2.3) can be rewritten as

$$(6.2.9)\quad L(\boldsymbol{\eta},\sigma^2)=\text{const}\,\sigma^{-N}\exp\left(-\frac{(\mathbf{x}-Z_1'C_2'\boldsymbol{\eta})'V^{-1}(\mathbf{x}-Z_1'C_2'\boldsymbol{\eta})}{2\sigma^2}\right)$$

and using the above method, we get

$$(6.2.10)\qquad \operatorname*{Max}_{\Omega_0} L(\eta,\sigma^2)=\text{const}\left(\frac{s_0^2}{N}\right)^{-N/2}e^{-N/2},$$

where

$$(6.2.11)\quad \begin{aligned} s_0^2&=\mathbf{x}'V^{-1}\mathbf{x}-\mathbf{x}'V^{-1}Z_1'C_2'\left(C_2Z_1V^{-1}Z_1'C_2'\right)^{-1}C_2Z_1V^{-1}\mathbf{x}\\ &=s^2+\mathbf{x}'V^{-1}Z_1'\left[\left(Z_1V^{-1}Z_1'\right)^{-1}-C_2'\left(C_2Z_1V^{-1}Z_1'C_2'\right)^{-1}C_2\right]\\ &\quad\times Z_1V^{-1}\mathbf{x}.\end{aligned}$$

Using Corollary 1.9.2, we get

(6.2.12) $$s_0^2 = s^2 + \omega' C_1' G^{-1} C_1 \omega,$$

where $\omega$ and $s^2$ are defined in Eq. (6.2.5) and

(6.2.13) $$G = C_1(Z_1 V^{-1} Z_1')^{-1} C_1' = C(ZV^{-1}Z')^{-} C'.$$

Equations (6.2.10) and (6.2.6) give the likelihood ratio

$$(s^2/s_0^2)^{N/2} = \lambda^{N/2},$$

and the test procedure is equivalent to the following:

(6.2.14) $$\text{reject} \quad C\boldsymbol{\beta} = 0 \quad \text{if} \quad (N-r)\omega' C_1' G^{-1} C_1 \omega / ls^2 \geqslant c_\alpha,$$

where $c_\alpha$ is a constant to be determined from

(6.2.15) $$P(F_{l,N-r} \geqslant c_\alpha) = \alpha.$$

To obtain the power function of the test procedure, we observe from Corollary 2.11.2 and Corollary 2.11.4 that

(6.2.16) $$\frac{s^2}{\sigma^2} = \frac{\mathbf{x}'\left[V^{-1} - V^{-1}Z'(ZV^{-1}Z')^{-} ZV^{-1}\right]\mathbf{x}}{\sigma^2} \sim \chi^2_{N-r},$$

(6.2.17)
$$\frac{\omega' C_1' G^{-1} C_1 \omega}{\sigma^2} = \frac{\mathbf{x}'\left[V^{-1}Z'(ZV^{-1}Z')^{-} C' G^{-1} C(ZV^{-1}Z')^{-} ZV^{-1}\right]\mathbf{x}}{\sigma^2}$$
$$\sim \chi_l^2(\delta),$$

and $s^2/\sigma^2$ and $\omega' C_1' G^{-1} C_1 \omega/\sigma^2$ are independently distributed, where

(6.2.18) $$\delta = \boldsymbol{\beta}' C' G^{-1} C \boldsymbol{\beta}/\sigma^2.$$

From Eqs. (6.2.16) and (6.2.17),

(6.2.19) $$(N-r)\omega' C_1' G^{-1} C_1 \omega / ls^2 \sim F_{l,N-r}(\delta).$$

For the test procedure, we observe that

$$C_1 \omega = C(ZV^{-1}Z')^{-} ZV^{-1}\mathbf{x} \sim N(C\boldsymbol{\beta}, G).$$

Note that in the above derivation, we have taken $C: l \times m$ to be of rank $l \leqslant r$. If $\rho(C) < l$, then the above derivation still goes through with the minor change of replacing $G^{-1}$ by $G^{-}$ and $l$ by $\rho(C) = \rho(C_1)$.

### *6.2.2 Union–Intersection Test Procedure when $V$ Is Nonsingular and $\rho(C) = l$*

We observe from Eq. (6.2.2) that for all nonnull vectors $\mathbf{a}$,

$$H \in \Omega_0 = \bigcap_{\mathbf{a}} H_{\mathbf{a}}, \qquad H_{\mathbf{a}} = \{\mathbf{a}' C_1 \boldsymbol{\beta}_1 = 0\}$$

and

$$A \in \Omega = \bigcup_{\mathbf{a}} A_{\mathbf{a}}, \qquad A_{\mathbf{a}} = \{\mathbf{a}' C_1 \boldsymbol{\beta}_1 \neq 0\}.$$

Given $\mathbf{a}$, we want to test $H_{\mathbf{a}}$ vs $A_{\mathbf{a}}$. To accomplish this purpose, we observe from Eq. (6.2.3) that $\omega$ is the maximum-likelihood estimate, that $\omega \sim N\big(\boldsymbol{\beta}_1, \sigma^2 (Z_1 V^{-1} Z_1')^{-1}\big)$ and that it is distributed independently of $s^2$. Hence, by univariate results,

$$\text{reject} \quad H_{\mathbf{a}} \quad \text{if} \quad \frac{(N-r)(\mathbf{a}' C_1 \omega)^2}{(\mathbf{a}' G \mathbf{a}) s^2} \geqslant c,$$

where $G$ is defined by Eq. (6.2.13) and $c$ depends on the size of the test. Using the union–intersection principle, we get the test procedure

$$\text{(6.2.20)} \qquad \text{reject} \quad H \in \Omega_0 \quad \text{if} \quad \sup_{\mathbf{a}} \frac{(N-r)(\mathbf{a}' C_1 \omega)^2}{(\mathbf{a}' G \mathbf{a}) s^2} \geqslant c.$$

But from Corollary 1.10.1,

$$\sup_{\mathbf{a}} \frac{(\mathbf{a}' C_1 \omega)^2}{(\mathbf{a}' G \mathbf{a})} = \omega' C_1' G^{-1} C_1 \omega.$$

Hence, from Eq. (6.2.20), we get the same test procedure as given by Eq. (6.2.14).

### *6.2.3 A Test Procedure Based on* BLUE *when $V$ Is Singular*

When $V$ is nonsingular, we have seen in Subsection 6.2.1 that the likelihood-ratio test procedure (6.2.14) is based on the BLUE for $C\boldsymbol{\beta}$. When $V$ is singular or nonsingular, the BLUE $C\hat{\beta}$ for $C\boldsymbol{\beta}$ is given by

$$C\hat{\beta} = C(ZV_1^- Z')^- ZV_1^- \mathbf{x},$$

where $V_1 = V + Z'RZ$, $R$ being any symmetric matrix such that $\rho(V_1) = \rho(V, Z')$.

Let $B = V_1^- - V_1^- Z'(ZV_1^- Z')^- ZV_1^-$ and $B_1 = Z'(ZV_1^- Z')^- ZV_1^-$. Then

$$V_1 V_1^- V = V, \qquad V_1 V_1^- Z' = Z', \qquad B_1 Z' = Z',$$

$$BZ' = 0, \qquad BV = BV_1, \qquad VB = V_1 B,$$

$$VBV = V_1 - Z'(ZV_1^- Z')^- Z = VBVBV,$$

$$\rho(VBV) = \operatorname{tr} BV = \operatorname{tr} V_1 V_1^- - \operatorname{tr}(ZV_1^- Z')^-(ZV_1^- Z')$$
$$= \rho(V_1) - \rho(ZV_1^- Z') = \rho(V, Z') - \rho(Z),$$

and

$$B_1 VB = B_1 V_1 B = Z'(ZV_1^- Z')^- ZB = 0.$$

Hence, by Corollaries 2.11.2 and 2.11.4,

$$B_1\mathbf{x} \quad \text{and} \quad \mathbf{x}'B\mathbf{x} \quad \text{are independently distributed,}$$

$$B_1\mathbf{x} \sim N\left(Z'\boldsymbol{\beta}, \sigma^2\left[Z'(ZV_1^- Z')^- Z - Z'RZ\right]\right)$$

and

$$s^2 = \mathbf{x}'B\mathbf{x} \sim \sigma^2\chi_n^2, \qquad n = \rho(V, Z') - \rho(Z).$$

These results show that for estimable functions $C\boldsymbol{\beta}$,

$$n(\hat{\beta}-\beta)'C'Q^- C(\hat{\beta}-\beta)/ls^2, \tag{6.2.21}$$

where

$$Q = C(ZV_1^- Z')^- C' - CRC',$$

is distributed as $F_{l,n}$ with $l = \rho(Q)$. This can be used in obtaining a test procedure for testing $C\boldsymbol{\beta} = \mathbf{0}$, and the simultaneous confidence bounds on the parametric functions $C\boldsymbol{\beta}$ are given by

$$\mathbf{a}'C\hat{\boldsymbol{\beta}} - \left[c(l/n)s^2\mathbf{a}'Q\mathbf{a}\right]^{1/2} \leqslant \mathbf{a}'C\boldsymbol{\beta} \leqslant \mathbf{a}'C\hat{\boldsymbol{\beta}} + \left[c(l/n)s^2\mathbf{a}'Q\mathbf{a}\right]^{1/2}$$

for all nonnull vectors $\mathbf{a}$ such that $\mathbf{a}'$, $Q\mathbf{a} \neq 0$ and $P(F_{l,n} \leqslant c) = 1 - \alpha$. Such a test procedure has been proposed by Rao (1973a, b) by a different approach. As in Subsection 6.2.2, the test statistic given by Eq. (6.2.21) can be shown to be a union–intersection test procedure.

## 6.3 Multivariate Linear Model

Before considering the model, we introduce some notation for the sake of convenience of presentation.

For a $p \times N$ matrix $X$, $(\operatorname{vec} X)'$ denotes its row-vector representation when the elements of $X$ are stretched out by taking each row of $X$ one after the other in order. That is, if $X = (x_{ij})$,

$$(\operatorname{vec} X)' = (x_{11}, \ldots, x_{1N}, x_{21}, \ldots, x_{2N}, \ldots, x_{p1}, \ldots, x_{pN})$$

is a $1 \times pN$ row vector. Suppose $X \sim N_{p,N}(\mu, \Sigma, V)$, that is, suppose its pdf is given by

$$(2\pi)^{-\frac{1}{2}pN}|\Sigma|^{-\frac{1}{2}N}|V|^{-\frac{1}{2}p}\operatorname{etr}\left[-\tfrac{1}{2}\Sigma^{-1}(X-\mu)V^{-1}(X-\mu)\right].$$

Then, clearly, $\operatorname{vec} X \sim N_{pN}(\operatorname{vec}\mu, \Sigma \otimes V)$. Hence $X \sim N_{p,N}(\mu, \Sigma, V)$ iff $\operatorname{vec} X \sim N_{pN}(\operatorname{vec}\mu, \Sigma \otimes V)$. Since

$$(\operatorname{vec} A)'(\operatorname{vec} X) \sim N((\operatorname{vec} A)'(\operatorname{vec}\mu), (\operatorname{vec} A)'(\Sigma \otimes V)(\operatorname{vec} A)),$$

and since $(\operatorname{vec} A)'(\operatorname{vec} X) = \operatorname{tr} A'X$ and $(\operatorname{vec} A)'(\Sigma \otimes V)(\operatorname{vec} A) = \operatorname{tr} A'\Sigma AV$, it follows that $X \sim N_{p,N}(\mu, \Sigma, V)$ iff for every nonnull matrix $A : p \times N$, $\operatorname{tr} A'X$

$\sim N(\operatorname{tr} A'\mu, \operatorname{tr} A'\Sigma A V)$. This shows that for any matrices $L: l \times p$ and $M: N \times m$, $LXM \sim N_{l,m}(L\mu M, L\Sigma L', M'VM)$. These results will be frequently used in this section.

We now describe the model and the testing problem. Let

$$X = \beta Z + \varepsilon, \tag{6.3.1}$$

where $Z$ is a known $m \times N$ matrix, $\beta$ is an unknown $p \times m$ matrix, and $\varepsilon$ is a $p \times N$ random matrix such that

$$\varepsilon \sim N_{p,N}(0, \Sigma, V). \tag{6.3.2}$$

Here, $\Sigma$ is assumed to be an unknown p.d. matrix, and $V$ is a known $N \times N$ p.s.d. matrix. We are interested in testing the hypothesis $H: \beta C' = 0$ vs $A: \beta C' \neq 0$, where $C: l \times m$, $l \leqslant m$, and $\rho(C) = l \leqslant r = \rho(Z)$.

This model is similar to that considered in Section 6.2 when $p = 1$. Let $\mathbf{a}$ be any nonnull vector. Then for $\mathbf{a} \neq 0$,

$$\mathbf{a}'X = \mathbf{a}'\beta Z + \mathbf{a}'\varepsilon, \qquad \varepsilon'\mathbf{a} \sim N_N(\mathbf{0}, \mathbf{a}'\Sigma\mathbf{a} V), \tag{6.3.3}$$

and the BLUE for $Z'\beta'$ is

$$Z'\hat{\beta}' = Z'(ZV_1^- Z')^- ZV_1^- X', \tag{6.3.4}$$

where $V_1 = V + Z'RZ$ and $\rho(V_1) = \rho(V, Z')$. When $V$ is nonsingular, $R = 0$. We shall first consider the situation when $V$ is nonsingular.

### *6.3.1 Likelihood-Ratio Test Procedure when V Is Nonsingular*

Using the transformation (6.3.3) and the results of Subsection 6.2.1, the maximum-likelihood estimates of $Z'\beta'\mathbf{a}$ and $\mathbf{a}'\Sigma\mathbf{a}$ are respectively

$$Z'\hat{\beta}'\mathbf{a} = Z'(ZV^{-1}Z')^- ZV^{-1}X'\mathbf{a} \tag{6.3.5}$$

and

$$\mathbf{a}'\hat{\Sigma}\mathbf{a} = \mathbf{a}'X\left[V^{-1} - V^{-1}Z'(ZV^{-1}Z')^- ZV^{-1}\right]X'\mathbf{a}/N. \tag{6.3.6}$$

Since this is true for all nonnull vectors $\mathbf{a}$, the maximum-likelihood estimates for $Z'\beta'$ and $\Sigma$ are

$$Z'\hat{\beta}' = Z'(ZV^{-1}Z')^- ZV^{-1}X' \quad \text{and} \quad \hat{\Sigma} = \frac{XV^{-1}X' - XV^{-1}Z'\hat{\beta}'}{N}, \tag{6.3.7}$$

where

$$\hat{\beta}Z \sim N_{p,N}\left(\beta Z, \Sigma, Z'(ZV^{-1}Z')^- Z\right), \tag{6.3.8}$$

$$N\hat{\Sigma} \sim W_p(\Sigma, N - r),$$

and are independently distributed. Here $r = \rho(Z)$.

The likelihood function for $\beta$ and $\Sigma$ is given by

$$L(\beta,\Sigma)=\text{const}\,|\Sigma|^{-N/2}\,\text{etr}\Big[-\tfrac{1}{2}\Sigma^{-1}(X-\beta Z)V^{-1}(X-\beta Z)'\Big], \tag{6.3.9}$$

and using Eq. (6.3.7), we get

$$\operatorname*{Sup}_{\Omega} L(\beta,\Sigma)=\text{const}\,|\hat{\Sigma}|^{-N/2}e^{-Np/2}. \tag{6.3.10}$$

In the same way, we can obtain the supremum of the likelihood under the hypothesis $\beta C'=0$ where $\rho(Z,C')=\rho(Z)$, and it is given by

$$\operatorname*{Sup}_{\Omega_0} L(\beta,\Sigma)=\text{const}\,|\hat{\Sigma}_0|^{-N/2}e^{-Np/2}, \tag{6.3.11}$$

where

$$\begin{aligned} N\hat{\Sigma}_0&=N\hat{\Sigma}+XV^{-1}Z'(ZV^{-1}Z')^{-}C'G^{-1}C(ZV^{-1}Z')^{-}Z'V^{-1}X' \\ &=S_e+S_H \quad \text{(say).} \end{aligned} \tag{6.3.12}$$

Here $G$ is the same as defined in Eq. (6.2.13). From Eqs. (6.3.11) and (6.3.12), the likelihood-ratio test procedure is equivalent to

$$\text{reject } H(\beta C'=0) \quad \text{if} \quad \lambda=\frac{|S_e|}{|S_e+S_H|}\leqslant c, \tag{6.3.13}$$

where $c$ is determined from $P(\lambda\leqslant c|H)=\alpha$. Using Eq. (6.3.8), it is easy to verify that $S_H$ and $S_e$ are independently distributed, $S_e\sim W_p(\Sigma,N-r)$ and $S_H\sim W_p(\Sigma,l,\Omega)$ with $\beta C'G^{-1}C\beta'=\Omega$. If $\rho(C)<l$, instead of $G^{-1}$ we shall have $G^{-}$, and instead of $l$ we shall have $\rho(C)$.

### *6.3.2 Union–Intersection Test Procedure when V Is Nonsingular and $\rho(C)=l$*

We observe that for all nonnull vectors **a** and **b**,

$$H\in\Omega_0=\bigcap_{\mathbf{a}}\bigcap_{\mathbf{b}}H_{\mathbf{a},\mathbf{b}}, \qquad H_{\mathbf{a},\mathbf{b}}=\{\mathbf{a}'\beta C\mathbf{b}=0,\ \mathbf{a}'\Sigma\mathbf{a}>0\},$$

and

$$A\in\Omega=\bigcup_{\mathbf{a}}\bigcup_{\mathbf{b}}A_{\mathbf{a},\mathbf{b}}, \qquad A_{\mathbf{a},\mathbf{b}}=\{\mathbf{a}'\beta C\mathbf{b}\neq 0,\ \mathbf{a}'\Sigma\mathbf{a}>0\}.$$

Given **a** and **b**, we want to test $H_{\mathbf{a},\mathbf{b}}$ vs $A_{\mathbf{a},\mathbf{b}}$. For this, we use the test procedure given in Subsection 6.2.2, namely,

$$\text{reject } H_{\mathbf{a},\mathbf{b}} \quad \text{if} \quad (N-r)\frac{(\mathbf{a}'\hat{\beta}C'\mathbf{b})^2}{(\mathbf{b}'G\mathbf{b})(\mathbf{a}'S_e\mathbf{a})}\geqslant c,$$

where $c$ depends on the size of the test,

$$\hat{\beta}C'=XV^{-1}Z'(ZV^{-1}Z')^{-}C', \qquad G=C(ZV^{-1}Z')^{-}C',$$

and

$$S_e = X\left[V^{-1} - V^{-1}Z'(ZV^{-1}Z')^{-}ZV^{-1}\right]X'.$$

Hence, by the union–intersection principle, we get the test procedure

$$\text{reject} \quad H \in \Omega_0 \quad \text{if} \quad \underset{\mathbf{a},\mathbf{b}}{\text{Sup}}\,(N-r)\frac{(\mathbf{a}'\hat{\beta}C'\mathbf{b})^2}{(\mathbf{b}'G\mathbf{b})(\mathbf{a}'S_e\mathbf{a})} \geqslant c.$$

Using Corollary 1.10.1, we see that

$$\underset{\mathbf{b}}{\text{Sup}}\,\frac{(\mathbf{a}'\hat{\beta}C'\mathbf{b})^2}{\mathbf{b}'G\mathbf{b}} = \mathbf{a}'S_H\mathbf{a}$$

and

$$\underset{\mathbf{a}}{\text{Sup}}\,\frac{\mathbf{a}'S_H\mathbf{a}}{\mathbf{a}'S_e\mathbf{a}} = \text{ch}_1\left(S_H S_e^{-1}\right),$$

where $\text{ch}_1(S_H S_e^{-1})$ denotes the maximum characteristic root of $S_H S_e^{-1}$. Hence, the test is given by

$$\text{(6.3.14)} \qquad \text{reject} \quad H \in \Omega_0 \quad \text{if} \quad (N-r)\,\text{ch}_1\left(S_H S_e^{-1}\right) \geqslant c,$$

where $c$ is to be determined from $P\left((N-r)\,ch_1\left(S_H S_e^{-1}\right) \geqslant c \,|\, H \in \Omega_0\right) = \alpha$. This test procedure was given by Roy (1957). This test can be used to give simultaneous confidence bounds on the parameters $\beta C'$ as, with probability $\geqslant (1-\alpha)$,

$$\text{(6.3.15)} \qquad \mathbf{a}'\hat{\beta}C'\mathbf{b} - \left(\frac{c(\mathbf{b}'G\mathbf{b})(\mathbf{a}'S_e\mathbf{a})}{N-r}\right)^{1/2}$$

$$\leqslant \mathbf{a}'\beta C'\mathbf{b} \leqslant \mathbf{a}'\hat{\beta}C'\mathbf{b} + \left(\frac{c(\mathbf{b}'G\mathbf{b})(\mathbf{a}'S_e\mathbf{a})}{N-r}\right)^{1/2}$$

for all nonnull vectors **a** and **b** such that $\mathbf{b}'G\mathbf{b} \neq 0$.

### *6.3.3 Testing Equality of k Mean Vectors*

This problem has already been considered. In this section, however, we shall show that it is a special case of the linear regression model. Let $\mathbf{x}_{i\alpha}$ be independently distributed as $N_p(\boldsymbol{\mu}_i, \Sigma)$, $\alpha = 1, 2, \ldots, N_i$, $i = 1, 2, \ldots, k$. Writing

$$X = (\mathbf{x}_{11}, \ldots, \mathbf{x}_{1N_1}, \mathbf{x}_{21}, \ldots, \mathbf{x}_{2N_2}, \ldots, \mathbf{x}_{k1}, \ldots, \mathbf{x}_{kN_k}) : p \times N,$$

$$\beta = (\boldsymbol{\mu}_1, \boldsymbol{\mu}_2, \ldots, \boldsymbol{\mu}_k) : p \times k,$$

$$N = N_1 + N_2 + \cdots + N_k$$

and

$$Z=\begin{bmatrix} \mathbf{e}'_{N_1} & \mathbf{0}' & \mathbf{0}' & \cdots & \mathbf{0}' \\ \mathbf{0} & \mathbf{e}'_{N_2} & \mathbf{0}' & \cdots & \mathbf{0}' \\ \vdots & \vdots & \vdots & & \vdots \\ \mathbf{0} & \mathbf{0} & \mathbf{0} & \cdots & \mathbf{e}'_{N_k} \end{bmatrix} : k\times N,$$

where $\mathbf{e}_r$ denotes an $r$ vector of ones, we get $X \sim N_{p,N}(\beta Z, \Sigma, I_N)$. Suppose the problem is to test the hypothesis

$$H: \boldsymbol{\mu}_1 = \boldsymbol{\mu}_2 = \cdots = \boldsymbol{\mu}_k$$

against the alternative $A \neq H$. We shall now reduce this problem to a regression model. Let

$$C=(I_{k-1}, -\mathbf{e}_{k-1}).$$

Then the hypothesis is equivalent to the problem of testing the hypothesis

$$H: C\beta' = 0$$

in the linear regression model $X \sim N_{p,N}(\beta Z, \Sigma, I_N)$.

### 6.3.4 *A Test Procedure Based on* BLUE *when $V$ Is Singular*

When $V$ is singular, we use the results of Section 6.2.3 and a BLUE given by Eq. (6.3.4). With this notation, a test procedure for testing $H: C\beta' = 0$ is given by

$$\text{(6.3.16)} \qquad \text{reject } H \text{ if } \lambda = \frac{|S_e|}{|S_e + S_H|} \leqslant c,$$

where $c$ is to be determined from $P(\lambda \leqslant c | H) = \alpha$,

$$\text{(6.3.17)} \qquad S_H = XV_1^- Z'(ZV_1^- Z')^- C'Q^- C(ZV_1^- Z')^- ZV_1^- X' \sim W_p(\Sigma, l, \Omega),$$

$$\text{(6.3.18)} \qquad S_e = X[V_1^- - V_1^- Z'(ZV_1^- Z')^- ZV_1^-]X' \sim W_p(\Sigma, n),$$

and $S_H$ and $S_e$ are independently distributed. Here,

$$V_1 = V + Z'RZ, \qquad \rho(V_1) = \rho(V, Z'),$$
$$\Omega = \beta C'Q^- C\beta,$$
$$Q = C(ZV_1^- Z')^- C' - CRC'$$
$$l = \rho[C(I - RZV_1^- Z')], \qquad n = \rho(V, Z') - \rho(Z).$$

The union–intersection test procedure can be given as

(6.3.19) $$\text{reject} \quad H:\beta C'=0 \quad \text{if} \quad n\,\text{ch}_1\left(S_H S_e^{-1}\right) \geqslant c,$$

where $c$ is a constant to be determined from $P\left(n\,ch_1\left(S_H S_e^{-1}\right) \geqslant c|H\right)=\alpha$. The simultaneous confidence bounds on the parametric functions of $\beta C'$ are similar to Eq. (6.3.15).

### *6.3.5 Moments of* $\lambda$ *Under* $H:\beta C'=0$

The statistics $\lambda$ defined in Eqs. (6.3.13) and (6.3.16) are similar. Hence we shall obtain the moments of $\lambda$ given by Eq. (6.3.16), because it reduces to Eq. (6.3.13) when $V$ is nonsingular. By Eq. (6.3.17), $S_H \sim W_p(\Sigma, l, \Omega)$, and so we can write

(6.3.20) $$S_H = YY', \qquad Y \sim N_{p,l}(\nu, \Sigma), \quad \text{and} \quad \Omega = \nu\nu'.$$

The joint density function of $Y$ and $S_e$ under $H:\nu=0$ is given by

(6.3.21) $$\left[2^{\frac{1}{2}p(n+l)}\pi^{\frac{1}{2}pl}\Gamma_p\left(\tfrac{1}{2}n\right)|\Sigma|^{\frac{1}{2}(n+l)}\right]^{-1}|S_e|^{\frac{1}{2}(n-p-1)}$$
$$\text{etr}\left[-\tfrac{1}{2}\Sigma^{-1}(S_e+YY')\right].$$

Use the transformations

$$S_e + YY' = S, \quad W = T^{-1}Y \qquad \text{with} \quad S = TT'.$$

Using Theorem 1.11.2 and Corollary 1.11.2, the Jacobian of the transformation is

$$J(S_e, Y \to S, W) = J(S_e \to S)J(Y \to W|S \text{ is fixed})$$
$$= |T|^l = |S|^{\frac{1}{2}l}.$$

Hence, the joint density of $S$ and $W$ is given by

(6.3.22) $$\left\{2^{\frac{1}{2}p(n+l)}\Gamma_p\left[\tfrac{1}{2}(n+l)\right]|\Sigma|^{\frac{1}{2}(n+l)}\right\}^{-1}|S|^{\frac{1}{2}(n+l-p-1)}\text{etr}\left(-\tfrac{1}{2}\Sigma^{-1}S\right)$$
$$\times \frac{\Gamma_p\left[\frac{1}{2}(n+l)\right]}{\Gamma_p\left(\frac{1}{2}n\right)}|I-WW'|^{\frac{1}{2}(n-p-1)},$$

and this shows that $W$ and $S$ are independently distributed, $S \sim W_p(\Sigma, n+l)$, and the density of $W$ is given by

(6.3.23) $$\frac{\Gamma_p\left[\frac{1}{2}(n+l)\right]}{\Gamma_p\left(\frac{1}{2}n\right)}|I-WW'|^{\frac{1}{2}(n-p-1)}.$$

We observe that

$$\lambda = |I - WW'|,$$

and hence the $h$th moment of $\lambda$ is given by

$$E\lambda^h = \frac{\Gamma_p\left[\frac{1}{2}(n+l)\right]}{\Gamma_p\left(\frac{1}{2}n\right)} \int |I - WW'|^{\frac{1}{2}(n+2h-p-1)} dW.$$

Replacing $n$ by $n+2h$ in Eq. (6.3.23), we get

$$E\lambda^h = \frac{\Gamma_p\left[\frac{1}{2}(n+l)\right]\Gamma_p\left(\frac{1}{2}n+h\right)}{\Gamma_p\left[\frac{1}{2}(n+l)+h\right]\Gamma_p\left(\frac{1}{2}n\right)}, \tag{6.3.24}$$

or

$$E\lambda^h = \frac{\Gamma_l\left[\frac{1}{2}(n+l)\right]\Gamma_l\left[\frac{1}{2}(n+l-p)+h\right]}{\Gamma_l\left[\frac{1}{2}(n+l)+h\right]\Gamma_l\left[\frac{1}{2}(n+l-p)\right]}, \tag{6.3.25}$$

the last equality being obtained from the fact that if $l \leqslant p$ (reversing the demonstration if $p \leqslant l$), then

$$\Gamma_p\left(\tfrac{1}{2}n\right) = \pi^{\frac{1}{2}l(p-l)}\Gamma_l\left(\tfrac{1}{2}n\right)\Gamma_{p-l}\left[\tfrac{1}{2}(n-l)\right]$$

and

$$\Gamma_p\left(\tfrac{1}{2}n\right) = \pi^{\frac{1}{2}l(p-l)}\Gamma_l\left[\tfrac{1}{2}(n+l-p)\right]\Gamma_{p-l}\left(\tfrac{1}{2}n\right).$$

Since $0 \leqslant \lambda \leqslant 1$, the moments determine the distribution uniquely. Denote the random variable $\lambda$ by $\lambda_{p,l,n}$ if $p \leqslant l$ and $E\lambda^h$ is given by Eq. (6.3.24), but denote it by $\lambda_{l,p,n+l-p}$ if $p \geqslant l$ and $E\lambda^h$ is given by Eq. (6.3.25). Note that $E\lambda^h$ is the same in both cases. Hence, we get

**Theorem 6.3.1** *Under $H$, the distribution of $\lambda_{p,l,n}$ is the same as that of $\lambda_{l,p,n+l-p}$.*

This theorem is very useful in obtaining the exact distribution in several special cases. Recalling the definition of $\Gamma_p(\cdot)$, we get

$$E\left(\lambda_{p,l,n}^h\right) = \prod_{i=1}^{p} \frac{\Gamma\left[\frac{1}{2}(n+1-i)+h\right]\Gamma\left[\frac{1}{2}(n+l+1-i)\right]}{\Gamma\left[\frac{1}{2}(n+1-i)\right]\Gamma\left[\frac{1}{2}(n+l+1-i)+h\right]} \tag{6.3.26}$$

$$= \prod_{i=1}^{p} E\left(x_i^h\right),$$

where $x_1,\ldots,x_p$ are independently distributed as beta random variables with parameters $\left(\frac{1}{2}(n+1-i), \frac{1}{2}l\right)$, $i=1,2,\ldots,p$. As usual, a random variable $x$ is said to have a beta distribution with parameters $(a,b)$ if its pdf is given by

$$\frac{\Gamma[(a+b)]}{\Gamma(a)\Gamma(b)}x^{a-1}(1-x)^{b-1}, \qquad 0<x<1,$$

and we shall write $x \sim \beta(a,b)$. Thus, we get

**Theorem 6.3.2** *The distribution of $\lambda_{p,l,n}$ is the distribution of the product of $p$ independent beta random variables with parameters $\left(\frac{1}{2}(n+1-i),\frac{1}{2}l\right)$, $i=1,2,\ldots,p$.*

Some further simplification is possible. Suppose $p$ is even, that is, $p=2r$. Then using the duplication formula

$$\Gamma\left(\gamma+\tfrac{1}{2}\right)\Gamma(\gamma+1)=\pi^{1/2}2^{-2\gamma}\Gamma(2\gamma+1),$$

we get the $h$th moment of $\lambda$ as

$$E(\lambda^h)=\prod_{i=1}^{r}\frac{\Gamma(l+n+1-2i)\Gamma(n+1-2i+2h)}{\Gamma(l+n+1-2i+2h)\Gamma(n+1-2i)}.$$

Hence,

$$E(\lambda^h)=\prod_{i=1}^{r}E\left(y_i^2\right)^h,$$

where $y_i^2 \sim \beta(n+1-2i,l)$ and are independent.

Similarly, if $p=2s+1$, then

$$E(\lambda^h)=E\left(\prod_{i=1}^{s}z_i^2 z_{s+1}\right)^h,$$

where the $z_i$'s are independent, $z_i \sim \beta(n+1-2i,l)$ for $i=1,2,\ldots,s$, and $z_{s+1} \sim \beta(\frac{1}{2}(n+1-p),\frac{1}{2}l)$. Thus we get

**Theorem 6.3.3** *$\lambda_{2r,l,n}$ is distributed as $\prod_{i=1}^{r}y_i^2$, where the $y_i^2$'s are independently distributed as $\beta(n+1-2i,l)$. And $\lambda_{2s+1,l,n}$ is distributed as $\prod_{i=1}^{s}z_i^2 z_{s+1}$, where the $z_i$'s and $z_{s+1}$ are independently distributed with $z_i^2 \sim \beta(n+1-2i,l)$, $i=1,2,\ldots,s$ and $z_{s+1} \sim \beta\left(\frac{1}{2}(n+1-p),\frac{1}{2}l\right)$.*

### 6.3.6 *Exact Distribution of $\lambda_{p,l,n}$*

First, we consider a few special cases.

*Case* 1. $p=1$ (and results available from Theorem 6.3.1) If $p=1$, then we get from Eq. (6.3.26) that

$$\lambda_{1,l,n} \sim \beta\left(\tfrac{1}{2}n,\tfrac{1}{2}l\right).$$

Hence,

$$\frac{1-\lambda_{1,l,n}}{\lambda_{1,l,n}}\frac{n}{l}\sim F_{l,n}.$$

Using Theorem 6.3.1, we get

$$\frac{1-\lambda_{p,1,n}}{\lambda_{p,1,n}}\frac{n+1-p}{p}\sim F_{p,n+1-p}.$$

Thus, we get

**Theorem 6.3.4** *We have the distributions*

$$\frac{n}{l}\frac{1-\lambda_{1,l,n}}{\lambda_{1,l,n}}\sim F_{l,n}$$

*and*

$$(n+1-p)\frac{1-\lambda_{p,1,n}}{\lambda_{p,1,n}}\sim F_{p,n+1-p}.$$

*Case* 2. $p=2$ When $p=2$, we get from Theorem 6.3.3 that

$$\lambda_{2,l,n}^{1/2}\sim\beta(n-1,l).$$

Hence

$$\frac{n-1}{l}\frac{1-\lambda_{2,l,n}^{1/2}}{\lambda_{2,l,n}^{1/2}}\sim F_{2l,2(n-1)}.$$

Using Theorem 6.3.1, we get

$$\frac{n+1-p}{p}\frac{1-\lambda_{p,2,n}^{1/2}}{\lambda_{p,2,n}^{1/2}}\sim F_{2p,2(n+1-p)}.$$

Thus, we get

**Theorem 6.3.5** *We have the distributions*

$$\frac{n-1}{l}\frac{1-\lambda_{2,l,n}^{1/2}}{\lambda_{2,l,n}^{1/2}}\sim F_{2l,2(n-1)}$$

*and*

$$\frac{n+1-p}{p}\frac{1-\lambda_{p,2,n}^{1/2}}{\lambda_{p,2,n}^{1/2}}\sim F_{2p,2(n+1-p)}.$$

To obtain results for $p \geqslant 3$, we need the following results:

**Lemma 6.3.1** *Let $v = v_1 + v_2$ where $v_1$ and $v_2$ are independently distributed with* pdf *given by*

$$\frac{a^{k+1}}{\Gamma(k+1)} v_1^k e^{-av_1}, \qquad v_1 > 0, \quad k \text{ a nonnegative integer},$$

*and*

$$be^{-bv_2}, \qquad v_2 > 0, \quad b > 0,$$

*respectively. Then the* pdf *of $v$ for $a \neq b$ is given by (integrating by parts)*

$$(6.3.27) \qquad \frac{ba^{k+1}}{\Gamma(k+1)}$$

$$\times \left[ e^{-bv} \left( \sum_{\gamma=1}^{k+1} (-1)^{\gamma+1} \frac{k!}{(k-\gamma+1)!} \frac{v^{k-\gamma+1}}{(b-a)^{\gamma}} \right) + e^{-av}(a-b)^{-k-1} k! \right].$$

**Lemma 6.3.2** *If the random variable $x_i$ has a beta distribution with* pdf *given by*

$$\beta\left(x_i; \tfrac{1}{2}(n-i+1), \tfrac{1}{2}l\right) = K_i x_i^{\frac{1}{2}(n-i-1)}(1-x_i)^{\frac{1}{2}(l-2)},$$
$$= n \geqslant i, \quad 0 < x_i < 1, \quad l > 0,$$

*and if $l$ is an even integer, then if $b \equiv \frac{1}{2}(l-2)$,*

$$\beta\left(x_i; \tfrac{1}{2}(n-i+1), \tfrac{1}{2}l\right) = K_i \sum_{\gamma=0}^{b} (-1)^{\gamma} \binom{b}{\gamma} x_i^{\frac{1}{2}(n-i-1+2\gamma)}.$$

**Corollary 6.3.1** *If $y_i = -\ln x_i$, then the* pdf *of $y_i$ is given by ($b \equiv \frac{1}{2}(l-2)$ is an integer)*

$$(6.3.28) \qquad K_i \sum_{\gamma=0}^{b} (-1)^{\gamma} \binom{b}{\gamma} \exp\left[ -\tfrac{1}{2} y_i (n-i+1+2\gamma) \right], \qquad y_i > 0,$$

*where*

$$(6.3.29) \qquad K_i^{-1} = \beta\left(\tfrac{1}{2}(n-i+1), \tfrac{1}{2}l\right) = \frac{\Gamma\left[\frac{1}{2}(n-i+1)\right]\Gamma\left(\frac{1}{2}l\right)}{\Gamma\left[\frac{1}{2}(l+n-i+1)\right]}.$$

Now, we note that if we set $y_i = -\ln x_i$, then, from Theorem 6.3.2,

$$-\ln\lambda_{p,l,n} = -\sum_{i=1}^{p} \ln x_i = \sum_{i=1}^{p} y_i,$$

where the $y_i$ are mutually independently distributed with density functions given in Eq. (6.3.28). Thus the distribution problem has been transformed to that of a sum of independently distributed random variables. It can be solved by successive convolutions, provided that the integrals at each stage can be evaluated. We now have

**Theorem 6.3.6** *When $l$ is an even integer, the density function of $\lambda_{p,l,n}$ is of the form*

$$f(\lambda) = \left(\prod_{i=1}^{p} K_i\right) \sum_{j=1}^{m} c_j \lambda^{\frac{1}{2}(n-r_j)} (-\ln\lambda)^{k_j}, \tag{6.3.30}$$

*where $K_i$ is defined in Eq.* (6.3.29), *and the constants $c_j$ and integers $m$, $r_j$, and $k_j$ are determined from $p, l, n$.*

PROOF. It can easily be proved by induction. For, when $p=1$, $f(\lambda_{1,l,n})$ is the pdf of a beta random variable with parameters $(\frac{1}{2}n, \frac{1}{2}l)$ according to Theorem 6.3.2, and since $l$ is an even integer, this function can easily be seen to be of the required form for the theorem with $m = b+1$,

$$c_j = (-1)^{j-1}\binom{b}{j-1}, \qquad r_j = 4-2j, \qquad k_j = 0,$$

and $b = \frac{1}{2}(l-2)$.

Suppose that $f(\lambda)$ is given by Eq. (6.3.30) for $p = t$. Then if $y = -\ln\lambda$, the pdf of $y$ is given by

$$g(y_{t,l,n}) = \left(\prod_{i=1}^{b} K_i\right)\left(\sum_{j=1}^{m} c_j \exp\left[-\tfrac{1}{2}y(n-r_j+2)\right]\right) y^{k_j}. \tag{6.3.31}$$

Hence, when $p = t+1$, the pdf of $y_{t+1,l,n}$ will be the convolution of Eq. (6.3.31) and

$$K_{t+1} \sum_{\gamma=0}^{b} (-1)^{\gamma} \binom{b}{\gamma} \exp\left[-\tfrac{1}{2}y_{t+1}(n-t+2\gamma)\right],$$

which can be evaluated. But each of these is of the form (6.3.27), so that $f(\lambda_{t+1,l,n})$ is again of the form (6.3.30). □

**Corollary 6.3.2** *When $l$ is odd and $p$ is even, Theorem* 6.3.6 *is still valid, since the role of $p$ and $l$ can be interchanged due to Theorem* 6.3.1.

Theorem 6.3.6 does not indicate explicitly how to find the values of the constants $m$, $c_j$, $r_j$, and $k_j$, a task that is by no means easy for large values of $p$ or $l$. However, the theorem provides the basis for a recursive algorithm, which has been programmed by Schatzoff (1966) and run on an IBM 7094. Thus in effect the computer derives the density and distribution functions at successive stages of the convolution process. This enabled Schatzoff (1966) to tabulate the factors for converting $\chi^2_{pl}$ percentiles to exact percentiles of $-[n-\frac{1}{2}(p-l+1)]\ln\lambda$ for $p=3(1)8$ and $pl\leqslant 70$.

In some special cases of $p=3,4,5,6$ the density and the cdf can be derived explicitly with the help of Theorems 6.3.1–6.3.3. We give here the pdf for $p=3$ and $p=4$, without proof [for proof refer to Pillai and Gupta (1969)].

**Theorem 6.3.7** *When $p=3$ and $l$ is even, the* pdf *of $\lambda_{3,l,n}$ is given by*

$$\left[2\beta(n-1,l)\beta\left(\tfrac{1}{2}(n-2),\tfrac{1}{2}l\right)\right]^{-1}$$

$$\times\left[\sum_{m=1}^{\frac{1}{2}l-1}(-1)^{m-1}\binom{l-1}{2m-1}\binom{\frac{1}{2}l-1}{m}(-\ln\lambda)\lambda^{\frac{1}{2}(n+2m-4)}\right]$$

$$+2\sum_{\substack{r=0\\ r\neq 2m-}}^{l-1}\sum_{m=0}^{\frac{1}{2}l-1}\frac{(-1)^{r+m}}{2m-r-1}\binom{l-1}{r}\binom{\frac{1}{2}l-1}{m}\left(\lambda^{\frac{1}{2}(n+r-3)}-\lambda^{\frac{1}{2}(n+2m-4)}\right).$$

**Theorem 6.3.8** *When $p=4$, the* pdf *of $\lambda_{4,l,n}$ is given by*

$$\prod_{i=1}^{2}\frac{1}{2\beta(n-2i+1,l)}\left[\sum_{r=0}^{l-3}\binom{l-1}{r}\binom{l-1}{r+2}(-\ln\lambda)\lambda^{\frac{1}{2}(n+r-3)}\right.$$

$$\left.+2\sum_{\substack{r=0\\ l\neq m-2}}^{l-1}\sum_{m=0}^{l-1}\frac{(-1)^{r+m}}{m-r-2}\binom{l-1}{r}\binom{l-1}{m}\left(\lambda^{\frac{1}{2}(n+r-3)}-\lambda^{\frac{1}{2}(n+m-s)}\right)\right]$$

These theorems helped Pillai and Gupta (1969) to extend Schatzoff's table. For alternative expressions, see Khatri and Bhargava (1975), and Mathai and Rathie (1971).

### 6.3.7 Asymptotic Distribution

Since

$$E(e^{h\ln\lambda}) = E(\lambda^h),$$

the $h$th moment of $\lambda$ is just the moment generating function of $\ln\lambda$. However, it is difficult to invert it to get the pdf of $\lambda$. Alternatively, one may use it to approximate the cdf. This has been the basis of several approximations. The asymptotic approximations for the distribution of the likelihood-ratio statistic have been considered by several authors. For example, Bartlett (1938) obtained a chi-square approximation to $-n\ln\lambda$; Wald and Brookner (1941) developed an asymptotic expansion, which was further modified by Rao (1948) to obtain the first three terms of a more rapidly convergent series for the cdf of $-[n-\frac{1}{2}(p-l+1]\ln\lambda$. Further, Box (1949) gave asymptotic approximations to functions of general likelihood-ratio statistics that include Rao's approximation as a special case. In addition, Rao (1952) has given a second approximation as a series of beta functions. In this, we follow Box's method, in which we essentially use an expansion formula for the gamma function [Barnes (1899, p. 64)] which is asymptotic in $x$ for bounded $h$:

$$\ln\Gamma(x+h) = \ln\sqrt{2\pi} + \left(x+h-\tfrac{1}{2}\right)\ln x - x$$

$$-\sum_{r=1}^{m}(-1)^r\frac{B_{r+1}(h)}{r(r+1)x^r} + R_{m+1}(x),$$

where $R_n(x) = O(x^{-n})$, that is, $|x^nR_n(x)|$ is bounded as $|x|\to\infty$, and $B_r(h)$ is the Bernoulli polynomial of degree $r$ and order unity defined by

$$\frac{\delta e^{\delta h}}{e^\delta - 1} = \sum_{r=0}^{\infty}\frac{\delta^r}{r!}B_r(h), \qquad \text{or} \quad \delta e^{\delta h} = (e^\delta - 1)\sum_{r=0}^{\infty}\frac{\delta^r}{r!}B_r(h).$$

Thus by expanding and comparing the coefficients of $\delta^r$, $r = 1,2,3$, on both sides, we get

$$B_0(h) = 1, \qquad B_1(h) = h - \tfrac{1}{2}, \qquad B_2(h) = h^2 - h + \tfrac{1}{6},$$

and

$$B_3(h) = h^3 - \tfrac{3}{2}h^2 + \tfrac{1}{2}h.$$

Other terms can be obtained similarly.

Let

$$m = \rho(Z), \quad m = m_1 + m_2, \quad l = m_2, \quad n = N - m, \quad \text{and} \quad N = r + 2\alpha,$$

where $\alpha$ is so chosen that the term of order $r^{-1}$ cancels out in the asymptotic expansion of the cdf of $-r\ln\lambda$. The moment generating function of $-r\ln\lambda$ is given by [see Eq. (6.3.24)]

$$\phi_r(t)\equiv E(e^{-rt\ln\lambda})=E(\lambda^{-rt})$$

$$=\frac{\Gamma_p\left[\frac{1}{2}(N-m_1-m_2)-rt\right]\Gamma_p\left[\frac{1}{2}(r+2\alpha-m_1)\right]}{\Gamma_p\left[\frac{1}{2}(r+2\alpha-m_1-m_2)\right]\Gamma_p\left[\frac{1}{2}(r+2\alpha-m_1)-rt\right]}$$

$$=\frac{\Gamma_p\left[\frac{1}{2}r(1-2t)+\alpha-\frac{1}{2}(m_1+m_2)\right]\Gamma_p\left[\frac{1}{2}r+\alpha-\frac{1}{2}m_1\right]}{\Gamma_p\left[\frac{1}{2}r+\alpha-\frac{1}{2}(m_1+m_2)\right]\Gamma_p\left[\frac{1}{2}r(1-2t)+\alpha-\frac{1}{2}m_1\right]}.$$

Using the above asymptotic expansion of $\ln\Gamma(x+h)$ for bounded $h$, we get

$$\ln\Gamma_p\left[\tfrac{1}{2}r(1-2t)+\alpha-\tfrac{1}{2}m\right]-\tfrac{1}{4}p(p-1)\ln\pi$$

$$=\sum_{k=1}^{p}\ln\Gamma\left[\tfrac{1}{2}r(1-2t)+\alpha-\tfrac{1}{2}(m+k-1)\right]$$

$$=\tfrac{1}{2}p\ln 2\pi+\sum_{k=1}^{p}\left\{\left[\tfrac{1}{2}r(1-2t)+\alpha-\tfrac{1}{2}(m+k-1)-\tfrac{1}{2}\right]\ln\tfrac{1}{2}r(1-2t)\right.$$

$$\left.-\tfrac{1}{2}r(1-2t)-\sum_{j=1}^{l}\frac{(-1)^jB_{j+1}\left[\alpha-\frac{1}{2}(m+k-1)\right]}{j(j+1)\left[\frac{1}{2}r(1-2t)\right]^j}\right\}+O(r^{-l-1})$$

and

$$\ln\Gamma_p\left(\tfrac{1}{2}r+\alpha-\tfrac{1}{2}m\right)-\tfrac{1}{4}p(p-1)\ln\pi$$

$$=\tfrac{1}{2}p\ln 2\pi+\sum_{k=1}^{p}\left\{\left[\tfrac{1}{2}r+\alpha-\tfrac{1}{2}(m+k-1)-\tfrac{1}{2}\right]\ln\tfrac{1}{2}r\right.$$

$$\left.-\tfrac{1}{2}r-\sum_{j=1}^{l}\frac{(-1)^jB_{j+1}\left[\alpha-\frac{1}{2}(m+k-1)\right]}{j(j+1)\left(\frac{1}{2}r\right)^j}\right\}+O(r^{-l-1}).$$

Hence

$$a_1(t)\equiv\ln\Gamma_p\left[\tfrac{1}{2}r(1-2t)+\alpha-\tfrac{1}{2}m\right]-\ln\Gamma_p\left[\tfrac{1}{2}r+\alpha-\tfrac{1}{2}m\right]$$

$$=\sum_{k=1}^{p}\left\{\left[\tfrac{1}{2}r(1-2t)+\alpha-\tfrac{1}{2}(m+k-1)-\tfrac{1}{2}\right]\ln\tfrac{1}{2}r(1-2t)\right.$$

$$-\left[\tfrac{1}{2}r+\alpha-\tfrac{1}{2}(m+k-1)-\tfrac{1}{2}\right]\ln\tfrac{1}{2}r+rt$$

$$\left.+\sum_{j=1}^{l}\left[1-(1-2t)^{-j}\right]\frac{(-1)^jB_{j+1}\left[\alpha-\frac{1}{2}(m+k-1)\right]}{j(j+1)\left(\frac{1}{2}r\right)^j}\right\}+O(r^{-l-1}).$$

Similarly

$$a_2(t) \equiv \ln \Gamma_p\left[\tfrac{1}{2}r(1-2t)+\alpha-\tfrac{1}{2}m_1\right]-\ln \Gamma_p\left[\tfrac{1}{2}r+\alpha-\tfrac{1}{2}m_1\right]$$

can be obtained from the above by substituting $m_1$ for $m$. Hence

$$\begin{aligned}
\ln \phi_r(t) &= a_1(t)-a_2(t) \\
&= \sum_{k=1}^{p}\Bigg\{\left(-\tfrac{1}{2}m+\tfrac{1}{2}m_1\right)\ln \tfrac{1}{2}r(1-2t)+\tfrac{1}{2}(m-m_1)\ln \tfrac{1}{2}r \\
&\quad + \sum_{j=1}^{l}\frac{(-1)^j\{B_{j+1}[\alpha-(m+k-1)]-B_{j+1}[\alpha-(m_1+k-1)]\}}{\left[1-(1-2t)^{-j}\right]^{-1}j(j+1)\left(\tfrac{1}{2}r\right)^j}\Bigg\} \\
&\quad + O(r^{-l-1}) \\
&= -\tfrac{1}{2}pm_2\ln(1-2t) \\
&\quad + \sum_{k=1}^{p}\sum_{j=1}^{l}(-1)^j\left[\frac{B_{j+1}\left[\alpha-\tfrac{1}{2}(m+k-1)\right]-B_{j+1}\left[\alpha-\tfrac{1}{2}(m_1+k-1)\right]}{\left[1-(1-2t)^{-j}\right]^{-1}j(j+1)\left(\tfrac{1}{2}r\right)^j}\right] \\
&\quad + O(r^{-l-1}).
\end{aligned}$$

But

$$\begin{aligned}
&B_2\left[\alpha-\tfrac{1}{2}(k-1)-\tfrac{1}{2}m\right] \\
&\quad = \left[\alpha-\tfrac{1}{2}(k-1)-\tfrac{1}{2}m\right]^2-\left[\alpha-\tfrac{1}{2}(k-1)-\tfrac{1}{2}m\right]+\tfrac{1}{6} \\
&\quad = \left[\alpha-\tfrac{1}{2}(k-1)\right]^2-(m+1)\left[\alpha-\tfrac{1}{2}(k-1)\right]+\tfrac{1}{4}m^2+\tfrac{1}{2}m+\tfrac{1}{6}.
\end{aligned}$$

Hence,

$$\begin{aligned}
&B_2\left[\alpha-\tfrac{1}{2}(k-1)-\tfrac{1}{2}m\right]-B_2\left[\alpha-\tfrac{1}{2}(k-1)-\tfrac{1}{2}m_1\right] \\
&\quad = -m_2\left[\alpha-\tfrac{1}{2}(k-1)\right]+\tfrac{1}{4}\left(m^2-m_1^2\right)+\tfrac{1}{2}m_2,
\end{aligned}$$

and

$$\begin{aligned}
&\sum_{k=1}^{p}\left\{B_2\left[\alpha-\tfrac{1}{2}(k-1)-\tfrac{1}{2}m\right]-B_2\left[\alpha-\tfrac{1}{2}(k-1)-\tfrac{1}{2}m_1\right]\right\} \\
&\quad = -m_2p\alpha+\tfrac{1}{4}m_2p(p-1)+\tfrac{1}{2}m_2p+\tfrac{1}{4}m_2(m+m_1)p \\
&\quad = -m_2p\alpha+\tfrac{1}{4}m_2p(p+1)+\tfrac{1}{4}m_2(2m_1+m_2)p.
\end{aligned}$$

Thus, in order for the term of order $r^{-1}$ to cancel out, we need (by equating the above term to zero)

$$2\alpha = m_1 + \tfrac{1}{2}(p + m_2 + 1).$$

Hence,

$$B_3\left[\alpha - \tfrac{1}{2}(k-1) - \tfrac{1}{2}m\right] = B_3\left[\tfrac{1}{4}(p+1) - \tfrac{1}{2}(k-1) - \tfrac{1}{4}m_2\right],$$

and

$$B_3\left[\alpha - \tfrac{1}{2}(k-1) - \tfrac{1}{2}m_1\right] = B_3\left[\tfrac{1}{4}(p+1) - \tfrac{1}{2}(k-1) + \tfrac{1}{4}m_2\right].$$

Consequently

$$\begin{aligned}
B_3&\left[\alpha - \tfrac{1}{2}(k-1) - \tfrac{1}{2}m\right] - B_3\left[\alpha - \tfrac{1}{2}(k-1) - \tfrac{1}{2}m_1\right] \\
&= B_3\left[\tfrac{1}{4}(p+1) - \tfrac{1}{2}(k-1) - \tfrac{1}{4}m_2\right] - B_3\left[\tfrac{1}{4}(p+1) - \tfrac{1}{2}(k-1) + \tfrac{1}{4}m_2\right] \\
&= -6\left[\tfrac{1}{4}(p+1) - \tfrac{1}{2}(k-1)\right]^2\left(\tfrac{1}{4}m_2\right) - 2\left(\tfrac{1}{4}m_2\right)^3 \\
&\quad + \tfrac{3}{2}(4)\left[\tfrac{1}{4}(p+1) - \tfrac{1}{2}(k-1)\right]\left(\tfrac{1}{4}m_2\right) \\
&= -\tfrac{3}{2}\left[\tfrac{1}{16}(p+1)^2 - \tfrac{1}{4}(p+1)(k-1) + \tfrac{1}{4}(k-1)^2\right]m_2 - \tfrac{1}{32}m_2^3 \\
&\quad + \tfrac{3}{2}\left[\tfrac{1}{4}(p+1) - \tfrac{1}{2}(k-1)\right]m_2 - \tfrac{1}{4}m_2 \\
&= \left[-\tfrac{3}{32}(p+1)^2 + \tfrac{3}{8}(p+1)(k-1) - \tfrac{3}{8}(k-1)^2\right]m_2 - \tfrac{1}{32}m_2^3 \\
&\quad + \left[\tfrac{3}{8}(p+1) - \tfrac{3}{4}(k-1) - \tfrac{1}{4}\right]m_2.
\end{aligned}$$

This gives

$$\begin{aligned}
\sum_{k=1}^{p} &\left\{B_3\left[\alpha - \tfrac{1}{2}(k-1) - \tfrac{1}{2}m\right] - B_3\left[\alpha - \tfrac{1}{2}(k-1) - \tfrac{1}{2}m_1\right]\right\} \\
&= \left[-\tfrac{3}{32}p(p+1)^2 + \tfrac{3}{16}(p+1)p(p-1) - \tfrac{3}{48}(p-1)p(2p-1)\right]m_2 - \tfrac{1}{32}m_2^3 p \\
&\quad + \left[\tfrac{3}{8}p(p+1) - \tfrac{3}{8}p(p-1) - \tfrac{1}{4}p\right]m_2 \\
&= -\tfrac{1}{32}pm_2\left(p^2 + m_2^2 - 5\right).
\end{aligned}$$

Calculating the other $B$'s similarly, we find that

$$\begin{aligned}
\ln\phi_r(t) = &-f\ln(1-2t) + r^{-2}\gamma_2\left[(1-2t)^{-2} - 1\right] \\
&+ r^{-4}\omega_4\left[(1-2t)^{-4} - 1\right] + O(r^{-6}),
\end{aligned}$$

where

$$f=pm_2,$$

$$\gamma_2=\frac{f}{48}(p^2+m_2-5),$$

$$\omega_4=\frac{f}{1920}\left[3p^4+3m_2^4+10p^2m_2^2-50(p^2+m_2^2)+159\right].$$

Hence,

$$\begin{aligned}\phi_r(t)&=E\left[e^{-rt\ln\lambda}\right]\\&=(1-2t)^{-\frac{1}{2}f}\left\{1+r^{-2}\gamma_2\left[(1-2t)^{-2}-1\right]+r^{-4}\omega_4\left[(1-2t)^{-4}-1\right]\right.\\&\quad\left.+\tfrac{1}{2}r^{-4}\gamma_2^2\left[(1-2t)^{-4}-2(1-2t)^{-2}+1\right]\right\}\\&=(1-2t)^{-\frac{1}{2}f}\left(1+r^{-2}\gamma_2\left[(1-2t)^{-2}-1\right]\right.\\&\quad\left.+r^{-4}\left\{\gamma_4\left[(1-2t)^{-4}-1\right]-\gamma_2^2\left[(1-2t)^{-2}-1\right]\right\}+O(r^{-6})\right),\end{aligned}$$

and

$$\begin{aligned}&P(-r\ln\lambda\leqslant z)\\&\quad=P(\chi_f^2\leqslant z)+r^{-2}\gamma_2\left[P(\chi_{f+4}^2\leqslant z)-P(\chi_f^2\leqslant z)\right]\\&\qquad+r^{-4}\left\{\gamma_4\left[P(\chi_{f+8}^2\leqslant z)-P(\chi_f^2\leqslant z)\right]-\gamma_2^2\left[P(\chi_{f+4}^2\leqslant z)-P(\chi_f^2\leqslant z)\right]\right\}\\&\qquad+O(m^{-6}),\end{aligned}$$

where

$$\gamma_4=\tfrac{1}{2}\gamma_2^2+\omega_4 \qquad\text{and}\qquad r=n-\tfrac{1}{2}(p-m_2+1).$$

For tabulated values, see Lee (1972b).

### 6.3.8 *Distribution of* $\mathrm{ch}_1(S_HS_e^{-1})$ *Under H*

In previous sections, we have obtained results by the likelihood-ratio method. Now we shall give some results by Roy's union–intersection principle based on $\mathrm{ch}_1(S_HS_e^{-1})$ where the distributions of $S_H$ and $S_e$ are given by Eqs. (6.3.20) and (6.3.18), and they are independently distributed. Under $H_0$, we have $S_H=YY'$, $Y\sim N_{p,l}(0,\Sigma,I_l)$, and $S_e\sim W_p(\Sigma,n)$. Since $\mathrm{ch}_1(S_HS_e^{-1})=\mathrm{ch}_1\left((\Sigma^{-1/2}S_e\Sigma^{-1/2})^{-1}(\Sigma^{-1/2}S_H\Sigma^{-1/2})\right)$, we can take $\Sigma=I$. Using the results of Section 3.6, the joint distribution of the nonzero ch roots $l_1>l_2>\cdots>l_t>0$, $t=\mathrm{Min}(p,l)$, of $S_HS_e^{-1}$ is given by

$$\frac{\Gamma_t\left[\frac{1}{2}(\tilde{n}+u)\right]}{\pi^{\frac{1}{2}t(u-t)}\Gamma_t\left(\frac{1}{2}\tilde{n}\right)\Gamma_t\left(\frac{1}{2}t\right)\Gamma_t\left(\frac{1}{2}u\right)}\,\frac{\left(\prod_{i=1}^{t}l_i\right)^{\frac{1}{2}(u-t-1)}}{\left(\prod_{i=1}^{t}(1+l_i)\right)^{\frac{1}{2}(\tilde{n}+u)}}\prod_{i=1}^{t-1}\prod_{j=i+1}^{t}(l_i-l_j),$$

where $ut=pl$ and $\tilde{n}=n-p+t$. Hence, we have either $(t=p,\ u=l,\ \tilde{n}=n)$ or $(t=l,\ u=p,\ \tilde{n}=n+l-p)$. The distribution of $l_1$ will be obtained after integrating over $l_2, l_3, \ldots, l_t$. This is done by Roy (1957), and Pillai (1955). The percentage points for the distribution of $l_1/(1+l_1)$ are given by Pillai (1967) and Heck (1960). For an extensive bibliography, the reader is referred to Krishnaiah (1978).

### *6.3.9 Confidence Bounds on Estimable Functions $\beta C'$*

This has been given in Eq. (6.3.15). In this section we give some further results. Let $c$ be such that

$$P(l_1 \leqslant c)=1-\alpha,$$

where the distribution of $l_1$ is given in Subsection 6.3.8. Note that

$$\operatorname*{Sup}_{\mathbf{a},\mathbf{b}} \frac{\left[\mathbf{a}'(\hat{\beta}-\beta)C'\mathbf{b}\right]^2}{(\mathbf{b}'G\mathbf{b})(\mathbf{a}'S_e\mathbf{a})}=l_1,$$

where

$$\hat{\beta}C'=XV_1^-Z'(ZV_1^-Z')^-C', \qquad G=C(ZV_1^-Z')^-C'-CRC',$$

$$\mathbf{b}'G\mathbf{b}\neq 0, \qquad S_e=X\left[V_1^- - V_1^-Z'(ZV_1^-Z')^-ZV_1^-\right]X',$$

with

$$V_1=V+Z'RZ \quad \text{and} \quad \rho(V_1)=\rho(V,Z').$$

Hence, simultaneous confidence bounds on $\beta C'$ with confidence greater than or equal to $1-\alpha$ for all nonnull vectors $\mathbf{a}$ and $\mathbf{b}$ (with $\mathbf{b}'G\mathbf{b}\neq 0$) are given [as in Eq. (6.3.15), $c$ being differently defined here] by

(6.3.32)
$$\mathbf{a}'\hat{\beta}C'\mathbf{b}-\left\{c(\mathbf{a}'S_e\mathbf{a})(\mathbf{b}'G\mathbf{b})\right\}^{1/2}$$
$$\leqslant \mathbf{a}'\beta C'\mathbf{b} \leqslant \mathbf{a}'\hat{\beta}C'\mathbf{b}+\left\{c(\mathbf{a}'S_e\mathbf{a})(\mathbf{b}'G\mathbf{b})\right\}^{1/2}.$$

Now, for an arbitrary p.d. matrix $B_1$ of order $p\times p$, we have

$$\operatorname*{Sup}_{\mathbf{a}} \frac{(\mathbf{a}'\beta C'\mathbf{b})}{(\mathbf{a}'B_1\mathbf{a})^{1/2}} \leqslant \operatorname*{Sup}_{\mathbf{a}} \left[\frac{\mathbf{a}'\hat{\beta}C'\mathbf{b}+\left[c(\mathbf{a}'S_e\mathbf{a})(\mathbf{b}'G\mathbf{b})\right]^{1/2}}{(\mathbf{a}'B_1\mathbf{a})^{1/2}}\right]$$

$$\leqslant \operatorname{Min}\left\{\left[\operatorname*{Sup}_{\mathbf{a}}\left(\frac{\mathbf{a}'\hat{\beta}C'\mathbf{b}}{(\mathbf{a}'S_e\mathbf{a})^{1/2}}+(c\mathbf{b}'G\mathbf{b})^{1/2}\right)\right]\operatorname*{Sup}_{\mathbf{a}}\left(\frac{\mathbf{a}'S_e\mathbf{a}}{\mathbf{a}'B_1\mathbf{a}}\right)^{1/2},\right.$$

$$\left.\operatorname*{Sup}_{\mathbf{a}}\left(\frac{\mathbf{a}'\hat{\beta}C'\mathbf{b}}{(\mathbf{a}'B_1\mathbf{a})^{1/2}}\right)+\operatorname*{Sup}_{\mathbf{a}}\left(\frac{\mathbf{a}'S_e\mathbf{a}}{\mathbf{a}'B_1\mathbf{a}}\right)^{1/2}(c\mathbf{b}'G\mathbf{b})^{1/2}\right\},$$

which can be written as

(6.3.33)
$$(\mathbf{b}'C\beta'B_1^{-1}\beta C'\mathbf{b})^{1/2} \leqslant \text{Min}\Big\{\big[\text{ch}_1(S_eB_1^{-1})\big]^{1/2}\big[(\mathbf{b}'C\hat{\beta}'S_e^{-1}\hat{\beta}C'\mathbf{b})^{1/2}+(c\mathbf{b}'G\mathbf{b})^{1/2}\big], (\mathbf{b}'C\hat{\beta}'B_1^{-1}\hat{\beta}C'\mathbf{b})^{1/2}+\big[c(\mathbf{b}'G\mathbf{b})\,\text{ch}_1(S_eB_1^{-1})\big]^{1/2}\Big\} \equiv\gamma_2(\mathbf{b}) \quad \text{(say)}.$$

Similarly, since

$$\operatorname*{Sup}_{\mathbf{a}}\left\{\frac{\mathbf{a}'\hat{\beta}C'\mathbf{b}-\big[c(\mathbf{b}'G\mathbf{b})(\mathbf{a}'S_e\mathbf{a})\big]^{1/2}}{(\mathbf{a}'B_1\mathbf{a})^{1/2}}\right\}\leqslant \operatorname*{Sup}_{\mathbf{a}}\frac{\mathbf{a}'\beta C'\mathbf{b}}{\mathbf{a}'B_1\mathbf{a}},$$

we get

(6.3.34)
$$(\mathbf{b}'C\beta'B_1^{-1}\beta C'\mathbf{b})^{1/2} \geqslant \text{Max}\Big\{\big[\text{ch}_p(B_1^{-1}S_e)\big]^{1/2}\big[(\mathbf{b}'C\hat{\beta}'S_e^{-1}\hat{\beta}C'\mathbf{b})^{1/2}-(c\mathbf{b}'G\mathbf{b})^{1/2}\big], (\mathbf{b}'C\hat{\beta}'B_1^{-1}\hat{\beta}C'\mathbf{b})^{1/2}-\big[c(\mathbf{b}'G\mathbf{b})\,\text{ch}_1(S_eB_1^{-1})\big]^{1/2}\Big\} \equiv\gamma_1(\mathbf{b}) \quad \text{(say)},$$

where $\text{ch}_p(S_eB_1^{-1})$ is the minimum characteristic root of $S_eB_1^{-1}$. Combining Eqs. (6.3.33) and (6.3.34), a simultaneous confidence bound with confidence greater than or equal to $1-\alpha$ on $\mathbf{b}'C\beta'B_1^{-1}\beta C'\mathbf{b}$ for all nonnull vectors with $\mathbf{b}'G\mathbf{b}\neq 0$ is

(6.3.35)
$$\gamma_1(\mathbf{b})\leqslant(\mathbf{b}'C\beta'B_1^{-1}\beta C'\mathbf{b})^{1/2}\leqslant\gamma_2(\mathbf{b}),$$

where $\gamma_1(\mathbf{b})$ and $\gamma_2(\mathbf{b})$ are defined in Eqs. (6.3.34) and (6.3.33) respectively. Similarly, a simultaneous confidence bound with confidence greater than or equal to $1-\alpha$ on $\mathbf{a}'\beta C'B_2^{-1}C\beta'\mathbf{a}$ for all nonnull vectors $\mathbf{a}$ is

(6.3.36)
$$\gamma_3(\mathbf{a})\leqslant(\mathbf{a}'\beta C'B_2^{-1}C\beta'\mathbf{a})^{1/2}\leqslant\gamma_4(\mathbf{a}),$$

where $B_2$ is an arbitrary p.d. matrix,

(6.3.37)
$$\gamma_3(\mathbf{a})=\text{Max}\Big[(\text{ch}_l GB_2^{-1})^{1/2}\{(\mathbf{a}'S_H\mathbf{a})^{1/2}-(c\mathbf{a}'S_e\mathbf{a})^{1/2}\}, \{(\mathbf{a}'\hat{\beta}C'B_2^{-1}C\hat{\beta}'\mathbf{a})^{1/2}-(c(\mathbf{a}'S_e\mathbf{a})(\text{ch}_1 GB_2^{-1}))^{1/2}\}\Big],$$

and

(6.3.38)
$$\gamma_4(\mathbf{a})=\text{Min}\Big[(\text{ch}_1(GB_2^{-1}))^{1/2}\{(\mathbf{a}'S_H\mathbf{a})^{1/2}+(c\mathbf{a}'S_e\mathbf{a})^{1/2}\}, \{(\mathbf{a}'\hat{\beta}C'B_2^{-1}C\hat{\beta}'\mathbf{a})^{1/2}+(c(\mathbf{a}'S_e\mathbf{a})(\text{ch}_1 GB_2^{-1}))^{1/2}\}\Big],$$

where $S_H=\hat{\beta}C'GC\hat{\beta}'$, and $\text{ch}_1(GB_2^{-1})$ denotes the largest nonzero characteristic root of $GB_2^{-1}$. From Eqs. (6.3.36) and (6.3.35), proceeding as above, we get simultaneous confidence bounds with confidence greater than or equal to $1-\alpha$ on the characteristic roots of $(B_1^{-1}\beta C'B_2^{-1}C\beta')$:

(6.3.39) $$\gamma_1 \leqslant \left\{\text{all ch}\left(B_1^{-1}\beta C'B_2^{-1}C\beta'\right)^{1/2}\right\} \leqslant \gamma_2,$$

where

(6.3.40) $$\begin{aligned}\gamma_1=\text{Max}\Big\{&\left(\text{ch}_p\left(B_1^{-1}S_e\right)\right)^{1/2}\left(\text{ch}_l\left(B_2^{-1}G\right)\right)^{1/2}\left(l_1^{1/2}-c^{1/2}\right),\\ &\left(\text{ch}_l\left(B_2^{-1}G\right)\right)^{1/2}\left\{\left(\text{ch}_1\left(B_1^{-1}S_H\right)\right)^{1/2}-\left(c\,\text{ch}_1B_1^{-1}S_e\right)^{1/2}\right\},\\ &\left(\text{ch}_p\left(B_1^{-1}S_e\right)\right)^{1/2}\left[\left(\text{ch}_1\left(S_e^{-1}\hat{\beta}C'B_2^{-1}C\hat{\beta}\right)\right)^{1/2}-\left(c\,\text{ch}_1GB_2^{-1}\right)\right)^{1/2}\Big],\\ &\left[\left(\text{ch}_1\left(B_1^{-1}\hat{\beta}C'B_2^{-1}D\hat{\beta}'\right)\right)^{1/2}-\left(c\left(\text{ch}_1\left(B_1^{-1}S_e\right)\right)\left(\text{ch}_1\left(GB_2^{-1}\right)\right)\right)^{1/2}\right]\Big\}\end{aligned}$$

and

(6.3.41) $$\begin{aligned}\gamma_2=\text{Min}\Big\{&\left(\text{ch}_1\left(B_1^{-1}S_e\right)\right)^{1/2}\left(\text{ch}_1\left(GB_2^{-1}\right)\right)^{1/2}\left(l_1^{1/2}+c^{1/2}\right),\\ &\left(\text{ch}_1\left(B_2^{-1}G\right)\right)^{1/2}\left[\left(\text{ch}_1\left(B_1^{-1}S_H\right)^{1/2}+\left(c\left(\text{ch}_1\left(B_1^{-1}S_e\right)\right)\right)^{1/2}\right],\\ &\left(\text{ch}_1\left(B_1^{-1}S_e\right)\right)^{1/2}\left[\left(\text{ch}_1\left(S_e^{-1}\hat{\beta}C'B_2^{-1}C\hat{\beta}'\right)\right)^{1/2}+\left(c\text{ch}_1B_2^{-1}G\right)\right)^{1/2}\Big],\\ &\left(\text{ch}_1\left(B_1^{-1}\hat{\beta}C'B_2^{-1}C\hat{\beta}'\right)\right)^{1/2}+\left(c\left(\text{ch}_1\left(B_1^{-1}S_e\right)\right)\left(\text{ch}_1\left(B_2^{-1}G\right)\right)\right)^{1/2}\Big\},\end{aligned}$$

with $l_1=\text{ch}_1(S_e^{-1}S_H)$. All the above results are true for the submatrices of $\beta C'$, and they can be obtained by choosing the vectors $\mathbf{a}$ and $\mathbf{b}$; for example, if $\mathbf{a}'=(\mathbf{a}_1',\mathbf{0})$ and $\mathbf{b}'=(\mathbf{b}_1',\mathbf{0})$, then $\mathbf{a}'\beta C\mathbf{b}=\mathbf{a}_1'\beta_{11}\mathbf{b}$, where

$$\beta C=\begin{pmatrix}\beta_{11} & \beta_{12}\\ \beta_{21} & \beta_{22}\end{pmatrix}.$$

Thus, we can obtain the confidence bounds on $\beta_{11}$ from Eq. (6.3.32), and various other types like Eqs. (6.3.35), (6.3.36), and (6.3.39). These are not mentioned explicitly.

### *6.3.10 Canonical Reduction for the Noncentral Distributions*

From Eqs. (6.3.17) and (6.3.18), we can write

(6.3.42) $$S_H=YY', \qquad S_e=Y_1Y_1',$$

where $Y\sim N_{p,l}(\nu,\Sigma,I_l)$, $\nu\nu'=\Omega$, $Y_1\sim N_{p,n}(0,\Sigma,I_n)$, and $Y$ and $Y_1$ are independently distributed. Since $\Sigma$ is p.d. and $\nu$ is a $p\times l$ matrix, we can find a nonsingular matrix $F$ and an orthogonal matrix $H$ such that

$$\Sigma = FF' \quad \text{and} \quad \nu = F\begin{pmatrix} D_\omega & 0 \\ 0 & 0 \end{pmatrix} H,$$

where $D_\omega = \text{diag}(\omega_1, \omega_2, \ldots, \omega_s)$, $\rho(\nu) = \rho(\Omega) = s$. Then, using the transformations

$$Y_2 = F^{-1}YH' \quad \text{and} \quad Y_3 = F^{-1}Y_1,$$

we find that the $\omega_i^2$'s are the ch roots of $\Sigma^{-1}\Omega$,

(6.3.43a) $$Y_2 \sim N_{p,l}\left(\begin{pmatrix} D_\omega & 0 \\ 0 & 0 \end{pmatrix}, I_p, I_l\right), \qquad Y_3 \sim N_{p,n}(0, I_p, I_n),$$

and

(6.3.43b) $Y_2$ and $Y_3$ are independently distributed.

Further, it is easy to verify that

(6.3.44) $$\lambda = \frac{|Y_3Y_3'|}{|Y_2Y_2' + Y_3Y_3'|}$$

and

(6.3.45) $$l_1 = \text{ch}_1\left((Y_3Y_3')^{-1}Y_2Y_2'\right).$$

Thus, the distributions of $\lambda$ and $l_1$ involve nothing but the characteristic roots $\omega_1^2, \ldots, \omega_s^2$ of $\Sigma^{-1}\Omega$ as noncentrality parameters.

### *6.3.11 Other Tests*

If $l_1 > l_2 > \cdots > l_t$, $t = \text{Min}(p, l)$, denote the nonzero characteristic roots of $(Y_3Y_3')^{-1}Y_2Y_2'$, where $Y_2$ and $Y_3$ are defined in Eq. (6.3.43), then the likelihood-ratio test $\lambda$ is based on

$$\prod_{i=1}^{t} (1 + l_i)^{-1},$$

while the union–intersection test statistic is $l_1$, which is the Roy's maximum-root statistic.

But, even if we restrict our attention to invariant tests, there are several possible candidates. For example, Lawley (1938) and Hotelling (1947) suggested a test based on the statistic

$$\sum_{i=1}^{t} l_i = \text{tr}\left[(Y_2Y_2')(Y_3Y_3')^{-1}\right],$$

which is distributed approximately [see Pillai (1955)] as

$$\frac{t^2 u}{t(\tilde{n}-t-1)+2} F_{tu,t(\tilde{n}-t-1)+2}$$

with

$$tu = pl \quad \text{and} \quad \tilde{n} = n-p+t.$$

Also, Nanda (1950) suggested a test, now known as Pillai's (1955) trace test, based on the statistic

$$V^{(t)} = \sum_{i=1}^{p} \frac{l_i}{1+l_i} = \operatorname{tr}\left[(Y_2 Y_2')(Y_2 Y_2' + Y_3 Y_3')^{-1}\right],$$

where approximately [see Pillai (1955)]

$$\frac{V^{(t)}}{t - V^{(t)}} \frac{\tilde{n}}{u} \sim F_{ut,\tilde{n}t},$$

with $ut = pl$ and $\tilde{n} = n-p+t$. For better approximations and tables, see Davis (1970a, b).

## 6.4 Generalized Linear Models (Growth-Curve Models)

### *6.4.1 Introduction*

Let $X$ be a $p \times N$ matrix of random variables. For given matrices $B: p \times q$ and $A: m \times N$, if $E(X) = B\xi A$, where $\xi: q \times m$ is the matrix of regression parameters, we get as in Chapter 5, the generalized linear model (more commonly called the "growth-curve" model) of Potthoff and Roy (1964) [see also Rao (1959)]; its principal application has been in the analysis of growth curves. We will assume that $X \sim N_{p,N}(B\xi A, \Sigma, I_N)$. For the convenience of presentation, we shall assume that $B$ and $A$ are of full rank, $q$ ($\leqslant p$) and $m$ ($< N$) respectively. If $A$ and $B$ are not of full rank, we reduce the problem to that of full rank as explained in Subsection 6.2.1.

### *6.4.2 Maximum-Likelihood Estimates of $\xi$ and $\Sigma$*

The likelihood function of $\xi$ and $\Sigma$ is

$$L(\xi, \Sigma) = (2\pi)^{-\frac{1}{2}pN} |\Sigma|^{-\frac{1}{2}N} \operatorname{etr}\left[-\tfrac{1}{2}\Sigma^{-1}(X - B\xi A)(X - B\xi A)'\right].$$

When $\xi$ is known, we find from Theorem 1.10.4, that

$$\text{(6.4.1)} \qquad L(\xi, \Sigma) \leqslant L(\xi, N^{-1}(X - B\xi A)(X - B\xi A)') = (2\pi)^{-\frac{1}{2}pN} |\hat{\Sigma}_\xi|^{-\frac{1}{2}N} e^{-\frac{1}{2}pN},$$

and the equality holds iff

$$\hat{\Sigma}_{\xi} = N^{-1}(X - B\xi A)(X - B\xi A)'.$$

From Theorem 1.10.3, it follows that

$$|\hat{\Sigma}_{\xi}| \geqslant |\hat{\Sigma}_{\hat{\xi}}| \equiv |\hat{\Sigma}|, \tag{6.4.2}$$

and the equality holds iff

$$\xi = \hat{\xi} = (B'S^{-1}B)^{-1}B'S^{-1}XA'(AA')^{-1}, \tag{6.4.3}$$

where

$$S = N^{-1}X[I - A'(AA')^{-1}A]X' = N^{-1}XX' - S_1. \tag{6.4.4}$$

We observe that

$$\begin{aligned} \hat{\Sigma}_{\hat{\xi}} \equiv \hat{\Sigma} &= N^{-1}(X - B\hat{\xi}A)(X - B\hat{\xi}A)' \\ &= N^{-1}(X - TS^{-1}XA'(AA')^{-1}A)(X - TS^{-1}XA'(AA')^{-1}A)' \\ &= S + (I - TS^{-1})S_1(I - S^{-1}T), \end{aligned} \tag{6.4.5}$$

where

$$T = B(B'S^{-1}B)^{-1}B'.$$

Using Eq. (6.4.2) in Eq. (6.4.1), we get

$$\begin{aligned} L(\xi, \Sigma) \leqslant L(\xi, \hat{\Sigma}_{\xi}) &\leqslant L(\hat{\xi}, \hat{\Sigma}) \\ &= (2\pi)^{-\frac{1}{2}pN}|\hat{\Sigma}|^{-\frac{1}{2}N}e^{-\frac{1}{2}pN}, \end{aligned}$$

and the equality holds iff

$$\xi = \hat{\xi} \quad \text{and} \quad \Sigma = \hat{\Sigma},$$

where $\hat{\xi}$ and $\hat{\Sigma}$ have been defined in Eqs. (6.4.3) and (6.4.5). These are maximum-likelihood estimates of $\xi$ and $\Sigma$. These estimates were obtained by Khatri (1966a).

### 6.4.3 *Canonical Reduction of the Model*

Let $\Gamma: N \times N$ and $\Delta: p \times p$ be two known orthogonal matrices defined by

$$\Delta' = p(\overset{q}{B(B'B)^{-1/2}}, \overset{p-q}{B_0}) \quad \text{and} \quad \Gamma = N(\overset{m}{A'(AA')^{-1/2}}, \overset{N-m}{A_0}), \tag{6.4.6}$$

where $B_0$ and $A_0$ are chosen such that $\Delta: p \times p$ and $\Gamma: N \times N$ are orthogonal. Since $\Delta$ and $\Gamma$ are known (orthogonal) matrices, observing $X$ is equivalent to observing

$$(6.4.7)\qquad \Delta X\Gamma \equiv p(\overset{m}{\dot{Z}}, \overset{N-m}{Z}) \equiv \begin{matrix} q \\ p-q \end{matrix}\overset{\begin{matrix} m & N-m \end{matrix}}{\left[\begin{array}{c:c} Z_1 & \\ & Z \\ Z_2 & \end{array}\right]}$$

and $\Delta X\Gamma \sim N_{p,N}(\Delta B\xi A\Gamma, \Lambda, I_N)$, where

$$(6.4.8)\qquad \Lambda = \Delta\Sigma\Delta'.$$

Hence $\dot{Z}$ and $Z$ are independently normally distributed. Since

$$(6.4.9)\qquad \Delta B\xi A\Gamma = \left[\begin{array}{c:c} (B'B)^{1/2}\xi(AA')^{1/2} & \\ & 0 \\ 0 & \end{array}\right] \equiv \left[\begin{array}{c:c} \mu_1 & \\ & 0 \\ 0 & \end{array}\right],$$

where

$$\mu_1 \equiv (B'B)^{1/2}\xi(A'A)^{1/2},$$

we find that

$$(6.4.10)\qquad \dot{Z} \equiv \begin{pmatrix} Z_1 \\ Z_2 \end{pmatrix} \sim N_{p,m}(\dot{\mu}, \Lambda, I_m) \quad \text{and} \quad Z \sim N_{p,n}(0, \Lambda, I_n),$$

where

$$n \equiv N - m \quad \text{and} \quad \dot{\mu} = (\mu_1', 0')'.$$

### 6.4.4 Likelihood-Ratio Test for $\xi = 0$

Let $X \sim N_{p,N}(B\xi A, \Sigma, I)$, where $B: p\times q$ and $A: m\times N$ are of full ranks. Suppose we wish to test the hypothesis $H: \xi = 0$ against the alternative $A \neq H$. In terms of the canonical variables $\Delta X\Gamma$ of Subsection 6.4.3, we are given independent random matrices, $\dot{Z}$ and $Z$, where $\dot{Z} \sim N_{p,m}(\dot{\mu}, \Lambda, I)$ and $Z \sim N_{p,n}(0, \Lambda, I)$, and we wish to test the hypothesis that $\mu_1 \equiv (B'B)^{1/2}\xi(A'A)^{1/2} = 0$ against the alternative that $\mu_1 \neq 0$. The likelihood function is

$$L(\mu_1, \Lambda) = \text{const}\,|\Lambda|^{-\frac{1}{2}N} \operatorname{etr}\left\{-\tfrac{1}{2}\Lambda^{-1}\left[ZZ' + (\dot{Z} - \dot{\mu})(\dot{Z} - \dot{\mu})'\right]\right\}.$$

Hence, the ratio is given by (see Chapter 4)

$$(6.4.11)\qquad \frac{\operatorname{Sup}_H L(\mu_1, \Lambda)}{\operatorname{Sup}_A L(\mu_1, \Lambda)} = \left[\frac{\operatorname*{Inf}_{\mu_1} |ZZ' + (\dot{Z} - \dot{\mu})(\dot{Z} - \dot{\mu})'|}{|ZZ' + \dot{Z}\dot{Z}'|}\right]^{\frac{1}{2}N}$$

$$= \left(\frac{|I + Z_2' V_{22}^{-1} Z_2|}{|I + \dot{Z}' V^{-1} \dot{Z}|}\right)^{\frac{1}{2}N}$$

$$= \lambda^{\frac{1}{2}N},$$

where

$$V \equiv \begin{pmatrix} V_{11} & V_{12} \\ V_{12}' & V_{22} \end{pmatrix} \equiv ZZ'. \tag{6.4.12}$$

In terms of the original variables, this is given by

$$\lambda = \frac{|S+(I-TS^{-1})S_1(I-S^{-1}T)|}{|S+S_1|}. \tag{6.4.13}$$

### 6.4.5 *h*th *Moment of* $\lambda$ *Under H*

From Eq. (6.4.11) we have

$$\lambda = \frac{|I+Z_2'V_{22}^{-1}Z_2|}{|I+Z'V^{-1}\dot{Z}|} = \frac{|V||V_{22}+Z_2Z_2'|}{|V_{22}||V+\dot{Z}\dot{Z}'|},$$

$$= \frac{|V_{11}-V_{12}V_{22}^{-1}V_{12}'|}{|(V_{11}+Z_1Z_1')-(V_{12}+Z_1Z_2')(V_{22}+Z_2Z_2')^{-1}(V_{12}+Z_1Z_2')|}$$

$$= \frac{|V_{1.2}|}{|V_{1.2}+W_1|},$$

where

$$V_{1.2} = V_1 - V_{12}V_{22}^{-1}V_{12}'$$

and

$$W_1 = (Z_1 - V_{12}V_{22}^{-1}Z_2)(I+Z_2'V_{22}^{-1}Z_2)^{-1}(Z_1 - V_{12}V_{22}^{-1}Z_2)'.$$

Since $V \sim W_p(\Lambda, n)$, then from Theorem 3.3.5, $V_{1.2}$ and $(V_{12}', V_{22})$ are independently distributed, $V_{1.2} \sim W_q(\Lambda_{1.2}, n-p+q)$, $V_{22} \sim W_{p-q}(\Lambda_{22}, n)$, and given $V_{22}$, $V_{12}V_{22}^{-1} \sim N_{q,p-q}(\Lambda_{12}\Lambda_{22}^{-1}, \Lambda_{1.2}, V_{22}^{-1})$, where

$$\Lambda = \left(\begin{array}{c|c} \Lambda_{11} & \Lambda_{12} \\ \hline \Lambda_{12}' & \Lambda_{22} \end{array}\right) \quad \text{and} \quad \Lambda_{1.2} = \Lambda_{11} - \Lambda_{12}\Lambda_{22}^{-1}\Lambda_{12}'.$$

Further, $\dot{Z} \sim N_{p,m}(\dot{\mu}, \Lambda, I_m)$, and $\dot{Z}$ and $V$ are independently distributed. Hence, given $Z_2$, $Z_1 \sim N_{q,m}(\mu_1 + \Lambda_{12}\Lambda_{22}^{-1}Z_2, \Lambda_{1.2}, I_m)$. From these, given $Z_2$ and $V_{22}$, $Z_1 - V_{12}V_{22}^{-1}Z_2 \sim N_{q,m}(\mu_1, \Lambda_{1.2}, I_m + Z_2'V_{22}^{-1}Z_2)$, and it is independent of $V_{1.2}$. Hence, we have the following:

Under $H$, $\mu_1 = 0$, $Y = (Z_1 - V_{12}V_{22}^{-1}Z_2)(I+Z_2'V_{22}^{-1}Z_2)^{-1/2}$ and $V_{1.2}$ are independently distributed, $V_{1.2} \sim W_q(\Lambda_{1.2}, n-p+q)$, $Y \sim N_{q,m}(0, \Lambda_{1.2}, I_m)$, and $\lambda = |V_{1.2}|/|V_{1.2}+YY'|$. Hence, using Theorem 6.3.1, the distributions of $\lambda$ is the same as that of $\lambda_{q,m,n-p+q}$ or $\lambda_{m,q,n-p+m}$ and its distribution is given in Subsection 6.3.6.

REMARK 6.4.1. As in Chapter 4, the likelihood-ratio test is the conditional test, and this test remains the same for testing the hypothesis ($\boldsymbol{\mu}_1 = \Lambda_{12}\Lambda_{22}^{-1}\boldsymbol{\mu}_2$, $\Lambda > 0$), where

$$\dot{\mu}' = (\boldsymbol{\mu}_1', \boldsymbol{\mu}_2') \quad \text{and} \quad \Lambda = \begin{pmatrix} \Lambda_{11} & \Lambda_{12} \\ \Lambda_{12}' & \Lambda_{22} \end{pmatrix},$$

instead of those defined in Eq. (6.4.10). Another test has been proposed by Rao (1966) for testing $H: \xi = 0, \Sigma > 0$. The test procedure is

$$\text{reject } H_0 \quad \text{if} \quad \frac{|V + \dot{Z}\dot{Z}'|}{|V|} - \frac{|V_{22} + \dot{Z}_2\dot{Z}_2'|}{|V_{22}|} \geq c,$$

but it is difficult to obtain the distribution for the above test.

REMARK 6.4.2. Khatri (1966a) considered the extended hypothesis mentioned in Remark 6.4.1 and he also established the likelihood-ratio test mentioned in this section for both the types of hypotheses. Khatri (1966a) proposed the following test procedure:

$$\text{reject } H_0: \xi = 0 \quad \text{if} \quad \text{ch}_1(W_1 V_{1.2}^{-1}) \geq c,$$

where $c$ is a constant to be determined from $P(ch_1(W_1 V_{1.2}^{-1}) \geq c | H) = \alpha$. The distribution of $\text{ch}_1(W_1 V_{1.2}^{-1})$ under $H$ is the same as $l_1$ given in Subsection 6.3.8 when $n \to n - p + q$, $p \to q$, $m_2 \to m$. The purpose of defining the above test procedure is to obtain simultaneous confidence bounds on $\xi$, and these can be obtained as in Subsection 6.3.9 by the following changes:

$$\beta C' \to \mu_1 = (B'B)^{1/2}\xi(A'A)^{1/2}, \qquad \hat{\beta}C' \to (Z_1 - V_{12}V_{22}^{-1}Z_2),$$

$$S_e \to V_{1.2}, \qquad (Z_2 G_1 Z_2')^{1} \to I_m + Z_2' V_{22}^{-1} Z_2.$$

These types of confidence bounds were given by Khatri (1966a).

## Problems

**6.1** Let $X \sim N_{p,N}(\beta Z, \Sigma, I)$, where $\beta: p \times m$, $Z: m \times N$, and $Z$ is of full rank. Let $\beta = (\beta_1, \beta_2)$, where $\beta_1: p \times m_1$, $\beta_2: p \times m_2$, and $m_1 + m_2 = m$. Find the likelihood-ratio test for $H: \beta_1 = 0$ against $A \neq H$. Give its asymptotic distribution.

**6.2** Show that the likelihood-ratio test for testing the equality of means in Subsection 6.3.3 is based on the statistic

$$\lambda = \frac{\left| \sum_{i,\alpha} (\mathbf{y}_\alpha^{(i)} - \bar{\mathbf{y}}^{(i)})(\mathbf{y}_\alpha^{(i)} - \bar{\mathbf{y}}^{(i)})' \right|}{\left| \sum_{i,\alpha} (\mathbf{y}_\alpha^{(i)} - \bar{\mathbf{y}}^{(i)})(\mathbf{y}_\alpha^{(i)} - \bar{\mathbf{y}}^{(i)})' + \sum N_i (\bar{\mathbf{y}}^{(i)} - \bar{\mathbf{y}})(\bar{\mathbf{y}}^{(i)} - \bar{\mathbf{y}})' \right|}$$

where $\bar{\mathbf{y}}^{(i)} = N_i^{-1}\Sigma_{\alpha=1}^{N_i} y_\alpha^{(i)}$ and $\bar{\mathbf{y}} = N^{-1}\Sigma N_i \bar{\mathbf{y}}^{(i)}$. Give its asymptotic distribution. Also, give simultaneous confidence bounds on the location parameters using the union–intersection principle.

**6.3** Let $\mathbf{y}_{ij}$ be distributed independently as $N(\boldsymbol{\mu}_{ij}, \Sigma)$, $i=1,2,\ldots,r$, $j=1,2,\ldots,c$, where

$$\boldsymbol{\mu}_{ij} = \boldsymbol{\mu} + \mathbf{r}_i + \mathbf{c}_j, \qquad \sum_{i=1}^{r} \mathbf{r}_i = \sum_{j=1}^{c} \mathbf{c}_j = 0.$$

Express this problem as a regression model. Find the likelihood-ratio and union–intersection tests for the hypothesis $H: \mathbf{r}_i = 0$, $i=1,2,\ldots,r$ against the alternative $A \neq H$. Find the simultaneous confidence bounds on the parameters.

**6.4** (Continuation.) Generalize the above problem to Latin squares.

**6.5** When $l$ is an even integer, show that the cdf of $\lambda_{p,l,n}$ is of the form

$$F(u) = \left\{ \prod_{i=1}^{p} K_i \right\} \sum_{j=1}^{m} \left\{ c_j u^{\frac{1}{2}(n-r_j)} \sum_{s_j=1}^{k_j+1} \frac{[2/(n-r_j)]^{s_j} (-\ln u)^{k_j - s_j + 1} k_j}{(k_j - s_j + 1)!} \right\}$$

$$= \left\{ \prod_{i=1}^{k} K_i \right\} \sum_{h=1}^{M} \left\{ d_h u^{\frac{1}{2}(n-r_h)} \left( \frac{2}{n-r_h} \right)^{s_h} (-\ln u)^{t_h} \right\},$$

where $F(u) = P\{\lambda_{p,l,n} \leqslant u\}$ and the pdf of $\lambda_{p,l,n}$ is given in Theorem 6.3.6.

**6.6** Show the equivalence of the statistics in Eqs. (6.4.11) and (6.4.13).

**6.7** Let $X \sim N_{p,N}(B\xi A + B_1 \xi_1 A, \Sigma, I)$ where $B_1 : p \times (p-q)$ is of rank $p-q$ and $\xi_1$ is unknown. Show that the problem of testing $H: C\eta V = 0$ against $H \neq A$, where

$$\eta = \xi - (B'\Sigma^{-1}B)^{-1}(B'\Sigma^{-1}B_1)\xi_1$$

is equivalent to the problem in Subsection 6.4.4.

**6.8** Derive an asymptotic cdf for the likelihood-ratio test for $\xi = 0$ by using an asymptotic expansion of $\ln\Gamma(x+h)$ for bounded $h$.

**6.9** Let $X \sim N_{p,N}(B\xi A, \Sigma, I_N)$, where $B$ and $A$ are $p \times q$ and $m \times N$ matrices of ranks $q$ and $m$ respectively. Let

$$\xi = \begin{pmatrix} \xi_1 & \xi_2 \\ \xi_3 & \xi_4 \end{pmatrix} \qquad \text{with} \quad \xi_1 = 0.$$

Find the maximum value of the likelihood function. Obtain the likelihood-ratio test procedure for testing $H_0: \xi_1 = 0$ vs $A: \xi_1 \neq 0$. Extend this result to testing $H_0: C\xi D = 0$ vs $A: C\xi D \neq 0$. Establish the distribution under $H_0$. [Hint: Let $B = (B_1, B_2)$, $A' = (A_1', A_2')$,

$$\eta_1 = \xi_3 \quad \text{and} \quad \eta = \begin{pmatrix} \xi_2 \\ \xi_4 \end{pmatrix}.$$

Then $B\xi A = B_2\eta_1 A_1 + B\eta A_2$. Given $\eta_1$, minimize $|(X - B_2\eta_1 A_1 - B\eta A_2)(X - B_2\eta_1 A_1 - B\eta A_2)'|$ with respect to $\eta$, using Theorem 1.10.3. This minimum value will be

$$|S_{\eta_1}||I + S_{1\eta_1}\left(S_{\eta_1}^{-1} - S_{\eta_1}^{-1}B\left(B'S_{\eta_1}^{-1}B\right)^{-1}B'S_{\eta_1}^{-1}\right)|$$

with

$$S_{\eta_1} = (X - B_2\eta_1 A_1)(X - B_2\eta_1 A_1)' - S_{1\eta_1}$$

and

$$S_{1\eta_1} = (X - B_2\eta_1 A_1)\left[I - A_2'(A_2A_2')^{-1}A_2\right](X - B_2\eta_1 A_1)'.$$

Observe that from Corollary 1.9.2

$$S_{\eta_1}^{-1} - S_{\eta_1}^{-1}B\left(B'S_{\eta_1}^{-1}B\right)^{-1}B'S_{\eta_1}^{-1} = B_0(B_0'S_{\eta_1}B_0)^{-1}B_0',$$

where $B_0'B = 0$ and $\rho(B) + \rho(B_0) = p$. Then show that

$$|I + S_{1\eta_1}\left(S_{\eta_1}^{-1} - S_{\eta_1}^{-1}B\left(B'S_{\eta_1}^{-1}B\right)^{-1}B'S_{\eta_1}^{-1}\right)|$$

does not depend on $\eta_1$. Now minimize $|S_{\eta_1}|$ with respect to $\eta_1$ using Theorem 1.10.3. Then use the technique of Subsection 6.4.2 to obtain maximum-likelihood estimates, and directly obtain the likelihood-ratio test procedure. Observe that $\eta = C\xi D$ can be written as

$$\xi = \begin{pmatrix} \eta & \xi_2 \\ \xi_3 & \xi_4 \end{pmatrix}$$

with some parameters $\xi_2$, $\xi_3$, and $\xi_4$ making a one-to-one mapping.]

# 7

# Inference on Covariances

## 7.1 Introduction

In this chapter we derive likelihood-ratio tests and their distributions (mostly asymptotic) for testing problems on covariances and mean vectors. Some other tests based on invariance are also considered. The monotonicity and unbiasedness of some of these tests are considered in Chapter 10. As in earlier chapters, attention will be confined to sufficient statistics only. For example, if $\mathbf{x}_1, \mathbf{x}_2, \ldots, \mathbf{x}_N$ is a random sample from $N_p(\boldsymbol{\theta}, \Sigma)$, then we shall consider only tests based on $\mathbf{y} \equiv N^{1/2}\bar{\mathbf{x}}$ and $V = \Sigma_{\alpha=1}^{N}(\mathbf{x}_\alpha - \bar{\mathbf{x}})(\mathbf{x}_\alpha - \bar{\mathbf{x}})'$, where $\mathbf{y} \sim N_p(\boldsymbol{\mu}, \Sigma)$ and $V \sim W_p(\Sigma, n)$ are independently distributed; $n \equiv N-1$ and $\boldsymbol{\mu} \equiv N^{1/2}\boldsymbol{\theta}$.

## 7.2 Testing That the Covariance Matrix Is an Identity Matrix

Let $\mathbf{y} \sim N_p(\boldsymbol{\mu}, \Sigma)$ and $V \sim W_p(\Sigma, n)$, $\Sigma > 0$, be independently distributed. Suppose we wish to test the hypothesis

$$H: \Sigma = I \quad \text{vs} \quad A: \Sigma \neq I. \tag{7.2.1}$$

It may be noted that the problem of testing $\Sigma = \Sigma_0$ vs $\Sigma \neq \Sigma_0$ where $\Sigma_0$ is a specified matrix reduces to the above problem, since $\mathbf{z} \equiv \Sigma_0^{-1/2}\mathbf{y} \sim N_p(\boldsymbol{\nu}, \Lambda)$ and $W \equiv \Sigma_0^{-1/2} V \Sigma_0^{-1/2} \sim W_p(\Lambda, n)$, where $\Lambda = \Sigma_0^{-1/2}\Sigma\Sigma_0^{-1/2}$ and $\boldsymbol{\nu} = \Sigma_0^{-1/2}\boldsymbol{\mu}$, and observing $\mathbf{y}$ and $V$ is equivalent to observing $\mathbf{z}$ and $W$. The problem now reduces to that of testing $\Lambda = I$ vs $\Lambda \neq I$.

The likelihood function is given by

$$L(\boldsymbol{\mu}, \Sigma) = c|\Sigma|^{-\frac{1}{2}N}|V|^{\frac{1}{2}(n-p-1)}\operatorname{etr}\left\{-\tfrac{1}{2}\Sigma^{-1}\left[V + (\mathbf{y}-\boldsymbol{\mu})(\mathbf{y}-\boldsymbol{\mu})'\right]\right\}, \tag{7.2.2}$$

where $c$ is a constant and $N=n+1$. Hence

$$\sup_H L(\boldsymbol{\mu},\Sigma)=c|V|^{\frac{1}{2}(n-p-1)}\operatorname{etr}\left(-\tfrac{1}{2}V\right), \tag{7.2.3}$$

and

$$\sup_A L(\boldsymbol{\mu},\Sigma)=c|N^{-1}V|^{-\frac{1}{2}N}|V|^{\frac{1}{2}(n-p-1)}e^{-\frac{1}{2}Np}. \tag{7.2.4}$$

Thus, the likelihood ratio is given by

$$\lambda=(e/N)^{\frac{1}{2}pN}|V|^{\frac{1}{2}N}\operatorname{etr}\left(-\tfrac{1}{2}V\right). \tag{7.2.5}$$

Now the test based on the likelihood ratio is not *unbiased* [see Sugiura and Nagao (1968)], whereas the test based on the modified likelihood ratio in which $N$ is replaced by $n$ (the number of degrees of freedom) is unbiased. Therefore we will consider the latter. The modified likelihood ratio $\lambda^*$ is given by

$$\lambda^*=(e/n)^{\frac{1}{2}pn}|V|^{\frac{1}{2}n}\operatorname{etr}\left(-\tfrac{1}{2}V\right). \tag{7.2.6}$$

The test based on $\lambda^*$ not only is unbiased, but has a monotone power function in each root of $\Sigma$ (see Chapter 10).

### 7.2.1 *Invariance*

The problem of testing $H$ against $A$ remains invariant under the additive and orthogonal groups of transformations

$$\mathbf{y}\to\Gamma\mathbf{y}+\mathbf{a} \text{ and } V\to\Gamma V\Gamma' \qquad \text{where} \quad \Gamma\Gamma'=I.$$

It can be shown that the roots of $V$ are maximal invariants. Hence, any test based on the characteristic roots of $V$ is invariant. In particular, the likelihood-ratio test ($\lambda$ or $\lambda^*$) is invariant, and so is the test based on the largest and the smallest root of $V$ (see Subsection 7.2.5) suggested by S. N. Roy (1957). Also, the distribution of any invariant test will depend only on the characteristic roots of $\Sigma$.

### 7.2.2 *Moments of* $\lambda^*$

Since $V\sim W_p(\Sigma,n)$, the $h$th moment of $\lambda^*$ is given by

$$E(\lambda^{*h})=\left(\frac{e}{n}\right)^{\frac{1}{2}pnh}|\Sigma|^{-\frac{1}{2}n}C(p,n)\int|V|^{\frac{1}{2}(nh+n-p-1)}\operatorname{etr}\left[-\tfrac{1}{2}V(hI+\Sigma^{-1})\right]dV$$

$$=\left(\frac{e}{n}\right)^{\frac{1}{2}pnh}|\Sigma^{-1}+hI|^{-\frac{1}{2}(nh+n)}|\Sigma|^{-\frac{1}{2}n}\frac{C(p,n)}{C(p,nh+n)},$$

where

$$C(p,n)=\left[2^{\frac{1}{2}pn}\pi^{\frac{1}{4}p(p-1)}\prod_{i=1}^{p}\Gamma\left[\tfrac{1}{2}(n+1-i)\right]\right]^{-1}. \tag{7.2.7}$$

Hence,

$$E(\lambda^{*h})=\left(\frac{2e}{n}\right)^{\frac{1}{2}pnh}\frac{|\Sigma|^{\frac{1}{2}nh}}{|I+h\Sigma|^{\frac{1}{2}n(1+h)}}\frac{\prod_{i=1}^{p}\Gamma\left[\frac{1}{2}(n+nh+1-i)\right]}{\prod_{i=1}^{p}\Gamma\left[\frac{1}{2}(n+1-i)\right]}. \tag{7.2.8}$$

$$\equiv g_1(h)g_2(h),$$

where

$$g_1(h)=\left(\frac{2e}{n}\right)^{\frac{1}{2}pnh}(1+h)^{-\frac{1}{2}pn(1+h)}\frac{\Gamma_p\left[\frac{1}{2}n(1+h)\right]}{\Gamma_p\left(\frac{1}{2}n\right)}, \tag{7.2.9}$$

and

$$g_2(h)=(1+h)^{\frac{1}{2}pn(1+h)}|\Sigma|^{\frac{1}{2}nh}|I+h\Sigma|^{-\frac{1}{2}n(1+h)}. \tag{7.2.10}$$

Note that when the hypothesis $H$ is true,

$$E(\lambda^{*h})=g_1(h) \quad \text{and} \quad g_2(h)=1. \tag{7.2.11}$$

### *7.2.3 Asymptotic Expansion of the* cdf *of* $-2\ln\lambda^*$ *Under H*

When the hypothesis $H$ is true, we get from Eq. (7.2.11)

$$E(\lambda^{*h})=g_1(h).$$

Using the asymptotic expansion of $\ln\Gamma(x+c)$ for bounded $c$, we get

$$\begin{aligned}g_1(h)=(1+h)^{-\frac{1}{2}f}\Big(&1+n^{-1}B_2\left[(1+h)^{-1}-1\right]\\&+\frac{1}{6n^2}\left[(3B_2^2-4B_3)(1+h)^{-2}-6B_2^2(1+h)^{-1}\right.\\&\left.+(3B_2^2+4B_3)\right]\\&+\frac{1}{6n^3}\left[(4B_1-4B_2B_3+B_2^3)(1+h)^{-3}\right.\\&+B_2(4B_3-3B_2^2)(1+h)^{-2}\\&+B_2(4B_3+3B_2^2)(1+h)^{-1}\\&\left.-(1B_4+4B_2B_3+B_2^3)\right]+O(n^{-4})\Big),\end{aligned} \tag{7.2.12}$$

where

(7.2.13)
$$f=\tfrac{1}{2}p(p+1),$$
$$B_2=\tfrac{1}{24}p(2p^2+3p-1),$$
$$B_3=-\tfrac{1}{32}p(p-1)(p+1)(p+2),$$
$$B_4=\tfrac{1}{480}p(6p^4+15p^3-10p^2-30p+3).$$

Thus, as a first approximation, $-2\ln\lambda^*$ is asymptotically distributed as a chi square with $f$ d.f. Hence, we get

**Theorem 7.2.1** *Let* $\mathbf{y}\sim N_p(\boldsymbol{\mu},\Sigma)$ *and* $V\sim W_p(\Sigma,n)$ *be independently distributed. Then as* $n\to\infty$, $-2\ln\lambda^*$, *where* $\lambda^*$ *is the modified likelihood ratio for testing* $H:\Sigma=I$ vs $A:\Sigma\neq I$, *has, under* $H$, *a chi-square distribution with* $f=\tfrac{1}{2}p(p+1)$ d.f.

An asymptotic distribution up to $O(n^{-2})$ is given in

**Theorem 7.2.2** *For the above problem of testing* $H$ vs $A$ *an asymptotic expansion of* $-2\ln\lambda^*$ *up to* $O(n^{-2})$ *is given by*

$$P(-2\ln\lambda^*\leqslant z)=P(\chi_f^2\leqslant z)+n^{-1}B_2P\left[(\chi_{f+2}^2\leqslant z)-P(\chi_f^2\leqslant z)\right]+O(n^{-2}).$$

PROOF. Since $0<\lambda^*<1$, all the moments of $\lambda^*$ exist and determine the distribution uniquely. The moment-generating function of $-2\ln\lambda^*$ is given by

$$E(e^{-2h\ln\lambda^*})=E(\lambda^{*-2h})=g_1(-2h),$$

whose asymptotic expansion is given by Eq. (7.2.12) with $h$ replaced by $-2h$. The result now follows by noting that $(1-2h)^{-\frac{1}{2}r}$ is the moment-generating function of a $\chi_r^2$. □

Asymptotic expansions up to $O(n^{-3})$ and $O(n^{-4})$ can be obtained from Eq. (7.2.13).

### 7.2.4 *Asymptotic Expansion of the* cdf *of* $-2\ln\lambda^*$ *Under Alternatives Close to the Hypothesis* $H$

We assume in this subsection that the alternative is close to the hypothesis in the following sense:

(7.2.14) $$I-\Sigma^{-1}=(2/n)^{1/2}P \qquad \text{where } P \text{ is fixed as } n\to\infty.$$

Under this assumption, we get

(7.2.15) $g_2(h)$

$$=(1+h)^{\frac{1}{2}pn(1+h)}|\Sigma^{-1}|^{\frac{1}{2}n}|\Sigma^{-1}+hI|^{-\frac{1}{2}n(1+h)}$$

$$=(1+h)^{\frac{1}{2}pn(1+h)}\left|I-\left(\frac{2}{n}\right)^{1/2}P\right|^{\frac{1}{2}n}\left|I-\left(\frac{2}{n}\right)^{1/2}P+hI\right|^{-\frac{1}{2}n(1+h)}$$

$$=(1+h)^{\frac{1}{2}pn(1+h)}\left|I-\left(\frac{2}{n}\right)^{1/2}P\right|^{\frac{1}{2}n}\left|(1+h)I-(2/n)^{1/2}P\right|^{-\frac{1}{2}n(1+h)}$$

$$=\left|I-\left(\frac{2}{n}\right)^{1/2}P\right|^{\frac{1}{2}n}\left|I-\left(\frac{2}{n}\right)^{1/2}(1+h)^{-1}P\right|^{-\frac{1}{2}n(1+h)}.$$

Hence, with the help of Problem 1.25(i), we get

(7.2.16) $\ln g_2(h)$

$$=\tfrac{1}{2}n\ln\left|I-\left(\frac{2}{n}\right)^{1/2}P\right|-\tfrac{1}{2}n(1+h)\ln\left|I-\left(\frac{2}{n}\right)^{1/2}(1+h)^{-1}P\right|$$

$$=\tfrac{1}{2}n(1+h)\left[\left(\frac{2}{n}\right)^{1/2}(1+h)^{-1}\operatorname{tr}P+\frac{1}{2}\left(\frac{2}{n}\right)(1+h)^{-2}\operatorname{tr}P^2\right.$$

$$\left.+\frac{1}{3}\left(\frac{2}{n}\right)^{3/2}(1+h)^{-3}\operatorname{tr}P^3\right]-\tfrac{1}{2}n\left[\left(\frac{2}{n}\right)^{1/2}\operatorname{tr}P+\frac{1}{2}\left(\frac{2}{n}\right)\operatorname{tr}P^2\right.$$

$$\left.+\frac{1}{3}\left(\frac{2}{n}\right)^{3/2}\operatorname{tr}P^3\right]+O(n^{-2})$$

$$=\sigma_2\left[(1+h)^{-1}-1\right]+\left(\frac{2}{n}\right)^{1/2}\sigma_3\left[(1+h)^{-2}-1\right]$$

$$+\left(\frac{2}{n}\right)\sigma_4\left[(1+h)^{-3}-1\right]+\left(\frac{2}{n}\right)^{3/2}\sigma_5\left[(1+h)^{-4}-1\right]+O(n^{-2}),$$

where

(7.2.17) $$\sigma_i=\operatorname{tr}P^i/i,\qquad i=2,3,4,5.$$

Since $g_2(h)=\exp[\ln g_2(h)]$, we get

(7.2.18) $g_2(h)$

$$=\left[\exp\left(\frac{-2h\sigma_2}{1+h}\right)\right]\left[1+\left(\frac{2}{n}\right)^{1/2}\sigma_3[(1+h)^{-2}-1]\right.$$

$$+\frac{2}{n}\left(\sigma_4[(1+h)^{-3}-1]+\tfrac{1}{2}\sigma_3^2[(1+h)^{-2}-1]^2\right)$$

$$+\left(\frac{2}{n}\right)^{3/2}\left(\sigma_5[(1+h)^{-4}-1]\right.$$

$$\left.+\sigma_3\sigma_4[(1+h)^{-2}-1][(1+h)^{-3}-1]+\tfrac{1}{6}\sigma_3^3[(1+h)^{-2}-1]^3\right)$$

$$\left.+O(n^{-2})\right].$$

Using the expansion for $g_1(h)$ in Eq. (7.2.12) and for $g_2(h)$ in Eq. (7.2.18), we get the following (Khatri and Srivastava, 1974c):

**Theorem 7.2.3** *Let* $\mathbf{y}\sim N_p(\boldsymbol{\mu},\Sigma)$ *and* $V\sim W_p(\Sigma,n)$ *be independently distributed. Then under the alternatives* (7.2.14) *an asymptotic expansion up to* $O(n^{-2})$ *of* $P(-2\ln\lambda^*\leq z)$, *where* $\lambda^*$ *is the modified likelihood ratio for testing* $H:\Sigma=I$ vs $A:\Sigma\neq I$, *is given by*

(7.2.19) $P(-2\ln\lambda^*\leq z)$

$$=P[\chi_f^2(\sigma_2)\leq z]+\left(\frac{2}{n}\right)^{1/2}\sigma_3\{P[\chi_{f+4}^2(\sigma_2)\leq z]$$

$$-P[\chi_f^2(\sigma_2)\leq z]\}+\frac{2}{n}\sum_{i=0}^{4}b_iP[\chi_{f+2i}^2(\sigma_2)\leq z]$$

$$+\left(\frac{2}{n}\right)^{3/2}\sum_{i=0}^{6}c_iP[\chi_{f+2i}^2(\sigma_2)\leq z]+O(n^{-2}),$$

*where* $\chi_r^2(\delta)$ *denotes a noncentral chi-square random variable with* $r$ d.f.

*and noncentrality parameter* $\delta$, *and*

(7.2.20)
$$f=\tfrac{1}{2}p(p+1),\qquad \sigma_i=\operatorname{tr}p^i/i,\quad i=2,3,4,5,$$
$$b_0=\tfrac{1}{2}\sigma_3^2-\tfrac{1}{2}B_2-\sigma_4,$$
$$b_1=\tfrac{1}{2}b_2,\quad b_2=-\sigma_3^2,\quad b_3=\sigma_4,\quad b_4=\tfrac{1}{2}\sigma_3^2,$$
$$c_0=\tfrac{1}{2}\sigma_3B_2-\sigma_5+\sigma_3\sigma_4-\tfrac{1}{6}\sigma_3^3,\qquad c_1=-\tfrac{1}{2}\sigma_3B_2,$$
$$c_2=-\tfrac{1}{2}\sigma_3B_2-\sigma_3\sigma_4+\tfrac{1}{2}\sigma_3^3,\qquad c_3=\tfrac{1}{2}\sigma_3B_2-\sigma_3\sigma_4,$$
$$c_4=\sigma_5-\tfrac{1}{2}\sigma_3^3,\qquad c_5=\sigma_3\sigma_4,\qquad c_6=\tfrac{1}{6}\sigma_3^3,$$
$$B_2=\tfrac{1}{24}p(2p^2+3p-1).$$

As a first approximation, we get the following:

**Theorem 7.2.4** *Under the alternative* (7.2.14), $-2\ln\lambda^*$ *has asymptotically noncentral chi square with* $f=\frac{1}{2}p(p+1)$ d.f. *and noncentrality parameter* $\sigma_2=\frac{1}{2}\operatorname{tr}P^2$.

### *7.2.5 Union–Intersection Test Procedure and Confidence Bounds*

We observe that

$$H:\Sigma=I \Leftrightarrow \bigcap_{\mathbf{a}} H_{\mathbf{a}},\qquad \text{where}\quad H_{\mathbf{a}}:\mathbf{a}'\Sigma\mathbf{a}=\mathbf{a}'\mathbf{a},$$

and

$$A:\Sigma\neq I \Leftrightarrow \bigcup_{\mathbf{a}} A_{\mathbf{a}},\qquad \text{where}\quad A_{\mathbf{a}}:\mathbf{a}'\Sigma\mathbf{a}\neq\mathbf{a}'\mathbf{a},$$

where $\mathbf{a}$ is any nonnull vector. Given $\mathbf{a}$, $\mathbf{a}'V\mathbf{a}/\mathbf{a}'\Sigma\mathbf{a}\sim\chi_n^2$; and hence using univariate results, the test procedure for testing $H_{\mathbf{a}}$ vs $A_{\mathbf{a}}$ is

$$\text{reject } H_{\mathbf{a}}\ \text{ if}$$
$$\mathbf{a}'V\mathbf{a}/\mathbf{a}'\mathbf{a}\geqslant c_2\quad\text{or}\quad \mathbf{a}'V\mathbf{a}/\mathbf{a}'\mathbf{a}\leqslant c_1,$$

where the constants $c_1$ and $c_2$ depend on $n$ and $\alpha$, and $c_2\geqslant c_1$. Hence, from the union–intersection principle we get

$$\text{reject } H \text{ vs } A\ \text{ if}$$
$$\operatorname*{Sup}_{\mathbf{a}}\frac{\mathbf{a}'V\mathbf{a}}{\mathbf{a}'\mathbf{a}}\geqslant c_2\quad\text{or}\quad \operatorname*{Inf}_{\mathbf{a}}\frac{\mathbf{a}'V\mathbf{a}}{\mathbf{a}'\mathbf{a}}\leqslant c_1.$$

Using Theorem 1.10.1, we get

$$\text{reject } H \text{ vs } A\ \text{ if}$$
$$\operatorname{ch}_1(V)\geqslant c_2\quad\text{or}\quad \operatorname{ch}_p(V)\leqslant c_1,$$

and the acceptance region is

$$c_1 < \mathrm{ch}_p(V) \leqslant \mathrm{ch}_1(V) < c_2,$$

where $c_1$ and $c_2$ are to be determined from $P[c_1 < \mathrm{ch}_p(V) \leqslant \mathrm{ch}_1(V) < c_2 | H] = 1-\alpha$. This test may fail to be unbiased unless $c_1$ and $c_2$ are properly chosen. When $p=1$, this is done by using unbiased conditions.

This test procedure can be used to obtain simultaneous confidence bounds as follows:

If $V \sim W_p(\Sigma, n)$, then

$$P\left[c_1 < \mathrm{ch}_p(\Sigma^{-1}V) \leqslant \mathrm{ch}_1(\Sigma^{-1}V) < c_2 | A\right] = 1-\alpha,$$

and hence with confidence coefficient equal to $1-\alpha$, the simultaneous confidence bounds are

$$\frac{\mathbf{a}'V\mathbf{a}}{c_2} \leqslant \mathbf{a}'\Sigma\mathbf{a} \leqslant \frac{\mathbf{a}'V\mathbf{a}}{c_1} \qquad \text{for all nonnull vectors} \quad \mathbf{a}.$$

From the above, we can get various types of confidence bounds on $\Sigma$ by choosing the vector **a**. For example

$$\frac{v_{ii}}{c_2} \leqslant \sigma_{ii} \leqslant \frac{v_{ii}}{c_1}, \qquad \frac{\mathrm{ch}_i(V)}{c_2} \leqslant \mathrm{ch}_i(\Sigma) \leqslant \frac{\mathrm{ch}_i(V)}{c_1}$$

$$\text{for} \quad i = 1, 2, \ldots, p$$

gives the simultaneous confidence bounds on all variances and all the characteristic roots of $\Sigma$ with confidence coefficient greater than or equal to $1-\alpha$.

Suppose one is interested only in the confidence bounds on $\mathrm{ch}_i(\Sigma)$, $i = 1, 2, \ldots, p$. The above types of confidence bounds are wide; narrower ones can be obtained as follows [see Khatri (1965c)]: Let $c_3$ and $c_4$ be chosen such that

$$P(\chi_n^2 \geqslant c_3) P(\chi_{n-p+1}^2 \leqslant c_4) = 1-\alpha.$$

Since $V \sim W_p(\Sigma, n)$, we have $W = \Gamma V \Gamma' \sim W_p(D_\lambda, n)$, where $D_\lambda = \mathrm{diag}(\lambda_1, \ldots, \lambda_p)$, $\lambda_1 \geqslant \lambda_2 \geqslant \cdots \geqslant \lambda_p$, and $\Gamma$ is an orthogonal matrix such that $\Sigma = \Gamma' D_\lambda \Gamma$. Notice that $w_{11}/\lambda_1$ and $(w^{pp}/\lambda_p)^{-1}$ are independently distributed, where $w_{11}/\lambda_1 \sim \chi_n^2$ and $(w^{pp}/\lambda_p)^{-1} \sim \chi_{n-p+1}^2$. Therefore, $P(w_{11}/\lambda_1 \geqslant c_3)$ and $(w^{pp}/\lambda_p)^{-1} \leqslant c_4) = 1-\alpha$, and the confidence bounds with confidence coefficient $1-\alpha$ on $\lambda_1$ and $\lambda_p$ are given by

$$(w^{pp}c_4)^{-1} \leqslant \lambda_p \leqslant \lambda_1 \leqslant w_{11}/c_3.$$

Since we don't know $\Gamma$, we cannot calculate $w_{11}$ and $w^{pp}$, but using Problem 1.39, we have

$$\mathrm{ch}_1(V) \geqslant \left(v_{ii}, (v^{jj})^{-1}\right) \geqslant \mathrm{ch}_p(V),$$

and hence we have the simultaneous confidence bounds with confidence greater than or equal to $1-\alpha$ on the characteristic roots of $\Sigma$ as

$$\mathrm{ch}_p(V)/c_4 \leqslant \mathrm{ch}_p(\Sigma) \leqslant \mathrm{ch}_1(\Sigma) \leqslant \mathrm{ch}_1(V)/c_3.$$

Suppose one is only interested in the confidence bounds on variances. Then, from Problem 2.9, we have

$$P(v_{ii} \leqslant \beta_i \sigma_{ii},\ i=1,2,\ldots,p) \geqslant \prod_{i=1}^{p} P(v_{ii} \leqslant \beta_i \sigma_{ii}).$$

Now, let us choose $\beta_1,\ldots,\beta_p$ such that $\prod_{i=1}^{p} P(\chi_n^2 \leqslant \beta_i) = 1-\alpha$. Then

$$P(v_{ii}/\sigma_{ii} \leqslant \beta_i,\ i=1,2,\ldots,p) \geqslant 1-\alpha,$$

and hence with confidence greater than or equal to $1-\alpha$, we have

$$\sigma_{ii} \geqslant v_{ii}/\beta_i \qquad \text{for} \quad i=1,2,\ldots,p$$

which gives one-sided confidence bounds.

### 7.2.6 *Some Other Tests*

From Eq. (7.2.8), we can write down the moment-generating function of $-2n^{-1/2}\ln\lambda^*$ and its asymptotic expansion. From this, one can easily obtain that asymptotically, the statistic

$$-2n^{-1/2}\ln\lambda^* - n^{1/2}\{\mathrm{tr}(\Sigma - I) - \ln|\Sigma|\} \tag{7.2.21}$$

is normally distributed with mean zero and variance

$$\tau_1^2 = 2\,\mathrm{tr}(I-\Sigma)^2. \tag{7.2.22}$$

Thus, $\tau_1^2$ may be regarded as a measure of departure from the null hypothesis $H$. Replacing $\Sigma$ by its unbiased estimate $n^{-1}V$, Nagao (1973a) proposed an alternative test statistic

$$T_1^2 = \tfrac{1}{2} n\,\mathrm{tr}(n^{-1}V - I)^2. \tag{7.2.23}$$

However, it is not even unbiased, as can be seen for $p=1$.

## 7.3 Testing That the Covariance Matrix Is an Identity Matrix and the Mean Vector Is Zero

Let $\mathbf{y} \sim N_p(\boldsymbol{\mu}, \Sigma)$ and $V \sim W_p(\Sigma, n)$ be independently distributed. Suppose we wish to test the hypothesis

$$H: \boldsymbol{\mu} = 0,\ \Sigma = I \quad \text{vs} \quad A \neq H. \tag{7.3.1}$$

It may be noted that the problem of testing $\boldsymbol{\mu} = \boldsymbol{\mu}_0$ and $\Sigma = \Sigma_0$ can be reduced to the above problem.

The likelihood function is given by Eq. (7.2.2). Hence $\operatorname{Sup}_A L(\boldsymbol{\mu},\Sigma)$ is given by Eq. (7.2.4) and

$$\sup_H L(\boldsymbol{\mu},\Sigma)=c|V|^{\frac{1}{2}(n-p-1)}\operatorname{etr}\left[-\tfrac{1}{2}(V+\mathbf{y}\mathbf{y}')\right]. \tag{7.3.2}$$

Thus, the likelihood ratio is the ratio of Eq. (7.3.2) to Eq. (7.2.4) and is given by

$$\lambda=(e/N)^{\frac{1}{2}pN}|V|^{\frac{1}{2}N}\operatorname{etr}\left[-\tfrac{1}{2}(V+\mathbf{y}\mathbf{y}')\right]. \tag{7.3.3}$$

This (unmodified) likelihood-ratio test is unbiased. (See Chapter 10.)

It is left as an exercise to the reader in Problem 7.2 to show that

$$E(\lambda^h)=g_1(h)g_2(h)g_3(h), \tag{7.3.4}$$

where

$$g_1(h)=\left(\frac{2e}{N}\right)^{\frac{1}{2}pNh}(1+h)^{-\frac{1}{2}pN(1+h)}\frac{\Gamma_p\left(\frac{1}{2}N(1+h)\right)}{\Gamma_p\left(\frac{1}{2}N\right)}, \tag{7.3.5}$$

$$g_2(h)=(1+h)^{\frac{1}{2}pN(1+h)}|\Sigma|^{\frac{1}{2}Nh}|I+h\Sigma|^{-\frac{1}{2}N(1+h)},$$

$$g_3(h)=\operatorname{etr}\left\{-\tfrac{1}{2}h\boldsymbol{\mu}\boldsymbol{\mu}'\left[I-h(\Sigma^{-1}+hI)^{-1}\right]\right\}.$$

It may be noted that under $H$, we have $g_2(h)=1$ and $g_3(h)=1$. Hence, the asymptotic distribution of $-2\ln\lambda$, under $H$, is given by Theorems 7.2.1 and 7.2.2 with $n$ replaced by $N$. The asymptotic distribution of $-2\ln\lambda$ under alternatives close to the hypothesis can be obtained on the same lines as in Subsection 7.2.2 and is given as an exercise to the reader as Problems 7.3 and 7.4 (Khatri and Srivastava, 1974c).

### 7.3.1 *Invariance*

The problem of testing $H$ vs $A$ remains invariant under the orthogonal group of transformations $\mathbf{y}\to\Gamma\mathbf{y}$ and $V\to\Gamma V\Gamma'$, $\Gamma\Gamma'=I$. It can easily be checked that the likelihood-ratio test $\lambda$ is invariant under the above transformations.

## 7.4 Test for Sphericity

Let $\mathbf{y}\sim N_p(\boldsymbol{\mu},\Sigma)$ and $V\sim W_p(\Sigma,n)$ be independently distributed. Suppose we wish to test the hypothesis

$$H:\Sigma=\sigma^2 I,\ \sigma^2>0, \quad \text{vs} \quad A\neq H. \tag{7.4.1}$$

The hypothesis $H$ is called the *sphericity hypothesis*. The likelihood function is given by [the same as Eq. (7.2.2)]

$$L(\boldsymbol{\mu}, \Sigma) = c|\Sigma|^{-\frac{1}{2}N}|V|^{\frac{1}{2}(n-p-1)}\operatorname{etr}\left\{-\tfrac{1}{2}\Sigma^{-1}\left[V+(\mathbf{y}-\boldsymbol{\mu})(\mathbf{y}-\boldsymbol{\mu})'\right]\right\}.$$

Hence,

$$\text{(7.4.2)} \qquad \operatorname*{Sup}_{H} L(\boldsymbol{\mu}, \Sigma) = c|V|^{\frac{1}{2}(n-p-1)}\left(\frac{\operatorname{tr} V}{Np}\right)^{-\frac{1}{2}Np} e^{-\frac{1}{2}Np},$$

and

$$\text{(7.4.3)} \qquad \operatorname*{Sup}_{A} L(\boldsymbol{\mu}, \Sigma) = c|V|^{\frac{1}{2}(n-p-1)}|N^{-1}V|^{-\frac{1}{2}N}e^{-\frac{1}{2}Np}.$$

Thus, the likelihood ratio is given by

$$\text{(7.4.4)} \qquad \lambda = \frac{|V|^{\frac{1}{2}N}}{(p^{-1}\operatorname{tr} V)^{\frac{1}{2}Np}} \equiv (\lambda^*)^{\frac{1}{2}N},$$

where

$$\text{(7.4.5)} \qquad \lambda^* = |V|/(p^{-1}\operatorname{tr} V)^p.$$

For the monotonicity of the power function see Chapter 10.

### 7.4.1 *Invariance*

The problem of testing $H$ vs $A$ remains invariant under the following transformations: $\mathbf{y}\to c\Gamma\mathbf{y}+\mathbf{a}$ and $V\to c^2\Gamma V\Gamma'$, $c\neq 0$, $\Gamma\Gamma'=I$. It can easily be checked that the likelihood-ratio test $\lambda^*$ is invariant under the above group of transformations. A set of maximal invariants is $(l_1/l_p, l_2/l_p, \ldots, l_{p-1}/l_p)$, where $l_1, \ldots, l_p$ are the roots of $V$. A set of maximal invariants on the parameter space is $(\sigma_1^2/\sigma_p^2, \ldots, \sigma_{p-1}^2/\sigma_p^2)$, where $\sigma_1^2, \ldots, \sigma_p^2$ are the roots of $\Sigma$.

### 7.4.2 *Moments of* $\lambda^*$ *Under* $H$

The $h$th moment of $\lambda^*$ under the alternative $A$ has been given by Khatri and Srivastava (1971), but is beyond the scope of this book. We will thus derive the $h$th moment of $\lambda^*$ under the hypothesis $H$, when $V\sim W_p(\sigma^2 I, n)$. Since the likelihood-ratio test $\lambda^*$ is invariant under the scale transformation $\{\mathbf{y}\to c\mathbf{y}$ and $V\to c^2V\}$, we may assume without loss of generality that

$\sigma^2=1$. Hence, under $H$, $V\sim W_p(I,n)$ and

$$(7.4.6)\quad E(\lambda^{*h})=p^{ph}\int\frac{|W|^{\frac{1}{2}(2h+n-p-1)}\operatorname{etr}\left(-\frac{1}{2}V\right)}{\Gamma_p\left(\frac{1}{2}n\right)2^{\frac{1}{2}pn}(\operatorname{tr}V)^{ph}}\,dV$$

$$=\frac{\Gamma_p\left(\frac{1}{2}n+h\right)p^{ph}2^{\frac{1}{2}p(n+2h)}}{\Gamma_p\left(\frac{1}{2}n\right)2^{\frac{1}{2}pn}}$$

$$\times\int\left(\sum_{i=1}^{p}v_{ii}\right)^{-\frac{1}{2}ph}\frac{|V|^{\frac{1}{2}(2h+n-p-1)}\operatorname{etr}\left(-\frac{1}{2}V\right)}{\Gamma_p\left(\frac{1}{2}n+h\right)2^{\frac{1}{2}p(n+2h)}}\,dV$$

$$=(2p)^{ph}\frac{\Gamma_p\left(\frac{1}{2}n+h\right)}{\Gamma_p\left(\frac{1}{2}n\right)}E\left[\left(\sum_{i=1}^{p}v_{ii}\right)^{-ph}\right],$$

where $V=(v_{ij})$ and $V\sim W_p(I,n+2h)$. Note that when $V\sim W_p(I,n+2h)$, the $v_{ii}$'s are independently distributed as chi square with $n+2h$ d.f. Hence under $H$, $\Sigma_{i=1}^{p}v_{ii}$ has a chi-square distribution with $p(n+2h)$ d.f. Thus

$$(7.4.7)\qquad E(\lambda^{*h})=p^{ph}\frac{\Gamma_p\left(\frac{1}{2}n+h\right)}{\Gamma_p\left(\frac{1}{2}n\right)}E\left[\frac{1}{2}\sum_{i=1}^{p}v_{ii}\right]^{-ph}$$

$$=p^{ph}\frac{\Gamma_p\left(\frac{1}{2}n+h\right)}{\Gamma_p\left(\frac{1}{2}n\right)}\frac{\Gamma\left[p\left(\frac{1}{2}n+h\right)-ph\right]}{\Gamma\left[p\left(\frac{1}{2}n+h\right)\right]}$$

$$=p^{ph}\frac{\Gamma_p\left(\frac{1}{2}n+h\right)}{\Gamma_p\left(\frac{1}{2}n\right)}\frac{\Gamma\left(\frac{1}{2}np\right)}{\Gamma\left(\frac{1}{2}np+hp\right)}.$$

Using Gauss's multiplication formula

$$(7.4.8)\qquad \Gamma(n\alpha)=(2\pi)^{-\frac{1}{2}(n-1)}n^{n\alpha-\frac{1}{2}}\prod_{r=0}^{n-1}\Gamma(\alpha+n^{-1}r),$$

we get

$$(7.4.9)\qquad E(\lambda^{*h})=\prod_{i=1}^{p-1}\frac{\Gamma\left[\frac{1}{2}(n-i)+h\right]\Gamma\left[\frac{1}{2}n+p^{-1}i\right]}{\Gamma\left[\frac{1}{2}(n-i)\right]\Gamma\left[\frac{1}{2}n+h+p^{-1}i\right]}.$$

### *7.4.3 Exact Distribution Under H*

From Eq. (7.4.9) we get that $\lambda^*$ is distributed as the product of $p-1$ independent beta random variables $Y_i$, with parameters $\frac{1}{2}(n-i)$ and $i\left(\frac{1}{2}+p^{-1}\right)$, $i=1,2,\ldots,p-1$. Thus, when $p=2$, we get the following:

**Theorem 7.4.1** *When $p=2$, the likelihood-ratio test statistic $\lambda^* = \lambda^{2/N}$ given in Eq. (7.4.6) for testing the sphericity has, under the hypothesis $H$, a beta distribution with parameters $\frac{1}{2}(n-1)$ and 1. Thus the* pdf *of $\lambda^*$ is given by*

$$f(\lambda^*) = \tfrac{1}{2}(n-1)\lambda^{*\frac{1}{2}(n-1)-1}, \qquad 0<\lambda^*<1,$$

*and*

$$P(\lambda^* \leqslant z) = z^{\frac{1}{2}(n-1)}, \qquad 0<z\leqslant 1.$$

The exact pdf (and cdf) of $\lambda^*$ has been given by Consul (1967b, 1969) and Mathai and Rathie (1970) under $H$ and by Khatri and Srivastava (1971) under the alternative $A$. Even under $H$, the algebra is heavy and involves Meijer's $G$-functions [see Erdelyi (1953)] except for $p=2,3,4,6$. The exact pdf under $H$ for $p=3$ and $p=4$ is given in the following theorems without proofs; the proofs may be found in Consul (1967b), where the exact pdf for $p=6$ is given as well.

**Theorem 7.4.2** *For $p=3$, the exact* pdf *of $\lambda^*$ under $H$ is given by*

$$f(\lambda^*) = K(n)\lambda^{*\frac{1}{2}n-2}(1-\lambda^*)^{3/2}{}_2F_1\left(\tfrac{5}{6},\tfrac{7}{6};\tfrac{5}{2};1-\lambda^*\right), \qquad 0<\lambda^*<1,$$

*where*

$$K(n) = 2^{n+1}\Gamma\left(\tfrac{3}{2}n\right)\left[\Gamma(n-1)\Gamma\left(\tfrac{1}{2}n-1\right)3^{\frac{1}{2}(3n+1)}\right]^{-1}.$$

For alternative expressions for the exact pdf in this case see John (1972) and Nagarsenker and Pillai (1973); in the latter reference tables of percentage points are also provided.

**Theorem 7.4.3** *For $p=4$, the exact* pdf *of $\lambda^*$ under $H$ is given by*

$$f(\lambda^*) = K(n)\lambda^{*\frac{1}{2}(n-5)}(1-\lambda^{*1/2})^{7/2}{}_2F_1\left(1,\tfrac{3}{2};\tfrac{9}{2};1-\lambda^{*1/2}\right), \qquad 0<\lambda^*<1,$$

*where*

$$K(n) = \frac{2(n-1)\Gamma\left(n+\frac{1}{2}\right)}{7\left[2\Gamma(n-3)\Gamma\left(\frac{7}{2}\right)\right]}.$$

### *7.4.4 Asymptotic Expansion of the* cdf *Under H*

In this section we give an asymptotic expansion of the cdf of $-m\ln\lambda^*$ under $H$ where

(7.4.10) $$n = m + 2\alpha$$

and $\alpha$ is chosen so that the term of order $m^{-1}$ is zero. From Eq. (7.4.7), we get

(7.4.11) $$\phi(t)\equiv E(e^{-mt\ln\lambda^*})=E(\lambda^{*-mt})$$
$$=p^{-pmt}\frac{\Gamma_p\left(\frac{1}{2}m-mt+\alpha\right)\Gamma\left(\frac{1}{2}mp+\alpha p\right)}{\Gamma_p\left(\frac{1}{2}m+\alpha\right)\Gamma\left(\frac{1}{2}mp+\alpha p-mtp\right)}.$$

Using the asymptotic expansion of $\ln(x+h)$ for bounded $h$ and choosing $\alpha$ such that the term of order $m^{-1}$ is zero, we get

(7.4.12) $$\alpha=\frac{2p^2+p+2}{12p},$$

and

(7.4.13) $$P(-m\ln\lambda^*\leqslant z)$$
$$=P(\chi_f^2\leqslant z)+\frac{a}{m^2}\left[P(\chi_{f+4}^2\leqslant z)-P(\chi_f^2\leqslant z)\right]+O(m^{-3}),$$

where

(7.4.14) $$f=\tfrac{1}{2}p(p+1)-1$$

and

(7.4.15) $$a=(p+2)(p-1)(p-2)(2p^3+6p^2+3p+2)/288p^2.$$

The expansion for alternatives close to the hypothesis has been given by Khatri and Srivastava (1974a).

### *7.4.5 An Alternative Test*

Although beyond the scope of this book, we can, as in Subsection 7.2.5, write down the moment-generating function of $-m^{1/2}\ln\lambda^*$ by using the noncentral results of Khatri and Srivastava (1971) and show that the statistic

(7.4.16) $$m^{1/2}\left[-\ln\lambda^*-\ln\frac{(\operatorname{tr}\Sigma/p)^p}{|\Sigma|}\right]$$

is asymptotically normally distributed with mean zero and variance

(7.4.17) $$\tau_2^2=2p\left[p\frac{\operatorname{tr}\Sigma^2}{(\operatorname{tr}\Sigma)^2}-1\right]$$
$$=2p^2\operatorname{tr}\left[\Sigma(\operatorname{tr}\Sigma)^{-1}-p^{-1}I\right]^2.$$

Thus, $\tau_2^2$ may be regarded as a measure of departure from the null hypothesis $H$. Replacing $\Sigma$ by its unbiased estimate $n^{-1}V$, Nagao (1973a)

proposed the statistic

$$T_2=\tfrac{1}{2}p^2 n\,\mathrm{tr}\left[V(\mathrm{tr}\,V)^{-1}-p^{-1}I\right]^2 \tag{7.4.18}$$

for testing the sphericity. From another point of view, this statistic has been proposed by John (1971) and Sugiura (1973) [see also Giri (1968)], who showed that the test based on this statistic is the locally most powerful invariant. However, while other properties of this test still remain to be investigated (except for $p=2$, when the two tests $T_2$ and $\lambda^*$ coincide), Carter and Srivastava (1976) have shown that the tests based on $T_2$ and $\lambda^*$ have the same power up to $O(n^{-3/2})$ for local alternatives—alternatives for which $T_2$ is supposed to be superior. Thus, any further investigation of $T_2$ appears to be redundant.

### *7.4.6 Union–Intersection Test Procedures*

We observe that

$$H(\Sigma=\sigma^2 I)=\bigcap_{\substack{\mathbf{a}_1,\mathbf{a}_2\\ \mathbf{a}_1'\mathbf{a}_2=0}} H_{\mathbf{a}_1,\mathbf{a}_2},$$

where

$$H_{\mathbf{a}_1,\mathbf{a}_2}:\left(\frac{\mathbf{a}_1'\Sigma\mathbf{a}_1}{\mathbf{a}_1'\mathbf{a}_1}=\frac{\mathbf{a}_2'\Sigma\mathbf{a}_2}{\mathbf{a}_2'\mathbf{a}_2},\ \mathbf{a}_1'\Sigma\mathbf{a}_2=0\right),$$

and

$$A(\Sigma\neq\sigma^2 I)=\bigcup_{\substack{\mathbf{a}_1,\mathbf{a}_2\\ \mathbf{a}_1'\mathbf{a}_2=0}} A_{\mathbf{a}_1,\mathbf{a}_2},$$

where

$$A_{\mathbf{a}_1,\mathbf{a}_2}:\left(\frac{\mathbf{a}_1'\Sigma\mathbf{a}_1}{\mathbf{a}_1'\mathbf{a}_1}\neq\frac{\mathbf{a}_2'\Sigma a_2}{\mathbf{a}_2'\mathbf{a}_2}\ \text{ or }\ \mathbf{a}_1'\Sigma\mathbf{a}_2\neq 0\right),$$

where $\mathbf{a}_1$ and $\mathbf{a}_2$ are any nonnull orthogonal vectors. We observe that $H_{\mathbf{a}_1,\mathbf{a}_2}$ and $A_{\mathbf{a}_1,\mathbf{a}_2}$ are composite hypotheses, and hence we shall use the likelihood-ratio test procedure based on $V$. Using Eq. (7.4.5) for $p=2$, for given $\mathbf{a}_1$ and $\mathbf{a}_2$ such that $\mathbf{a}_1'\mathbf{a}_2=0$,

$$\text{reject } H_{\mathbf{a}_1,\mathbf{a}_2} \text{ vs } A_{\mathbf{a}_1,\mathbf{a}_2} \quad \text{if}$$

$$\frac{4\left[(\mathbf{a}_1'V\mathbf{a}_1)(\mathbf{a}_2'V\mathbf{a}_2)-(\mathbf{a}_1'V\mathbf{a}_2)^2\right]}{(\mathbf{a}_1'V\mathbf{a}_1+\mathbf{a}_2'V\mathbf{a}_2)^2}\leqslant c,$$

where $c$ is a constant depending on $n$ and $\alpha$. Hence, by the union–intersection test procedure,

$$\text{reject } H \text{ vs } A \quad \text{if}$$

$$\sup_{\substack{\mathbf{a}_1,\mathbf{a}_2\\ \mathbf{a}_1'\mathbf{a}_2=0}} \left\{ \frac{(\mathbf{a}_1'V\mathbf{a}_1+\mathbf{a}_2'V\mathbf{a}_2)^2}{4\left[(\mathbf{a}_1'V\mathbf{a}_1)(\mathbf{a}_2'V\mathbf{a}_2)-(\mathbf{a}_1'V\mathbf{a}_2)^2\right]} \right\} \geqslant c.$$

It is easy to verify that the supremum value of

$$f(\mathbf{a}_1,\mathbf{a}_2)\equiv \frac{(\mathbf{a}_1'V\mathbf{a}_1+\mathbf{a}_2'V\mathbf{a}_2)^2}{4\left[(\mathbf{a}_1'V\mathbf{a}_1)(\mathbf{a}_2'V\mathbf{a}_2)-(\mathbf{a}_1'V\mathbf{a}_2)^2\right]}$$

will be obtained when $\mathbf{a}_1$ and $\mathbf{a}_2$ are the characteristic vectors of $V$. Hence,

$$\sup_{\substack{\mathbf{a}_1,\mathbf{a}_2\\ \mathbf{a}_1'\mathbf{a}_2=0}} f(\mathbf{a}_1,\mathbf{a}_2)=\max_{\substack{i,j\\ i\neq j}} \frac{(l_i+l_j)^2}{4l_il_j}=\frac{(l_1+l_p)^2}{4l_1l_p},$$

where $l_1>l_2>\cdots>l_p>0$ are the characteristic roots of $V$. Thus, the test procedure for testing $H$ against $A$ is

(7.4.19) $$\text{reject } H \quad \text{if} \quad (l_1+l_p)^2/4l_1l_p \geqslant c,$$

where $c$ is determined from $P[(l_1+l_p)^2/4l_1l_p \geqslant c|H]=\alpha$.

A different approach than the above one given by Venables (1973) is useful in obtaining confidence bounds and extending the test procedure. With this point in mind, we reestablish Eq. (7.4.19) by Venables' method. First, we observe that for any nonnull vector $\mathbf{a}$, the inequality

(7.4.20) $$(\mathbf{a}'\mathbf{a})^2 \leqslant (\mathbf{a}'\Sigma\mathbf{a})(\mathbf{a}'\Sigma^{-1}\mathbf{a}) \leqslant (\mathbf{a}'\mathbf{a})^2\frac{(\lambda_1+\lambda_p)^2}{4\lambda_1\lambda_p}$$

holds, where $\lambda_1 \geqslant \lambda_2 \geqslant \cdots \geqslant \lambda_p$ are the characteristic roots of $\Sigma$. Hence,

$$(H:\Sigma=\sigma^2I)=\bigcap_{\mathbf{a}:\mathbf{a}'\mathbf{a}=1} H_{\mathbf{a}},$$

where

$$H_{\mathbf{a}}:(\mathbf{a}'\Sigma\mathbf{a})(\mathbf{a}'\Sigma^{-1}\mathbf{a})=1,$$

and

$$(A:\Sigma\neq\sigma^2I)=\bigcup_{\mathbf{a}:\mathbf{a}'\mathbf{a}} A_{\mathbf{a}},$$

where

$$A_{\mathbf{a}}:(\mathbf{a}'\Sigma\mathbf{a})(\mathbf{a}'\Sigma^{-1}\mathbf{a})>1.$$

If $K=(\mathbf{a}, K_2)$ is an orthogonal matrix, then by Corollary 1.9.2,

$$\Sigma=\Sigma K_2(K_2'\Sigma K_2)^{-1}K_2'\Sigma+\frac{\mathbf{a}\mathbf{a}'}{\mathbf{a}'\Sigma^{-1}\mathbf{a}}$$

and hence

(7.4.21) $$(\mathbf{a}'\Sigma^{-1}\mathbf{a})(\mathbf{a}'\Sigma\mathbf{a})=1 \quad \Leftrightarrow \quad \mathbf{a}'\Sigma K_2=\mathbf{0}.$$

Thus, given $\mathbf{a}$, $H_{\mathbf{a}}$ is equivalent to testing the hypothesis that $\mathbf{a}'\Sigma K_2=\mathbf{0}$. Hence, the likelihood-ratio test procedure for testing $H_a$ is

$$\text{reject } H_{\mathbf{a}} \text{ vs } A_{\mathbf{a}} \quad \text{if} \quad (\mathbf{a}'Va)(\mathbf{a}'V^{-1}\mathbf{a})\geq c.$$

Thus, by the union–intersection principle, we get the same test procedure as Eq. (7.4.19), namely,

$$\text{reject } H \text{ vs } A \quad \text{if} \quad (l_1+l_p)^2/4l_1l_p\geq c,$$

where $c$ is determined from $P[(l_1+l_p)^2/4l_1l_p\geq c|H]=\alpha$.

To generalize the above method, consider an orthogonal matrix $K=(K_1,K_2)$ where $K_1$ is a $p\times q$ matrix. Then we have the following inequality:

(7.4.22) $$1\leq|K_1'\Sigma K_1||K_1'\Sigma^{-1}K_1|\leq\prod_{i=1}^{q}\frac{(\lambda_i+\lambda_{p-i+1})^2}{4\lambda_i\lambda_{p-i+1}}$$

for all matrices $K_1:p\times q$ such that $p\geq 2q$ and $K_1'K_1=I_q$. Then

$$(H:\Sigma=\sigma^2I)=\bigcap_{K_1:K_1'K_1=I_q} H_{K_1},$$

where

$$H_{K_1}:|K_1'\Sigma K_1||K_1'\Sigma^{-1}K_1|=1$$

and

$$(A:\Sigma\neq\sigma^2I)=\bigcup_{K_1:K_1'K_1=I_q} A_{K_1},$$

where

$$A_{K_1}:|K_1'\Sigma K_1||K_1'\Sigma^{-1}K_1|>1.$$

As before, we observe that the hypothesis $H_{K_1}$ is equivalent to testing the hypothesis that

(7.4.23) $$K_1'\Sigma K_2=0.$$

From the likelihood-ratio test [see Eq. (7.5.8) of the next subsection] for independence, we get the test procedure for testing $H_{K_1}$ vs $A_{K_1}$ as

$$\text{reject } H_{K_1} \quad \text{if} \quad |K_1'VK_1||K_1'V^{-1}K_1|\geq c,$$

because

$$K'VK=\begin{pmatrix} V_{11} & V_{12} \\ V_{12}' & V_{22} \end{pmatrix},$$

with $V_{11}=K_1'VK_1$ and $V^{11}=K_1'V^{-1}K_1$.

Hence, the union–intersection test procedure is

(7.4.24) $$\text{reject } H \text{ vs } A \quad \text{if} \quad \prod_{i=1}^{q} \frac{(l_i+l_{p-i+1})^2}{4l_il_{p-i+1}} \geqslant c,$$

where $c$ is to be determined from

$$P\left\{ \prod_{i=1}^{q} \frac{(l_i+l_{p-i+1})^2}{4l_il_{p-i+1}} \geqslant c \,\middle|\, H \right\} = \alpha,$$

where $q=1,2,\ldots,[p/2]$, $[x]$=greatest integer contained in $x$. This is the test derived by Venables (1973). However, the distribution of the statistic in Eq. (7.4.24) is not yet available, and hence $c$ cannot be determined except when $q=1$, the statistic in Eq. (7.4.19). This case can be handled as follows:

$$\begin{aligned} &P\left\{ (l_1+l_p)^2/l_1l_p \geqslant 4c \,\middle|\, H \right\} \\ &= P\left\{ \left[ \left( \frac{l_1}{l_p} \right)^{1/2} + \left( \frac{l_p}{l_1} \right)^{1/2} \right]^2 \geqslant 4c \,\middle|\, H \right\} \\ &= P\left\{ \left( \frac{l_1}{l_p} \right)^{1/2} + \left( \frac{l_p}{l_1} \right)^{1/2} \geqslant 2c^{1/2} \,\middle|\, H \right\} \\ &= P\left\{ z+z^{-1} > 2c^{1/2} | H \right\} \qquad (\text{where} \quad z=l_1/l_p) \\ &= P\left\{ z^2-2c^{1/2}z+1>0 | H \right\} \\ &\equiv P\left\{ h_1(c)<z<h_2(c) | H \right\}, \quad \text{say,} \end{aligned}$$

where the upper percentage points of $z$ have been given by Krishnaiah and Schuurmann (1974). Thus, with the help of their tables, values of $c$ can be obtained. For some other test procedures, refer to Khatri (1978b).

### *7.4.7 Simultaneous Confidence Bounds*

Let us define $c$ such that

$$P\left\{ \frac{\left[\operatorname{ch}_1(\Sigma^{-1}V)+\operatorname{ch}_p(\Sigma^{-1}V)\right]^2}{4\operatorname{ch}_1(\Sigma^{-1}V)\operatorname{ch}_p(\Sigma^{-1}V)} \leqslant c \,\middle|\, A \right\} = 1-\alpha.$$

Then, as in Subsection 7.4.6, we have with probability $1-\alpha$

$$\frac{(\mathbf{a}'V\mathbf{a})(\mathbf{a}'\Sigma V^{-1}\Sigma\mathbf{a})}{(\mathbf{a}'\Sigma\mathbf{a})^2} \leqslant c \qquad \text{for all} \quad \mathbf{a}\neq\mathbf{0}.$$

Since for any $\mathbf{a}\neq 0$ and $\mathbf{b}\neq 0$,

$$\frac{(\mathbf{a}'\Sigma\mathbf{b})^2}{\mathbf{b}'V\mathbf{b}} \leqslant \mathrm{ch}_1(\Sigma\mathbf{a}\mathbf{a}'\Sigma V^{-1}) = \mathbf{a}'\Sigma V^{-1}\Sigma\mathbf{a},$$

the above inequality implies

$$\frac{(\mathbf{a}'V\mathbf{a})(\mathbf{a}'\Sigma\mathbf{b})^2}{(\mathbf{a}'\Sigma\mathbf{a})^2(\mathbf{b}'V\mathbf{b})} \leqslant c.$$

Hence,

(7.4.25) $$-\left(c\frac{\mathbf{b}'V\mathbf{b}}{\mathbf{a}'V\mathbf{a}}\right)^{1/2} \leqslant \frac{\mathbf{a}'\Sigma\mathbf{b}}{\mathbf{a}'\Sigma\mathbf{a}} \leqslant \left(c\frac{\mathbf{b}'V\mathbf{b}}{\mathbf{a}'V\mathbf{a}}\right)^{1/2}$$

for all $\mathbf{a}\neq\mathbf{0}$ and $\mathbf{b}\neq\mathbf{0}$, with confidence coefficient greater than or equal to $1-\alpha$.

From Eq. (7.4.25) various particular cases can be obtained, which are left to the reader. Now we shall consider the test procedure given by Eq. (7.4.24). Let us find $c$ such that

$$P\left\{\prod_{i=1}^{q}\frac{(\tilde{l}_i+\tilde{l}_{p-i+1})^2}{4\tilde{l}_i\tilde{l}_{p-i+1}} \geqslant c \,\middle|\, A\right\} = 1-\alpha,$$

where $\tilde{l}_i = \mathrm{ch}_i(\Sigma^{-1}V)$. Hence, using Eq. (7.4.22), we have with probability $1-\alpha$

(7.4.26) $$\frac{|B'VB||B'\Sigma V^{-1}\Sigma B|}{|B'\Sigma B|^2} \leqslant c$$

for all matrices $B:p\times q$ of rank $q$. Let $Q$ be any $p\times q$ matrix of rank $q$. Then, it is easy to see that

(7.4.27) $$|B'\Sigma Q|^2/|Q'VQ| \leqslant |B'\Sigma V^{-1}\Sigma B|,$$

because $V^{-1} = Q(Q'VQ)^{-1}Q' + V^{-1}Q_1(Q_1'VQ_1)^{-1}Q_1'V^{-1}$ for $Q'Q_1=0$ and $\operatorname{rank} Q_1 = p-q$. Hence, the simultaneous confidence bounds with confidence greater than or equal to $1-\alpha$ on $\Sigma$ are given by

(7.4.28) $$\left(\frac{|B'\Sigma Q|}{|B'\Sigma B|}\right)^2 \leqslant c\frac{|Q'VQ|}{|B'VB|}$$

for all matrices $B:p\times q$ and $Q:p\times q$ of rank $q$. Equation (7.4.25) can be obtained from (7.4.28) when $q=1$. Some special results can be obtained from (7.4.28), but are left to the reader.

## 7.5 Test for Independence

Let $\mathbf{y}\sim N_p(\boldsymbol{\mu},\Sigma)$ and $V\sim W_p(\Sigma,n)$ be independently distributed, where $\mathbf{y}=(\mathbf{y}_1',\mathbf{y}_2',\ldots,\mathbf{y}_k')'$, $\boldsymbol{\mu}=(\boldsymbol{\mu}_1',\boldsymbol{\mu}_2',\ldots,\boldsymbol{\mu}_k')'$, and

$$\Sigma=\begin{bmatrix}\Sigma_{11} & \Sigma_{12} & \cdots & \Sigma_{1k}\\ \Sigma_{12}' & \Sigma_{22} & \cdots & \Sigma_{2k}\\ \vdots & \vdots & \ddots & \vdots\\ \Sigma_{1k}' & \Sigma_{2k}' & \cdots & \Sigma_{kk}\end{bmatrix}>0. \tag{7.5.1}$$

The $\mathbf{y}_i$'s and $\boldsymbol{\mu}_i$'s are $p_i$-vectors, and the $\Sigma_{ij}$'s are $p_i\times p_j$ matrices, $i,j=1,2,\ldots,k$, where $p=\Sigma_{i=1}^k p_i$. Suppose we wish to test the hypothesis that all the subvectors $\mathbf{y}_i$ are independently distributed. Thus, the hypothesis is

$$H:(\Sigma_{ij}=0_{ij},\ i\neq j,\ i,j=1,2,\ldots,k), \tag{7.5.2}$$

and the alternative is

$$A\neq H. \tag{7.5.3}$$

Here $0_{ij}$ denote zero matrices of order $p_i\times p_j$, $i\neq j$, $i,j=1,2,\ldots,k$.

The likelihood function is given by

$$L(\boldsymbol{\mu},\Sigma)=c|\Sigma|^{-\frac{1}{2}N}|V|^{\frac{1}{2}(n-p-1)}\operatorname{etr}\left\{-\tfrac{1}{2}\Sigma^{-1}\left[V+(\mathbf{y}-\boldsymbol{\mu})(\mathbf{y}-\boldsymbol{\mu})'\right]\right\}. \tag{7.5.4}$$

Hence,

$$\operatorname*{Sup}_H L(\boldsymbol{\mu},\Sigma)=c|V|^{\frac{1}{2}(n-p-1)}\prod_{i=1}^k|N^{-1}V_{ii}|^{-\frac{1}{2}N}e^{-\frac{1}{2}Np},$$

where the partitioning of $V$ is like that of $\Sigma$ given in Eq. (7.5.1), that is,

$$V=\begin{bmatrix}V_{11} & V_{12} & \cdots & V_{1k}\\ V_{12}' & V_{22} & \cdots & V_{2k}\\ \vdots & \vdots & \ddots & \vdots\\ V_{1k}' & V_{2k}' & \cdots & V_{kk}\end{bmatrix}. \tag{7.5.5}$$

Similarly,

$$\operatorname*{Sup}_A L(\boldsymbol{\mu},\Sigma)=c|V|^{\frac{1}{2}(n-p-1)}|N^{-1}V|^{-\frac{1}{2}N}e^{-\frac{1}{2}Np}. \tag{7.5.6}$$

Hence, the likelihood ratio is given by

$$\lambda = \frac{\sup_H L(\boldsymbol{\mu}, \Sigma)}{\sup_A L(\boldsymbol{\mu}, \Sigma)} \tag{7.5.7}$$

$$= \left[ \frac{|V|}{\prod_{i=1}^{k} |V_{ii}|} \right]^{\frac{1}{2}N}$$

$$= (\lambda^*)^{\frac{1}{2}N}.$$

Thus, the hypothesis $H$ is rejected for small values of $\lambda^*$, where

$$\lambda^* = \frac{|V|}{\prod_{i=1}^{k} |V_{ii}|}. \tag{7.5.8}$$

When

$$k=2 \quad \text{and} \quad V^{-1} = \begin{pmatrix} V^{11} & V^{12} \\ V^{12'} & V^{22} \end{pmatrix},$$

then $\lambda^{*-1} = |V_{11}||V^{11}|$ and the test based on $\lambda^*$ has monotone power function (see Chapter 10).

### 7.5.1 Invariance

Let $B$ be a nonsingular matrix of the form

$$\begin{bmatrix} B_1 & & 0 & \\ & B_2 & & \\ 0 & & \ddots & \\ & & & B_k \end{bmatrix}. \tag{7.5.9}$$

Then the hypothesis $H$ and the alternative $A$ remain invariant under the group of transformations of the form

$$\mathbf{y} \to B\mathbf{y} + \mathbf{a} \quad \text{and} \quad V \to BVB', \tag{7.5.10}$$

where $\mathbf{a}$ is an element of an additive group of transformations. It can easily be checked that the likelihood-ratio test $\lambda^*$ is invariant under the above group of transformations.

### 7.5.2 Moments of $\lambda^*$ Under $H$

Under the hypothesis $H$, $\Sigma$ is of the form

$$\Sigma_0 = \begin{bmatrix} \Sigma_{11} & & & 0 \\ & \Sigma_{22} & & \\ 0 & & \ddots & \\ & & & \Sigma_{kk} \end{bmatrix}. \tag{7.5.11}$$

And since the likelihood-ratio test $\lambda^*$ is invariant under the group of transformations $V \to BVB'$, where $B$ is of the form given in Eq. (7.5.9), we may assume without any loss of generality that $\Sigma_0 = I$. Thus, we need to find

$$E(\lambda^{*h}) = E\left[\frac{|V|^h}{\prod_{i=1}^{k} |V_{ii}|^h}\right], \tag{7.5.12}$$

where $V \sim W_p(I,n)$. Hence

$$\begin{aligned} E(\lambda^{*h}) &= C(p,n)\int |V|^{\frac{1}{2}(n-p-1)+h}\left(\prod_{i=1}^{k}|V_i|^{-h}\right)\operatorname{etr}\left(-\tfrac{1}{2}V\right)dV \\ &= \frac{C(p,n)}{C(p,n+2h)}E\left(\prod_{i=1}^{k}|V_{ii}|^{-h}\right), \end{aligned} \tag{7.5.13}$$

where the expectation is taken with respect to $V$, which is distributed as $W_p(I,n+2h)$. But when $V \sim W_p(I,n+2h)$, the $V_{ii}$'s are independently distributed as $W_{p_i}(I_{p_i},n+2h)$. Hence

$$\begin{aligned} E(\lambda^{*h}) &= \frac{C(p,n)}{C(p,n+2h)}\prod_{i=1}^{k} E|V_{ii}|^{-h} \\ &= \frac{C(p,n)}{C(p,n+2h)}\prod_{i=1}^{k}\frac{C(p_i,n+2h)}{C(p_i,n)} \\ &= \prod_{\alpha=1}^{p}\frac{\Gamma\left[\frac{1}{2}(n+1-\alpha)+h\right]}{\Gamma\left[\frac{1}{2}(n+1-\alpha)\right]}\prod_{i=1}^{k}\prod_{j=1}^{p_i}\frac{\Gamma\left[\frac{1}{2}(n+1-j)\right]}{\Gamma\left[\frac{1}{2}(n+1-j)+h\right]} \\ &= \prod_{\alpha=p_1+1}^{p}\frac{\Gamma\left[\frac{1}{2}(n+1-\alpha)+h\right]}{\Gamma\left[\frac{1}{2}(n+1-\alpha)\right]}\prod_{i=2}^{k}\prod_{j=1}^{p_i}\frac{\Gamma\left[\frac{1}{2}(n+1-j)\right]}{\Gamma\left[\frac{1}{2}(n+1-j)+h\right]}. \end{aligned} \tag{7.5.14}$$

Since

$$\prod_{p_1+1}^{p} = \prod_{\bar{p}_2+1}^{\bar{p}_3} \prod_{\bar{p}_3+1}^{\bar{p}_4} \cdots \prod_{\bar{p}_k+1}^{p},$$

where

$$\bar{p}_i = p_1 + \cdots + p_{i-1}, \qquad i = 1,2,\ldots,p+1,$$

with

$$\bar{p}_1 = 0 \quad \text{and} \quad \bar{p}_{k+1} = p,$$

and since

$$\prod_{\alpha=\bar{p}_{i-1}+1}^{\bar{p}_i} \frac{\Gamma\left[\frac{1}{2}(n+1-\alpha)+h\right]}{\Gamma\left[\frac{1}{2}(n+1-\alpha)\right]} = \prod_{\alpha=1}^{p_i} \frac{\Gamma\left[\frac{1}{2}(n+1-\bar{p}_i-\alpha)+h\right]}{\Gamma\left[\frac{1}{2}(n+1-\bar{p}_i-\alpha)\right]},$$

we can rewrite Eq. (7.5.14) as

$$(7.5.15) \quad E(\lambda^{*h}) = \prod_{i=2}^{k} \prod_{j=1}^{p_i} \frac{\Gamma\left[\frac{1}{2}(n+1-\bar{p}_i-j)+h\right]\Gamma\left[\frac{1}{2}(n+1-j)\right]}{\Gamma\left[\frac{1}{2}(n+1-\bar{p}_i-j)\right]\Gamma\left[\frac{1}{2}(n+1-j)+h\right]},$$

$$= \prod_{i=2}^{k} \prod_{j=1}^{p_i} \frac{\Gamma\left[\frac{1}{2}(n+1-j)\right]\Gamma\left[\frac{1}{2}(n+1-\bar{p}_i-j)+h\right]\Gamma\left(\frac{1}{2}\bar{p}_i\right)}{\Gamma\left[\frac{1}{2}(n+1-\bar{p}_i-j)\right]\Gamma\left(\frac{1}{2}\bar{p}_i\right)\Gamma\left[\frac{1}{2}(n+1-j)+h\right]}$$

$$= \prod_{i=2}^{k} \prod_{j=1}^{p_i} E\left(x_{ij}^{h}\right),$$

where the $x_{ij}$'s are independent beta random variables with parameters $\frac{1}{2}(n+1-\bar{p}_i-j)$, $\frac{1}{2}\bar{p}_i$.

Since $0<\lambda^*<1$, these moments determine the distribution of $\lambda^*$ uniquely. Hence $\lambda^*$ is distributed as the product of independent beta random variables. However, an explicit closed-form pdf for $\lambda^*$ is difficult to obtain except in a few special cases given in the next subsection.

### *7.5.3 Exact Distribution Under H*

In this section we give expressions for the exact pdf of $\lambda^*$ for a few special cases; some of these have been worked out by Wilks (1932), some by Consul (1967a), and others follow from Chapter 5.

*Case* 1. k=2 In this case the $h$th moment of $\lambda^*$ is given by

$$E(\lambda^*) = \prod_{\alpha=1}^{p} \frac{\Gamma\left[\frac{1}{2}(n+1-\alpha)+h\right]}{\Gamma\left[\frac{1}{2}(n+1-\alpha)\right]} \prod_{i=1}^{2} \prod_{j=1}^{p_i} \frac{\Gamma\left[\frac{1}{2}(n+1-j)\right]}{\Gamma\left[\frac{1}{2}(n+1-j)+h\right]} \tag{7.5.16}$$

$$= \prod_{\alpha=p_2+1}^{p} \frac{\Gamma\left[\frac{1}{2}(n+1-\alpha)+h\right]}{\Gamma\left[\frac{1}{2}(n+1-\alpha)\right]} \prod_{j=1}^{p_1} \frac{\Gamma\left[\frac{1}{2}(n+1-j)\right]}{\Gamma\left[\frac{1}{2}(n+1-j)+h\right]}$$

$$= \prod_{i=1}^{p_1} \frac{\Gamma\left[\frac{1}{2}(n+1-p_2-i)+h\right]\Gamma\left[\frac{1}{2}(n+1-i)\right]}{\Gamma\left[\frac{1}{2}(n+1-p_2-i)\right]\Gamma\left[\frac{1}{2}(n+1-i)+h\right]}.$$

Thus the distribution of $\lambda^*$ can be obtained from Chapter 5. Hence the distribution of $\lambda^*$ is that of $\lambda^*_{p_1,p_2,n-p_2}$, given in Theorem 5.4.2. It is the product of $p_1$ independent beta random variables with parameters $\left(\frac{1}{2}(n+1-p_2-i), \frac{1}{2}p_2\right)$, $i=1,2,\ldots,p_1$. Thus if $p_1=1$, the distribution of $\lambda^*$ is given by Theorem 5.4.4, namely

$$f(\lambda^*) = \frac{\Gamma\left(\frac{1}{2}n\right)}{\Gamma\left[\frac{1}{2}p_2\right]\Gamma\left[\frac{1}{2}(n-p_2)\right]} \lambda^{*\frac{1}{2}(n-p_2)-1}(1-\lambda^*)^{\frac{1}{2}p_2-1}. \tag{7.5.17}$$

Other special cases can be obtained from Chapter 5.

*Case* 2. k=3, $p_1=1$, $p_2=1$ The exact pdf of $\lambda^*$ in this case is given by (7.5.18)

$$f(\lambda^*) = c\lambda^{*\frac{1}{2}(n-p_3-3)}(1-\lambda^*)^{p_3-\frac{1}{2}}\,{}_2F_1\left(\tfrac{1}{2}p_3, \tfrac{1}{2}p_3; p_3+\tfrac{1}{2}; 1-\lambda^*\right),$$

where

$$c = \frac{\left[\Gamma\left(\frac{1}{2}n\right)\right]^2}{\Gamma\left[\frac{1}{2}(n-p_3)\right]\Gamma\left[\frac{1}{2}(n-p_3-1)\right]\Gamma\left[p_3+\frac{1}{2}\right]}. \tag{7.5.19}$$

This pdf has been obtained by Consul (1967a) with the help of Mellin's inversion theorem. Consul (1967a) has also given the exact pdf of $\lambda^*$ when $p_1=1$, $p_2=2$; $p_1=2$, $p_2=2$; $p_1=1$, $p_2=3$; $p_1=2$, $p_2=3$, $p_3$ even; and $p_1=2$, $p_2=4$.

### *7.5.4 Asymptotic Distribution Under H*

Let

$$N = m + 2\alpha, \tag{7.5.20}$$

where $\alpha$ is so chosen that the term of order $m^{-1}$ in the expansion of the cdf of $-m\ln\lambda^*$ cancels out. From Eq. (7.5.14) we get the moment-generating function of $-m\ln\lambda^*$ as

$$E(e^{-mt\ln\lambda^*}) = E(\lambda^{*-mt}) = \prod_{i=1}^{p} \frac{\Gamma\left[\frac{1}{2}(N-i)-mt\right]}{\Gamma\left[\frac{1}{2}(N-i)\right]} \prod_{i=1}^{k} \prod_{j=1}^{p_i} \frac{\Gamma\left[\frac{1}{2}(N-j)\right]}{\Gamma\left[\frac{1}{2}(N-j)-mt\right]}. \tag{7.5.21}$$

Expanding $\ln\Gamma(x+c)$ for bounded $c$ and choosing $\alpha$ such that the term of $O(m^{-1})$ cancels out, we get

$$\alpha = \frac{\left(p^3 - \sum_{i=1}^{k} p_i^3\right) + 9\left(p^2 - \sum_{i=1}^{k} p_i^2\right)}{6\left(p^2 - \sum_{i=1}^{k} p_i^2\right)}, \tag{7.5.22}$$

$$E(e^{-mt\ln\lambda^*}) = (1-2t)^{-\frac{1}{2}f}\left\{1 + m^{-2}\gamma_2\left[(1-2t)^{-2} - 1\right] + O(m^{-3})\right\}, \tag{7.5.23}$$

where

$$f = \tfrac{1}{2}\left(p^2 - \sum_{i=1}^{k} p_i^2\right), \tag{7.5.24}$$

$$\gamma_2 = \frac{p^4 - \sum_{i=1}^{k} p_i^4}{48} - \frac{5\left(p^2 - \sum_{i=1}^{k} p_i^2\right)}{96} - \frac{\left(p^3 - \sum_{i=1}^{k} p_i^3\right)^2}{72\left(p^2 - \sum_{i=1}^{k} p_i^2\right)}.$$

Hence, we get the following

**Theorem 7.5.1**

$$P(-m\ln\lambda^* \leqslant z) = P(\chi_f^2 \leqslant z) + m^{-2}\gamma_2\left[P(\chi_{f+4}^2 \leqslant z) - P(\chi_f^2 \leqslant z)\right] + O(m^{-3}).$$

### 7.5.5 *An Alternative Test*

By the same considerations as in Subsections 7.2.5 and 7.4.5, Nagao (1973a) proposed a test

$$T_3 = \tfrac{1}{2}n\,\mathrm{tr}\left(VV_D^{-1} - I\right)^2, \tag{7.5.25}$$

where

$$V_D = \begin{bmatrix} V_{11} & & & 0 \\ & V_{22} & & \\ & & \ddots & \\ 0 & & & V_{kk} \end{bmatrix};$$

the $V_{ii}$'s have been defined in Eq. (7.5.5).

An asymptotic expansion for the cdf of $T_3$ is given in

**Theorem 7.5.2** *The null distribution of the statistic $T_3$ given by Eq.* (7.5.25) *is expanded asymptotically for large n as*

$$P(T_3 \leqslant x) = P_f + n^{-1}\Big[\tfrac{1}{12}(p^3 - 3p\bar{p}_2 + 2\bar{p}_3)P_{f+6}$$
$$+\tfrac{1}{8}(-2p^3 + 4p\bar{p}_2 - 2\bar{p}_3 - p^2 + \bar{p}_2)P_{f+4}$$
$$+\tfrac{1}{4}(p^3 - p\bar{p}_2 + p^2 - \bar{p}_3)P_{f+2}$$
$$+\tfrac{1}{24}(-2p^3 + 2\bar{p}_3 - 3p^2 + 3\bar{p}_2)P_f\Big] + O(n^{-2}),$$

*where*

$$\bar{p}_r = \sum_{\alpha=1}^{k} p_\alpha^r, \qquad f = \tfrac{1}{2}(p^2 - \bar{p}_2), \quad \textit{and} \quad P_f = P(x_j^2 \leqslant x).$$

### *7.5.6 Union–Intersection Test Procedure when $k = 2$*

Let us consider the situation when $k=2$. We may observe that

$$H(\Sigma_{12} = 0) = \bigcap_{\mathbf{a}_1, \mathbf{a}_2} H_{\mathbf{a}_1, \mathbf{a}_2},$$

where

$$H_{\mathbf{a}_1, \mathbf{a}_2} : \mathbf{a}_1' \Sigma_{12} \mathbf{a}_2 = 0,$$

and

$$A(\Sigma_{12} \neq 0) = \bigcup_{\mathbf{a}_1, \mathbf{a}_2} A_{\mathbf{a}_1, \mathbf{a}_2},$$

where

$$A_{\mathbf{a}_1, \mathbf{a}_2} : \mathbf{a}_1' \Sigma_{12} \mathbf{a}_2 \neq 0,$$

$\mathbf{a}_1$ and $\mathbf{a}_2$ being any nonnull vectors. Given $\mathbf{a}_1$ and $\mathbf{a}_2$,

$$\begin{pmatrix} \mathbf{a}_1' V_{11} \mathbf{a}_1 & \mathbf{a}_1' V_{12} \mathbf{a}_2 \\ \mathbf{a}_2' V_{12} \mathbf{a}_1 & \mathbf{a}_2' V_{22} \mathbf{a}_2 \end{pmatrix} \sim W_2\left( \begin{pmatrix} \mathbf{a}_1 \Sigma_{11} \mathbf{a}_1 & \mathbf{a}_1' \Sigma_{12} \mathbf{a}_2 \\ \mathbf{a}_1 \Sigma_{12} \mathbf{a}_2 & \mathbf{a}_2' \Sigma_{22} \mathbf{a}_2 \end{pmatrix}, n \right),$$

and the test procedure for testing $H_{\mathbf{a}_1, \mathbf{a}_2}$ vs $A_{\mathbf{a}_1, \mathbf{a}_2}$ is

$$\text{reject } H_{\mathbf{a}_1, \mathbf{a}_2} \quad \text{if} \quad \frac{(\mathbf{a}_1' V_{12} \mathbf{a}_2)^2}{(\mathbf{a}_1' V_{11} \mathbf{a}_1)(\mathbf{a}_2' V_{22} \mathbf{a}_2)} \geqslant c,$$

where $c$ is a constant depending on $n$ and $\alpha$. Hence, by the union–intersection principle,

$$\text{reject } H \text{ vs } A \quad \text{if} \quad \sup_{\mathbf{a}_1, \mathbf{a}_2} \frac{(\mathbf{a}_1' V_{12} \mathbf{a}_2)^2}{(\mathbf{a}_1' V_{11} \mathbf{a}_1)(\mathbf{a}_2' V_{22} \mathbf{a}_2)} \geqslant c,$$

and by using Problem 1.18(ii), the test procedure is

$$\text{reject } H \quad \text{if} \quad \text{ch}_1\left(V_{12}V_{22}^{-1}V_{12}'V_{11}^{-1}\right) \geq c \ \text{ or } \ \text{ch}_1\left(V_{1.2}^{-1}V_{12}V_{22}^{-1}V_{12}'\right) \geq c_1,$$

where $V_{1.2} = V_{11} - V_{12}V_{22}^{-1}V_{12}'$, and where $c_1 = c/(1-c)$, and $c$ is a constant to be determined from

$$P\left[\text{ch}_1\left(V_{12}V_{22}^{-1}V_{12}'V_{11}^{-1}\right) \geq c \,|\, H\right] = \alpha.$$

The distribution of $\text{ch}_1(V_{12}V_{22}^{-1}V_{12}'V_{1.2}^{-1})$ and the confidence bounds on $\Sigma_{12}\Sigma_{22}^{-1} = \beta_1$ (or $\Sigma_{12}'\Sigma_{11}^{-1} = \beta_2$) are the same as those given in Subsections 6.3.2 and 6.3.9 provided we make the following changes:

$$l \to p_1, \quad n \to n - p_2,$$

$$S_e \to V_{1.2}, \quad \hat{\beta}C' \to V_{12}V_{22}^{-1},$$

and

$$G \to V_{22}$$

$$\left(\text{or} \qquad S_e \to V_{22} - V_{12}'V_{11}^{-1}V_{12}, \qquad \hat{\beta}C' = V_{12}'V_{11}^{-1}, \qquad G \to V_{11}\right).$$

When $k > 2$, we do not get manageable test procedures, and that case is not considered at this stage.

## 7.6 Test for the Equality of Covariances

Let $\{\mathbf{y}_1, V_1\}, \{\mathbf{y}_2, V_2\}, \ldots, \{\mathbf{y}_k, V_k\}$ be independently distributed, where $\mathbf{y}_\alpha \sim N_p(\boldsymbol{\mu}_\alpha, \Sigma_\alpha)$ and $V_\alpha \sim W_p(\Sigma_\alpha, n_\alpha)$ are also independently distributed, $\alpha = 1, 2, \ldots, k$.

In this section we consider the following problem:

$$H: \Sigma_1 = \Sigma_2 = \cdots = \Sigma_k \quad \text{vs} \quad A \neq H.$$

The likelihood function is given by

(7.6.1)
$$L(\boldsymbol{\mu}_i, \Sigma_i, i = 1, \ldots, k)$$
$$= c \prod |\Sigma_i|^{-\frac{1}{2}N_i} |V_i|^{\frac{1}{2}(n_i - p - 1)} \text{etr}\left\{-\tfrac{1}{2}\Sigma_i^{-1}\left[V_i + (\mathbf{y}_i - \boldsymbol{\mu}_i)(\mathbf{y}_i - \boldsymbol{\mu}_i)'\right]\right\}.$$

Hence

(7.6.2)
$$\sup_H L(\boldsymbol{\mu}_i, \Sigma_i, i = 1, \ldots, k)$$
$$= c\left(\prod^k |V_i|^{\frac{1}{2}(n_i - p - 1)}\right)|N^{-1}V|^{-\frac{1}{2}N} e^{-\frac{1}{2}Np},$$

where

(7.6.3)
$$V = \sum_{i=1}^k V_i \quad \text{and} \quad N = \Sigma N_i, \quad N_i = n_i + 1.$$

Similarly

(7.6.4)
$$\sup_A L(\boldsymbol{\mu}_i, \Sigma_i, i=1,\ldots,k) = c \prod_{i=1}^{k} |N_i^{-1}V_i|^{-\frac{1}{2}N} |V_i|^{\frac{1}{2}(n_i-p-1)} e^{-\frac{1}{2}Np}.$$

Thus, the likelihood ratio $\lambda$ is given by

(7.6.5)
$$\lambda = \frac{\sup_H L(\boldsymbol{\mu}_i, \Sigma_i, i=1,\ldots,k)}{\sup_A L(\boldsymbol{\mu}_i, \Sigma_i, i=1,\ldots,k)} = \frac{\prod_{i=1}^{k} |N_i^{-1}V_i|^{\frac{1}{2}N_i}}{|N^{-1}V|^{\frac{1}{2}N}} = \frac{\prod_{i=1}^{k} |V_i|^{\frac{1}{2}N_i}}{|V|^{\frac{1}{2}N}} \frac{N^{\frac{1}{2}pN}}{\prod_{i=1}^{k} N_i^{\frac{1}{2}pN_i}}.$$

This test, however, is not unbiased. The modified likelihood-ratio test based on

(7.6.6)
$$\lambda' = \frac{\prod_{i=1}^{k} |V_i|^{\frac{1}{2}n_i}}{|V|^{\frac{1}{2}n}} \frac{n^{\frac{1}{2}pn}}{\prod_{i=1}^{k} n_i^{\frac{1}{2}pn_i}},$$

in which the $N_i$'s and $N$ are replaced by $n_i$ and $n = \sum_{i=1}^{k} n_i$ (the numbers of degrees of freedom associated with $V_i$ and $V$ respectively), is unbiased. Further, when $k=2$ the power function has a monotonicity property (see Chapter 10).

### *7.6.1 Invariance*

The hypothesis $H$, the alternative $A$, and the likelihood ratio ($\lambda$ or $\lambda'$) remain invariant under the following group of transformations:

(7.6.7)
$$\mathbf{y}_i \to C\mathbf{y}_i + \mathbf{a}, \qquad V_i \to CV_iC',$$

where $C$ is an arbitrary nonsingular matrix and $\mathbf{a}$ is an arbitrary vector.

### 7.6.2 *Moments of* $\lambda'$ *Under* $H$

Under $H$, the $V_i$'s are independently distributed as $W_p(\Sigma, n)$, where $\Sigma$ is the common covariance matrix. Since the modified likelihood-ratio test $\lambda'$ is invariant under the transformation $V_i \to CV_iC'$, where $C$ is an arbitrary nonsingular matrix, we may assume without any loss of generality that $\Sigma = I$. Hence

$$E(\lambda'^h) = \frac{n^{\frac{1}{2}pnh}}{\prod_{i=1}^k n_i^{\frac{1}{2}n_i ph}} E\left[\frac{\prod_{i=1}^k |V_i|^{\frac{1}{2}n_i h}}{|V|^{\frac{1}{2}nh}}\right]$$

$$= \prod_{i=1}^k C(p,n_i) \frac{n^{\frac{1}{2}pnh}}{\prod_{i=1}^k n_i^{\frac{1}{2}pn_i h}}$$

$$\times \int |V|^{-\frac{1}{2}nh} \prod_{i=1}^k \left(|V_i|^{\frac{1}{2}n_i h + \frac{1}{2}(n_i - p + 1)} \operatorname{etr}\left(-\tfrac{1}{2}V_i\right)\right) dV_i$$

$$= \frac{n^{\frac{1}{2}pnh}}{\prod_{i=1}^k n_i^{\frac{1}{2}n_i ph}} \frac{\prod_{i=1}^k C(p,n_i)}{\prod_{i=1}^k C(p,n_i + n_i h)} E\left(|V|^{-\frac{1}{2}nh}\right)$$

where the expectation is with respect to $k$ independent random Wishart matrices $V_i$ that are distributed as $W_p(I, n_i + n_i h)$. But $V = \Sigma_{i=1}^k V_i$ has a Wishart distribution with mean $(n + nh)I$ and $n + nh$ d.f., where $n = \Sigma n_i$. Hence

$$E(\lambda'^h) = \frac{n^{\frac{1}{2}nph}}{\prod_{i=1}^k n_i^{\frac{1}{2}n_i ph}} \frac{\prod_1^k C(p,n_i)}{\prod_1^k C(p,n_i + n_i h)} \cdot \frac{C(p, n+nh)}{C(p,n)} \tag{7.6.8}$$

$$= \frac{n^{\frac{1}{2}nph}}{\prod_{i=1}^k n_i^{\frac{1}{2}n_i ph}} \prod_{i=1}^k \frac{\Gamma_p\left(\frac{1}{2}(n_i + hn_i)\right)}{\Gamma_p\left(\frac{1}{2}n_i\right)} \frac{\Gamma_p\left(\frac{1}{2}n\right)}{\Gamma_p\left(\frac{1}{2}(n+hn)\right)}.$$

Let

$$\lambda^* = (\lambda')^{2/n}. \tag{7.6.9}$$

In the next subsection, we give an asymptotic cdf of $-m \ln \lambda^*$, where

$$n = m + 2\alpha \tag{7.6.10}$$

and $\alpha$ is so chosen that the term of $O(m^{-1})$ cancels out.

### *7.6.3 Asymptotic* cdf *of* $-m\ln\lambda^*$ *Under H*

Letting

$$r_i=\frac{n_i}{n},\qquad i=1,2,\ldots,k, \tag{7.6.11}$$

we find from Eq. (7.6.8) that the moment-generating function of $-m\ln\lambda^*$ is given by

$$E(e^{-mh\ln\lambda^*})=E(\lambda^{*-mh})=E(\lambda'^{-2mh/n}) \tag{7.6.12}$$

$$=\frac{n^{\frac{1}{2}mph}}{\prod_{i=1}^{2}n_i^{\frac{1}{2}mr_iph}}\frac{\Gamma_p\left[\frac{1}{2}m+\alpha\right]}{\Gamma_p\left[\frac{1}{2}m(1-2h)+\alpha\right]}\prod_{i=1}^{k}\frac{\Gamma_p\left[\frac{1}{2}r_im(1-2h)+\alpha r_i\right]}{\Gamma_p\left[\frac{1}{2}r_im+\alpha r_i\right]},$$

which exists for $h<\frac{1}{2}$. For the asymptotic expansion, we assume that

$$\lim_{n\to\infty} r_i>0. \tag{7.6.13}$$

Next, we give the asymptotic expansion for $k=2$.

*Case* 1. k=2 Expanding $\ln\Gamma(x+c)$ for bounded $c$, we get from Eq. (7.6.12) under the condition (7.6.13)

$$\{\text{coefficient of } m^0\}=mph(r_1\ln r_1+r_2\ln r_2) \tag{7.6.14}$$

$$+\left[\sum_{i=1}^{p}\left(\tfrac{1}{2}m+\alpha-\tfrac{1}{2}i\right)\ln\tfrac{1}{2}m\right.$$

$$\left.-\sum_{i=1}^{p}\left[\tfrac{1}{2}m(1-2h)+\alpha-\tfrac{1}{2}i\right]\ln\left[\tfrac{1}{2}m(1-2h)\right]\right]$$

$$+\sum_{j=1}^{2}\left[\sum_{i=1}^{p}\left[\tfrac{1}{2}r_jm(1-2h)+r_j\alpha-\tfrac{1}{2}i\right]\ln\left[\tfrac{1}{2}r_jm(1-2h)\right]\right.$$

$$\left.-\sum_{i=1}^{p}\left(\tfrac{1}{2}r_jm+r_j\alpha-\tfrac{1}{2}i\right)\ln\tfrac{1}{2}r_jm\right]$$

$$=-\tfrac{1}{2}f\ln(1-2h),$$

where

(7.6.15) $$f=\tfrac{1}{2}p(p+1);$$

(7.6.16) $$\{\text{coefficient of } m^{-1}[(1-2h)^{-1}-1]\}$$

$$= \sum_{i=1}^{p}\left[\left(\alpha-\frac{i-1}{2}\right)^2 - r_1^{-1}\left(r_1\alpha-\frac{i-1}{2}\right)^2 - r_2^{-1}\left(r_2\alpha-\frac{i-1}{2}\right)^2\right]$$

$$- \sum_{i=1}^{p}\left[\left(\alpha-\frac{i-1}{2}\right) - r_1^{-1}\left(r_1\alpha-\frac{i-1}{2}\right) - r_2^{-1}\left(r_2\alpha-\frac{i-1}{2}\right)\right]$$

$$+\tfrac{1}{6}p(1-r_1^{-1}-r_2^{-1})$$

$$=\tfrac{1}{2}\alpha p(p+1)-\tfrac{1}{24}(r_1^{-1}+r_2^{-1}-1)p(2p^2+3p-1);$$

(7.6.17) $$\{\text{coefficient of } m^{-2}(\tfrac{2}{3})[(1+h)^{-2}-1]\}$$

$$= \sum_{i=1}^{p}\Bigg[-\left(\alpha-\frac{i-1}{2}\right)^3+\frac{3}{2}\left(\alpha-\frac{i-1}{2}\right)^2-\frac{1}{2}\left(\alpha-\frac{i-1}{2}\right)$$

$$-r_1^{-2}\left[\left(r_1\alpha-\frac{i-1}{2}\right)^3-\frac{3}{2}\left(r_1\alpha-\frac{i-1}{2}\right)^2+\frac{1}{2}\left(r_1\alpha-\frac{i-1}{2}\right)\right]$$

$$-r_2^{-2}\left[\left(r_2\alpha-\frac{i-1}{2}\right)^3-\frac{3}{2}\left(r_2\alpha-\frac{i-1}{2}\right)^2+\frac{1}{2}\left(r_2\alpha-\frac{i-1}{2}\right)\right]\Bigg]$$

$$=\tfrac{3}{4}\alpha^2p(p+1)-\tfrac{1}{8}\alpha(r_1^{-1}+r_2^{-1}-1)p(2p^2+3p-1)$$

$$+\tfrac{1}{32}(r_1^{-2}+r_2^{-2}-1)p(p-1)(p^2+3p+2).$$

In order for the term of $m^{-1}$ to cancel out, we need

(7.6.18) $$\alpha=(r_1^{-1}+r_2^{-1}-1)(2p^2+3p-1)/12(p+1).$$

Hence

(7.6.19) $$E(e^{-mh\ln\lambda^*})=(1-2h)^{-\frac{1}{2}f}|1+m^{-2}\gamma_2[(1-2h)^{-2}-1]+O(m^{-3})|,$$

where

$$f=\tfrac{1}{2}p(p+1),$$

and

(7.6.20) $$\gamma_2=\tfrac{1}{48}p(p+1)[(p-1)(p+2)(r_1^{-2}+r_2^{-2}-1)-24\alpha^2].$$

Thus,

$$P(-m\ln\lambda^* \leqslant z) = P(x_f^2 \leqslant z) + m^{-2}\gamma_2\left[P(x_{f+4}^2 \leqslant z) - P(x_f^2 \leqslant z)\right] + O(m^{-3}). \tag{7.6.21}$$

*Case* 2. General $k$ Proceeding as above, we get in this case

$$P(-m\ln\lambda^* \leqslant z) = P(x_f^2 \leqslant z) + m^{-2}\gamma_2\left[P(x_{f+4}^2 \leqslant z) - P(x_f^2 \leqslant z)\right] + O(m^{-3}), \tag{7.6.22}$$

where

$$f = \tfrac{1}{2}(k-1)p(p+1), \tag{7.6.23}$$

$$\alpha = \left(\sum_{j=1}^{k} r_j^{-1} - 1\right)\frac{2p^2+3p-1}{12(p+1)(k-1)},$$

$$\gamma_2 = \tfrac{1}{48}p(p+1)\left[(p-1)(p+2)\left(\Sigma r_j^{-2} - 1\right) - 24(k-1)\alpha^2\right].$$

REMARK 7.6.1. A problem of considerable practical interest is to test the hypothesis $H_0: \Sigma_1 = \cdots = \Sigma_k = \Sigma_0$, $\Sigma_0$ unknown against the alternatives $H_1: \Sigma_j = \Sigma_0$ for $j \neq i$, $\Sigma_i = \Sigma_0 + \Delta$, $\Delta$ unknown, $i = 1,2,\ldots,k$. The problem is to find which of the $(k+1)$ hypotheses is true. This is known as a slippage problem. For a solution of this problem, the reader is referred to Srivastava (1966).

REMARK 7.6.2. Another problem of practical interest is the case of missing observations. When these missing observations are of monotone type, Bhargava (1962) obtained the likelihood-ratio test and gave its null distribution.

### *7.6.4 An Alternative Test*

For testing the homogeneity of covariance, Nagao (1973a) suggests

$$T_4 = \frac{1}{2}\sum_{\alpha=1}^{k} n_\alpha \operatorname{tr}\left[\frac{S_\alpha}{n_\alpha}\left(\frac{S}{n}\right)^{-1} - I\right]^2 \tag{7.6.24}$$

as an alternative test statistic to $\lambda^*$. An asymptotic expansion for the cdf of $T_4$ is given in

**Theorem 7.6.1** *The null distribution of the test statistic $T_4$ given by Eq. (7.6.24), for large $n=\Sigma_{\alpha=1}^{k} n_\alpha$, is*

$$P(T_4 \leqslant x)=P_f+n^{-1}\Big\{\tfrac{1}{12}\big[\bar{r}p(4p^2+9p+7)-3k^2p(p+1)^2$$
$$-(3k-2)p(p^2+3p+4)\big]P_{f+6}$$
$$+\tfrac{1}{8}\big[-\bar{r}p(6p^2+13p+9)+4k^2p(p+1)^2$$
$$+(2k-1)p(2p^2+5p+5)\big]P_{f+4}$$
$$+\tfrac{1}{4}(2\bar{r}-k^2-k)p(p+1)^2P_{f+2}$$
$$+\tfrac{1}{24}(1-\bar{r})p(2p^2+3p-1)P_f\Big\}+O(n^{-2}),$$

*where*

(7.6.25)
$$r_\alpha=n_\alpha/n,$$
$$\bar{r}=\sum_{\alpha=1}^{k} r_\alpha^{-1},$$
$$P_f=P\big(\chi_f^2\leqslant x\big)\quad\text{with}\quad f=\tfrac{1}{2}(k-1)p(p+1).$$

For another alternative test when $k=2$, see Dempster (1964).

### *7.6.5 Union–Intersection Test Procedure and Confidence Bounds*

#### 7.6.5.1 $k=2$

First we shall consider $k=2$. We observe that for any nonnull vector $\mathbf{a}$,

$$H(\Sigma_1=\Sigma_2)=\bigcap_{\mathbf{a}} H_{\mathbf{a}},\qquad \text{where}\quad H_{\mathbf{a}}:\mathbf{a}'\Sigma_1\mathbf{a}=\mathbf{a}'\Sigma_2\mathbf{a},$$

and

$$A(\Sigma_1\neq\Sigma_2)=\bigcup_{\mathbf{a}} A_{\mathbf{a}},\qquad \text{where}\quad A_{\mathbf{a}}:\mathbf{a}'\Sigma_1\mathbf{a}\neq\mathbf{a}'\Sigma_2\mathbf{a}.$$

Given $\mathbf{a}$, $\mathbf{a}'V_i\mathbf{a}/\mathbf{a}'\Sigma_i\mathbf{a}$ $(i=1,2)$ are independently distributed as chi squares with $n_i$ $(i=1,2)$ d.f. Hence we get the test procedure for testing $H_{\mathbf{a}}$ vs $A_{\mathbf{a}}$ as

$$\text{reject } H_{\mathbf{a}} \quad\text{if}$$
$$\frac{\mathbf{a}'V_1\mathbf{a}}{\mathbf{a}'V_2\mathbf{a}}\geqslant c_2 \quad\text{or}\quad \frac{\mathbf{a}'V_1\mathbf{a}}{\mathbf{a}'V_2\mathbf{a}}\leqslant c_1,$$

where $c_1<c_2$, and $c_1,c_2$ are constants depending on $n_1$, $n_2$, and $\alpha$. Hence, by the union–intersection principle,

$$\text{reject } H \quad\text{if}$$
$$\operatorname*{Sup}_{\mathbf{a}}\frac{\mathbf{a}'V_1\mathbf{a}}{\mathbf{a}'V_2\mathbf{a}}\geqslant c_2 \quad\text{or}\quad \operatorname*{Inf}_{\mathbf{a}}(\mathbf{a}'V_1\mathbf{a})(\mathbf{a}'V_2\mathbf{a})\leqslant c_1.$$

Using Corollary 1.10.1, the test procedure is

$$\text{reject } H \quad \text{if}$$

$$\text{ch}_1(V_1V_2^{-1}) \geqslant c_2 \text{ or } \text{ch}_p(V_1V_2^{-1}) \leqslant c_1,$$

where $\text{ch}_1(V_1V_2^{-1}) \geqslant \cdots \geqslant \text{ch}_p(V_1V_2^{-1})$, $\text{ch}_j(V_1V_2^{-1}) = j$th largest characteristic root of $V_1V_2^{-1}$, and $c_1$, $c_2$ are determined from

$$P\left[c_1 < \text{ch}_p(V_1V_2^{-1}) < \text{ch}_1(V_1V_2^{-1}) < c_2 \middle| H\right] = 1-\alpha.$$

The joint density of $l_i = \text{ch}_i(V_1V_2^{-1})$, $i=1,2,\ldots,p$, under $H$ is given in Subsection 5.4.8 if $(t,u,\tilde{n})$ is replaced by $(p,n_1,n_2)$, and one can obtain $c_1$ and $c_2$ by integrating over $l_1,\ldots,l_p$ over the domain $c_1 < l_p < \cdots < l_1 < c_2$. This is not easy, and the test may or may not be unbiased.

For the simultaneous confidence bounds on the elements of $\Sigma_1\Sigma_2^{-1}$, let us write $V_{(1)} = B^{-1}V_1B'^{-1}$ and $V_{(2)} = B^{-1}V_2B'^{-1}$, where $B$ is a nonsingular matrix such that $\Sigma_1 = BD_\lambda B'$, $\Sigma_2 = BB'$, and write $D_\lambda = \text{diag}(\lambda_1,\ldots,\lambda_p)$, $\lambda_1 \geqslant \lambda_2 \geqslant \cdots \geqslant \lambda_p > 0$ being the ch roots of $\Sigma_1\Sigma_2^{-1}$ (see Theorem 1.9.2 for the existence of $B$). Then $V_{(1)} \sim W_p(D_\lambda, n_1)$, $V_{(2)} \sim W_p)(I, n_2)$, and they are independent. We may observe that

$$1-\alpha = P\left\{c_1 \leqslant \left[\text{all ch}(D_\lambda^{-1/2}V_1D_\lambda^{-1/2})\right] \leqslant c_2 \middle| A\right\}$$
$$= P\left\{c_1 \leqslant \left[\text{all ch}(Q^{-1}V_1^{1/2}V_2^{-1}V_1^{1/2})\right] \leqslant c_2 \middle| A\right\}, \tag{7.6.26}$$

where $Q = V_1^{1/2}B'^{-1}D_\lambda^{1/2}B'V_1^{-1}BD_\lambda^{1/2}B^{-1}V_1^{1/2} \equiv P'P$. Hence from Problem 1.18,

$$\text{ch}_p(Q) \leqslant \left[\text{ch}_i(P)\right]^2 = \lambda_i \leqslant \text{ch}_1(Q) \qquad \text{for} \quad i=1,2,\ldots,p. \tag{7.6.27}$$

From Eq. (7.6.26), the simultaneous confidence bound on $Q$ is

$$\frac{\mathbf{a}'V_1^{1/2}V_2^{-1}V_1^{1/2}\mathbf{a}}{c_2} \leqslant \mathbf{a}'Q\mathbf{a} \leqslant \frac{\mathbf{a}'V_1^{1/2}V_2^{-1}V_1^{1/2}\mathbf{a}}{c_1} \tag{7.6.28}$$

for all nonnull vectors $\mathbf{a}$ and with confidence coefficient equal to $1-\alpha$. Notice that $Q$ is not explicitly in terms of unknown parameters, and in order to use Eq. (7.6.27) we obtain the confidence bounds in terms of the ch roots. From Eq. (7.6.28), it is easy to see that

$$\frac{\text{ch}_p(V_1V_2^{-1})}{c_2} \leqslant \text{ch}_p(Q) \leqslant \lambda_i \leqslant \text{ch}_1(Q) \leqslant \frac{\text{ch}_1(V_1V_2^{-1})}{c_1} \tag{7.6.29}$$

with confidence coefficient greater than or equal to $1-\alpha$. Hence, we have the simultaneous confidence bounds on $\Sigma_1$ and $\Sigma_2$ as

$$\frac{\text{ch}_p(V_1V_2^{-1})}{c_2} \leqslant \frac{\mathbf{a}'\Sigma_1\mathbf{a}}{\mathbf{a}'\Sigma_2\mathbf{a}} \leqslant \frac{\text{ch}_1(V_1V_2^{-1})}{c_1} \tag{7.6.30}$$

for all nonnull vectors $\mathbf{a}$ and for confidence coefficient greater than or equal to $1-\alpha$. This type of confidence bound was given by Roy (1957). Notice by Eq. (7.6.29) that we have obtained the confidence bounds on $\mathrm{ch}(\Sigma_1\Sigma_2^{-1})$ only, and if we pose the problem of obtaining confidence bounds only on the $\lambda_i$'s, then narrower confidence bounds can be obtained by using only the $F$ distribution as given in Anderson (1965) and Khatri (1965c). We follow Khatri's method (1965c) and use Problem 1.39, namely

$$\mathrm{ch}_p\left(V_1V_2^{-1}\right)=\mathrm{ch}_p\left(V_{(1)}V_{(2)}^{-1}\right)\leqslant\left(v_{(1)}^{pp}\boldsymbol{\delta}_p'V_{(2)}\boldsymbol{\delta}_p\right)^{-1}$$

and

$$\mathrm{ch}_1\left(V_1V_2^{-1}\right)=\mathrm{ch}_1\left(V_{(1)}V_{(2)}^{-1}\right)\geqslant\left(v_{11(1)}\boldsymbol{\delta}_1'V_{(2)}^{-1}\boldsymbol{\delta}_1\right),$$

where $V_{(1)}=(v_{ij(1)})$, $V_{(1)}^{-1}=(v_{(1)}^{ij})$, and $\boldsymbol{\delta}_1$ and $\boldsymbol{\delta}_p$ denote, respectively, the characteristic vectors corresponding to the maximum and the minimum characteristic roots of $V_{(1)}$. Note that $(\lambda_p v_{(1)}^{pp})^{-1}$, $\boldsymbol{\delta}_p'V_{(2)}\boldsymbol{\delta}_p$, $v_{11(1)}/\lambda_1$, $(\boldsymbol{\delta}'V_{(2)}^{-1}\boldsymbol{\delta}_1)^{-1}$ are independent chi-square variates. Defining

(7.6.31)
$$F_{n_1-p+1,n_2}^{(2)}=\left(\lambda_p v_{(1)}^{pp}\boldsymbol{\delta}_p'V_{(2)}\boldsymbol{\delta}_p\right)^{-1}$$

$$F_{n_1,n_2-p+1}^{(1)}=\frac{v_{11(1)}\boldsymbol{\delta}_1'V_{(2)}^{-1}\boldsymbol{\delta}_1}{\lambda_1},$$

we have the simultaneous confidence bound on $\lambda_1$ and $\lambda_p$ with confidence coefficient equal to $1-\alpha$ as

$$F_2\left(v_{(1)}^{pp}\boldsymbol{\delta}_p'V_{(2)}\boldsymbol{\delta}_p\right)^{-1}\leqslant\lambda_p\leqslant\lambda_1\leqslant v_{11(1)}\boldsymbol{\delta}_1'V_{(2)}^{-1}\boldsymbol{\delta}_1/F_1,$$

where $F_1$ and $F_2$ are determined from

$$P\left(F_{n_1,n_2-p+1}^{(1)}\geqslant F_1\right)P\left(F_{n_1-p+1,n_2}^{(2)}\leqslant F_2\right)=1-\alpha\qquad\text{with}\quad F_2>F_1.$$

The above confidence bounds contain the unknown matrix $B$, and hence, using Eq. (7.6.27), we get the simultaneous confidence bound with confidence coefficient greater than or equal to $1-\alpha$ as

(7.6.32)
$$\frac{\mathrm{ch}_p\left(V_1V_2^{-1}\right)}{F_2}\leqslant\left[\text{all ch}\left(\Sigma_1\Sigma_2^{-1}\right)\right]\leqslant\frac{\mathrm{ch}_1\left(V_1V_2^{-1}\right)}{F_1},$$

where $F_1$ and $F_2$ satisfy Eq. (7.6.31) with $F_2>F_1$. These bounds were shown to be optimum by Anderson (1965).

### 7.6.5.2 $k>2$

Let us consider the case when $k>2$. We observe that for any nonnull vector $\mathbf{a}$,

$$H(\Sigma_1=\Sigma_2=\cdots=\Sigma_k)=\bigcap_{\mathbf{a}}H_{\mathbf{a}},$$

where $H_{\mathbf{a}}:\mathbf{a}'\Sigma_1\mathbf{a}=\cdots=\mathbf{a}'\Sigma_k\mathbf{a}$, and

$$A(\text{at least one } \Sigma_i\neq\Sigma_j)=\bigcup_{\mathbf{a}} A_{\mathbf{a}},$$

where $A_{\mathbf{a}}$ : at least one $\mathbf{a}'\Sigma_i\mathbf{a}\neq\mathbf{a}'\Sigma_j\mathbf{a}$, $i\neq j$. Notice that the hypothesis $H_{\mathbf{a}}$ is the hypothesis of equality of variances in $k$-populations when $\mathbf{a}'V_j\mathbf{a}/\mathbf{a}'\Sigma_j\mathbf{a}$, $j=1,2,\ldots,k$, are independently distributed as chi squares with $n_j$ d.f. $(j=1,2,\ldots,k)$. The likelihood ratio for $H_{\mathbf{a}}$ vs $A_{\mathbf{a}}$ is

$$\lambda_{\mathbf{a}}=\prod_{i=1}^{k}\left[\left(\frac{\mathbf{a}'V_i\mathbf{a}}{\mathbf{a}'V\mathbf{a}}\right)^{\frac{1}{2}n_i}\left(\frac{n}{n_i}\right)^{\frac{1}{2}n_i}\right],\qquad V=\sum_{i=1}^{k}V_i.$$

Notice that the distribution of $\lambda_{\mathbf{a}}$ under $H$ depends on $n_1,n_2,\ldots,n_k$. Hence, using the union–intersection principle, the test procedure is

$$\text{reject } H \quad \text{if} \quad \operatorname*{Sup}_{\mathbf{a}}\lambda_{\mathbf{a}}^{-1}\geqslant c,$$

where $c$ is to be determined from $P[\operatorname{Sup}_{\mathbf{a}}\lambda_{\mathbf{a}}^{-1}\geqslant c|H]=\alpha$. Notice that

$$\begin{aligned}\operatorname*{Sup}_{\mathbf{a}}\lambda_{\mathbf{a}}^{-1}&\leqslant\prod_{i=1}^{k}\left(\frac{n_i}{n}\right)^{\frac{1}{2}n_i}\left(\operatorname*{Sup}_{\mathbf{a}}\frac{\mathbf{a}'V\mathbf{a}}{\mathbf{a}V_i\mathbf{a}}\right)^{\frac{1}{2}n_i}\\&=\prod_{i=1}^{k}\left(\frac{n_i}{n}\right)^{\frac{1}{2}n_i}\left[\operatorname{ch}_1(VV_i^{-1})\right]^{\frac{1}{2}n_i},\end{aligned}$$

and hence

$$\begin{aligned}(7.6.33)\qquad P\left(\operatorname*{Sup}_{\mathbf{a}}\lambda_{\mathbf{a}}^{-1}\geqslant c\right)&\leqslant\alpha\\&=P\left[\prod_{i=1}^{k}\left(\frac{n_i}{n}\right)^{\frac{1}{2}n_i}\left[\operatorname{ch}_1(VV_i^{-1})\right]^{\frac{1}{2}n_i}\geqslant c\right].\end{aligned}$$

Thus, we can use the test procedure as

$$(7.6.34)\qquad \text{reject } H \text{ vs } A \quad \text{if } \prod_{i=1}^{k}\left(\frac{n_i}{n}\right)^{\frac{1}{2}n_i}\left[\operatorname{ch}_1(VV_i^{-1})\right]^{\frac{1}{2}n_i}\geqslant c,$$

where $c$ is determined from Eq. (7.6.33).

Using different types of test procedures for $H_{\mathbf{a}}$, we can obtain different types of test procedures; this task is left to the reader.

## 7.7 Testing $\Sigma=I$ in the Growth-Curve Model

Let $X\sim N_{p,N}(B\xi A,\Sigma,I_N)$, where $B:p\times q$ and $A:m\times n$ are given matrices of full rank. We wish to test the hypothesis

$$(7.7.1)\qquad H:\Sigma=I \quad \text{vs} \quad A\neq H.$$

In the canonical form, we get from Eq. (5.5.3) that observing $X$ is equivalent to observing

$$(7.7.2)\qquad Y=\Delta X\Gamma\equiv p(\overset{m}{\dot{Z}},\overset{N-m}{Z})\equiv\begin{bmatrix} & \overset{m}{} & \overset{N-m}{} \\ q & Z_1 & \\ & & Z \\ p-q & Z_2 & \end{bmatrix},$$

where $\Delta:p\times p$ and $\Gamma:N\times N$ are known orthogonal matrices defined by

$$(7.7.3)\qquad \Delta'=(\overset{q}{B(B'B)^{-1/2}},\overset{p-q}{B_0})\quad\text{and}\quad \Gamma=\left(\overset{m}{A'(AA')^{-1/2}},\overset{N-m}{A_0}\right).$$

Since $X\sim N_{p,N}(B\xi A,\Sigma,I_N)$, it follows that

$$(7.7.4)\qquad Y\sim N_{p,N}(\Delta B\xi A\Gamma,\Lambda,I_N),$$

where

$$(7.7.5)\qquad \Lambda=\Delta\Sigma\Delta'.$$

Letting

$$(7.7.6)\qquad \mu_1=(B'B)^{1/2}\xi(A'A')^{1/2}\quad\text{and}\quad n=N-m,$$

we find that

$$(7.7.7)\qquad \Delta B\xi A\Gamma=\begin{bmatrix} \overset{m}{\mu_1} & \overset{n}{} \\ & 0 \\ 0 & \end{bmatrix}\equiv(\overset{m}{\dot{\mu}},\overset{n}{0}).$$

Thus the problem of testing $H$ vs $A$ is equivalent to

$$(7.7.8)\qquad H':\Lambda=I\quad\text{vs}\quad A'\neq H,$$

where $\dot{Z}\sim N_{p,m}(\dot{\mu},\Lambda,I_m)$ and $Z\sim N_{p,n}(0,,I_n)$ are independently distributed. The likelihood function is given by

$$(7.7.9)\qquad L(\dot{\mu},\Sigma)=c|\Lambda|^{-\frac{1}{2}N}\operatorname{etr}\left\{-\tfrac{1}{2}\Lambda^{-1}\left[ZZ'+(\dot{Z}-\dot{\mu})(\dot{Z}-\dot{\mu})'\right]\right\}$$

Hence

$$(7.7.10)\qquad \operatorname*{Sup}_{H'}L(\dot{\mu},\Sigma)=c\operatorname{etr}\left(-\tfrac{1}{2}ZZ'\right)\operatorname{etr}\left(-\tfrac{1}{2}Z_2'Z_2\right),$$

$$(7.7.11)\qquad \operatorname*{Sup}_{A'}L(\dot{\mu},\Sigma)=c\left(\frac{N}{e}\right)^{\frac{1}{2}Np}|ZZ'|^{-\frac{1}{2}N}|I+Z_2'V_{22}^{-1}Z_2|^{-\frac{1}{2}N}.$$

Thus, the likelihood ratio is given by

$$\lambda = \frac{\operatorname*{Sup}_{H'} L(\mu_1, \Lambda)}{\operatorname*{Sup}_{A'} L(\mu_1, \Lambda)} \tag{7.7.12}$$

$$= (e/N)^{\frac{1}{2}Np} |V_{1.2}|^{\frac{1}{2}N} |V_{22} + Z_2 Z_2'|^{\frac{1}{2}N} \operatorname{etr}\left(-\tfrac{1}{2} V_{11}\right) \times \operatorname{etr}\left[-\tfrac{1}{2}(V_{22} + Z_2 Z_2')\right],$$

where

$$V = \begin{pmatrix} V_{11} & V_{12} \\ V_{12} & V_{22} \end{pmatrix} = ZZ' \sim W_p(\Lambda, n). \tag{7.7.13}$$

However, the test based on $\lambda$ will not be unbiased, as was shown by Sugiura and Nagao (1968) in the special case of $q=p$ and $m=1$. Thus, we shall modify $\lambda$ to $\lambda^*$ by replacing $N$ with $n \equiv N-m$, the number of d.f. The modified likelihood-ratio statistic is then given by

$$\lambda^* = \left(\frac{e}{n}\right)^{\frac{1}{2}np} |V_{1.2}|^{\frac{1}{2}n} |V_{22} + Z_2 Z_2'|^{\frac{1}{2}n} \operatorname{etr}\left(-\tfrac{1}{2} V_{11}\right) \times \operatorname{etr}\left[-\tfrac{1}{2}(V_{22} + Z_2 Z_2')\right]. \tag{7.7.14}$$

In terms of the original observation, it is given by

$$\lambda^* = \left(\frac{e}{n}\right)^{\frac{1}{2}np} |S + (I - TS^{-1}) S_1 (I - S^{-1}T)|^{\frac{1}{2}n} \times \operatorname{etr}\left(-\tfrac{1}{2}\{S + [I - B(B'B)^{-1}B'] S_1\}\right), \tag{7.7.15}$$

where $S$ and $T$ have been defined in Eq. (5.5.2), that is,

$$S = X[I - A'(AA')^{-1}A]X' \quad \text{and} \quad T = B(B'S^{-1}B)^{-1}B'. \tag{7.7.16}$$

### *7.7.1* **h**th *Moment of* $\lambda^*$ *Under* $H$

Since under the hypothesis $H$, $V \sim W_p(I, n)$, we find that $V_{1.2}$ and $(V_{12}, V_{22})$ are independently distributed, where $V_{1.2} \sim W_q(I, n-p+q)$, $V_{22} \sim W_{p-q}(I, n)$, and $V_{12}$ given $V_{22} \sim N_{q,p-q}(0, I, V_{22})$. Thus, since $V$ and $Z_2$ are independently distributed with $Z_2 \sim N_{p-q,m}(0, I, I_m)$, we get

$$E(\lambda^{*h}) = (e/n)^{\frac{1}{2}pnh} E\left[|V_{1.2}|^{\frac{1}{2}nh} \operatorname{etr}\left(-\tfrac{1}{2} V_{1.2}\right)\right] \times E\left\{|V_{22} + Z_2 Z_2'|^{\frac{1}{2}nh} \operatorname{etr}\left[-\tfrac{1}{2}h(V_{22} + Z_2 Z_2')\right] \operatorname{etr}\left[-\tfrac{1}{2}h V_{12}' V_{22}^{-1} V_{12}\right]\right\} \equiv (e/n)^{\frac{1}{2}pnh} g_1(h) g_2(h), \tag{7.7.17}$$

where the expectations are denoted by $g_1(h)$ and $g_2(h)$, respectively. Hence

$$g_1(h)=\frac{2^{\frac{1}{2}nhq}\Gamma_q\left[\frac{1}{2}\{n(1+h)-p+q\}\right]}{(1+h)^{\frac{1}{2}q[n(1+h)-p+q]}\Gamma_q\left[\frac{1}{2}(n-p+q)\right]}. \tag{7.7.18}$$

Next, we evaluate $g_2(h)$. We have

(7.7.19) $g_2(h)$

$$\begin{aligned}
&=E\left\{|V_{22}+Z_2Z_2'|^{\frac{1}{2}nh}\operatorname{etr}\left[-\tfrac{1}{2}h(V_{22}+Z_2Z_2')\right]E\left[\operatorname{etr}\left(-\tfrac{1}{2}hV_{12}'V_{22}^{-1}V_{12}\right)\Big|V_{22}\right]\right\}\\
&=E\left\{|V_{22}+Z_2Z_2'|^{\frac{1}{2}nh}\operatorname{etr}\left[-\tfrac{1}{2}h(V_{22}+Z_2Z_2')\right](1+h)^{-\frac{1}{2}q(p-q)}\right\}\\
&=(1+h)^{-\frac{1}{2}q(p-q)}E\left[|U|^{\frac{1}{2}nh}\operatorname{etr}\left(-\tfrac{1}{2}hU\right)\right],
\end{aligned}$$

where $U\sim W_{p-q}(I,n+m)$. Hence

(7.7.20) $g_2(h)=$

$$2^{\frac{1}{2}nh(p-q)}(1+h)^{\frac{1}{2}[n(1+h)+m-q](p-q)}\frac{\Gamma_{p-q}\left[\frac{1}{2}n(1+h)+\frac{1}{2}m\right]}{\Gamma_{p-q}\left[\frac{1}{2}(n+m)\right]}.$$

Hence,

$$E(\lambda^{*h})=\frac{\left(\frac{2e}{n}\right)^{\frac{1}{2}nhp}\Gamma_q\left(\frac{n(1+h)-p+q}{2}\right)\Gamma_{p-q}\left(\frac{n(1+h)+m}{2}\right)}{(1+h)^{-\frac{1}{2}n(1+h)p-\frac{1}{2}m(p-q)}\Gamma_q\left(\frac{n-p+q}{2}\right)\Gamma_{p-q}\left(\frac{n+m}{2}\right)}. \tag{7.7.21}$$

For further testing problems on covariances in a growth-curve model, see Khatri (1973). The above result is due to Khatri and Srivastava (1974b).

## Problems

**7.1** Let $\lambda^*$ be given by Eq. (7.2.6). Then, under the alternative that $I-\Sigma^{-1}=(2/n)P$, where $P$ is fixed as $n\to\infty$, show that

$$P(-2\ln\lambda^*\leqslant z)=P(\chi_f^2\leqslant z)+\frac{2a}{n}\left[P(\chi_{f+2}^2\leqslant z)-P(\chi_f^2\leqslant z)\right]$$

$$+\left(\frac{2}{n}\right)^2\sum_{i=0}^{2}b_iP(\chi_{f+2i}^2\leqslant z)+\left(\frac{2}{n}\right)^3\sum_{i=0}^{3}c_iP(\chi_{f+2i}^2\leqslant z)+O(n^{-4}),$$

where

$$a = B_2 + 2\sigma_2,$$

$$b_0 = \tfrac{1}{6}(3B_2^2 + 4B_3) + 2B_2\sigma_2 + 2\sigma_2^2 + 4\sigma_3,$$

$$b_1 = -B_2^2 - 4B_2\sigma_2 - 4\sigma_2^2,$$

$$b_2 = \tfrac{1}{6}(3B_2^2 - 4B_3) + 2B_2\sigma_2 + 2\sigma_2^2 + 4\sigma_3,$$

$$c_0 = \tfrac{1}{6}(4B_4 + 4B_2B_3 + B_2^3) - 8\sigma_4 + 16\sigma_2\sigma_3 - \tfrac{4}{3}\sigma_2^3 - \tfrac{1}{6}(3B_2^2 + 4B_3) + 4B_2\sigma_3 - 2B_2\sigma_2^2,$$

$$c_1 = \tfrac{1}{3}(3B_2^2 + 4B_3)\left(\tfrac{1}{2}B_2 + \sigma_2\right) + 2B_2\sigma_2(B_2 + 3\sigma_2) - 4\sigma_3(B_2 + 4\sigma_2) + 4\sigma_2^3,$$

$$c_2 = \tfrac{1}{3}(4B_3 - 3B_2^2)\left(\tfrac{1}{2}B_2 + \sigma_2\right) - 4\sigma_3(B_2 + 4\sigma_2) - 2B_2\sigma_2(B_2 + 3\sigma_2) - 4\sigma_2^3,$$

$$c_3 = \tfrac{1}{6}(4B_4 - 4B_2B_3 + B_2^3) + 8\sigma_4 + 16\sigma_2\sigma_3 + \tfrac{4}{3}\sigma_2^3 + \tfrac{1}{3}\sigma_2(3B_2^2 - 4B_3) + 4B_2\sigma_3 + 2\sigma_2^2B_2.$$

**7.2** Derive Eq. (7.3.4).

**7.3** Let $\lambda$ be given by Eq. (7.3.3). Then, under the alternative that $I - \Sigma^{-1} = (2/N)P$, where $P$ is fixed as $N \to \infty$, show that

$$P(-2\ln\lambda^* \leqslant z) = P\left(\chi_f^2(\delta) \leqslant z\right) + N^{-1}\sum_{i=0}^{2} a_i P\left(\chi_f^2(\delta) \leqslant z\right) + N^{-2}\sum_{i=0}^{4} b_i P\left[\chi_f^2(\delta) \leqslant z\right] + O(N^{-3}),$$

where

$$\delta = \tfrac{1}{2}\boldsymbol{\mu}'\Sigma^{-1}\boldsymbol{\mu},$$

$$B_2' = \tfrac{1}{24}p(2p^2 + 9p + 11), \qquad B_3' = -\tfrac{1}{32}p(p+1)(p+2)(p+3),$$

$$B_4' = \tfrac{1}{480}p(6p^4 + 45p^3 + 110p^2 + 90p + 3),$$

$$a_0 = -(B_2' + 2\sigma_2), \qquad R = \boldsymbol{\mu}\boldsymbol{\mu}'\Sigma^{-1},$$

$$a_1 = -a_0 \operatorname{tr} RP, \qquad a_2 = \operatorname{tr} RP,$$

$$b_0 = \tfrac{1}{6}(3B_2'^2 + 4B_3') + 2B_2'\sigma_2 - 4\sigma_4,$$

$$b_1 = -a_0a_2 - B_2'^2 - 4B_2'\sigma_2 - 4\sigma_2^2,$$

$$b_3 = -a_0a_2 + 2\operatorname{tr} RP^2 - a_2^2, \qquad b_4 = \tfrac{1}{2}a_2^2,$$

$$b_2 = 2a_0a_2 + \tfrac{1}{6}(3B_2'^2 - 4B_3') + 2B_2'\sigma_2 + 2\sigma_2^2 + 4\sigma_3 - 2\operatorname{tr} RP^2 + \tfrac{1}{2}a_2^2,$$

$$f = \tfrac{1}{2}p(p+1),$$

and

$$\sigma_i = \operatorname{tr} P^i / i, \qquad i = 2,3,4.$$

**7.4** Let $\lambda$ be given by Eq. (7.3.3). Then under the alternative that $I - \Sigma^{-1} = (2/N)^{1/2} P$, where $P$ is fixed as $N \to \infty$, show that

$$P[-2\ln\lambda^* \leqslant z] = P\left[\chi_f^2(\delta^*) \leqslant z\right] + \left(\frac{2}{N}\right)^{1/2} \sum_{j=0}^{2} a_j^* P\left[\chi_{f+2j}^2(\delta^*) \leqslant z\right]$$

$$+ \frac{2}{N} \sum_{j=0}^{4} b_j^* P\left[\chi_{f+2j}^2(\delta^*) \leqslant z\right]$$

$$+ \left(\frac{2}{N}\right)^{3/2} \sum_{j=0}^{6} c_j^* P\left[\chi_{f+2j}^2(\delta^*) \leqslant z\right] + O(N^{-2}),$$

with

$$\delta^* = \delta + \sigma^2,$$

$$a_0^* = -\sigma_3, \qquad a_1^* = -\operatorname{tr} RP, \qquad a_2^* = -(a_0^* + a_1^*),$$

$$b_0 = -\sigma_4 + \tfrac{1}{2}\sigma_3^2 - \tfrac{1}{2}B_2', \qquad b_1^* = \tfrac{1}{2}B_2' - \sigma_3 a_1^*,$$

$$b_2^* = -\sigma_3^2 - \tfrac{1}{2}\operatorname{tr} RP^2 + \tfrac{1}{8}a_1^{*2} + \sigma_3 a_1^*,$$

$$b_3^* = \sigma_4 + \tfrac{1}{2}\operatorname{tr} RP^2 - \tfrac{1}{4}a_1^{*2} + \sigma_3^2,$$

$$b_4^* = \tfrac{1}{2}{\sigma_3}^2 + \tfrac{1}{8}a_1^* - \sigma_3 a_1^*,$$

$$c_0^* = -\tfrac{1}{18}\sigma_3^3 + \sigma_3\sigma_4 - \sigma_5 + \tfrac{1}{2}B_2'\sigma_3,$$

$$c_1^* = -\tfrac{1}{4}B_2' a_1^* + \tfrac{1}{4}a_1^*\sigma_3^2 - \tfrac{1}{2}B_2'\sigma_3,$$

$$c_2^* = \tfrac{1}{2}B_2' a_1^* - \tfrac{1}{4}a_1^*\sigma_3^2 + \tfrac{1}{2}\sigma_4 a_1^* + \tfrac{1}{6}\sigma_3^2 - \sigma_2\sigma_3 - \tfrac{1}{2}B_2'\sigma_3,$$

$$c_3^* = -\tfrac{1}{4}B_2' a_1^* - \tfrac{1}{2}a_1^*\sigma_3^2 - \tfrac{1}{4}a_1^* \operatorname{tr} RP^2 - \tfrac{1}{2}\operatorname{tr} RP^3 - \sigma_3\sigma_4 + \tfrac{1}{2}B_2'\sigma_3,$$

$$c_4^* = \tfrac{1}{2}a_1^*\sigma_3^2 + \tfrac{1}{2}a_1^*\sigma_4 + \tfrac{1}{2}a_1^* \operatorname{tr} RP^2 + \tfrac{1}{2}\operatorname{tr} RP^3 + \sigma_3,$$

$$c_5^* = \tfrac{1}{4}a_1^*\sigma_3^2 - \tfrac{1}{2}a_1^*\sigma_4 + \tfrac{1}{4}\operatorname{tr} RP \operatorname{tr} RP^2 + \sigma_3\sigma_4,$$

$$c_6^* = -\tfrac{1}{4}a_1^*\sigma_3^2.$$

**7.5** Let $\mathbf{y} \sim N_p(\boldsymbol{\mu}, \Sigma)$ and $V \sim W_p(\Sigma, n)$. Obtain the likelihood-ratio test for testing $H: (\boldsymbol{\mu} = \mathbf{0},\ \Sigma = \sigma^2 I,\ \sigma^2 > 0)$ vs $A \neq H$. Find the $h$th moment of the criterion under $H$, and give its asymptotic distribution.

**7.6** Let $V \sim W_p(\sigma^2 I, n)$, and let $U = (\operatorname{tr} V^2)/(\operatorname{tr} V)^2$. Show that

$$E(U^\theta) = E\left[(\Sigma\lambda_i^2)^\theta\right] / E(\Sigma\lambda_i)^{2\theta},$$

where $\lambda_1, \lambda_2, \ldots, \lambda_p$ are the roots of $V$.

**7.7** (Continuation) Show that

(i) $E[(\Sigma\lambda_i)^2]=(np+2)np\sigma^4$,

(ii) $E[(\Sigma\lambda_i)^4]=(np+6)(np+4)(np+2)np\sigma^8$,

(iii) $E(\Sigma\lambda_i^2)=np(n+p+1)\sigma^4$,

(iv) $E[(\Sigma\lambda_i^2)^2]=[pn^3+(2p^2+2p+8)n^2+(p^3+2p^2+21p+20)n+8p^2+20p+20]np\sigma^8$.

**7.8**

(i) Let $\mathbf{y}_1,\mathbf{y}_2,\ldots,\mathbf{y}_k, V_1,V_2,\ldots,V_k$ be independently distributed, where $\mathbf{y}_\alpha \sim N_p(\boldsymbol{\mu}_\alpha,\Sigma_\alpha)$ and $V_\alpha \sim W_p(\Sigma_\alpha,n_\alpha)$, $\alpha=1,2,\ldots,k$. Show that the likelihood-ratio test for testing $H:\boldsymbol{\mu}_1=\cdots=\boldsymbol{\mu}_k$, $\Sigma_1=\cdots=\Sigma_k$ against $A\neq H$ is given by

$$\lambda=\prod_{i=1}^{k}\frac{|V_i|^{\frac{1}{2}N_i}}{N_i^{\frac{1}{2}pN_i}}\;\frac{N^{\frac{1}{2}pN}}{\left|\sum_{i=1}^{k}[V_i+(\mathbf{y}_i-\bar{\mathbf{y}})(\mathbf{y}_i-\bar{\mathbf{y}})]^{\frac{1}{2}N}\right|}$$

The modified likelihood ratio $\lambda'$ is obtained from $\lambda$ by replacing $N_i$ with $n_i$ and $N$ with $n$.

(ii) Show that the $h$th moment of $\lambda'$ is given by

$$E(\lambda'^h)=\frac{N^{\frac{1}{2}pN}}{\prod_{i=1}^{k}N_i^{\frac{1}{2}pn_i}}\;\frac{\Gamma_p\left[\frac{1}{2}(n+k-1)\right]}{\Gamma_p\left[\frac{1}{2}(n+hn+k-1)\right]}\prod_{i=1}^{k}\frac{\Gamma_p\left[\frac{1}{2}(n_i+hn_i)\right]}{\Gamma_p\left[\frac{1}{2}n_i\right]}.$$

(iii) Show that if $\lim_{n\to\infty}(n_i/n)>0$, then

$$P(-m\ln\lambda^*\leqslant z)=P(\chi_f^2\leqslant z)+m^{-2}\gamma_2\left[P(\chi_{f+4}^2\leqslant z)-P(\chi_f^2\leqslant z)\right]+O(m^{-3}),$$

where

$$\lambda^*=(\lambda')^{2/n},$$

$$f=\tfrac{1}{2}(k-1)p(p+3),$$

$$m=n+2\alpha,$$

$$\alpha=(\Sigma r_i^{-1}-1)\frac{2p^2+3p-1}{12(k-1)(p+3)}+\frac{p-k+2}{p+3},$$

$$\gamma_2=\frac{p}{288}\Bigg[6(\Sigma r_i^{-2}-1)(p+1)(p-1)(p+2)$$

$$-(\Sigma r_i^{-1}-1)^2\frac{(2p^2+3p-1)^2}{(k-1)(p+3)}-12(\Sigma r_i^{-1}-1)\frac{(2p^2+3p-1)(p-k+2)}{p+3}$$

$$-36\frac{(k-1)(p-k+2)^2}{p+3}-12(k-1)(-2k^2+7k+3pk-2p^2-6p-4)\Bigg].$$

**7.9** Find the asymptotic distribution of $-2\ln\lambda^*$, where $E(\lambda^{*h})$ is given in Eq. (7.7.21).

**7.10** Let $X \sim N_{p,N}(B\xi A, \Sigma, I_N)$. Find the likelihood-ratio test for testing the hypothesis $H:(\Sigma=I,\ \xi=0)$ vs $A \neq H$. Find the $h$th moment of the modified likelihood-ratio statistic (in which $N$ is replaced by $n=N-m$, the number of d.f. under $H$. Give an asymptotic expansion for its cdf.

**7.11** For the above growth-curve model, find the likelihood-ratio test for testing the hypothesis (i) $H_1:\Sigma=\sigma^2 I$ vs $A_1 \neq H_1$, (ii) $H_2:(\Sigma=\sigma^2 I,\ \xi=0)$ vs $A_2 \neq H_2$. Find the $h$th moments of the modified likelihood-ratio statistics and their asymptotic distributions under $H_1$ and $H_2$.

**7.12** Let $S \sim CW_p(\Sigma, n)$, where $\Sigma=\Sigma_1+i\Sigma_2$ is Hermitian p.d. Show the likelihood-ratio test for testing the hypothesis $H:\Sigma_2=0$ vs $A \neq H$ is based on the statistic

$$\lambda^* = |S_1+iS_2|/|S_1|,$$

where $S=S_1+iS_2$. Find the $h$th moment and an (appropriate) asymptotic distribution under $H$.

**7.13** Let $V \sim W_p(\Sigma, n)$, where

$$\Sigma=\begin{pmatrix}\Sigma_1 & \Sigma_{12}' \\ \Sigma_{12} & \Sigma_2\end{pmatrix},$$

$\Sigma_1:r\times r$ and $\Sigma_2:s\times s$, $r\leqslant s$, $r+s=p$. Suppose we wish to test the hypothesis $H:\Sigma_{12}=0$ vs $A:\Sigma_{12}\neq 0$. Show that the problem remains invariant under the transformation

$$V \rightarrow \begin{pmatrix}A_1 & 0 \\ 0 & A_2\end{pmatrix} V \begin{pmatrix}A_1 & 0 \\ 0 & A_2\end{pmatrix}',$$

where $A_1:r\times r$ and $A_2:s\times s$ are nonsingular matrices. Find a set of maximal invariants. Suppose $r=1$. In this case give a uniformly most powerful invariant test.

# 8

# Classification and Discrimination

## 8.1 Introduction

Consider the situation in which a doctor is diagnosing a patient's illness. Perhaps there are a group of diseases which are often confused, and the doctor wishes to record those symptoms and take those measurements that will best enable him to correctly identify the illness from which the patient is suffering. The doctor will also be interested in the best utilization of the measurements taken and in knowing how successfully the distinction between illnesses can be made.

This is essentially a problem of classification. One wishes to classify an individual with a particular population and obtain a measure of how often the classification is likely to be incorrect.

Suppose that two illnesses have been characterized by two $k$-dimensional probability distributions and that samples from each of these distributions will overlap. The doctor is interested in a quantity that distinguishes as well as possible between the two distributions, and at the outset he wishes to know the probability of making an error in a decision based on the use of that quantity.

The first clear statement and solution of this kind of problem was given by Fisher, who was consulted by Barnard (1935) in classifying skeletal remains. Fisher (1936) introduced the discriminant function for distinguishing between two multivariate normal distributions with common covariance matrix.

This chapter is devoted to the problem of classification and the theory associated with its implementation.

## 8.2 The Problem of Classifying into Two Known Normals with Common Covariance

The problem of classifying an individual into one of two categories may be described as follows. Let $\pi_0$, $\pi_1$, and $\pi_2$ denote the three populations. It is known that $\pi_0 = \pi_i$ for exactly one $i \in (1,2)$. The problem is to find for which $i$ this is true. In the case of multivariate normal populations with common nonsingular covariance matrix $\Sigma$, the problem reduces to that of finding (or testing) whether $\boldsymbol{\mu}_0 = \boldsymbol{\mu}_1$ or $\boldsymbol{\mu}_0 = \boldsymbol{\mu}_2$, where $\boldsymbol{\mu}_0$, $\boldsymbol{\mu}_1$, and $\boldsymbol{\mu}_2$ are respectively the mean vectors of $\pi_0$, $\pi_1$, and $\pi_2$. Thus if an observation $\mathbf{x}_0$ (a subject with measurements $\mathbf{x}_0$ from $\pi_0$) is to be classified in $\pi_1$ or $\pi_2$ when all the parameters $\boldsymbol{\mu}_1$, $\boldsymbol{\mu}_2$, and $\Sigma$ are known, we obtain the likelihood ratio as

$$\text{(8.2.1)} \qquad \begin{aligned} &\operatorname{etr}\left\{ -\tfrac{1}{2}\Sigma^{-1}\left[(\mathbf{x}_0-\boldsymbol{\mu}_1)(\mathbf{x}_0-\boldsymbol{\mu}_1)' - (\mathbf{x}_0-\boldsymbol{\mu}_2)(\mathbf{x}_0-\boldsymbol{\mu}_2)'\right]\right\} \\ &= \operatorname{etr}\left\{ -\tfrac{1}{2}\Sigma^{-1}\left[\boldsymbol{\mu}_1\boldsymbol{\mu}_1' - 2\boldsymbol{\mu}_1\mathbf{x}_0' + 2\boldsymbol{\mu}_2\mathbf{x}_0' - \boldsymbol{\mu}_2\boldsymbol{\mu}_2'\right]\right\} \\ &= \exp\left[(\boldsymbol{\mu}_1-\boldsymbol{\mu}_2)'\Sigma^{-1}\mathbf{x}_0 - \tfrac{1}{2}(\boldsymbol{\mu}_1-\boldsymbol{\mu}_2)'\Sigma^{-1}(\boldsymbol{\mu}_1+\boldsymbol{\mu}_2)\right]. \end{aligned}$$

Hence, from the Neyman–Pearson lemma, the best procedure is to classify $\mathbf{x}_0$ in $\pi_1$ or $\pi_2$ according as

$$\text{(8.2.2)} \qquad (\boldsymbol{\mu}_1-\boldsymbol{\mu}_2)'\Sigma^{-1}\mathbf{x}_0 - \tfrac{1}{2}(\boldsymbol{\mu}_1-\boldsymbol{\mu}_2)'\Sigma^{-1}(\boldsymbol{\mu}_1+\boldsymbol{\mu}_2) \gtrless C,$$

where $C$ is so chosen that one of the errors is fixed at some specified level. This procedure was given by Welch (1939). It can alternatively be rewritten as follows: classify $\mathbf{x}_0$ in $\pi_1$ if

$$\text{(8.2.3)} \qquad \boldsymbol{\delta}'\Sigma^{-1}(\mathbf{x}_0-\boldsymbol{\mu}_1) > -\tfrac{1}{2}\Delta^2 + C,$$

and classify $\mathbf{x}_0$ in $\pi_2$ if

$$\text{(8.2.4)} \qquad \boldsymbol{\delta}'\Sigma^{-1}(\mathbf{x}_0-\boldsymbol{\mu}_2) < \tfrac{1}{2}\Delta^2 + C,$$

where

$$\text{(8.2.5)} \qquad \boldsymbol{\delta} = (\boldsymbol{\mu}_1-\boldsymbol{\mu}_2),$$

$$\text{(8.2.6)} \qquad \Delta^2 = \boldsymbol{\delta}'\Sigma^{-1}\boldsymbol{\delta}.$$

$\Delta^2$ is called the Mahalanobis (squared) distance.

### *8.2.1 Errors of Misclassification*

Let $e_1$ denote the probability of misclassifying an individual from $\pi_1$ (into $\pi_2$) and $e_2$ denote the probability of misclassifying an individual from $\pi_2$ (into $\pi_1$). Then from Eq. (8.2.3)

$$\text{(8.2.7)} \qquad \begin{aligned} e_1 &= P\left[\boldsymbol{\delta}'\Sigma^{-1}(\mathbf{x}_0-\boldsymbol{\mu}_1) < -\tfrac{1}{2}\Delta^2 + C \,\middle|\, \mathbf{x}_0 \in \pi_1\right] \\ &= \Phi\left(\Delta^{-1}C - \tfrac{1}{2}\Delta\right), \end{aligned}$$

since when $\mathbf{x}_0 \in \pi_1$, $\boldsymbol{\delta}'\Sigma^{-1}(\mathbf{x}_0-\boldsymbol{\mu}_1) \sim N(0,\Delta^2)$. Here $\Phi$ denotes the cumulative distribution function of a standard normal random variable $N(0,1)$. Similarly, we get from Eq. (8.2.4)

$$e_2 = 1-\Phi\left(\Delta^{-1}C+\tfrac{1}{2}\Delta\right)=\Phi\left(-\Delta^{-1}C-\tfrac{1}{2}\Delta\right). \tag{8.2.8}$$

Thus, if we wish to have the two errors of misclassification equal, then

$$C \equiv 0 \quad \text{and} \quad e_1 = e_2 = \Phi\left(-\tfrac{1}{2}\Delta\right). \tag{8.2.9}$$

### *8.2.2 Fisher's Discriminant Function*

The above procedure in Eq. (8.2.2) is linear in $\mathbf{x}_0$. The linear function

$$L(\mathbf{x}_0) = \boldsymbol{\delta}'\Sigma^{-1}\mathbf{x}_0 \equiv (\boldsymbol{\mu}_1-\boldsymbol{\mu}_2)'\Sigma^{-1}\mathbf{x}_0 \tag{8.2.10}$$

is called Fisher's linear discriminant function; he obtained this result in 1936 from a different viewpoint as follows.

Let $\mathbf{x}_0 \sim N_p(\boldsymbol{\mu}_0, \Sigma)$. Then $\mathbf{a}'\mathbf{x}_0 \sim N(\mathbf{a}'\boldsymbol{\mu}_0, \mathbf{a}'\Sigma\mathbf{a})$ for any nonnull vector $\mathbf{a}$. Fisher suggested that the vector $\mathbf{a}$ be chosen so that the distance between populations $\pi_1$ and $\pi_2$ is a maximum—that is, so that

$$\frac{(\mathbf{a}'\boldsymbol{\mu}_1-\mathbf{a}'\boldsymbol{\mu}_2)^2}{\mathbf{a}'\Sigma\mathbf{a}} = \frac{(\mathbf{a}'\boldsymbol{\delta})^2}{\mathbf{a}'\Sigma\mathbf{a}} \tag{8.2.11}$$

is a maximum. By the Cauchy–Schwartz inequality (see Corollary 1.10.2),

$$\frac{(\mathbf{a}'\boldsymbol{\delta})^2}{\mathbf{a}'\Sigma\mathbf{a}} \leqslant \Delta^2 = \boldsymbol{\delta}'\Sigma^{-1}\boldsymbol{\delta}$$

and the equality holds iff $\mathbf{a} = \lambda\Sigma^{-1}\boldsymbol{\delta}$ where $\lambda$ is constant of proportionality. Without loss of generality, one can take $\lambda = 1$. Hence, the linear discriminant function is given by Eq. (8.2.10).

An interesting property of $L(\mathbf{x}_0)$ is given in

**Theorem 8.2.1** *The statistic $L(\mathbf{x}_0)$ is sufficient (for $\beta$) for a class of normal densities $N_p(\alpha\boldsymbol{\mu}_1+\beta\boldsymbol{\mu}_2, \Sigma)$, where $\alpha+\beta=1$, $\alpha \geqslant 0$, $\beta \geqslant 0$, and where $\boldsymbol{\mu}_1$, $\boldsymbol{\mu}_2$, and $\Sigma$ are known parameters.*

PROOF. The pdf of $\mathbf{x}_0$ can be written as

$$(2\pi)^{-\frac{1}{2}p}|\Sigma|^{-1/2}\operatorname{etr}\left[-\tfrac{1}{2}\Sigma^{-1}(\mathbf{x}_0-\boldsymbol{\mu}_1+\beta\boldsymbol{\delta})(\mathbf{x}_0-\boldsymbol{\mu}_1+\beta\boldsymbol{\delta})'\right],$$

since $\alpha+\beta=1$. Here, as before, $\boldsymbol{\delta}=\boldsymbol{\mu}_1-\boldsymbol{\mu}_2$. Since

$$\begin{aligned}(\mathbf{x}_0-\boldsymbol{\mu}_1+\beta\boldsymbol{\delta})'\Sigma^{-1}(\mathbf{x}_0-\boldsymbol{\mu}_1+\beta\boldsymbol{\delta})\\ =(\mathbf{x}_0-\boldsymbol{\mu}_1)'\Sigma^{-1}(\mathbf{x}_0-\boldsymbol{\mu}_1)+2\beta\boldsymbol{\delta}'\Sigma^{-1}(\mathbf{x}_0-\boldsymbol{\mu}_1)+\beta^2\boldsymbol{\delta}'\Sigma^{-1}\boldsymbol{\delta},\end{aligned}$$

we can rewrite the density of $\mathbf{x}_0$ as

$$(2\pi)^{-\frac{1}{2}p}|\Sigma|^{-1/2}\exp\left[-\tfrac{1}{2}B(\mathbf{x}_0)\right]\exp\left[-\beta L(\mathbf{x}_0)+H(\beta)\right],$$

where

$$B(\mathbf{x}_0)=(\mathbf{x}_0-\boldsymbol{\mu}_1)'\Sigma^{-1}(\mathbf{x}_0-\boldsymbol{\mu}_1) \quad\text{and}\quad H(\beta)=\beta\boldsymbol{\delta}'\Sigma^{-1}\boldsymbol{\mu}_1-\tfrac{1}{2}\beta^2\boldsymbol{\delta}'\Sigma^{-1}\boldsymbol{\delta}.$$

Hence, from the factorization criterion, $L(\mathbf{x}_0)$ is sufficient for $\beta$. This result is due to C. R. Rao (1973a) and is a generalization of a result of Smith (1947) who proved it for $\beta=0$ and $\beta=1$. □

From the above result, not only can we construct a test to determine whether $\mathbf{x}_0\in\pi_1$ or $\mathbf{x}_0\in\pi_2$, as in Eq. (8.2.2), but we can construct a test for $H:\mathbf{x}_0\sim N_p(\alpha\boldsymbol{\mu}_1+\beta\boldsymbol{\mu}_2,\Sigma)$, $\alpha+\beta=1$, $\alpha\geqslant 0$, $\beta\geqslant 0$, against $A\neq H$; the mean of this distribution lies on the line joining the means of $\pi_1$ and $\pi_2$. A test for this $H$ vs $A$ has been proposed by C. R. Rao (1973a) as follows:

**Theorem 8.2.2** *Let* $\mathbf{x}_0$ *be a p-dimensional random vector. For testing* $H:\mathbf{x}_0\sim N_p(\alpha\boldsymbol{\mu}_1+\beta\boldsymbol{\mu}_2,\Sigma)$, $\alpha\geqslant 0$, $\beta\geqslant 0$, $\alpha+\beta=1$, *against* $A\neq H$, *a criterion is given by*

$$Q(\mathbf{x}_0)=(\mathbf{x}_0-\boldsymbol{\mu}_1)'\Sigma^{-1}(\mathbf{x}_0-\boldsymbol{\mu}_1)-\Delta^{-2}\left[(\mathbf{x}_0-\boldsymbol{\mu}_1)'\Sigma^{-1}\boldsymbol{\delta}\right]^2,$$

*which has, under H, a chi-square distribution with* $p-1$ d.f. *Here we assume that the parameters* $\mu_1$, $\mu_2$, *and* $\Sigma$ *are* known, *and the only* unknown *parameter is* $\beta$ *(or* $\alpha$*).*

For a proof, refer to Problem 2.27.

## 8.3 Classifying into Two Normals with Known Common Covariance

Here the two normal populations are not known except for the common covariance matrix $\Sigma$, which is assumed to be known. On the basis of independent observations $\mathbf{x}_1,\ldots,\mathbf{x}_{N_1}$, $\mathbf{x}_{N_1+1},\ldots,\mathbf{x}_{N_1+N_2}$, where the first $N_1$ observations are from $\pi_1$ and the last $N_2$ observations are from $\pi_2$, we wish to classify $\mathbf{x}_0$ (a subject with measurements $\mathbf{x}_0$) into $\pi_i$, $i=1,2$. The set of sufficient statistics for $\boldsymbol{\mu}_0$, $\boldsymbol{\mu}_1$, and $\boldsymbol{\mu}_2$ is $\{\mathbf{x}_0,\bar{\mathbf{x}}_1,\bar{\mathbf{x}}_2\}$, where $\bar{\mathbf{x}}_1=N_1^{-1}\Sigma_{i=1}^{N_1}\mathbf{x}_i$ and $\bar{\mathbf{x}}_2=N_2^{-1}\Sigma_{N_1+1}^{N_1+N_2}\mathbf{x}_i$. When $\boldsymbol{\mu}_0=\boldsymbol{\mu}_1$, the likelihood function is given by

$$L(\boldsymbol{\mu}_1,\boldsymbol{\mu}_2|\boldsymbol{\mu}_0=\boldsymbol{\mu}_1)=c\ \mathrm{etr}\left\{-\tfrac{1}{2}\Sigma^{-1}\left[(\mathbf{x}_0-\boldsymbol{\mu}_1)(\mathbf{x}_0-\boldsymbol{\mu}_1)',\right.\right.$$
$$\left.\left.+N_1(\bar{\mathbf{x}}_1-\boldsymbol{\mu}_1)(\bar{\mathbf{x}}_1-\boldsymbol{\mu}_1)'+N_2(\bar{\mathbf{x}}_2-\boldsymbol{\mu}_2)(\bar{\mathbf{x}}_2-\boldsymbol{\mu}_2)'\right]\right\},$$

where

$$c=(2\pi)^{-\frac{1}{2}p(N_1+N_2+1)}|\Sigma|^{-\frac{1}{2}(N_1+N_2+1)}\operatorname{etr}\left[-\tfrac{1}{2}\Sigma^{-1}(V_1+V_2)\right],$$

with

$$V_1=\sum_{i=1}^{N_1}(\mathbf{x}_i-\bar{\mathbf{x}}_1)(\mathbf{x}_i-\bar{\mathbf{x}}_1)'$$

and

$$V_2=\sum_{i=N_1+1}^{N_1+N_2}(\mathbf{x}_i-\bar{\mathbf{x}}_2)(\mathbf{x}_i-\bar{\mathbf{x}}_2)'$$

and the maximum-likelihood estimates of $\boldsymbol{\mu}_1$ and $\boldsymbol{\mu}_2$ are $(N_1+1)^{-1}(N_1\bar{\mathbf{x}}_1+\mathbf{x}_0)$ and $\bar{\mathbf{x}}_2$, respectively. Hence

$$\operatorname*{Sup}_{\boldsymbol{\mu}_1,\boldsymbol{\mu}_2} L(\boldsymbol{\mu}_1,\boldsymbol{\mu}_2|\boldsymbol{\mu}_0=\boldsymbol{\mu}_1)=c\exp\left[-\tfrac{1}{2}\left(1+N_1^{-1}\right)^{-1}(\mathbf{x}_0-\bar{\mathbf{x}}_1)'\Sigma^{-1}(\mathbf{x}_0-\bar{\mathbf{x}}_1)\right].$$

Similarly,

$$\operatorname*{Sup}_{\boldsymbol{\mu}_1,\boldsymbol{\mu}_2} L(\boldsymbol{\mu}_1,\boldsymbol{\mu}_2|\boldsymbol{\mu}_0=\boldsymbol{\mu}_2)=c\exp\left[-\tfrac{1}{2}\left(1+N_2^{-1}\right)^{-1}(\mathbf{x}_0-\bar{\mathbf{x}}_2)'\Sigma^{-1}(\mathbf{x}_0-\bar{\mathbf{x}}_2)\right].$$

Hence, the *maximum-likelihood rule* is to classify $\mathbf{x}_0$ in $\pi_1$ or $\pi_2$ according as

$$(8.3.1)\qquad \left(1+N_1^{-1}\right)^{-1}(\mathbf{x}_0-\bar{\mathbf{x}}_1)'\Sigma^{-1}(\mathbf{x}_0-\bar{\mathbf{x}}_1) \lesseqgtr\left(1+N_2^{-1}\right)^{-1}(\mathbf{x}_0-\bar{\mathbf{x}}_2)'\Sigma^{-1}(\mathbf{x}_0-\bar{\mathbf{x}}_2)+C.$$

The so-called *minimum-distance rule* is to classify $\mathbf{x}_0$ in $\pi_1$ or $\pi_2$ according as

$$(8.3.2)\qquad (\mathbf{x}_0-\bar{\mathbf{x}}_1)'\Sigma^{-1}(\mathbf{x}_0-\bar{\mathbf{x}}_1)\lesseqgtr(\mathbf{x}_0-\bar{\mathbf{x}}_2)'\Sigma^{-1}(\mathbf{x}_0-\bar{\mathbf{x}}_2).$$

This rule can, however, be rewritten as: Classify $\mathbf{x}_0$ in $\pi_1$ or $\pi_2$ according as

$$(8.3.3)\qquad (\bar{\mathbf{x}}_1-\bar{\mathbf{x}}_2)'\Sigma^{-1}\mathbf{x}_0-\tfrac{1}{2}(\bar{\mathbf{x}}_1-\bar{\mathbf{x}}_2)'\Sigma^{-1}(\bar{\mathbf{x}}_1+\bar{\mathbf{x}}_2)\gtrless 0,$$

which is the same rule as Eq. (8.2.2) except that the unknown parameters have been replaced by their estimates and $C\equiv 0$, a rule suggested by Anderson (1951). Thus when $C\equiv 0$, the minimum-distance rule and Anderson's rule are the same and coincide with the maximum-likelihood rule when $N_1\equiv N_2$. Thus, the maximum-likelihood, the minimum-distance, and Anderson's rules are special cases of the following rule: Classify $\mathbf{x}_0$ in $\pi_1$ or $\pi_2$ according as

$$(8.3.4)\qquad b(\mathbf{x}_0-\bar{\mathbf{x}}_1)'\Sigma^{-1}(\mathbf{x}_0-\bar{\mathbf{x}}_1)\lesseqgtr(\mathbf{x}_0-\bar{\mathbf{x}}_2)'\Sigma^{-1}(\mathbf{x}_0-\bar{\mathbf{x}}_2)+K.$$

For example, $b=1$ with $-\frac{1}{2}K=C$ gives Anderson's rule, and $b=1$ with $K\equiv 0$ gives the minimum-distance rule. Similarly $b=(1+N_1^{-1})^{-1}(1+N_2^{-1})$ with $K=(1+N_2^{-1})C$ gives the maximum-likelihood rule.

### 8.3.1 Errors of Misclassification

In this subsection we evaluate $e_1$, the probability of misclassifying an individual from $\pi_1$ into $\pi_2$ for the rule (8.3.4). We have

$$(8.3.5)\qquad e_1=P\big[b(\mathbf{x}_0-\bar{\mathbf{x}}_1)'\Sigma^{-1}(\mathbf{x}_0-\bar{\mathbf{x}}_1)-(\mathbf{x}_0-\bar{\mathbf{x}}_2)'\Sigma^{-1}(\mathbf{x}_0-\bar{\mathbf{x}}_2) > K\,|\,\mathbf{x}_0\in\pi_1\big].$$

Let

$$(8.3.6)\qquad \begin{aligned} k_1^2&=b(1+N_1^{-1})+(1+N_2^{-1})-2b^{1/2},\\ k_2^2&=b(1+N_1^{-1})+(1+N_2^{-1})+2b^{1/2}. \end{aligned}$$

Define

$$(8.3.7)\qquad \begin{aligned} \mathbf{u}'&=k_1^{-1}\big[b^{1/2}(\mathbf{x}_0-\bar{\mathbf{x}}_1)-(\mathbf{x}_0-\bar{\mathbf{x}}_2)\big]'\Sigma^{-1/2},\\ \mathbf{v}'&=k_2^{-1}\big[b^{1/2}(\mathbf{x}_0-\bar{\mathbf{x}}_1)+(\mathbf{x}_0-\bar{\mathbf{x}}_2)\big]'\Sigma^{-1/2}. \end{aligned}$$

Then when $\boldsymbol{\mu}_0=\boldsymbol{\mu}_1$,

$$(8.3.8)\qquad \begin{pmatrix}\mathbf{u}\\ \mathbf{v}\end{pmatrix}\sim N_{2p}\left(\begin{pmatrix}-\boldsymbol{\delta}^*/k_1\\ +\boldsymbol{\delta}^*/k_2\end{pmatrix},\begin{pmatrix}I_p & \rho I_p\\ \rho I_p & I_p\end{pmatrix}\right),$$

where

$$\boldsymbol{\delta}^{*\prime}=(\boldsymbol{\mu}_1-\boldsymbol{\mu}_2)'\Sigma^{-1/2}\quad\text{and}\quad \rho=\frac{b(1+N_1^{-1})-(1+N_2^{-1})}{k_1k_2}.$$

Hence

$$(8.3.9)\qquad e_1=P(2\mathbf{u}'\mathbf{v}>C^*),\qquad \text{where}\quad C^*=2K/k_1k_2.$$

Thus we get

**Theorem 8.3.1** *For the classification rule* (8.3.4),

$$(8.3.10)\qquad \begin{aligned} e_1&=P\big[(\mathbf{u}+\mathbf{v})'(\mathbf{u}+\mathbf{v})-(\mathbf{u}-\mathbf{v})'(\mathbf{u}-\mathbf{v})>2C^*\big]\\ &=P(a_1w_1-a_2w_2>C^*), \end{aligned}$$

*where $w_1$ and $w_2$ are independently distributed as noncentral chi-square random variables with $p$* d.f. *The noncentrality parameters are respectively given by*

$$(8.3.11)\qquad \begin{aligned} \lambda_1&=(k_2^{-1}-k_1^{-1})^2\Delta^2/2a_1,\\ \lambda_2&=(k_2^{-1}+k_1^{-1})^2\Delta^2/2a_2, \end{aligned}$$

*where*

$$a_1=(1+\rho) \quad \text{and} \quad a_2=(1-\rho).$$

Hence, we get

**Theorem 8.3.2** *For the classification rule* (8.3.4) *with* $K=0$,

$$e_1=P\left[(w_1/w_2)>(a_2/a_1)\right],$$

*where* $w_1/w_2$ *has a doubly noncentral F distribution with* $(p,p)$ d.f. *and noncentrality parameters* $(\lambda_1,\lambda_2)$.

See Srivastava (1973a) for some tabulated values. A table for doubly noncentral F-distribution is given by Tiku (1975).

In another special case, when $\rho=0$, so that $b=(1+N_1^{-1})^{-1}(1+N_2^{-1})$ (that is, for the maximum-likelihood rule), we get

**Theorem 8.3.3** *For the maximum-likelihood rule, that is, for the classification rule* (8.3.4) *with* $b=(1+N_1^{-1})^{-1}(1+N_2^{-1})$, *we have*

$$e_1=P(w_1-w_2>C^*),$$

*where* $w_1$ *and* $w_2$ *are independent noncentral chi squares each with* $p$ d.f. *and noncentrality parameters* $\lambda_1$ *and* $\lambda_2$, *respectively, which are defined in Eq.* (8.3.11).

An expression for the pdf of $w_1-w_2$ can be obtained from Theorem 2.11.2 with $a_1=a_2=1$. [See also John (1961)].

### *8.3.2 Monotonicity of the Errors of Misclassification*

In this subsection we show that under some restrictions on $b$ and $K$ (which include most of the classification rules) $e_1$ is a monotone decreasing function of the Mahalanobis distance $\Delta$. This is contained in the following Theorem 8.3.4, which was established by Das Gupta (1974); a simpler proof is given here [see Srivastava and Selliah (1976)].

**Theorem 8.3.4** *For the classification rule* (8.3.4),

(i) *the probability of misclassification* $e_1$ *(misclassifying an individual from* $\pi_1$ *into* $\pi_2$*) is a monotone decreasing function of the Mahalanobis distance* $\Delta$ *if*

(8.3.12) $$b\geqslant\left(1+N_1^{-1}\right)^{-2} \quad \textit{and} \quad K\leqslant 0;$$

(ii) $e_2$ *is a monotone decreasing function of* $\Delta$ *if*

(8.3.13) $$0<b\leqslant\left(1+N_2^{-1}\right)^2 \quad\text{and}\quad K\geqslant 0.$$

PROOF. From Eq. (8.3.4), we get

$$\begin{aligned}P_1\equiv 1-e_1=P\big[&b(\mathbf{x}_0-\bar{\mathbf{x}}_1)'\Sigma^{-1}(\mathbf{x}_0-\bar{\mathbf{x}}_1)\\&<(\mathbf{x}_0-\bar{\mathbf{x}}_2)'\Sigma^{-1}(\mathbf{x}_0-\bar{\mathbf{x}}_2)+K|\mathbf{x}_0\in\pi_1\big].\end{aligned}$$

Letting

(8.3.14) $$\tilde{\mathbf{x}}_1=\frac{\mathbf{x}_0+N_1\bar{\mathbf{x}}_1}{N_1+1} \quad\text{and}\quad a=b\left(1+N_1^{-1}\right)^2-1,$$

we get

(8.3.15) $$\bar{\mathbf{x}}_1=\frac{(N_1+1)\tilde{\mathbf{x}}_1-\mathbf{x}_0}{N_1}, \qquad (\mathbf{x}_0-\bar{\mathbf{x}}_1)=\left(1+N_1^{-1}\right)(\mathbf{x}_0-\tilde{\mathbf{x}}_1),$$

and

$$\begin{aligned}P_1=P\big[&b\left(1+N_1^{-1}\right)^2(\mathbf{x}_0-\tilde{\mathbf{x}}_1)'\Sigma^{-1}(\mathbf{x}_0-\tilde{\mathbf{x}}_1)\\&<(\mathbf{x}_0-\tilde{\mathbf{x}}_1+\tilde{\mathbf{x}}_1-\bar{\mathbf{x}}_2)'\Sigma^{-1}(\mathbf{x}_0-\tilde{\mathbf{x}}_1+\tilde{\mathbf{x}}_1-\bar{\mathbf{x}}_2)+K|\mathbf{x}_0\in\pi_1\big]\\=P\big[&a(\mathbf{x}_0-\tilde{\mathbf{x}}_1)'\Sigma^{-1}(\mathbf{x}_0-\tilde{\mathbf{x}}_1)\\&-2(\mathbf{x}_0-\tilde{\mathbf{x}}_1)'\Sigma^{-1}(\tilde{\mathbf{x}}_1-\bar{\mathbf{x}}_2)-(\tilde{\mathbf{x}}_1-\bar{\mathbf{x}}_2)'\Sigma^{-1}(\tilde{\mathbf{x}}_1-\bar{\mathbf{x}}_2)\\&<K|\mathbf{x}_0\in\pi_1\big].\end{aligned}$$

Let

(8.3.16) $$\mathbf{u}=\Sigma^{-1/2}(\mathbf{x}_0-\tilde{\mathbf{x}}_1) \quad\text{and}\quad \mathbf{v}=\Sigma^{-1/2}(\tilde{\mathbf{x}}_1-\bar{\mathbf{x}}_2).$$

Then when $\mathbf{x}_0\in\pi_1$, it follows that $\mathbf{u}$, $\tilde{\mathbf{x}}_1$ and $\bar{\mathbf{x}}_2$ are independently normally distributed. Here $\mathbf{u}\sim N_p(\mathbf{0},(1+N_1^{-1})^{-1}I)$ and $\mathbf{v}\sim N_p(\boldsymbol{\delta}^*,[N_2^{-1}+(N_1+1)^{-1}]I)$. Hence

$$P_1=P\{a\mathbf{u}'\mathbf{u}-2\mathbf{u}'\mathbf{v}-\mathbf{v}'\mathbf{v}<K\}.$$

Let $\Gamma$ be an orthogonal matrix with first row $\mathbf{v}'/(\mathbf{v}'\mathbf{v})^{1/2}$. Let $\mathbf{w}=\Gamma\mathbf{u}$. Then when $\mathbf{x}_0\in\pi_1$, $\mathbf{w}\sim N_p(\mathbf{0},(1+N_1^{-1})^{-1}I)$ and is independently distributed of $\tilde{\mathbf{x}}_1$ and $\bar{\mathbf{x}}_2$ (or $\mathbf{v}$). Letting

(8.3.17) $$\mathbf{w}=(w_1,\ldots,w_p)' \quad\text{and}\quad \tilde{D}^2=\mathbf{v}'\mathbf{v}=(\tilde{\mathbf{x}}_1-\bar{\mathbf{x}}_2)'\Sigma^{-1}(\tilde{\mathbf{x}}_1-\bar{\mathbf{x}}_2),$$

we find that

(8.3.18) $$\begin{aligned}P_1&=P\big[a\mathbf{w}'\mathbf{w}-2w_1\tilde{D}-\tilde{D}^2<K\big]\\&=P\big[2w_1\geqslant a(\mathbf{w}'\mathbf{w})\tilde{D}^{-1}-\tilde{D}-K\tilde{D}^{-1}\big]\\&=E\big[P(g(\mathbf{w},\tilde{D})\geqslant 0|\mathbf{w})\big]=E\big[EI(g(\mathbf{w},\tilde{D}))|\mathbf{w}\big],\end{aligned}$$

where

$$g(\mathbf{w},\tilde{D}) = K\tilde{D}^{-1} - a(\mathbf{w}'\mathbf{w})\tilde{D}^{-1} + \tilde{D} + 2w_1, \tag{8.3.19}$$

and given $\mathbf{w}$,

$$I(g(\mathbf{w},\tilde{D})) = \begin{cases} 1 & \text{if} \quad g(\mathbf{w},\tilde{D}) \geq 0 \\ 0 & \text{otherwise.} \end{cases}$$

When $a > 0$ and $K \leq 0$, then for a given value of $\mathbf{w}$, it is easy to see that $g(\mathbf{w},\tilde{D})$ is a monotonic increasing function of $\tilde{D}$, and consequently $I(g(\mathbf{w},\tilde{D}))$ is a monotonic increasing function of $\tilde{D}$. The distribution of $\tilde{D}^2$ (except for a constant) is a noncentral chi square with $p$ d.f. and noncentral parameter $\Delta^2$ (except for a constant), and it has a monotone likelihood-ratio property in $\Delta^2$; i.e., if $f(\tilde{D}|\Delta^2)$ is the pdf of $\tilde{D}$ and if $\Delta_1^2 \geq \Delta_2^2$, then $r(\tilde{D}|\Delta_1^2,\Delta_2^2) = f(\tilde{D}|\Delta_1^2)/f(\tilde{D}|\Delta_2^2)$ is a monotonic increasing function of $\tilde{D}$. Hence, for a given value of $\mathbf{w}$ and $\Delta_1^2 \geq \Delta_2^2$,

$$\begin{aligned} &\int I(g(\mathbf{w},\tilde{D}))f(\tilde{D}|\Delta_1^2)\,d\tilde{D} \\ &\quad = \int I(g(\mathbf{w},\tilde{D}))r(\tilde{D}|\Delta_1^2,\Delta_2^2)f(\tilde{D}|\Delta_2^2)\,d\tilde{D} \\ &\quad \geq \left[\int I(g(\mathbf{w},\tilde{D}))f(\tilde{D}|\Delta_2^2)\,d\tilde{D}\right]\left[\int r(\tilde{D}|\Delta_1^2,\Delta_2^2)f(\tilde{D}|\Delta_2^2)\,d\tilde{D}\right] \end{aligned}$$

on account of Theorem 1.10.5. Noting that

$$\int r(\tilde{D}|\Delta_1^2,\Delta_2^2)f(\tilde{D}|\Delta_2^2)\,d\tilde{D} = 1,$$

we find that for a given value of $\mathbf{w}$ and $\Delta_1^2 \geq \Delta_2^2$,

$$\int I(g(\mathbf{w},\tilde{D}))f(\tilde{D}|\Delta_1^2)\,d\tilde{D} \geq \int I(g(\mathbf{w},\tilde{D}))f(\tilde{D}|\Delta_2^2)\,d\tilde{D}.$$

That is,

$$P[\,g(\mathbf{w},\tilde{D}) \geq 0|\mathbf{w}]$$

is a monotonic increasing function of $\Delta^2$. Hence integrating over $\mathbf{w}$,

$$P[\,g(\mathbf{w},\tilde{D}) \geq 0]$$

is a monotonic increasing function of $\Delta^2$. This proves part (i) of the theorem. The proof of (ii) is similar. □

**Corollary 8.3.1** *Let* $K \equiv 0$. *Then, if*

$$(1+N_1^{-1})^{-2} \leq b \leq (1+N_2^{-1})^2, \tag{8.3.20}$$

$e_1$ *and* $e_2$ *are both monotone decreasing functions of* $\Delta$.

**Corollary 8.3.2** *Let* $\boldsymbol{\mu}_1=\boldsymbol{\mu}_2$, $K\equiv 0$, $P_i=1-e_i$ $(i=1,2)$, *and* $b=(1+N_1^{-1})^{-1}(1+N_2^{-1})$. *Then*

$$P_1\equiv P_2\equiv \tfrac{1}{2}.$$

PROOF. Under the above conditions [for the classification rule (8.3.4)] we get

$$P_1=P\Big[\big(1+N_1^{-1}\big)^{-1}(\mathbf{x}_0-\bar{\mathbf{x}}_1)'\Sigma^{-1}(\mathbf{x}_0-\bar{\mathbf{x}}_1)$$
$$<\big(1+N_2^{-1}\big)^{-1}(\mathbf{x}_0-\bar{\mathbf{x}}_2)'\Sigma^{-1}(\mathbf{x}_0-\bar{\mathbf{x}}_2)|\mathbf{x}_0\in\pi_1\Big].$$

When $\boldsymbol{\mu}_1=\boldsymbol{\mu}_2$, the joint distribution of $\mathbf{u}=(1+N_1^{-1})^{-1/2}\Sigma^{-1/2}(\mathbf{x}_0-\bar{\mathbf{x}}_1)$ and $\mathbf{v}=(1+N_2^{-1})^{-1/2}\Sigma^{-1/2}(\mathbf{x}_0-\bar{\mathbf{x}}_2)$ is $N_{2p}(\mathbf{0},\Sigma_1)$, irrespective of whether $\mathbf{x}_0\in\pi_1$ or $\pi_2$, where

$$\Sigma_1=\begin{pmatrix} I_p & \rho I_p \\ \rho I_p & I_p\end{pmatrix} \quad\text{and}\quad \rho=\big[(1+N_1^{-1})(1+N_2^{-1})\big]^{-1}.$$

Hence, $P_1=P(\mathbf{u}'\mathbf{u}<\mathbf{v}'\mathbf{v})=P(\mathbf{u}'\mathbf{u}>\mathbf{v}'\mathbf{v})=1-P_1$: the joint distribution of $\mathbf{u}$ and $\mathbf{v}$ is invariant under the permutation of the elements of $(\mathbf{u}',\mathbf{v}')$. This proves that $P_1=\frac{1}{2}$. Similarly, $P_2=\frac{1}{2}$ can be established. □

**Corollary 8.3.3** *Let* $b=(1+N_1^{-1})^{-1}(1+N_2^{-1})$ *and* $K\equiv 0$. *Then*

$$P_1\geqslant \tfrac{1}{2} \quad\text{and}\quad P_2\geqslant\tfrac{1}{2}.$$

*That is, for the maximum-likelihood rule with* $K\equiv 0$, $P_1\geqslant\frac{1}{2}$ *and* $P_2\geqslant\frac{1}{2}$.

PROOF. Since $K\equiv 0$ and $(1+N_1^{-1})^{-2}\leqslant b\leqslant(1+N_2^{-1})^2$, it follows from Corollary 8.3.1 that $P_1$ and $P_2$ are both monotone increasing functions of $\Delta$. Since $\Delta=0$ iff $\boldsymbol{\mu}_1=\boldsymbol{\mu}_2$, the minimum value of $P_1$ and $P_2$ is at $\Delta=0$, and is thus $\frac{1}{2}$ by Corollary 8.3.2. The result now follows. □

**Corollary 8.3.4** *Let* $\boldsymbol{\mu}_1=\boldsymbol{\mu}_2$ *and* $K\equiv 0$. *Then* $P_1+P_2=1$ *and*

$$P_1 \substack{<\\=\\>} \tfrac{1}{2} \quad\text{according as}\quad b \substack{>\\=\\<} \big(1+N_1^{-1}\big)^{-1}\big(1+N_2^{-1}\big).$$

The proof follows from the fact that $P(\beta\mathbf{u}'\mathbf{u}<\mathbf{v}'\mathbf{v}) \substack{<\\=\\>} \frac{1}{2}$ according as $\beta \substack{>\\=\\<} 1$, where $\mathbf{u}$ and $\mathbf{v}$ are defined in the proof of Corollary 8.3.2.

That is, for the Anderson's classification statistic with $K\equiv 0$ (or for the minimum-distance rule), $P_1\leqslant\frac{1}{2}$ if $N_1 \leqslant N_2$ and $\Delta=0$. This in turn implies due to monotonicity that $P_1$ could be less than $\frac{1}{2}$ for some values of the parameter $\Delta$, an undesirable feature. *For this reason, the maximum-likelihood rule is recommended, preferably with* $K \equiv 0$, *since in this case* $e_1$ *and* $e_2$ *are both monotone decreasing functions of* $\Delta$, *and are* $\leqslant \frac{1}{2}$.

### *8.3.3 An Estimator of the Error of Misclassification,* $e_1$

Since $e_1$ involves the unknown parameter $\Delta$, it is often desirable to have an estimate of $e_1$. In this section, we give uniformly minimum-variance unbiased estimator (UMVUE) of $e_1$ when $\mathbf{x}_0 \in \pi_1$. Let

$$t_1 = \begin{cases} 1 & \text{if } \; b(\mathbf{x}_0 - \bar{\mathbf{x}}_1)'\Sigma^{-1}(\mathbf{x}_0 - \bar{\mathbf{x}}_1) \geqslant (\mathbf{x}_0 - \bar{\mathbf{x}}_2)'\Sigma^{-1}(\mathbf{x}_0 - \bar{\mathbf{x}}_2) + K, \\ 0 & \text{otherwise.} \end{cases} \tag{8.3.21}$$

Then

$$E(t_1 | \mathbf{x}_0 \in \pi_1) = e_1.$$

Hence $t_1$ is an unbiased estimator of $e_1$. We shall now use Lehmann–Scheffé theorem [see Lehmann (1959)] to obtain UMVUE. Since $\mathbf{x}_0 \in \pi_1$, the set of sufficient statistics for $\boldsymbol{\mu}_1$ and $\boldsymbol{\mu}_2$ are

$$\tilde{\mathbf{x}}_1 = \frac{\mathbf{x}_0 + N_1\bar{\mathbf{x}}_1}{N_1 + 1} \quad \text{and} \quad \bar{\mathbf{x}}_2, \tag{8.3.22}$$

which are complete. Hence the UMVUE of $e_1$ is given by

$$h(\tilde{\mathbf{x}}_1, \bar{\mathbf{x}}_2) \equiv E(t_1 | \tilde{\mathbf{x}}_1, \bar{\mathbf{x}}_2).$$

Thus, we get the following

**Theorem 8.3.5** *When* $\mathbf{x}_0$ *belongs to* $\pi_1$, *the uniformly minimum-variance unbiased estimator of* $e_1$ *is given by*

$$h(\tilde{\mathbf{x}}_1, \bar{\mathbf{x}}_2) = \int_0^{\gamma a^{-1}(\tilde{D}^2 + K) + \lambda^2} f_p(\chi^2 | \lambda^2)\, d\chi^2, \tag{8.3.23}$$

*where*

$$\lambda^2 = \gamma a^{-2}\tilde{D}^2, \qquad a = b(1 + N_1^{-1})^2 - 1,$$

$$\tilde{D}^2 = (\tilde{\mathbf{x}}_1 - \bar{\mathbf{x}}_2)'\Sigma^{-1}(\tilde{\mathbf{x}}_1 - \bar{\mathbf{x}}_2), \qquad \gamma = (1 + N_1^{-1}),$$

*and* $f_p(\chi^2 | \lambda^2)$ *denotes the* pdf *of a noncentral chi-square random variable with* $p$ d.f. *and noncentrality parameter* $\lambda^2$.

PROOF. As in Subsection 8.3.2, we get

$$h(\tilde{\mathbf{x}}_1,\bar{\mathbf{x}}_2)=P\big(a\mathbf{w}'\mathbf{w}-2w_1\tilde{D}\leqslant\tilde{D}^2+K|\tilde{\mathbf{x}}_1,\bar{\mathbf{x}}_2\big)$$
$$=P\left(\gamma\sum_{i=1}^{p}w_i^2-2w_1\gamma\frac{\tilde{D}}{a}\leqslant\gamma a^{-1}(\tilde{D}^2+K)\middle|\tilde{\mathbf{x}}_1,\bar{\mathbf{x}}_2\right)$$
$$=P\left(\gamma\sum_{i=2}^{p}w_i^2+\big(\gamma^{1/2}w_1-\lambda\big)^2\leqslant\gamma a^{-1}(\tilde{D}^2+K)+\lambda^2\middle|\tilde{\mathbf{x}}_1,\bar{\mathbf{x}}_2\right),$$

where $\lambda$ has been defined above. Since the $w_i$'s are independent $N(0,\gamma^{-1})$ and are independent of $\tilde{D}$, it follows that given $\tilde{D}$, $\gamma\Sigma_{i=2}^{p}w_i^2+(\gamma^{1/2}w_1-\lambda)^2$ has a noncentral $\chi^2_{p,\lambda^2}$. This proves Theorem 8.3.5. □

REMARKS. The estimator (8.3.23) will not be an unbiased estimator of $e_1$ if $\mathbf{x}_0$ does not belong to $\pi_1$. Another biased estimator of $e_1$ can be obtained from Eq. (8.3.23) by changing $\tilde{\mathbf{x}}_1\to\bar{\mathbf{x}}_1$, $a\to bN_1^{-2}-1$, and $\gamma=N_1^{-1}$.

## 8.4 Classifying into Two Completely Unknown Normals

As before, let $\mathbf{x}_1,\ldots,\mathbf{x}_{N_1}$, $\mathbf{x}_{N_1+1},\ldots,\mathbf{x}_{N_1+N_2}$ denote independent observations with the first $N_1$ from $\pi_1$ and the last $N_2$ from $\pi_2$. On the basis of measurements $\mathbf{x}_0$ on a subject from $\pi_0$, we wish to classify it into $\pi_i$, $i=1,2$. Let

$$\text{(8.4.1)}\quad \bar{\mathbf{x}}_1=N_1^{-1}\sum_{i=1}^{N_1}\mathbf{x}_i,\qquad \bar{\mathbf{x}}_2=N_2^{-1}\sum_{i=N_1+1}^{N_1+N_2}\mathbf{x}_i,\quad\text{and}\quad S=r^{-1}V,$$

where

$$V=\sum_{i=1}^{N_1}(\mathbf{x}_i-\bar{\mathbf{x}}_1)(\mathbf{x}_i-\bar{\mathbf{x}}_1)'+\sum_{i=N_1+1}^{N_1+N_2}(\mathbf{x}_i-\bar{\mathbf{x}}_2)(\mathbf{x}_i-\bar{\mathbf{x}}_2)',$$

and

$$r=N_1+N_2-2.$$

Then $\mathbf{x}_0$, $\bar{\mathbf{x}}_1$, $\bar{\mathbf{x}}_2$, and $V$ are independently distributed. The likelihood function when $\boldsymbol{\mu}_0=\boldsymbol{\mu}_1$ is given by

$$L(\boldsymbol{\mu}_1,\boldsymbol{\mu}_2,\Sigma|\boldsymbol{\mu}_0=\boldsymbol{\mu}_1)$$
$$=\text{const}\,|\Sigma|^{-\frac{1}{2}(N_1+N_2+1)}|V|^{\frac{1}{2}(r-p-1)}$$
$$\times\operatorname{etr}\left\{-\tfrac{1}{2}\Sigma^{-1}\left[V+(\mathbf{x}_0-\boldsymbol{\mu}_1)(\mathbf{x}_0-\boldsymbol{\mu}_1)'+N_1(\bar{\mathbf{x}}_1-\boldsymbol{\mu}_1)(\bar{\mathbf{x}}_1-\boldsymbol{\mu}_1)'\right.\right.$$
$$\left.\left.+N_2(\bar{\mathbf{x}}_2-\boldsymbol{\mu}_2)(\bar{\mathbf{x}}_2-\boldsymbol{\mu}_2)'\right]\right\},$$

and the maximum-likelihood estimates of $\boldsymbol{\mu}_1$, $\boldsymbol{\mu}_2$, and $\Sigma$ are, respectively,

$$(8.4.2)\qquad \tilde{\mathbf{x}}_1=(\mathbf{x}_0+N_1\bar{\mathbf{x}}_1)/(N_1+1),\ \bar{\mathbf{x}}_2,\quad \text{and}\quad (N_1+N_2+1)^{-1}V_1,$$

where

$$\begin{aligned} V_1&=V+(\mathbf{x}_0-\tilde{\mathbf{x}}_1)(\mathbf{x}_0-\tilde{\mathbf{x}}_1)'+N_1(\bar{\mathbf{x}}_1-\tilde{\mathbf{x}}_1)(\bar{\mathbf{x}}_1-\tilde{\mathbf{x}}_1)'\\ &=V+\left(1+N_1^{-1}\right)^{-1}(\mathbf{x}_0-\bar{\mathbf{x}}_1)(\mathbf{x}_0-\bar{\mathbf{x}}_1)'. \end{aligned}$$

Hence,

$$\begin{aligned} &\operatorname*{Sup}_{\boldsymbol{\mu}_1,\boldsymbol{\mu}_2,\Sigma} L(\boldsymbol{\mu}_1,\boldsymbol{\mu}_2,\Sigma|\boldsymbol{\mu}_0=\boldsymbol{\mu}_1)\\ &\quad=\text{const}\,|V|^{\frac{1}{2}(r-p-1)}|(N_1+N_2+1)^{-1}V_1|^{-\frac{1}{2}(N_1+N_2+1)}\exp\left[-\tfrac{1}{2}p(N_1+N_2)\right]. \end{aligned}$$

Similarly,

$$\begin{aligned} &\operatorname*{Sup}_{\boldsymbol{\mu}_1,\boldsymbol{\mu}_2,\Sigma} L(\boldsymbol{\mu}_1,\boldsymbol{\mu}_2,\Sigma|\boldsymbol{\mu}_0=\boldsymbol{\mu}_2)\\ &\quad=\text{const}\,|V|^{\frac{1}{2}(r-p-1)}|(N_1+N_2+1)^{-1}V_2|^{-\frac{1}{2}(N_1+N_2+1)}\exp\left[-\tfrac{1}{2}p(N_1+N_2)\right], \end{aligned}$$

where

$$(8.4.3)\qquad V_2=V+\left(1+N_2^{-1}\right)^{-1}(\mathbf{x}_0-\bar{\mathbf{x}}_2)(\mathbf{x}_0-\bar{\mathbf{x}}_2)'.$$

Thus, the likelihood ratio is given by

$$\frac{\operatorname*{Sup}_{\boldsymbol{\mu}_1,\boldsymbol{\mu}_2,\Sigma} L(\boldsymbol{\mu}_1,\boldsymbol{\mu}_2,\Sigma|\boldsymbol{\mu}_0=\boldsymbol{\mu}_1)}{\operatorname*{Sup}_{\boldsymbol{\mu}_1,\boldsymbol{\mu}_2,\Sigma} L(\boldsymbol{\mu}_1,\boldsymbol{\mu}_2,\Sigma|\boldsymbol{\mu}_0=\boldsymbol{\mu}_2)}=\left(\frac{|V_2|}{|V_1|}\right)^{\frac{1}{2}(N_1+N_2+1)}$$

Hence, classify in $\pi_1$ or $\pi_2$ according as

$$\frac{|V+\left(1+N_2^{-1}\right)^{-1}(\mathbf{x}_0-\bar{\mathbf{x}}_2)(\mathbf{x}_0-\bar{\mathbf{x}}_2)'|}{|V+\left(1+N_1^{-1}\right)^{-1}(\mathbf{x}_0-\bar{\mathbf{x}}_1)(\mathbf{x}_0-\bar{\mathbf{x}}_1)'|}\gtrless C,$$

or

$$\begin{aligned} &C+C\left(1+N_1^{-1}\right)^{-1}(\mathbf{x}_0-\bar{\mathbf{x}}_1)'V^{-1}(\mathbf{x}_0-\bar{\mathbf{x}}_1)\\ &\lesseqgtr 1+(1+N_2)^{-1}(\mathbf{x}_0-\bar{\mathbf{x}}_2)'V^{-1}(\mathbf{x}_0-\bar{\mathbf{x}}_2), \end{aligned}$$

where $C$ is some positive constant (which may depend on the size of either of the two errors of misclassification). Hence, the likelihood rule is given as follows: Classify into $\pi_1$ or $\pi_2$ according as

$$(8.4.4)\qquad C(1+N_1^{-1})^{-1}(\mathbf{x}_0-\bar{\mathbf{x}}_1)'V^{-1}(\mathbf{x}_0-\bar{\mathbf{x}}_1)$$
$$\lesseqgtr (1+N_2^{-1})^{-1}(\mathbf{x}_0-\bar{\mathbf{x}}_2)'V^{-1}(\mathbf{x}_0-\bar{\mathbf{x}}_2)+(1-C).$$

In the above rule, we could use $S=r^{-1}V$ in place of $V$.

Anderson (1951) suggested using the classification rule (8.2.2) with unknown parameters replaced by their estimates. Hence, the above likelihood rule, the minimum-distance rule, and Anderson's classification rule are special cases of the following rule: Classify into $\pi_1$ or $\pi_2$ according as

$$(8.4.5)\qquad b(\mathbf{x}_0-\bar{\mathbf{x}}_1)'S^{-1}(\mathbf{x}_0-\bar{\mathbf{x}}_1)\lesseqgtr(\mathbf{x}_0-\bar{\mathbf{x}}_2)'S^{-1}(\mathbf{x}_0-\bar{\mathbf{x}}_2)+K.$$

For reasons indicated in the previous section, we should use the likelihood rule (8.4.4) with $C=1$. It should be mentioned that Wald (1944) suggested to use Fisher's discriminant function with parameters replaced by their estimates. However, it has the unpleasant feature that its distribution depends not only on $\Delta$ but also on $\boldsymbol{\mu}_1$ and $\boldsymbol{\mu}_2$.

In the following subsections we investigate the properties of the classification rule (8.4.5).

### *8.4.1 Errors of Misclassification*

In this subsection, we calculate the errors of misclassification,

$$e_1=P\left[b(\mathbf{x}_0-\bar{\mathbf{x}}_1)'S^{-1}(\mathbf{x}_0-\bar{\mathbf{x}}_2)>(\mathbf{x}_0-\bar{\mathbf{x}}_2)'S^{-1}(\mathbf{x}_0-\bar{\mathbf{x}}_2)+K|\mathbf{x}_0\in\pi_1\right].$$

As in Section 8.3.2, we get

$$(8.4.6)\qquad e_1=P\left[a(\mathbf{x}_0-\tilde{\mathbf{x}}_1)'V^{-1}(\mathbf{x}_0-\tilde{\mathbf{x}}_1)-2(\mathbf{x}_0-\tilde{\mathbf{x}}_1)'V^{-1}(\tilde{\mathbf{x}}_1-\mathbf{x}_2)\right.$$
$$\left.-(\tilde{\mathbf{x}}_1-\bar{\mathbf{x}}_2)'V^{-1}(\tilde{\mathbf{x}}_1-\bar{\mathbf{x}}_2)>r^{-1}K|\mathbf{x}_0\in\pi_1\right]$$
$$=P\left[a\mathbf{u}'W^{-1}\mathbf{u}-2\mathbf{u}'W^{-1}\mathbf{v}-\mathbf{v}'W^{-1}\mathbf{v}>r^{-1}K|\mathbf{x}_0\in\pi_1\right],$$

where

$$(8.4.7)\qquad r=N_1+N_2-2,\qquad a=b(1+N_1^{-1})^2-1,\qquad \tilde{\mathbf{x}}_1=\frac{\mathbf{x}_0+N_1\bar{\mathbf{x}}_1}{N_1+1},$$
$$\mathbf{u}=\Sigma^{-1/2}(\mathbf{x}_0-\tilde{\mathbf{x}}_1),\qquad \mathbf{v}=\Sigma^{-1/2}(\tilde{\mathbf{x}}_1-\bar{\mathbf{x}}_2),\qquad W=\Sigma^{-1/2}V\Sigma^{-1/2}.$$

When $\mathbf{x}_0\in\pi_1$, then $\mathbf{u}\sim N_p(\mathbf{0},(1+N_1^{-1})^{-1}I)$ and $\mathbf{v}\sim N_p(\boldsymbol{\delta},[N_2^{-1}+(1+N_1)^{-1}]I)$, and they are independently distributed. Also $W\sim W_p(I,r)$, and it is independent of $\mathbf{u}$ and $\mathbf{v}$. Let

$$(8.4.8)\qquad \gamma=(1+N_1^{-1})\quad\text{and}\quad W_1=W+\gamma\mathbf{u}\mathbf{u}'.$$

Then, using Theorem 1.4.1(iv), we get

$$W^{-1}=W_1^{-1}+\gamma(1-\gamma\mathbf{u}'W_1^{-1}\mathbf{u})^{-1}W_1^{-1}\mathbf{u}\mathbf{u}'W_1^{-1} \tag{8.4.9}$$

and

$$W^{-1}\mathbf{u}=\frac{W_1^{-1}\mathbf{u}}{1-\gamma\mathbf{u}'W_1^{-1}\mathbf{u}}. \tag{8.4.10}$$

Then

$$\mathbf{u}'W^{-1}\mathbf{u}=(\mathbf{u}'W_1^{-1}\mathbf{u})(1-\gamma\mathbf{u}'W_1^{-1}\mathbf{u})^{-1},$$

$$\mathbf{u}'W^{-1}\mathbf{v}=(\mathbf{u}'W_1^{-1}\mathbf{v})(1-\gamma\mathbf{u}'W_1^{-1}\mathbf{u})^{-1},$$

and

$$\mathbf{v}'W^{-1}\mathbf{v}=\frac{\mathbf{v}'W_1^{-1}\mathbf{v}(1-\gamma\mathbf{u}'W_1^{-1}\mathbf{u})+\gamma(\mathbf{u}'W_1^{-1}\mathbf{v})^2}{1-\gamma\mathbf{u}'W_1^{-1}\mathbf{u}}.$$

Hence, from Eq. (8.4.6), we get

$$\begin{aligned} e_1 &= P\left\{\frac{a\mathbf{u}'W_1^{-1}\mathbf{u}-2\mathbf{u}'W_1^{-1}\mathbf{v}-\mathbf{v}'W_1^{-1}\mathbf{v}(1-\gamma\mathbf{u}'W_1^{-1}\mathbf{u})-\gamma(\mathbf{u}'W_1^{-1}\mathbf{v})^2}{1-\gamma\mathbf{u}'W_1^{-1}\mathbf{u}}\right. \\ &\qquad \left. >r^{-1}K|\mathbf{x}_0\in\pi_1\right\} \\ &= P\left\{(a+r^{-1}K\gamma)\mathbf{u}'W_1^{-1}\mathbf{u}-2\mathbf{u}'W_1^{-1}\mathbf{v}-(\mathbf{v}'W_1^{-1}\mathbf{v})\left[1-\gamma(\mathbf{u}'W_1^{-1}\mathbf{u})\right]\right. \\ &\qquad \left. -\gamma(\mathbf{u}'W_1^{-1}\mathbf{v})^2>r^{-1}K|\mathbf{x}_0\in\pi_1\right\}. \end{aligned} \tag{8.4.11}$$

Let

$$\begin{aligned} &\tilde{D}_1^2=r\mathbf{v}'W_1^{-1}\mathbf{v}, \qquad r=N_1+N_2-2, \\ &\mathbf{l}'=r^{1/2}\mathbf{v}'W_1^{-1/2}/\tilde{D}_1, \qquad \mathbf{l}'\mathbf{l}=1, \\ &\mathbf{y}'=\gamma^{1/2}\mathbf{u}'W_1^{-1/2}, \qquad q=\mathbf{y}'\mathbf{y}, \quad \text{and} \quad y_1=\mathbf{l}'\mathbf{y}. \end{aligned} \tag{8.4.12}$$

Then Eq. (8.4.11) can be written as

$$\begin{aligned} e_1 &= P\left[(a\gamma^{-1}+r^{-1}K+r^{-1}\tilde{D}_1^2)\mathbf{y}'\mathbf{y}-2(\gamma r)^{-1/2}\tilde{D}_1\mathbf{l}'\mathbf{y}-r^{-1}\tilde{D}_1^2\right. \\ &\qquad \left. -r^{-1}\tilde{D}_1^2(\mathbf{l}'\mathbf{y})^2>r^{-1}K|\mathbf{x}_0\in\pi_1\right] \\ &= P\left[(ar+K\gamma+\tilde{D}_1^2\gamma)q-2(\gamma r)^{1/2}\tilde{D}_1y_1-\gamma\tilde{D}_1^2-\gamma\tilde{D}_1^2y_1^2\right. \\ &\qquad \left. >\gamma K|\mathbf{x}_0\in\pi_1\right]. \end{aligned} \tag{8.4.13}$$

From Problem 3.3, we find that the joint pdf of $q$ and $y_1$ for $p \geqslant 2$ is given by

(8.4.14) $$f(q,y_1)=\frac{(1-q)^{\frac{1}{2}(r+1-p)-1}\left(q-y_1^2\right)^{\frac{1}{2}(p-1)-1}}{\beta\left(\frac{1}{2},\frac{1}{2}(p-1)\right)\beta\left(\frac{1}{2}p,\frac{1}{2}(r+1-p)\right)},$$

$$-1<y_1<1, \quad y_1^2<q<1,$$

since $W_1 \sim W_p(I,r+1)$. For $p=1$, $q=y_1^2$, the pdf of $y_1$ is given by

(8.4.15) $$f(y_1)=\frac{\left(1-y_1^2\right)^{\frac{1}{2}r-1}}{\beta\left(\frac{1}{2},\frac{1}{2}r\right)}, \qquad -1<y_1<1.$$

Thus the pdf of $(q,y_1)$ does not depend upon any unknown parameters $(\boldsymbol{\mu}_1,\boldsymbol{\mu}_2,\Sigma)$. Since the set of (when $\mathbf{x}_0\in\pi_1$) sufficient statistics $(\tilde{\mathbf{x}}_1,\bar{\mathbf{x}}_2,V_1)$ is complete, it follows from Basu's (1955) theorem that $(q,y_1)$ is independent of $\tilde{D}_1^2$. The distribution of

$$\frac{\left[N_2^{-1}+(N_1+1)^{-1}\right]^{-1}(r-p+2)}{pr}\tilde{D}_1^2$$

is a noncentral $F$ distribution with $p$ and $r-p+2$ d.f. and noncentrality parameter $[N_2^{-1}+(N_1+1)^{-1}]^{-1}\Delta^2$. Hence, we get the following [see also Sitgreaves (1952)]:

**Theorem 8.4.1** *For the classification rule* (8.4.5), *the probability of misclassifying an individual from $\pi_1$ (into $\pi_2$) is given by*

(8.4.16) $$e_1=\int p(\tilde{D}_1)\,d\tilde{D}_1\int_A f(q,y_1)\,dq\,dy_1,$$

*where*

(8.4.17) $$A=\left\{(q,y_1):\left(ar+K\gamma+\tilde{D}_1^2\gamma\right)q-2(\gamma r)^{1/2}\tilde{D}_1 y_1 - \gamma\left(1+y_1^2\right)\tilde{D}_1^2>\gamma K\right\},$$

*and $p(\tilde{D}_1)$ denotes the* pdf *of $\tilde{D}_1$.*

It appears difficult to evaluate $e_1$ even numerically. For Anderson's statistic, an asymptotic expansion for $e_1$ has been given by Okamoto (1963), Bowker and Sitgreaves (1961), and Anderson (1973).

### *8.4.2 Monotonicity of the Errors of Misclassification*

In this subsection we show that Theorem 8.3.5 is true even when the covariance matrix is unknown. Thus we shall prove

**Theorem 8.4.2** *For the classification rule* (8.4.5),

(i) *the probability of misclassification* $e_1$ *is a monotone decreasing function of the Mahalanobis distance* $\Delta$ *if*

$$b \geqslant \gamma^{-2}, \quad K \leqslant 0, \qquad \text{where} \quad \gamma = 1 + N_1^{-1}; \tag{8.4.18}$$

(ii) $e_2$ *is a monotonically decreasing function of* $\Delta$ *if*

$$0 < b \leqslant \left(1 + N_2^{-1}\right)^2 \quad \text{and} \quad K \geqslant 0. \tag{8.4.19}$$

This result was first obtained by Das Gupta (1974). However, the following proof is due to Srivastava and Selliah (1976).

PROOF. From Eq. (8.4.13) we get

$$P_1 = 1 - e_1 = P\left[\, g(y_1, q, \tilde{D}_1) \geqslant 0 \right],$$

where $g(y_1,q,\tilde{D}_1) = 2(\gamma r)^{1/2} y_1 + \gamma(1 - q + y_1^2)\tilde{D}_1 + [\gamma K(1-q) - arq]\tilde{D}_1^{-1}$ is a monotonic increasing function of $\tilde{D}_1$ given $(q, y_1)$ when $a \geqslant 0$ and $K \leqslant 0$. Notice that $\tilde{D}_1$ and $(q, y_1)$ are independent and the distribution of $\tilde{D}_1$ has a monotone likelihood-ratio property in $\Delta^2$. Then, arguing as in the proof of Theorem 8.3.5, we get the required result.

In the same way, we have (ii). All the corollaries given at the end of Theorem 8.3.5 hold true here as well. □

### 8.4.3 An Estimate of $e_1$

Let

$$\tilde{t}_1 = \begin{cases} 1 & \text{if} \quad b(\mathbf{x}_0 - \bar{\mathbf{x}}_1)' V^{-1}(\mathbf{x}_0 - \bar{\mathbf{x}}_1) \geqslant (\mathbf{x}_0 - \bar{\mathbf{x}}_2)' V^{-1}(\mathbf{x}_0 - \bar{\mathbf{x}}_2) + r^{-1}K, \\ 0 & \text{otherwise.} \end{cases}$$

Then

$$E\left(\tilde{t}_1 \mid \mathbf{x}_0 \in \pi_1\right) = e_1.$$

Since $\tilde{t}_1$ is an unbiased estimator of $e_1$ when $\mathbf{x}_0$ belongs to $\pi_1$, the UMVUE of $e_1$ is given by

$$h_1(\bar{\mathbf{x}}_1, \mathbf{x}_2, V) \equiv E\left(\tilde{t}_1 \mid \bar{\mathbf{x}}_1, \bar{\mathbf{x}}_2, V\right).$$

Hence, proceeding as in Theorem 8.3.6, we get

**Theorem 8.4.3** *If* $\mathbf{x}_0$ *belongs to* $\pi_1$, $p=1$, $K=0$, $r>1$, $N_1\geqslant 1$, $N_2\geqslant 1$, *then*

$$h_1(\tilde{\mathbf{x}}_1,\bar{\mathbf{x}}_2,V)=\begin{cases}1-P\left[\sqrt{\frac{\gamma}{r}}\,\tilde{D}_1\left(\frac{1-\sqrt{1+a}}{a}\right)<y_1<\sqrt{\frac{\gamma}{r}}\,\tilde{D}_1\left(\frac{1+\sqrt{1+a}}{a}\right)\right], & \text{if } a>0,\\ P\left[\sqrt{\frac{\gamma}{r}}\,\tilde{D}_1\left(\frac{1+\sqrt{1+a}}{a}\right)<y_1<\sqrt{\frac{\gamma}{r}}\,\tilde{D}_1\left(\frac{1-\sqrt{1+a}}{a}\right)\right], & \text{if } -1<a<0,\end{cases}$$

*where the density of* $y_1$ *is given by Eq.* (8.4.15).

**Theorem 8.4.4** *If* $\mathbf{x}_0$ *belongs to* $\pi_1$, $p>1$, $K=0$, $a>0$, $r>p$, $N_1\geqslant 1$, $N_2\geqslant 1$, *then*

$$h_1(\tilde{\mathbf{x}}_1,\bar{\mathbf{x}}_2,V)=P\left\{q\geqslant\frac{\tilde{D}_1^2y_1^2+2\sqrt{r/\gamma}\,\tilde{D}_1y_1+\tilde{D}_1^2}{(r/\gamma)a+\tilde{D}_1^2}\right\},$$

*where the density of* $(q,y_1)$ *is given by Eq.* (8.4.14).

Note that the above estimates are biased if $\mathbf{x}_0$ does not belong to $\pi_1$.

These results are due to Srivastava and Selliah (1976) and parallel those of Broffit and Williams (1973), who estimate $e_1(N_1-1,N_2)$ and give a biased truncated estimator of $e_1\equiv e_1(N_1,N_2)$.

The above expression may be difficult to evaluate. As an alternative, the expression in Eq. (8.3.23) is suggested, with $\Sigma$ replaced by

$$(N_1+N_2-2)^{-1}V.$$

## 8.5 Classifying into $k$ Normals with Common Covariance

Let $N_p(\boldsymbol{\mu}_i,\Sigma)$ be the distribution of $\pi_i$, $i=0,1,\ldots,k$. First, we assume that except for $\boldsymbol{\mu}_0$, all the parameters are known. It is also known that $\boldsymbol{\mu}_0=\boldsymbol{\mu}_i$ for exactly one $i\in(1,2,\ldots,k)$. On the basis of measurements $\mathbf{x}_0$ on a subject from $\pi_0$, we wish to find for which $i$, $\boldsymbol{\mu}_0=\boldsymbol{\mu}_i$. Let $f_j(\mathbf{x}_0)$ denote the pdf of $\mathbf{x}_0$ when $\boldsymbol{\mu}_0=\boldsymbol{\mu}_j$. Then, as in Section 8.2, we can use the functions

$$u_{jk}(\mathbf{x}_0)=\ln\frac{f_j(\mathbf{x}_0)}{f_k(\mathbf{x}_0)}=\left[\mathbf{x}_0-\tfrac{1}{2}(\boldsymbol{\mu}_j+\boldsymbol{\mu}_k)\right]'\Sigma^{-1}(\boldsymbol{\mu}_j-\boldsymbol{\mu}_k)\tag{8.5.1}$$

to find the classification regions. Thus $\mathbf{x}_0$ is classified in $\pi_i$ if

$$u_{ij}(\mathbf{x}_0)>c_{ij}\qquad\text{for all}\quad j\neq i, j=1,2,\ldots,k,\tag{8.5.2}$$

where $c_{ij}$ are constants which may be determined from *a priori* information. Alternatively, if we wish to have all the errors of misclassification equal, we can choose $c_{ij} \equiv 0$.

Next, we consider the case when all the parameters are unknown. Let $\mathbf{x}_i$, $i=0,1,\ldots,k$, be independently distributed as $N_p(\boldsymbol{\mu}_i, \alpha_i\Sigma)$, where $\alpha_0=1$, and $\alpha_i=N_i^{-1}$, $i=1,2,\ldots,k$. Let $V$ be independently distributed of the $\mathbf{x}_i$'s as $W_p(\Sigma,r)$, where $r=\Sigma(N_i-1)=\Sigma N_i-k$. The problem is to find for which $i$ $\boldsymbol{\mu}_0=\boldsymbol{\mu}_i$. It may be noted that in relation to Section 8.4, $\mathbf{x}_i$, $i=1,2,\ldots,k$, here represents the sample mean, based on $N_i$ observations of the $i$th population, and $S=r^{-1}V$, the pooled sample estimate of the covariance matrix $\Sigma$. Thus proceeding as in Section 8.4, we obtain *a* likelihood rule as follows: Classify $\mathbf{x}_0$ in $\pi_i$ iff $i$ is the smallest integer for which the minimum of

$$(1+N_i^{-1})^{-1}(\mathbf{x}_0-\mathbf{x}_i)'S^{-1}(\mathbf{x}_0-\mathbf{x}_i) \tag{8.5.3}$$

is attained. Thus the minimum-distance rule will coincide with the likelihood rule if the $N_i$'s are all equal. Alternatively, one can use Eq. (8.5.1) with parameters replaced by their estimates, as suggested by Anderson (1958). Kiefer and Schwartz (1965) have shown that the above likelihood rule is admissible.

## 8.6 Classifying into $k$ Normals when the Population Means Are Linearly Restricted

Let $\mathbf{x}_i$, $i=0,1,\ldots,k$, a $p$-dimensional vector, be independently normally distributed with mean vector $\boldsymbol{\mu}_i$, $i=0,1,\ldots,k$, and common unknown nonsingular covariance matrix $\Sigma$. The $\boldsymbol{\mu}_i$'s, $i=0,1,\ldots,k$, are unknown, but it is known that $\mu=\xi A$, i.e., we have the following model:

$$H:\left[E(\mathbf{x}_0)=\boldsymbol{\mu}_0,\quad E(\underset{p\times k}{X})=\underset{p\times k}{\mu}=\xi A\right], \tag{8.6.1}$$

where $X=(\mathbf{x}_1,\mathbf{x}_2,\ldots,\mathbf{x}_k)$, $\mu=(\boldsymbol{\mu}_1,\boldsymbol{\mu}_2,\ldots,\boldsymbol{\mu}_k)$, $\xi$ is a $p\times m$ matrix of unknown parameters, and $A$ is a known, $m\times k$ $(m\leqslant k)$ matrix of rank $m$. The case when $A$ is not of full rank can be dealt with similarly.

Suppose it is known that $\boldsymbol{\mu}_0=\boldsymbol{\mu}_i$ for exactly one $i\in(1,2,\ldots,k)$, i.e., we have the following model:

$$H_i:(\boldsymbol{\mu}_0=\boldsymbol{\mu}_i \text{ for only one } i,\ i=1,2,\ldots,k,\quad \mu=\xi A). \tag{8.6.2}$$

The problem is to decide for which $i$ this is true. Let $H_i$ be the hypothesis that $\boldsymbol{\mu}_0=\boldsymbol{\mu}_i$, and $D_i$ be the decision of taking $\boldsymbol{\mu}_0=\boldsymbol{\mu}_i$. The problem is thus to find a statistical decision procedure for selecting one of the $k$ decisions $(D_1,\ldots,D_k)$, which should be optimum in a certain sense.

In this section we derive the likelihood rule that was shown by Srivastava (1967b) to be admissible. In order to derive the likelihood rule, we need

**Lemma 8.6.1** *Let $V$ be the matrix of error, sum of squares, and products under the model $H$, and let $V_i$ $(i=1,2,\ldots,k)$ be the matrix of error, sum of squares, and products under model $H_i$. Then*

(i) $V=X[I_k-A'(AA')^{-1}A]X'$,

(ii) $V_i=(\mathbf{x}_0,X)\{I_{k+1}-(\mathbf{c}_i,A)'[(\mathbf{c}_i,A)(\mathbf{c}_i,A)']^{-1}(\mathbf{c}_i,A)\}(\mathbf{x}_0,X)'$, *and*

(iii) $V_i=V+\mathbf{y}_i\mathbf{y}_i'$,

*where*

(8.6.3)
$$\begin{aligned}A&=(\mathbf{c}_1,\ldots,\mathbf{c}_k)\\ \mathbf{y}_i&=(1+b_i)^{-1/2}(\mathbf{x}_0-X\mathbf{z}_i),\\ b_i&=\mathbf{c}_i'(AA')^{-1}\mathbf{c}_i,\\ \mathbf{z}_i&=A'(AA')^{-1}\mathbf{c}_i.\end{aligned}$$

For the proof of (i), the reader is referred to Chapter 5.

Under the model $H_i$, we note that

(8.6.4)
$$E_i(\mathbf{x}_0,X)=\xi(\mathbf{c}_i,A),$$

and hence (ii) follows from (i). For the proof of (iii) we note that from Theorem 1.4.1 (iv)

$$\begin{aligned}&(\mathbf{c}_i,A)'\left[(\mathbf{c}_i,A)(\mathbf{c}_i,A)'\right]^{-1}(\mathbf{c}_i,A)\\ &=(\mathbf{c}_i,A)'\left[\mathbf{c}_i\mathbf{c}_i'+AA'\right]^{-1}(\mathbf{c}_i,A)\\ &=\begin{pmatrix}\mathbf{c}_i'\\ A'\end{pmatrix}\left[(AA')^{-1}-\frac{(AA')^{-1}\mathbf{c}_i\mathbf{c}_i'(AA')^{-1}}{1+\mathbf{c}_i'(AA')^{-1}\mathbf{c}_i}\right](\mathbf{c}_i,A)\\ &=\begin{bmatrix}\mathbf{c}_i'(AA')^{-1}-\dfrac{\mathbf{c}_i'(AA')^{-1}\mathbf{c}_i\mathbf{c}_i'(AA')^{-1}}{1+b_i}\\ A'(AA')^{-1}-\dfrac{A'(AA')^{-1}\mathbf{c}_i\mathbf{c}_i'(AA')^{-1}}{1+b_i}\end{bmatrix}(\mathbf{c}_i,A)\\ &=\begin{bmatrix}b_i-\dfrac{b_i^2}{1+b_i} & \left(1-\dfrac{b_i}{1+b_i}\right)\mathbf{z}_i'\\ \left(1-\dfrac{b_i}{1+b_i}\right)\mathbf{z}_i & A'(AA')^{-1}A-\dfrac{1}{1+b_i}\mathbf{z}_i\mathbf{z}_i'\end{bmatrix}\\ &=(1+b_i)^{-1}\begin{pmatrix}b_i & \mathbf{z}_i'\\ \mathbf{z}_i & (1+b_i)A'(AA')^{-1}A-\mathbf{z}_i\mathbf{z}_i'\end{pmatrix}.\end{aligned}$$

Hence

$$V_i - V = (\mathbf{x}_0, X)\left\{ I_{k+1} - (\mathbf{c}_i, A)'[(\mathbf{c}_i, A)(\mathbf{c}_i, A)']^{-1}(\mathbf{c}_i, A) - \begin{pmatrix} 0 & 0' \\ 0 & I_k - A'(AA')^{-1}A \end{pmatrix}\right\}\begin{pmatrix} \mathbf{x}_0' \\ X' \end{pmatrix} \tag{8.6.5}$$

$$= (\mathbf{x}_0, X)\begin{bmatrix} 1 - \dfrac{b_i}{1+b_i} & -(1+b_i)^{-1}\mathbf{z}_i' \\ -(1+b_i)^{-1}\mathbf{z}_i & (1+b_i)^{-1}\mathbf{z}_i\mathbf{z}_i' \end{bmatrix}\begin{pmatrix} \mathbf{x}_0' \\ X' \end{pmatrix}$$

$$= (1+b_i)^{-1}(\mathbf{x}_0, X)\begin{pmatrix} 1 & -\mathbf{z}_i' \\ -\mathbf{z}_i & \mathbf{z}_i\mathbf{z}_i' \end{pmatrix}\begin{pmatrix} \mathbf{x}_0' \\ X' \end{pmatrix}$$

$$= (1+b_i)^{-1}(\mathbf{x}_0 - X\mathbf{z}_i, -(\mathbf{x}_0 - X\mathbf{z}_i)\mathbf{z}_i')(\mathbf{x}_0, X)'$$

$$= (1+b_i)^{-1}(\mathbf{x}_0 - X\mathbf{z}_i)(\mathbf{x}_0 - X\mathbf{z}_i)'$$

$$= \mathbf{y}_i\mathbf{y}_i'$$

Hence, as in Eq. (8.4.4) with $C = 1$, we get the following likelihood rule:

**Theorem 8.6.1** *The likelihood rule for choosing among the decisions* $(D_1, \ldots, D_k)$ *can be described as follows: Take decision* $D_i$ *(that is, classify* $\mathbf{x}_0$ *in* $\pi_i$*) if*

$$T_i = \operatorname*{Min}_{1 < j < k} T_j,$$

*where*

$$T_j = (1+b_j)^{-1}(\mathbf{x}_0 - X\mathbf{z}_j)'V^{-1}(\mathbf{x}_0 - X\mathbf{z}_j).$$

## 8.7 Classifying into Two Normals with Unequal Covariances

Let $\mathbf{x}_0$ be the measurements on a subject from $\pi_0$. It is known that $\mathbf{x}_0$ belongs to either $\pi_1$ or $\pi_2$, where $\pi_i \sim N_p(\boldsymbol{\mu}_i, \Sigma_i)$, $i = 1, 2$. If we assume that all the parameters $(\boldsymbol{\mu}_1, \boldsymbol{\mu}_2, \Sigma_1, \Sigma_2)$ are known, then following as in Section 8.2, the likelihood rule is given by: classify into $\pi_1$ or $\pi_2$ according as

$$(\mathbf{x}_0 - \boldsymbol{\mu}_1)'\Sigma_1^{-1}(\mathbf{x}_0 - \boldsymbol{\mu}_1) \lessgtr (\mathbf{x}_0 - \boldsymbol{\mu}_2)'\Sigma_2^{-1}(\mathbf{x}_0 - \boldsymbol{\mu}_2) + [\ln|\Sigma_2| - \ln|\Sigma_1| + K]. \tag{8.7.1}$$

The errors of misclassification are difficult to evaluate except when $\Sigma_1 = \gamma\Sigma_2$ ($\gamma$ known). In this case it can be obtained from the tables of noncentral chi square and is left as an exercise in Problem 8.7. In the same manner, one can obtain the errors of misclassification for the classification rule (8.7.1) when the parameters $\boldsymbol{\mu}_1$ and $\boldsymbol{\mu}_2$ are unknown and replaced by their estimates $\bar{\mathbf{x}}_1$ and $\bar{\mathbf{x}}_2$; and $\Sigma_1 = \gamma\Sigma_2$.

If the parameters $\boldsymbol{\mu}_1$, $\boldsymbol{\mu}_2$, $\Sigma_1$, and $\Sigma_2$ are unknown, and a sample $\mathbf{x}_1, \ldots, \mathbf{x}_{N_1}, \mathbf{x}_{N_1+1}, \ldots, \mathbf{x}_{N_1+N_2}$ is given, where the first $N_1$ observations are from $\pi_1$ and the last $N_2$ from $\pi_2$, then the likelihood function when $\mathbf{x}_0 \in \pi_1$ is given by

$$\begin{aligned} &L(\boldsymbol{\mu}_i, \Sigma_i, i=1,2 | \mathbf{x}_0 \in \pi_1) \\ &\quad = \text{const} |\Sigma_1|^{-(N_1+1)/2} |\Sigma_2|^{-(N_2)/2} \\ &\qquad \times \operatorname{etr}\left(-\tfrac{1}{2}\left\{\Sigma_1^{-1}\left[(\mathbf{x}_0 - \boldsymbol{\mu}_1)(\mathbf{x}_0 - \boldsymbol{\mu}_1)' + N_1(\bar{\mathbf{x}}_1 - \boldsymbol{\mu}_1)(\mathbf{x}_1 - \boldsymbol{\mu}_1)' + V_1\right]\right.\right. \\ &\qquad \left.\left. + \Sigma_2^{-1}\left[N_2(\bar{\mathbf{x}}_2 - \boldsymbol{\mu}_2)(\mathbf{x}_2 - \boldsymbol{\mu}_2)' + V_2\right]\right\}\right), \end{aligned}$$

where

$$\bar{\mathbf{x}}_1 = N_1^{-1} \sum_{i=1}^{N_1} \mathbf{x}_i, \qquad \bar{\mathbf{x}}_2 = N_2^{-1} \sum_{i=N_1+1}^{N_1+N_2} \mathbf{x}_i, \tag{8.7.2}$$

$$V_1 = \sum_{i=1}^{N_1} (\mathbf{x}_i - \bar{\mathbf{x}}_1)(\mathbf{x}_i - \bar{\mathbf{x}}_1)',$$

$$V_2 = \sum_{N_1+1}^{N_1+N_2} (\mathbf{x}_i - \bar{\mathbf{x}}_2)(\mathbf{x}_i - \bar{\mathbf{x}}_2)'.$$

Hence

$$\begin{aligned} &\sup_{\boldsymbol{\mu}_i, \Sigma_i, i=1,2} L(\boldsymbol{\mu}_i, \Sigma_i, i=1,2 | \mathbf{x}_0 \in \pi_1) \\ &\quad = \text{const} |(N_1+1)^{-1}\left[V_1 + (\mathbf{x}_0 - \tilde{\mathbf{x}}_1)(\mathbf{x}_0 - \tilde{\mathbf{x}}_1)'\right. \\ &\qquad \left. + N_1(\bar{\mathbf{x}}_1 - \tilde{\mathbf{x}}_1)(\bar{\mathbf{x}}_1 - \tilde{\mathbf{x}}_1)'\right]|^{-\frac{1}{2}(N_1+1)} |N_2^{-1}V_2|^{-\frac{1}{2}N_2}, \end{aligned}$$

where

$$\tilde{\mathbf{x}}_1 = \frac{\mathbf{x}_0 + N_1\bar{\mathbf{x}}_1}{N_1+1}. \tag{8.7.3}$$

Similarly

$$\begin{aligned} &\sup_{\boldsymbol{\mu}_i, \Sigma_i, i=1,2} L(\boldsymbol{\mu}_i, \Sigma_i, i=1,2 | \mathbf{x}_0 \in \pi_2) \\ &\quad = \text{const} |(N_2+1)^{-1}\left[V_2 + (\mathbf{x}_0 - \tilde{\mathbf{x}}_2)(\mathbf{x}_0 - \tilde{\mathbf{x}}_2)'\right. \\ &\qquad \left. + N_2(\bar{\mathbf{x}}_2 - \tilde{\mathbf{x}}_2)(\mathbf{x}_2 - \tilde{\mathbf{x}}_2)'\right]|^{-\frac{1}{2}(N_2+1)} |N_1^{-1}V_1|^{-\frac{1}{2}N_1}, \end{aligned}$$

where

$$\tilde{\mathbf{x}}_2 = \frac{\mathbf{x}_0 + N_2\bar{\mathbf{x}}_2}{N_2+1}. \tag{8.7.4}$$

Hence, the likelihood ratio is given by

$$\lambda = \frac{\operatorname{Sup} L(\boldsymbol{\mu}_i, \Sigma_i | \mathbf{x}_0 \in \pi_1)}{\operatorname{Sup} L(\boldsymbol{\mu}_i, \Sigma_i | \mathbf{x}_0 \in \pi_2)}$$

$$= \frac{|(N_1+1)^{-1}V_1|^{-\frac{1}{2}(N_1+1)}|N_2^{-1}V_2|^{-\frac{1}{2}N_2}}{|(N_2+1)^{-1}V_2|^{-\frac{1}{2}(N_2+1)}|N_1^{-1}V_1|^{-\frac{1}{2}N_1}}$$

$$\times \frac{\left[1+(1+N_1^{-1})^{-1}(\mathbf{x}_0-\bar{\mathbf{x}}_1)'V_1^{-1}(\mathbf{x}_0-\bar{\mathbf{x}}_1)\right]^{-\frac{1}{2}(N_1+1)}}{\left[1+(1+N_2^{-1})^{-1}(\mathbf{x}_0-\bar{\mathbf{x}}_2)'V_2^{-1}(\mathbf{x}_0-\bar{\mathbf{x}}_2)\right]^{-\frac{1}{2}(N_2+1)}}$$

$$= \frac{(N_2+1)^{\frac{1}{2}p(N_2+1)}N_1^{\frac{1}{2}N_1p}}{(N_1+1)^{\frac{1}{2}p(N_1+1)}N_2^{\frac{1}{2}N_2p}}\left(\frac{|V_2|}{|V_1|}\right)^{1/2}$$

$$\times \frac{\left[1+(1+N_2^{-1})^{-1}(\mathbf{x}_0-\bar{\mathbf{x}}_2)'V_2^{-1}(\mathbf{x}_0-\bar{\mathbf{x}}_2)\right]^{\frac{1}{2}(N_2+1)}}{\left[1+(1+N_1^{-1})^{-1}(\mathbf{x}_0-\bar{\mathbf{x}}_1)'V_1^{-1}(\mathbf{x}_0-\bar{\mathbf{x}}_1)\right]^{\frac{1}{2}(N_1+1)}}.$$

Thus $\mathbf{x}_0$ is classified into $\pi_1$ or $\pi_2$ according as

$$\lambda \gtrless C, \tag{8.7.5}$$

where $C$ is some constant. If we choose $C=1$, then $\mathbf{x}_0$ is classified in $\pi_1$ or $\pi_2$ according as

(8.7.6)

$$(N_1+1)^{\frac{1}{2}p(N_1+1)}N_2^{\frac{1}{2}N_2p}|V_1|^{1/2}\left[1+(1+N_1^{-1})^{-1}(\mathbf{x}_0-\bar{\mathbf{x}}_1)'V_1^{-1}(\mathbf{x}_0-\bar{\mathbf{x}}_1)\right]$$
$$\lessgtr (N_2+1)^{\frac{1}{2}p(N_2+1)}N_1^{\frac{1}{2}N_1p}|V_2|^{1/2}\left[1+(1+N_2^{-1})^{-1}(\mathbf{x}_0-\bar{\mathbf{x}}_2)'V_2^{-1}(\mathbf{x}_0-\bar{\mathbf{x}}_2)\right].$$

According to the minimum-distance rule, $\mathbf{x}_0$ is classified in $\pi_1$ or $\pi_2$ according as

$$(N_1-1)(\mathbf{x}_0-\bar{\mathbf{x}}_1)'V_1^{-1}(\mathbf{x}_0-\bar{\mathbf{x}}_1) \lessgtr (\mathbf{x}_0-\bar{\mathbf{x}}_2)'V_2^{-1}(\mathbf{x}_0-\bar{\mathbf{x}}_2)(N_2-1). \tag{8.7.7}$$

Alternatively, one can use Eq. (8.7.1) with all the parameters replaced by their estimates given in Eq. (8.7.2); that is, classify in $\pi_1$ or $\pi_2$ according as

$$(\mathbf{x}_0-\bar{\mathbf{x}}_1)'S_1^{-1}(\mathbf{x}_0-\bar{\mathbf{x}}_1) \lessgtr (\mathbf{x}_0-\bar{\mathbf{x}}_2)'S_2^{-1}(\mathbf{x}_0-\bar{\mathbf{x}}_2)+(\ln|S_2|-\ln|S_1|+K), \tag{8.7.8}$$

where $S_i = V_i/(N_i-1)$, $i=1,2$.

However, in contrast to the equal-covariance case, neither the rules (8.7.1) nor (8.7.5)–(8.7.8) are linear in $\mathbf{x}_0$; they are quadratic in $\mathbf{x}_0$. When the $\Sigma_i$'s and $\boldsymbol{\mu}_i$'s are known, Anderson and Bahadur (1962) suggest a linear discriminant function, which they show to be admissible, in place of Eq. (8.7.1). [See also Clunier-Ross and Riffenburgh (1960).] The procedure can be described as follows: Classify $\mathbf{x}_0$ in $\pi_1$ or $\pi_2$ according as

$$\mathbf{b}'\mathbf{x}_0 \lesseqgtr c, \tag{8.7.9}$$

where $\mathbf{b}$ and $c$ are chosen in some optimum way. From normal theory, we find that $e_1$, the probability of misclassifying an individual from $\pi_1$ (into $\pi_2$) is given by

$$\begin{aligned} e_1 &= P(\mathbf{b}'\mathbf{x}_0 > c \mid \mathbf{x}_0 \in \pi_1) \\ &= P\left( \frac{\mathbf{b}'\mathbf{x}_0 - \mathbf{b}'\boldsymbol{\mu}_1}{(\mathbf{b}'\Sigma_1\mathbf{b})^{1/2}} > \frac{c - \mathbf{b}'\boldsymbol{\mu}_1}{(\mathbf{b}'\Sigma_1\mathbf{b})^{1/2}} \,\middle|\, \mathbf{x}_0 \in \pi_1 \right) \\ &= 1 - \Phi(y_1), \end{aligned} \tag{8.7.10}$$

where

$$y_1 = \frac{c - \mathbf{b}'\boldsymbol{\mu}_1}{(\mathbf{b}'\Sigma_1\mathbf{b})^{1/2}}. \tag{8.7.11}$$

Similarly, the probability of misclassifying an individual from $\pi_2$ is given by

$$\begin{aligned} e_2 &= P(\mathbf{b}'\mathbf{x}_0 < c \mid \mathbf{x}_0 \in \pi_2) \\ &= P(-\mathbf{b}'\mathbf{x}_0 > -c \mid \mathbf{x}_0 \in \pi_2) \\ &= 1 - \Phi(y_2), \end{aligned} \tag{8.7.12}$$

where

$$y_2 = \frac{\mathbf{b}'\boldsymbol{\mu}_2 - c}{(\mathbf{b}'\Sigma_2\mathbf{b})^{1/2}}. \tag{8.7.13}$$

To minimize the two errors, the problem is to find a vector $\mathbf{b}$ and $c$ such that

$$y_1 = \frac{c - \mathbf{b}'\boldsymbol{\mu}_1}{(\mathbf{b}'\Sigma_1\mathbf{b})^{1/2}} \quad \text{and} \quad y_2 = \frac{\mathbf{b}'\boldsymbol{\mu}_2 - c}{(\mathbf{b}'\Sigma_2\mathbf{b})^{1/2}} \tag{8.7.14}$$

have maximum values for given values of $\boldsymbol{\mu}_1$, $\boldsymbol{\mu}_2$, $\Sigma_1$, and $\Sigma_2$. For given $\mathbf{b}$, the variation of $c$ decreases one and increases the other, and hence this cannot be achieved. Therefore, for variation over $\mathbf{b}$, we have to fix $c$ or one value of $y$. This leads us to maximize $y_1$ for a given value of $y_2$ or to maximize $y_2$ for a given value of $y_1$. Anderson and Bahadur (1962) have shown that the above method leads to the complete class of admissible solutions for **b**, from which an optimum can be chosen. We shall consider here in detail the case when $y_1=y_2=y$ (say). In this situation, eliminating $c$, we have

$$(8.7.15)\qquad y=\frac{\mathbf{b}'\boldsymbol{\delta}}{(\mathbf{b}'\Sigma_1\mathbf{b})^{1/2}+(\mathbf{b}'\Sigma_2\mathbf{b})^{1/2}},\qquad \boldsymbol{\delta}=\boldsymbol{\mu}_2-\boldsymbol{\mu}_1.$$

It can be easily seen that if **b** is changed to $c_1\mathbf{b}$, then $y$ is changed to $(\operatorname{sign} c_1)y$. The value of $y$ is positive at $\mathbf{b}=\boldsymbol{\delta}$. Hence the maximum value of $y$ must be positive. We see that $y$ is a differentiable function of **b** except possibly at $\mathbf{b}=\mathbf{0}$, and we are not looking for a solution at $\mathbf{b}=\mathbf{0}$. Hence, the maximum (or the minimum) value of $y$ will be obtained by setting the first derivative of $y$ equal to zero. For this,

$$(8.7.16)\qquad \frac{\partial y}{\partial \mathbf{b}}=\left[(\mathbf{b}'\Sigma_1\mathbf{b})^{1/2}+(\mathbf{b}'\Sigma_2\mathbf{b})^{1/2}\right]^{-1}\left[\boldsymbol{\delta}-(t_1\Sigma_1+t_2\Sigma_2)\mathbf{b}\right],$$

where $t_i\propto(\mathbf{b}'\boldsymbol{\delta})(\mathbf{b}'\Sigma_i\mathbf{b})^{-1/2}[(\mathbf{b}'\Sigma_1\mathbf{b})^{1/2}+(\mathbf{b}'\Sigma_2\mathbf{b})^{1/2}]^{-1}$. Equating (8.7.16) to zero gives

$$(8.7.17)\qquad y=t_1(\mathbf{b}'\Sigma_1\mathbf{b})^{1/2}=t_2(\mathbf{b}'\Sigma_2\mathbf{b})^{1/2}.$$

The solutions of $\partial y/\partial\mathbf{b}=\mathbf{0}$ cannot be obtained easily, and we shall have to solve the equations iteratively. Since we are looking for the maximum values, $t_1$ and $t_2$ must be positive. For the given positive values of $t_1$ and $t_2$, $t_1\Sigma_1+t_2\Sigma_2$ is p.d. and

$$(8.7.18)\qquad \mathbf{b}=(t_1\Sigma_1+t_2\Sigma_2)^{-1}\boldsymbol{\delta}.$$

On account of the invariance property of $y$, we can choose without loss of generality $t_2=1-t$, $t_1=t$, and $0<t<1$. Since $\Sigma_1$ and $\Sigma_2$ are p.d., by Theorem 1.9.2 we can write $\Sigma_1=C'D_\lambda C$ and $\Sigma_2=C'C$, where $C$ is nonsingular and $D_\lambda=\operatorname{diag}(\lambda_1,\lambda_2,\ldots,\lambda_p)$. Let $\boldsymbol{\delta}_1=C'^{-1}\boldsymbol{\delta}=(\delta_1,\delta_2,\ldots,\delta_p)'$. Then

$$(8.7.19)\qquad C\mathbf{b}=\left[I+t(D_\lambda-I)\right]^{-1}\boldsymbol{\delta}_1,$$

and

$$(8.7.20)\qquad y=t\left[\sum_{j=1}^{p}\frac{\lambda_j\delta_j^2}{(1-t+\lambda_j t)^2}\right]^{1/2}$$

$$=(1-t)\left[\sum_{j=1}^{p}\frac{\delta_j^2}{(1-t+\lambda_j t)^2}\right]^{1/2}.$$

The problem reduces to finding the value of $t$ that makes the two expressions of $y$ equal, and this can be done, because

(8.7.21) $$t^2 \sum_{j=1}^{p} \lambda_j \delta_j^2 (1-t+\lambda_j t)^{-2}$$

is zero at $t=0$ and is a monotonic increasing function of $t$, while

(8.7.22) $$(1-t)^2 \sum_{j=1}^{p} \delta_j^2 (1-t+\lambda_j t)^{-2}$$

is positive at $t=0$ and is a monotonic decreasing function of $t$. With this value of $t$, we get the optimum value of $y$ and the solution for $\mathbf{b}$. For this value of $\mathbf{b}$ and $t$, the value of $c$ should be

(8.7.23) $$c = \mathbf{b}'\boldsymbol{\mu}_2 - t\mathbf{b}'\boldsymbol{\delta} = \mathbf{b}'\boldsymbol{\mu}_1 + (1-t)\mathbf{b}'\boldsymbol{\delta}.$$

See also Banerjee and Marcus (1965).

## 8.8 Classifying into $k$ Normals with Known Unequal Covariances

Let $\mathbf{x}_0, \mathbf{x}_1, \ldots, \mathbf{x}_k$ be independently distributed as $N_p(\boldsymbol{\mu}_i, \alpha\Sigma_i)$, where $\alpha = N^{-1}$, $i = 0, 1, 2, \ldots, k$, and the $\Sigma_i$'s are known for $i = 1, 2, \ldots, k$. Thus the $\boldsymbol{\mu}_i$'s, $i = 0, 1, \ldots, k$, and $\Sigma_0$ are unknown. It is known that $\boldsymbol{\mu}_0 = \boldsymbol{\mu}_i$ and $\Sigma_0 = \Sigma_i$ for some $i \in (1, 2, \ldots, k)$. The problem is to find for which $i$ this is true. It has been shown by Srivastava (1964) that the following theorem holds.

**Theorem 8.8.1** *Under the restriction of translation invariance and with a simple loss function, the following procedure of selecting one of the k decisions $D_1, \ldots, D_k$ ($D_i : \boldsymbol{\mu}_0 = \boldsymbol{\mu}_i$, $\Sigma_0 = \Sigma_i$) is admissible*:

$$\phi_i(\mathbf{y}) = \begin{cases} 1 & \text{if } U_i(\mathbf{y}) = \min\limits_{1<j<k} U_j(\mathbf{y}), \\ 0 & \text{otherwise}, \end{cases}$$

*where*

$$U_j(\mathbf{y}) = \ln|\Sigma_j| + \tfrac{1}{4}(\mathbf{x}_j - \mathbf{x}_0)'\Sigma_j^{-1}(\mathbf{x}_j - \mathbf{x}_0) + \tfrac{1}{2}\mathbf{y}'\Sigma_{1j}^{-1}\mathbf{y},$$

$$y = \sqrt{N}\,(\mathbf{x}_1 - \mathbf{x}_0, \ldots, \mathbf{x}_k - \mathbf{x}_0)',$$

*and*

$$\Sigma_{1j} = \begin{bmatrix} \Sigma_1 + \Sigma_j & \Sigma_j & \cdots & \Sigma_j \\ \Sigma_j & \Sigma_2 + \Sigma_j & \cdots & \Sigma_j \\ \vdots & \vdots & & \vdots \\ \Sigma_j & \Sigma_j & \cdots & \Sigma_k + \Sigma_j \end{bmatrix}.$$

A similar problem when $\Sigma_0$ is known (and the problem is to find for which $i$, $\boldsymbol{\mu}_0=\boldsymbol{\mu}_i$) can be handled along the same lines; see Ellison (1962).

When the $\Sigma_i$'s are unknown, the above procedure is suggested with the $\Sigma_i$'s replaced by their estimates.

## 8.9 A Sequential Approach to Classification

From the results on the testing of hypotheses [see Dantzig (1940)], it is well known that there does not exist a fixed-sample procedure that can control both the errors at some specified level. To achieve this one needs to proceed sequentially. In this section we describe such a procedure for $k=2$ and when the populations are normally distributed. The sampling is carried out sequentially from all the three populations $\pi_0, \pi_1, \pi_2$. While in the following, for the sake of convenience of presentation, equal numbers of observations are taken sequentially, the procedure can be carried out in any proportion $1:b_1:b_2$ where $b_1$ and $b_2$ are specified. Now we describe the sequential procedure for controlling the two errors of misclassification at a specified level $\alpha$. Since we are sampling from all three populations, the notation will differ somewhat from the previous sections. Let

$$lV_l = \sum_{i=0}^{2} \sum_{j=1}^{n} (\mathbf{x}_{ij}-\bar{\mathbf{x}}_{in})(\mathbf{x}_{ij}-\bar{\mathbf{x}}_{in})', \tag{8.9.1}$$

where

$$l=3(n-1), \qquad L=3(N-1), \qquad n\bar{\mathbf{x}}_{in} = \sum_{j=1}^{n} \mathbf{x}_{ij}, \quad i=0,1,2, \tag{8.9.2}$$

and $\mathbf{x}_{ij}$ denotes the $j$th observation from the $i$th population, $i=0,1,2$. Then the stopping variable $N$ is defined by

$$N = \text{smallest integer } n \ (\geqslant n_0) \text{ such that } n \geqslant \frac{6a^2}{\boldsymbol{\delta}_n' V_l^{-1} \boldsymbol{\delta}_n}, \tag{8.9.3}$$

where $a$ is given by

$$\Phi(a) = 1-\alpha, \tag{8.9.4a}$$

and

$$\boldsymbol{\delta}_n = \bar{\mathbf{x}}_{1n} - \bar{\mathbf{x}}_{2n} \tag{8.9.4b}$$

Here $n_0$ could be chosen as the smallest sample size required for $V_l$ to be positive definitive with probability one.

When the sampling is stopped at $N=n$, classify in $\pi_1$ or $\pi_2$ according as

$$\left[\bar{\mathbf{x}}_{0n} - \tfrac{1}{2}(\bar{\mathbf{x}}_{1n}+\bar{\mathbf{x}}_{2n})\right]' V_l^{-1} \boldsymbol{\delta}_n \gtrless 0. \tag{8.9.5}$$

We shall now show that

$$\lim_{\Delta\to 0} e_1 = \lim_{\Delta\to 0} e_2 \leqslant \alpha, \tag{8.9.6}$$

where, as before, $\Delta^2 = \boldsymbol{\delta}'\Sigma^{-1}\boldsymbol{\delta}$, $\boldsymbol{\delta} = \boldsymbol{\mu}_1 - \boldsymbol{\mu}_2$. We have

$$e_1 = P\left\{\left[\overline{\mathbf{x}}_{0N} - \tfrac{1}{2}(\overline{\mathbf{x}}_{1N} + \overline{\mathbf{x}}_{2N})'\right] V_L^{-1}\boldsymbol{\delta}_N < 0 \middle| \mathbf{x}_0 \in \pi_1\right\}, \tag{8.9.7}$$

where $\boldsymbol{\delta}_N = \overline{\mathbf{x}}_{1N} - \overline{\mathbf{x}}_{2N}$. Since $\overline{\mathbf{x}}_{1N} + \overline{\mathbf{x}}_{2N}$ and $\overline{\mathbf{x}}_{1N} - \overline{\mathbf{x}}_{2N}$ are stochastically independent, we get

$$\begin{aligned} e_1 &= P\left\{\left(\frac{2N}{3}\right)^{1/2} \frac{\left[\overline{\mathbf{x}}_{0N} - \frac{1}{2}(\overline{\mathbf{x}}_{1N} + \overline{\mathbf{x}}_{2N})' - \frac{1}{2}\boldsymbol{\delta}\right]' V_L^{-1}\boldsymbol{\delta}_N}{(\boldsymbol{\delta}_N' V_L^{-1}\Sigma V_L^{-1}\boldsymbol{\delta}_N)^{1/2}} \right. \\ &\qquad \left. < -\left(\frac{N}{6}\right)^{1/2} \frac{\boldsymbol{\delta}' V_L^{-1}\boldsymbol{\delta}_N}{(\boldsymbol{\delta}_N' V_L^{-1}\Sigma V_L^{-1}\boldsymbol{\delta}_N)^{1/2}} \middle| x_0 \in \pi_1 \right\} \\ &= 1 - E\Phi\left[\left(\frac{N}{6}\right)^{1/2} \frac{\boldsymbol{\delta}' V_L^{-1}\boldsymbol{\delta}_N}{(\boldsymbol{\delta}_N' V_L^{-1}\Sigma V_L^{-1}\boldsymbol{\delta}_N)^{1/2}}\right]. \end{aligned} \tag{8.9.8}$$

Since $\Phi(\cdot) \leqslant 1$, $\boldsymbol{\delta}_N \to \boldsymbol{\delta}$ a.s. and $N\boldsymbol{\delta}_n' V_L^{-1}\boldsymbol{\delta}_n / 6a^2 \to 1$ a.s. as $\Delta \to 0$, it follows from the bounded-convergence theorem that

$$\lim_{\Delta\to 0} e_1 = 1 - \Phi(a) = \alpha.$$

Similarly, it can be shown that $\lim_{\Delta\to 0} e_2 = \alpha$. See Srivastava (1973c).

## 8.10 The Problem of Minimum Distance

For the problem of classification one assumes that the individual $\pi_0$ to be classified belongs to one of the several given populations $\pi_1, \pi_2, \ldots, \pi_k$. However, when the external evidence is slight, the classification problem is not only subject to the error due to the misclassification, but also to the error due to the false assumption that $\pi_0$ belongs to one of the several given populations. The best thing would be to test first whether $\pi_0$ belongs to any one of the several given populations. If so, we assign $\pi_0$ to the $\pi_i$ that corresponds to the hypothesis to be accepted at the highest level of significance. If we reject, we estimate the position of the new group relative to the others. Unfortunately, no such test criterion is available. Alternatively we might be interested to find which of the $k$ population is "closest" —in the sense of distance—to the individual to be classified. This raises a natural question: what measure of distance between two populations should be used. For multivariate populations, we shall use the Mahalanobis (1936) generalized squared distance. Thus we are led to the investigation of the following problem. Given $k+1$ populations $\pi_0, \pi_1, \ldots, \pi_k$, to find which of the $k$ populations $\pi_1, \ldots, \pi_k$ is nearest to $\pi_0$. We consider in this section the case when the $\pi_i$'s, $i=0,1,\ldots,k$, are multivariate normal with means $\boldsymbol{\mu}_i$ and common nonsingular covariance matrix $\Sigma$, i.e., $\pi_i : N(\boldsymbol{\mu}_i, \Sigma)$. The following example, given by Cacoullos (1965a), shows clearly the situation in which the above problem of nearest distance makes more sense than the classification approach.

EXAMPLE 8.10.1 A $p$-dimensional observation $\mathbf{x}$ (e.g., the set of scores of a battery of $p$ tests) is made on an individual; this individual is considered as a random observation from a certain category or population. A set of, say, $k$ other populations is available. Each population may be thought of as a representative of a certain profession, and is characterized by a probability distribution of the $p$ measurements. The question is: Which of the $k$ populations does the individual fit best? If we introduce a measure of similarity between two professions, we are led to consider the problem of "nearest" (best-fit) profession for the individual $\mathbf{x}$.

The problem of smallest distance stems from Rao (1954), who suggested intuitively the maximum-likelihood rule. When the mean $\boldsymbol{\mu}_0$ and the common covariance matrix $\Sigma$ are both known, Cacoullos (1965b) proved the admissibility of the likelihood rule in the restricted class of *symmetric* invariant procedures; a decision procedure $\psi=(\psi_1,\ldots,\psi_k)$ is said to be symmetric if

(i) $0 \leqslant \psi_i(\mathbf{T}) \leqslant 1$,

(ii) $\sum_{i=1}^{k} \psi_i(\mathbf{T}) = 1$,

(iii) $\psi_j(T_{\pi 1},\ldots,T_{\pi k}) = \psi_{\pi j}(T_1,\ldots,T_k)$

for all permutations $(1,2,\ldots,k) \to (\pi_1,\ldots,\pi_k)$ and all $j=1,2,\ldots,k$. This imposes the restriction that the sample sizes from all the population must be equal.

In the case when $\boldsymbol{\mu}_0$ and $\Sigma$ are both unknown, it is not yet known whether the likelihood rule is admissible or not. However, when $k=2$, it has been shown by Srivastava (1967c) that the maximum-likelihood rule is admissible in the whole class of procedures, without restriction to symmetric invariant procedures. The maximum-likelihood rule in the equal sample case is: Take the $i$th decision if $i$ is the smallest integer for which $\text{Min}_{1 \leqslant j \leqslant k} (\mathbf{x}_i - \mathbf{x}_0)' V^{-1} (\mathbf{x}_i - \mathbf{x}_0)$ is attained. Here $\mathbf{x}_i$, $i=0,1,\ldots,k$ are independent with distribution $N_p(\boldsymbol{\mu}_i, \alpha\Sigma)$ ($\alpha \equiv N^{-1}$, $i=0,1,2,\ldots,k$), and $V$ is distributed independently of the $\mathbf{x}_i$'s as $W_p(\Sigma, r)$, where $r=(k+1)(N-1)$. For the corresponding problems of ranking and slippage based on distance, see Srivastava and Taneja (1972, 1973).

## 8.11 Remarks

In Section 8.10, we considered the problem of minimum distance. A general approach to this problem has been considered by Hariton (1972) in an unpublished thesis. In this approach he assumes that the three distributions $F_0$, $F_1$, and $F_2$ (of $\pi_0$, $\pi_1$, and $\pi_2$) are related by

$$F_0(x)=pF_1(x)+(1-p)F_2(x), \qquad 0\leqslant p\leqslant 1, \tag{8.11.1}$$

and considers the following problems:

(i) A nonparametric test of the hypothesis that (8.11.1) holds for some $p$, $0\leqslant p\leqslant 1$. He gives the asymptotic distribution of the test criterion.
(ii) Assuming that Eq. (8.11.1) holds for some $p$, $0\leqslant p\leqslant 1$, he constructs a rank-type consistent estimator for $p$, and considers its asymptotic distributions.
(iii) Assuming that Eq. (8.11.1) holds for either $p=1$ or $p=0$, one must choose one of the two values of $p$. This is equivalent to classifying $\pi_0=\pi_1$ or $\pi_0=\pi_2$. A rank-type classification criterion is given, along with its distribution.
(iv) Assuming that Eq. (8.11.1) holds for some $p$, $0\leqslant p\leqslant 1$, and further assuming that $\pi_i$ is distributed as $N_r(\boldsymbol{\mu}_i,\Sigma)$, $i=1,2$, where $\boldsymbol{\mu}_1$, $\boldsymbol{\mu}_2$, and $\Sigma$ are unknown, he constructs an estimator of $p$, along with its asymptotic distribution.

In another unpublished thesis Leung (1977) considers the classification problem with covariants which includes the growth curve model and also the case when the two populations are not independent.

Finally, in many practical situations, it may be desirable to do step-wise classification. For details, see Srivastava (1973a).

## Problems

**8.1** Let $x_1,\ldots,x_{N_1},x_{N_1+1},\ldots,x_{N_1+N_2}$ be independently normally distributed, where the first $N_1$ observations are from $\pi_1:N(\mu_1,\sigma^2)$ and the last $N_2$ observations are from $\pi_2:N(\mu_2,\sigma^2)$, with $\mu_1$, $\mu_2$, and $\sigma^2$ unknown. Suppose an observation $x_0$ is from $\pi_0:N(\mu_0,\sigma^2)$, where $\mu_0=\mu_1$ or $\mu_0=\mu_2$. Write down the Anderson classification statistic corresponding to Eq. (8.2.2) with $C=0$, and obtain the errors of misclassifications.

**8.2** Let $\pi_0$, $\pi_1$, and $\pi_2$ denote the three normal populations $N_p(0,\sigma_i^2 I)$, $i=0,1,2$. Suppose $\sigma_1^2$ and $\sigma_2^2$ are known, and it is known that $\sigma_0^2=\sigma_1^2$ or $\sigma_0^2=\sigma_2^2$. On the basis of measurements $\mathbf{x}_0$ on an individual from $\pi_0$, we wish to classify it in $\pi_i$, $i=1,2$. Derive the likelihood-ratio test and errors of misclassification.

**8.3** Suppose in the above problem we have $N_p(\mathbf{0},\Sigma_i)$, where $\Sigma_1=(1-\rho)I_p+\rho\mathbf{ee}'$ and $\Sigma_2=\sigma^2[(1-\rho)I_p+\rho\mathbf{ee}']$. It is known that $\Sigma_0=\Sigma_1$ or $\Sigma_0=\Sigma_2$. Derive the likelihood-ratio test and errors of misclassification.

**8.4** In the linear discriminant function (8.2.10) let $\mathbf{a}=\Sigma^{-1}\boldsymbol{\delta}=\Sigma^{-1}(\boldsymbol{\mu}_1-\boldsymbol{\mu}_2)$. Let $\mathbf{a}'=(\mathbf{a}_1',\mathbf{a}_2')$, where $\mathbf{a}_1$ is an $r$ vector and $\mathbf{a}_2$ is an $s$ vector, with $r+s=p$. On the basis of $N_1$ observations $\mathbf{x}_1,\ldots,\mathbf{x}_{N_1}$ from $\pi_1\sim N(\boldsymbol{\mu}_1,\Sigma)$ and $N_2$ observations $\mathbf{x}_{N_1+1},\ldots,\mathbf{x}_{N_1+N_2}$ from $\pi_2\sim N(\boldsymbol{\mu}_2,\Sigma)$, derive the likelihood-ratio test for $H:\mathbf{a}_2=0$ against $A:\mathbf{a}_2\neq 0$. Give its distribution.

**8.5** Let $P_1(\Delta)$ denote the probability of correctly classifying an individual from $\pi_0$ into $\pi_1$ for the classification rule (8.2.2) with $C=0$ [that is, $P_1(\Delta)\equiv\Phi(\frac{1}{2}\Delta)$]. Let

$$\hat{P}_1(\hat{\Delta})\equiv\Phi\left(\tfrac{1}{2}\hat{\Delta}\right),\qquad \text{where}\quad \hat{\Delta}=[(\bar{\mathbf{x}}-\bar{\mathbf{y}})'\Sigma^{-1}(\bar{\mathbf{x}}-\bar{\mathbf{y}})]^{1/2},$$

and let $\hat{P}_1(\hat{\Delta})$ be the probability of correctly classifying an individual from $\pi_0$ into $\pi_1$ for the rule (8.3.3). Also, let

$$\begin{aligned} c_1(\hat{\Delta}) &\equiv \text{the proportion of the observation from } \pi_1\\ &= \text{correctly classified by the rule (8.3.3).}\end{aligned}$$

Then show that:

(i) $P_1(\hat{\Delta})<P_1(\Delta)<E[c_1(\hat{\Delta})]$ when $N_1=N_2$.

(ii) $E[\hat{P}_1(\hat{\Delta})<E[c_1(\hat{\Delta})]$.

(iii) $P_1(\hat{\Delta})<E[c_1(\hat{\Delta})]$ for $p=1$.

**8.6** (*Monotone likelihood ratio*) A real-parameter family of distributions $\{P_\theta:\theta\in\Omega\}$ is said to have a monotone likelihood ratio if there exists a real-valued function $T(x)$ such that for any $\theta<\theta'$ the distributions $P_\theta$ and $P_{\theta'}$ are distinct and the ratio of densities (with respect to $P_\theta$ and $P_{\theta'}$)

$$f(x|\theta')/f(x|\theta)$$

is a nondecreasing function of $T(x)$. Show that the noncentral chi square has a monotone likelihood ratio.

**8.7** Suppose $\Sigma_1=\gamma\Sigma_2$, $\gamma$ known, in the classification rule (8.7.1). Evaluate $e_1$.

**8.8** (Continuation) When the population means $\boldsymbol{\mu}_1$ and $\boldsymbol{\mu}_2$ are unknown, suppose one uses the classification rule (8.7.1) with the means $\boldsymbol{\mu}_1$ and $\boldsymbol{\mu}_2$ replaced by their estimates $\bar{\mathbf{x}}_1$ and $\bar{\mathbf{x}}_2$ respectively. Obtain the errors of misclassification when $\Sigma_1=\gamma\Sigma_2$, $\gamma$ known. Give an asymptotic expansion of $e_1$.

# 9

# Principal-Component Analysis

## 9.1 Introduction

When a large number of measurements are available, it is natural to inquire whether they could be replaced by a smaller number of measurements or of functions of them, without much loss of information. The method of principal components, essentially as put forward by Hotelling (1933), is suggested. It is a method of transforming the vector **x** into a set of orthogonal components. This, in effect, amounts to breaking down the covariance matrix of **x** into characteristic roots and vectors. The orthogonal components are the linear combinations of **x** with these characteristic vectors, and the characteristic roots correspond to the variances of these components. These new components have the property that the sum of their variances is the same as the sum of the variances of the components of **x**. Thus, if only a few components are to be used to summarize the data, then the ones with smaller variances (characteristic roots) can be ignored. Indeed, it is the aim of the principal-component analysis to explain or account for the covariance matrix with as small number of variates or factors as possible. *However, the method is useful only if all the variates $x_i$ of the vector $\mathbf{x}=(x_1,\ldots,x_p)'$ are measured in the same units.* (See Subsection 9.2.1).

In many exploratory studies where the measurements are in the same units, the principal-component analysis could be used very effectively and profitably to reduce the large and massive data before subjecting to close scrutiny. Since the sum of the variances of the original variables is the same as the sum of the variances of the new variables (principal components), it may be enough to study a smaller number of principal components that account for most of the total variability.

The selection of the principal components with the largest variances, however, may have the disadvantage that all of the original variables (or almost all of them) may enter into some of the selected principal components with nonzero weights.

The related topic of factor analysis will not be considered in this book; the reader is referred to an excellent treatment by Lawley and Maxwell (1963).

## 9.2 Definition

Let **x** be a $p$-component vector with covariance matrix $\Sigma$. Since we will be interested only in variances and covariances, we may assume without any loss of generality that **x** has zero mean vector. Let $\lambda_1 \geq \lambda_2 \geq \cdots \geq \lambda_p$ be the ordered characteristic roots of $\Sigma$, that is, the roots of

(9.2.1) $$|\Sigma - \lambda I| = 0.$$

Let $\gamma_1, \gamma_2, \ldots, \gamma_p$ be the normalized characteristic vectors of $\Sigma$, that is, the vectors satisfying

(9.2.2) $$\Sigma\gamma_i = \lambda_i \gamma_i,$$

(9.2.3) $$\gamma_i'\gamma_j = \delta_{ij},$$

where $\delta_{ij}$ is the Kronecker delta. If several characteristic roots of $\Sigma$ are equal, the corresponding vectors have some indeterminacy, but they can be taken so that Eq. (9.2.3) holds.

Let

(9.2.4) $$y_i = \gamma_i'\mathbf{x}, \qquad i = 1, 2, \ldots, p,$$

and

(9.2.5) $$\mathbf{y} = (y_1, y_2, \ldots, y_p)'.$$

These new sets of $y_i$'s have the property that they are uncorrelated, since $\gamma_i'\gamma_j = 0$ for $i \neq j$ and the variance of $y_i$ is given by $\lambda_i$. Thus the variate $y_1$ is the one with the largest variance $\lambda_1$ and is uncorrelated with all the remaining variables $y_2, \ldots, y_p$. Similarly, the variate $y_2$ is the one with the second largest variance $y_2$, and is uncorrelated with $y_1$ and also with all the remaining variables $y_3, \ldots, y_p$, and so on. Since $\Sigma_{i=1}^{p} \lambda_i = \operatorname{tr} \Sigma$, the sum of the variances of the original variables is the same as the sum of the variances of the new. Thus, the variables with smaller variances could be ignored without effecting the total variance, thereby reducing the number of variables from $p$ to (say) $q \leq p$. When $q < p$, there will be loss of information in replacing **x** by the first $q$ principal components $(\gamma_1'\mathbf{x}, \ldots, \gamma_q'\mathbf{x})$.

Naturally $q$ should be so chosen that this loss ($\operatorname{tr}\Sigma$ minus the sum of the variances of the first $q$ principal components) is minimum. This raises a natural question: Does there exist any other set of variables—say $\mathbf{z}=T'\mathbf{x}$, where $T:p\times q$—for which the above loss is smaller than for the first $q$ principal components? The answer is clearly no, since

$$\operatorname{tr}\Sigma-\operatorname{tr}T'\Sigma T \tag{9.2.6}$$

is minimized at

$$T=(\gamma_1,\gamma_2,\ldots,\gamma_q). \tag{9.2.7}$$

Not only do the first $q$ principal components minimize the loss (9.2.6), but also they minimize other loss functions, such as (Rao 1964)

$$\operatorname{tr}\left[\Sigma-\Sigma T(T'\Sigma T)^{-1}T'\Sigma\right], \tag{9.2.8}$$

or

$$\|\Sigma-\Sigma T(T'\Sigma T)^{-1}T'\Sigma\|, \tag{9.2.9}$$

where the symbol $\|A\|$ denotes the Euclidean norm, which is the square root of the sum of the squares of the elements of $A$. These loss functions arise when we wish to determine to what extent we can predict $\mathbf{x}$ knowing $\mathbf{z}$. The predictive efficiency of $\mathbf{z}$ for $\mathbf{x}$ depends on the residual covariance matrix of $\mathbf{x}$ after subtracting its best linear predictor in terms of $\mathbf{z}$. Since

$$\operatorname{Cov}(\mathbf{x},\mathbf{z})=\begin{pmatrix}\Sigma & \Sigma T\\ T'\Sigma & T'\Sigma T\end{pmatrix}, \tag{9.2.10}$$

the residual covariance matrix is

$$\Sigma-\Sigma T(T'\Sigma T)^{-1}T'\Sigma. \tag{9.2.11}$$

It is left as an exercise for the reader to show that Eqs. (9.2.8) and (9.2.9) are minimized at

$$T=(\gamma_1,\ldots,\gamma_q). \tag{9.2.12}$$

### *9.2.1 Effects of Units of Measurement*

Now we examine the effects of units of measurement. If the variates $x_i$ are not measured in the same units, then a change in the scale of measurements of some or all of the variates results in the covariance matrix being multiplied on both sides by a diagonal matrix. The effect of this on the characteristic roots and vectors is difficult to assess. However, it is clear that the roles of the important and unimportant variables might be reversed; those with smaller variance might become the ones with larger variances. Thus, the principal-component analysis should be used only when all the variables are measured in the same units or at least in comparable units.

A way out of this difficulty is to consider the principal components corresponding to the correlation matrix. However, the distribution problems then present difficulties.

## 9.3 Maximum-Likelihood Estimates

In this section we assume that the random vector $\mathbf{x}$ is normally distributed and a random sample of size $N$ ($>p$) is given. That is, the observation matrix $X \sim N_{p,N}(\boldsymbol{\mu}\mathbf{e}', \Sigma, I)$. Then from Chapter 2, it follows that the maximum-likelihood estimates of $\Sigma$ is $(n/N)S$, where

$$(9.3.1)\qquad S = n^{-1}V \equiv n^{-1}(XX' - N\bar{\mathbf{x}}\bar{\mathbf{x}}'), \qquad \bar{\mathbf{x}} = N^{-1}X\mathbf{e}, \quad n = N-1,$$

and the maximum-likelihood estimates of the $\lambda_i$'s, the ordered (assumed distinct) roots of $\Sigma$, are $(n/N)l_i$, where the $l_i$'s are the ordered characteristic roots of $S$:

$$(9.3.2)\qquad |S - lI| = 0.$$

It may be noted that the $l_i$'s are distinct with probability one, since $N > p$. Similarly, if $\mathbf{c}_1, \mathbf{c}_2, \ldots, \mathbf{c}_p$ are the normalized characteristic vectors of $S$ with $c_{ii} > 0$, $i = 1, 2, \ldots, p$, that is, the unique vectors satisfying

$$(9.3.3)\qquad S\mathbf{c}_i = l_i \mathbf{c}_i,$$

$$(9.3.4)\qquad \mathbf{c}_i'\mathbf{c}_j = \delta_{ij},$$

where $\delta_{ij}$ is the Kronecker delta, and if $\boldsymbol{\gamma}_1, \ldots, \boldsymbol{\gamma}_p$ are the normalized characteristic vectors of $\Sigma$ with $\gamma_{ii} > 0$, $i = 1, 2, \ldots, p$, where the roots $\lambda_i$ of $\Sigma$ are assumed distinct, then the $\mathbf{c}_i$'s are the maximum-likelihood estimates of the $\boldsymbol{\gamma}_i$'s. Thus, we get

**Theorem 9.3.1** *Let $\boldsymbol{\gamma}_1, \ldots, \boldsymbol{\gamma}_p$ be the characteristic vectors of $\Sigma$ corresponding to the distinct and ordered characteristic roots $\lambda_1, \ldots, \lambda_p$. Similarly, let $\mathbf{c}_1, \ldots, \mathbf{c}_p$ and $l_1, \ldots, l_p$ be the characteristic vectors and the ordered roots of $S$. Then the maximum likelihood estimates of the $\boldsymbol{\gamma}_i$'s are $\mathbf{c}_i$, and of the $\lambda_i$'s are $(n/N)l_i$, $i = 1, \ldots, p$.*

Next, we consider the case when the roots of $\Sigma$ are $\lambda_1, \ldots, \lambda_r$ with multiplicities $q_1, q_2, \ldots, q_r$ respectively. In this case we write

$$(9.3.5)\qquad \Sigma = \Gamma D_\lambda \Gamma' \equiv (\Gamma_1, \Gamma_2, \ldots, \Gamma_r)\begin{bmatrix} \lambda_1 I_{q_1} & & 0 \\ & \ddots & \\ 0 & & \lambda_r I_{q_r} \end{bmatrix}\begin{bmatrix} \Gamma_1' \\ \Gamma_2' \\ \vdots \\ \Gamma_r' \end{bmatrix}$$

where $\Gamma_i$ is $p\times q_i$. Similarly, we can write

$$C\equiv(C_1,C_2,\ldots,C_r) \tag{9.3.6}$$

such that

$$S=CD_lC', \tag{9.3.7}$$

where $c_{jj}>0$, $j=1,2,\ldots,p$, that is, $C$ is unique, and $l_1>l_2>\cdots>l_p>0$. However, the $\Gamma_i$'s, $i=1,2,\ldots,r$, are not unique, in the sense that $\Gamma_i$ multiplied by any $q_i\times q_i$ orthogonal matrix will also be a characteristic matrix. Thus the maximum-likelihood estimate of $\Gamma_i$ will not be just $C_i$, but $C_i$ multiplied on the right by any $q_i\times q_i$ orthogonal matrix such that $c_{jj}$ is still greater than zero, $j=1,2,\ldots,p$. The logarithm of the likelihood function after maximization with respect to $\boldsymbol{\mu}$ (except for a constant) is

$$\begin{aligned}
&-\tfrac{1}{2}N\ln|\Sigma|-\tfrac{1}{2}\operatorname{tr}\Sigma^{-1}nS\\
&\quad=-\tfrac{1}{2}N\ln|\Gamma D_\lambda\Gamma'|-\tfrac{1}{2}n\operatorname{tr}D_\lambda^{-1}\Gamma S\Gamma'\\
&\quad=\left(-\tfrac{1}{2}N\ln\prod_{i=1}^{r}\lambda_i^{q_i}\right)-\left(\tfrac{1}{2}n\operatorname{tr}\sum_{i=1}^{r}\lambda_i^{-1}\Gamma_i'S\Gamma_i\right)\\
&\quad\leqslant\left(-\tfrac{1}{2}N\ln\prod_{i=1}^{r}\lambda_i^{q_i}\right)-\tfrac{1}{2}n\left(\lambda_1^{-1}\sum_{i=1}^{q_1}l_i+\lambda_2^{-1}\sum_{i=q_1+1}^{q_{(2)}}l_i+\cdots+\lambda_r^{-1}\sum_{q_{(r-1)}+1}^{q_{(r)}}l_i\right),
\end{aligned}$$

with $q_{(i)}$ defined as in Theorem 9.3.2 below, since by Theorem 1.10.2(ii), $\operatorname{tr}AB^{-1}\geqslant\sum_{j=1}^{p}(\alpha_j/\beta_j)$, where $\alpha_1\geqslant\cdots\geqslant\alpha_p$ are the roots of $A$ and $\beta_1\geqslant\cdots\geqslant\beta_p$ are the roots of $B$. Hence, we get

**Theorem 9.3.2** *If the roots of the population covariance matrix $\Sigma$ are $\lambda_1,\ldots,\lambda_r$ with multiplicities $q_1,\ldots,q_r$, respectively, then the maximum-likelihood estimate of $\lambda_k$ is*

$$\hat{\lambda}_k=\frac{1}{q_k}\frac{n}{N}\sum_{j=q_{(k-1)}+1}^{q_{(k)}}l_j,\qquad k=1,2,\ldots,r, \tag{9.3.8}$$

*where $q_{(k)}=q_{(k-1)}+q_k$ with $q_{(0)}=0$ and $q_{(r)}=p$, and the maximum-likelihood estimate $\hat{\Gamma}_k$ of $\Gamma_k$ is $C_k$ multiplied on the right by any $q_k\times q_k$ orthogonal matrix such that $\hat{\gamma}_{jj}>0$.*

## 9.4 Asymptotic Distribution of the Roots and Vectors of a Wishart Matrix

Let $l_1>\cdots>l_p$ and $\mathbf{c}_1,\ldots,\mathbf{c}_p$ be respectively the characteristic roots and the *unique* characteristic vectors (that is, one element of each $\mathbf{c}_i$ is greater than zero) of $S$ [that is, satisfying Eqs. (9.3.3) and (9.3.4), where $S=n^{-1}V$

and $V \sim W_p(\Sigma, n)$, $n \geqslant p, \Sigma > 0$]. In Chapter 3, we derived the exact joint distribution of $nl_1, nl_2, \ldots, nl_p$ (also of $\mathbf{c}_1, \ldots, \mathbf{c}_p$) when $\Sigma = I$. However, the marginal distribution of $l_i$ is not easy to obtain except for the largest or the smallest root (see Chapter 7). Even the asymptotic distribution of $l_i$ presents difficulty unless $\Sigma \neq I$. While the exact distribution of these roots and vectors when $\Sigma \neq I$ is beyond the scope of this book, we consider in this section the asymptotic distribution of $l_1, \ldots, l_p$ and $\mathbf{c}_1, \ldots, \mathbf{c}_p$ when $\Sigma \neq I$. In the case of all the roots of $\Sigma$ being different, Girshick (1939) has given the asymptotic variances and covariances of $l_1, \ldots, l_p$ and of $l_1^{1/2}\mathbf{c}_1, \ldots, l_p^{1/2}\mathbf{c}_p$. In the case of the smallest $p-q$ roots of $\Sigma$ being equal and the others different, Lawley (1953) has given the asymptotic variances and covariances of $l_1^{1/2}\mathbf{c}_1, \ldots, l_q^{1/2}\mathbf{c}_q$ (the sample vectors corresponding to the different roots of $\Sigma$), but he did not argue asymptotic normality. Anderson (1963) considered this problem in general and obtained the distribution of $l_1, \ldots, l_p, \mathbf{c}_1, \ldots, \mathbf{c}_p$ when the roots of $\Sigma$ have any multiplicities. As a special case, the vectors Lawley treats are shown by him to be asymptotically normal (a result not following directly and easily from the usual theory of maximum-likelihood estimators).

These results are useful not only in the study of principal components but also for inferences on the roots of $\Sigma$, which arise in several situations. Bartlett (1950, 1951a, 1951b, 1954) has considered testing the hypothesis that $p-q$ roots of $\Sigma$ are equal. These and other related problems will be considered in Section 9.5.

### *9.4.1 Asymptotic Distribution up to $O(n^{-1})$*

Let $\lambda_1 > \cdots > \lambda_r > 0$ be the distinct characteristic roots of $\Sigma$, with the multiplicities $q_1, \ldots, q_r$ respectively as in Eq. (9.3.5). Let, as in Eqs. (9.3.5)–(9.3.7),

$$(9.4.1) \quad \Gamma'\Sigma\Gamma = D_\lambda, \qquad \Gamma' S \Gamma = E D_l E', \quad \text{and} \quad D_h = n^{1/2}(D_l - D_\lambda),$$

where $E' = C'\Gamma$, $S = CD_lC'$, $D_\lambda = \operatorname{diag}(\lambda_1 I_{q_1}, \ldots, \lambda_r I_{q_r})$, $D_l = \operatorname{diag}(L_1, \ldots, L_r)$, $L_j = \operatorname{diag}(l_{q_{(j-1)}+1}, \ldots, l_{q_{(j)}})$, $q_{(j)} = \sum_{i=1}^{j} q_i$, $q_{(0)} = 0$, and $D_h = \operatorname{diag}(H_1, H_2, \ldots, H_r)$ with $H_i = n^{1/2}(L_i - \lambda_i I_{q_i})$, $i = 1, 2, \ldots, r$. Let $E = (E_{ij})$ be partitioned in such a way that $E_{ij}$ is a $q_i \times q_j$ matrix, and let $F_{ij} = n^{1/2} E_{ij}$ $(i<j)$, $i,j = 1, 2, \ldots, r$. Then we have

**Theorem 9.4.1** *Let $nS \sim W_p(\Sigma, n)$. Then, in the above notation, $H_i = n^{1/2}(L_i - \lambda_i I_{q_i})$, $E_{ii}$, and $F_{ij} = n^{1/2}E_{ij}$ $(i<j)$, $i,j = 1, 2, \ldots, r$, are asymptotically independent in distribution. The asymptotic distributions of $H_i$, $E_{ii}$, and $F_{ij}$ are respectively given by $f_i^{(1)}(H_i)$, $[dE_{ii}]$, and $g_{ij}(F_{ij})$, where if $H_i =$*

$\operatorname{diag}(x_1, x_2, \ldots, x_{q_i})$ *and* $x_1 > x_2 > \cdots > x_{q_i}$, *then*

$$(9.4.2)\quad f_i^{(1)}(x_1,\ldots,x_{q_i}) = \left\{2^{\frac{1}{4}q_i(q_i+3)}\lambda_i^{\frac{1}{2}q_i(q_i+1)}\Gamma_{q_i}\left(\tfrac{1}{2}q_i\right)\right\}^{-1}$$

$$\times \pi^{\frac{1}{4}q_i(q_i-1)}\left[\prod_{j=1}^{q_i-1}\prod_{j'=j+1}^{q_i}(x_j - x_{j'})\right]\exp\left(-\sum_{j=1}^{q_i}\frac{x_j^2}{4\lambda_i^2}\right),$$

$$(9.4.3)\quad [dE_{ii}] = \pi^{-\frac{1}{2}q_i^2}\left\{\Gamma_{q_i}\left(\tfrac{1}{2}q_i\right)\right\}f_{q_i}(E_{ii})$$

$$= \text{unit invariant Haar measure,}$$

*and*

$$(9.4.4)\quad g_{ij}(F_{ij}) = \left(\frac{(\lambda_i-\lambda_j)^2}{2\pi\lambda_i\lambda_j}\right)^{q_iq_j/2}\exp\left(-\frac{(\lambda_i-\lambda_j)^2\operatorname{tr}F_{ij}F_{ij}'}{2\lambda_i\lambda_j}\right).$$

*Here* $f_q(A) = J(A'dA \to dA)$ *is defined in Theorem* 1.11.5.

PROOF. Since $nS \sim W_p(\Sigma, n)$,

$$T = \Gamma' S \Gamma \sim W_p\left(\frac{1}{n}D_\lambda, n\right).$$

Using Theorem 1.11.5 for the Jacobian of the transformation $T = ED_lE'$ and then transforming $D_l$ to $D_h$ by $D_h = n^{1/2}(D_l - D_\lambda)$, the joint density of $h_1 > h_2 > \cdots > h_p$ and $E$ is given by

$$(9.4.5)\quad \left(\tfrac{1}{2}n\right)^{\frac{1}{2}pn}\left[n^{\frac{1}{2}p}\Gamma_p\left(\tfrac{1}{2}n\right)\right]^{-1}|D_\lambda|^{\frac{1}{2}(p+1)}|I + n^{-1/2}D_\lambda^{-1}D_h|^{\frac{1}{2}(n-p-1)}$$

$$\times\prod_{i=1}^{p-1}\prod_{j=i+1}^{p}\left[n^{-1/2}(h_i - h_j) + \delta_i - \delta_j\right]\operatorname{etr}\left(-\tfrac{1}{2}nD_\lambda^{-1}EDE'\right)f_p(E),$$

where $f_p(E) = J(E'dE \to dE)$, and $\delta_{q_{(i-1)}+j} = \lambda_i$ for $j = 1, 2, \ldots, q_i$ and $i = 1, 2, \ldots, r$. Let

$$E_{ij} = n^{-1/2}F_{ij} \qquad \text{for} \quad i < j,$$

where $E = (E_{ij})$, and $E_{ij}$ $(i, j = 1, 2, \ldots, r)$ are $q_i \times q_j$ matrices. We can fix the random elements in $E$ so that $E_{ij}$ for $i < j$ has random elements, and the Jacobian of this transformation is

$$(9.4.6)\quad J(E_{ij}, i<j \to F_{ij}, i<j) = n^{\frac{1}{4}(\Sigma_{j=1}^r q_j^2 - p^2)},$$

and using the expression for $f_p(E)$ (see Theorem 1.11.5), it is easy to see that

$$(9.4.7)\quad f_p(E) = \prod_{i=1}^{r} f_{q_i}(E_{ii}) + O(n^{-1/2}).$$

Further, we can write

$$E = E_1 + n^{-1/2}F,$$

where $E_1 = \operatorname{diag}(E_{11}, E_{22}, \ldots, E_{rr})$ and $F = (F_{ij})$, with $F_{ii} = 0$ and $E_{ij} = n^{-1/2}F_{ij}$ for $i \neq j$. Now

$$\begin{aligned} ED_lE' &= \left(E_1 + n^{-1/2}F\right)\left(n^{-1/2}D_h + D_\lambda\right)\left(E_1' + n^{-1/2}F'\right) \\ &= E_1D_\lambda E_1' + n^{-1/2}(E_1D_hE_1' + FD_\lambda E_1' + E_1D_\lambda F') \\ &\quad + n^{-1}(FD_hE_1' + E_1D_hF' + FD_\lambda F') + n^{-3/2}FD_hF'. \end{aligned}$$

We note that $D_\lambda E_1 = E_1 D_\lambda$. Hence,

$$\begin{aligned} (9.4.8) \qquad & n \operatorname{tr} D_\lambda^{-1}ED_lE' \\ &= n \operatorname{tr} E_1E_1' + n^{1/2}\operatorname{tr}\left(E_1D_\lambda^{-1}D_hE_1' + FE_1' + E_1F'\right) \\ &\quad + \operatorname{tr}\left(FD_\lambda^{-1}D_hE_1' + E_1D_\lambda^{-1}D_hF' + D_\lambda^{-1}FD_\lambda F'\right) + O(n^{-1/2}). \end{aligned}$$

Using $I = E_1'E_1 + n^{-1/2}(E_1'F + F'E_1) + n^{-1}F'F$ in Eq. (9.4.8), we get

$$\begin{aligned} (9.4.9) \qquad & n \operatorname{tr} D_\lambda^{-1}ED_lE' \\ &= np + n^{1/2}\operatorname{tr} D_\lambda^{-1}D_h + \operatorname{tr}\left(D_\lambda^{-1}FD_\lambda F' - FF'\right) + O(n^{-1/2}). \end{aligned}$$

By determinant expansion,

$$\begin{aligned} (9.4.10) \qquad & |I + n^{-1/2}D_\lambda^{-1}D_h|^{\frac{1}{2}(n-p-1)} \\ &= \operatorname{etr}\left[\tfrac{1}{2}\sqrt{n}\, D_\lambda^{-1}D_h - \tfrac{1}{4}D_\lambda^{-2}D_h^2 + O(n^{-1/2})\right] \end{aligned}$$

and by Stirling's expansion,

$$(9.4.11) \quad \Gamma_p\left(\tfrac{1}{2}n\right) \simeq \pi^{\frac{1}{4}p(p-1)}(2\pi)^{\frac{1}{2}p}\left(\tfrac{1}{2}n\right)^{\frac{1}{2}pn - \frac{1}{4}p(p+1)}\exp\left[-\tfrac{1}{2}np + O(n^{-1})\right].$$

Lastly, we have

$$\begin{aligned} (9.4.12) \qquad & \prod_{i=1}^{p-1}\prod_{j=i+1}^{p}\left[n^{-1/2}(h_i - h_j) + \delta_i - \delta_j\right] \\ &= n^{-\frac{1}{4}\Sigma_{j=1}^{r}q_j(q_j-1)}\left\{\left[\prod_{i=1}^{r}\prod_{j=1}^{q_i-1}\prod_{j'=i+1}^{q_i}\left(h_{q_{(i-1)}+j} - h_{q_{(i-1)}+j'}\right)\right]\right. \\ &\quad \left.\times\left[\prod_{i=1}^{r-1}\prod_{j=i+1}^{r}(\lambda_i - \lambda_j)^{q_iq_j}\right] + O(n^{-1/2})\right\}. \end{aligned}$$

Using Eqs. (9.4.6), (9.4.7), (9.4.9), (9.4.10), (9.4.11), and (9.4.12) in Eq.

(9.4.4), we can write the joint density function of $E_{ii}$, $F_{ij}$ $(i<j)$, and $h_1>h_2>\cdots>h_p$ as

$$(9.4.13)\qquad (2\pi)^{-\frac{1}{4}p(p+1)}2^{-\frac{1}{2}p}|D_\lambda|^{\frac{1}{2}(p+1)}\left(\prod_{i=1}^{r-1}\prod_{j=i+1}^{r}(\lambda_i-\lambda_j)^{q_iq_j}\right)$$
$$\times\left(\prod_{i=1}^{r}\prod_{j=1}^{q_i-1}\prod_{j'=j+1}^{q_i}\left(h_{q_{(i-1)}+j}-h_{q_{(i-1)}+j'}\right)\right)$$
$$\operatorname{etr}\left[-\tfrac{1}{4}D_\lambda^{-2}D_h^2-\tfrac{1}{2}\left(D_\lambda^{-1}FD_\lambda F'-FF'\right)\right]$$
$$\times\prod_{i=1}^{r}f_{q_i}(E_{ii})+O(n^{-1/2}).$$

Now, we notice that

$$E_{ii}E_{ii}'=I_{q_i}+n^{-1}\sum_{\substack{j=1\\(j\neq r)}}^{r}F_{ij}F_{ij}'$$

and for $i<j$,

$$E_{ii}F_{ji}'+F_{ij}E_{jj}'+n^{-1/2}\sum_{\substack{k=1\\(k\neq i,j)}}^{p}F_{ik}F_{jk}'=0.$$

Hence the $E_{ii}$'s are asymptotically orthogonal and $E_{ii}F_{ji}'+F_{ij}E_{jj}'\simeq 0$. With this notation, the asymptotic joint distribution of $E_{ii}$ $(i=1,2,\ldots,r)$, $F_{ij}$ $(i<j=1,2,\ldots,r)$ and $h_1>h_2>\cdots>h_p$ is

$$(9.4.14)\qquad \left(\prod_{i=1}^{r}f_i^{(1)}(h_{q_{(i-1)}+1},\ldots,h_{q_{(i)}})\right)\left(\prod_{i=1}^{r-1}\prod_{j=i+1}^{r}g_{ij}(F_{ij})\right)\prod_{i=1}^{r}[dE_{ii}]$$

where $f_i^{(1)}$, $[dE_{ii}]$, and $g_{ij}(F_{ij})$ are defined in Eqs. (9.4.2)–(9.4.4) respectively. This proves Theorem 9.4.1. □

The above theorem was established by T. W. Anderson (1963) by a different approach.

Let us consider the results of Theorem 9.4.1 in terms of the original coordinates. As before, let $C=(C_1,C_2,\ldots,C_r)$ and $\Gamma=(\Gamma_1,\ldots,\Gamma_r)$. Then, since $C=\Gamma E$, we have

$$(9.4.15)\qquad (C_1,C_2,\ldots,C_r)=(\Gamma_1,\Gamma_2,\ldots,\Gamma_r)(E_{ij})$$
$$=\left(\sum_{i=1}^{r}\Gamma_iE_{i1},\ \sum_{i=1}^{r}\Gamma_iE_{i2},\ldots,\ \sum_{i=1}^{r}\Gamma_iE_{ir}\right).$$

Thus

$$(9.4.16)\qquad C_k=\Gamma_kE_{kk}+n^{-1/2}\sum_{\substack{i\neq k\\i=1}}^{r}\Gamma_iF_{ik},\qquad k=1,\ldots,r.$$

Hence, if $q_1 = 1$, then $C_1$ is a vector, say $\mathbf{c}_1$, the 1st column of $C$, and $E_{11} = e_{11}$. Hence, from Theorem 9.4.1, $n^{1/2}(e_{11}^2 - 1)$ converges stochastically to zero. Since at least one element of the vector $\mathbf{c}_1$ is greater than zero, $e_{11}$ converges stochastically to one, and from Eq. (9.4.16)

$$n^{1/2}(\mathbf{c}_1 - \boldsymbol{\gamma}_1) = \sum_{i=2}^{r} \Gamma_i F_{i1}, \tag{9.4.17}$$

which converges to singular normal distribution with mean zero and covariance matrix

$$\sum_{i=2}^{r} \frac{\lambda_i \lambda_1}{(\lambda_i - \lambda_1)^2} \Gamma_i \Gamma_i'. \tag{9.4.18}$$

Thus, we get

**Corollary 9.4.1** *Let the jth column of* $\Gamma$, *say* $\boldsymbol{\gamma}_j$, *correspond to the jth root of* $\Sigma$ *of multiplicity* 1. *Then* $n^{-1/2}(\mathbf{c}_j - \boldsymbol{\gamma}_j)$ *is asymptotically normally distributed with mean* 0 *and covariance matrix*

$$\sum_{\substack{i \neq j \\ i=1}}^{r} \frac{\lambda_i \lambda_j}{(\lambda_i - \lambda_j)^2} \Gamma_i \Gamma_i'.$$

Thus if all the roots of $\Sigma$ are simple (of multiplicity one), then $n^{1/2}(\mathbf{c}_j - \boldsymbol{\gamma}_j)$, $j = 1, 2, \ldots, p$, are asymptotically normally distributed with mean zero and covariance

$$\sum_{\substack{i \neq j \\ i=1}}^{p} \frac{\lambda_i \lambda_j}{(\lambda_i - \lambda_j)^2} \boldsymbol{\gamma}_i \boldsymbol{\gamma}_i', \qquad j = 1, 2, \ldots, p. \tag{9.4.19}$$

The asymptotic covariance between $\mathbf{c}_k$ and $\mathbf{c}_m$ is given by (note that $F_{mk}'$ and $-F_{km}$ have the same limiting distribution)

$$-\frac{\lambda_k \lambda_m}{(\lambda_k - \lambda_m)^2} \boldsymbol{\gamma}_k \boldsymbol{\gamma}_m'. \tag{9.4.20}$$

It may be noted that this expression for the covariance between $\mathbf{c}_k$ and $\mathbf{c}_m$ remains unchanged even if other roots of $\Sigma$ have multiplicity, so long as the $k$th and the $m$th root of $\Sigma$ are simple.

**Corollary 9.4.2** *If the roots of* $\Sigma$ *are simple, then* $n^{1/2}(l_i - \lambda_i)/(2\lambda_i^2)^{1/2}$, $i = 1, 2, \ldots, p$, *is asymptotically independently distributed as* $N(0, 1)$. *Even if other roots of* $\Sigma$ *except the ith one, which is simple, have multiplicity,*

$n^{1/2}(l_i - \lambda_i)/(2\lambda_i^2)^{1/2}$ *is asymptotically distributed as* $N(0,1)$.

From Eq. (9.4.16) and the fact that

$$\sum_{j=1}^{r} \Gamma_j \Gamma_j' = I, \qquad \Gamma_j' \Gamma_i = 0 \quad (j \neq i), \qquad \Gamma_j' \Gamma_j = I, \tag{9.4.21}$$

we get

$$\begin{aligned} n^{1/2}(I - \Gamma_k \Gamma_k') C_k &= n^{1/2} \left( \sum_{j \neq k}^{r} \Gamma_j \Gamma_j' \right) C_k \\ &= n^{1/2} \left( \sum_{j \neq k}^{r} \Gamma_j \Gamma_j' \right) \left( \Gamma_k E_{kk} + n^{-1/2} \sum_{j \neq k}^{r} \Gamma_j F_{jk} \right) \\ &= \sum_{\substack{j \neq k \\ j=1}}^{r} \Gamma_j F_{jk}. \end{aligned} \tag{9.4.22}$$

Hence, we get

**Corollary 9.4.3**

$$n^{1/2}(C_k - \Gamma_k \Gamma_k' C_k) \sim N_{p,q_k} \left[ 0, \sum_{\substack{i=1 \\ i \neq k}}^{r} \lambda_i \lambda_k (\lambda_i - \lambda_k)^{-1} \Gamma_i \Gamma_i', I_{q_k} \right].$$

Next, we consider another normalization for the asymptotic distribution —used, for example, by Lawley. Note that if $C = (c_{g_i})$ and $\Gamma = (\gamma_{g_i})$,

$$\begin{aligned} n^{1/2}\left[ l_i^{1/2} c_{g_i} - \lambda_i^{1/2} \gamma_{g_i} \right] &= n^{1/2} \lambda_i^{1/2} (c_{g_i} - \gamma_{g_i}) + n^{1/2} c_{g_i} \left( l_i^{1/2} - \lambda_i^{1/2} \right) \\ &= n^{1/2} \lambda_i^{1/2} (c_{g_i} - \gamma_{g_i}) + n^{1/2} c_{g_i} \frac{l_i - \lambda_i}{(l_i^{1/2} + \lambda_i^{1/2})}. \end{aligned} \tag{9.4.23}$$

Since $c_{g_i} \to \gamma_{g_i}$ a.s. and $l_i^{1/2} \to \lambda_i^{1/2}$ a.s., this is asymptotically equivalent to

$$n^{1/2} \lambda_i^{1/2} (c_{g_i} - \gamma_{g_i}) + n^{1/2} \gamma_{g_i} \frac{l_i - \lambda_i}{2\lambda_i^{1/2}} \tag{9.4.24}$$

in distribution. Both the terms are asymptotically independently normally distributed. The variances and covariances can be obtained from the above theorem.

Similarly, we can treat the estimates of

$$\gamma_{g_i} (\lambda_i - \lambda_r)^{1/2}, \tag{9.4.25}$$

where the last root $\lambda_r$ is of multiplicity $q_r$.

### 9.4.2 *An Asymptotic Expansion of $l_i$ when $\lambda_i$ Is Simple*

Let us denote $T=\Gamma' S T$, $\Gamma'\Sigma\Gamma = D_\lambda$, $\lambda_1 > \lambda_2 > \cdots > \lambda_p$, and $S \sim W_p(\Sigma/n, n)$. Then $T \sim W_p(D_\lambda/n, n)$. Let

(9.4.26) $$\delta T = (\delta t_{ij}) = T - D_\lambda.$$

Then the moment generating function of $\delta T$ is

(9.4.27) $$E[\exp(\operatorname{tr}\theta\,\delta T)] = \operatorname{etr}(-D_\lambda \theta)|I - 2D_\lambda\theta/n|^{-n/2},$$

where $\theta$ is a symmetric $p \times p$ matrix. From this, it is easy to see that

(9.4.28) $$E(\delta^2) = O(n^{-1}), \quad E\delta^3 = O(n^{-3/2}), \quad E\delta^4 = O(n^{-2}), \quad \text{etc.},$$

where $\delta^r$ denotes the function of $\delta t_{ij}$'s such that its lowest order is the product of $r$ such $\delta t_{ij}$'s. Now let $Q$ be a skew-symmetric matrix given by

(9.4.29) $$Q = (q_{ij}), \qquad q_{ii} = 0 \quad \text{and} \quad q_{ij} = \frac{\delta t_{ij}}{\lambda_i - \lambda_j}.$$

Since $|I + Q| \neq 0$, let

(9.4.30) $$B = (I+Q)T(I+Q)^{-1}.$$

Then, using

$$(I+Q)^{-1} = I - Q + Q^2 - Q^3 + Q^4(I+Q)^{-1}$$

in Eq. (9.4.30), we get

$$B = (D_\lambda + QD_\lambda + \delta T + Q\,\delta T)(I+Q)^{-1}$$

and

$$QD_\lambda - D_\lambda Q + \delta T = \operatorname{diag}(\delta t_{11}, \delta t_{22}, \ldots, \delta t_{pp}) = D_{\delta t} \quad \text{(say)}.$$

Hence, we can write

$$\begin{aligned} B &= D_\lambda + (D_{\delta t} + Q\,\delta T)[I - Q + Q^2(I+Q)^{-1}] \\ &= D_\lambda + D_{\delta t} + Q\,\delta T - D_{\delta t}Q + O(\delta^3), \end{aligned}$$

and this can be rewritten as

(9.4.31) $$b_{ii} = \lambda_i + \delta t_{ii} + \sum_j{}' \frac{(\delta t_{ij})^2}{\lambda_i - \lambda_j} + O(\delta^3)$$

for $i = 1, 2, \ldots, p$ and

(9.4.32) $$b_{ij} = \sum_k{}' \frac{(\delta t_{ik})(\delta t_{kj})}{\lambda_i - \lambda_k} + O(\delta^3)$$

for $i \neq j$, $i,j=1,2,\ldots,p$, where $\Sigma'_j$ and $\Sigma'_k$ indicate the summation over all values from 1 to $p$ except $i$. From Eq. (9.4.30), the eigenvalues of $B$ are the eigenvalues of $T$, which are $l_1, l_2, \ldots, l_p$, the eigenvalues of $S$. Notice that

$$|B-lI| = \prod_{j=1}^{p} (l_{jj}-l) + O(\delta^4).$$

Hence, $l_i = b_{ii} + O(\delta^4)$. Using this in Eq. (9.4.31), we get

$$l_i = \lambda_i + \delta t_{ii} + \sum_j{}' \frac{(\delta t_{ij})^2}{\lambda_i - \lambda_j} + O(\delta^3), \tag{9.4.33}$$

for $i=1,2,\ldots,p$.

The above type of expansion of $l_i$ when the roots are distinct was given by Lawley (1956a) when the population eigenvalues are all simple. There is still a different approach: obtaining the joint density function of $l_1, l_2, \ldots, l_p$. This is done by G. Anderson (1965), Bingham (1972), and James (1969). The marginal asymptotic distribution of $l_i$ is given by Muirhead and Chikuse (1975). All these results are given in the next subsection.

### *9.4.3 Asymptotic Expansion up to $O(n^{-3/2})$ when the Roots of $\Sigma$ Are Simple*

The asymptotic joint distribution of $l_1, l_2, \ldots, l_p$, the eigenvalues of $S$, is given by

**Theorem 9.4.2** *Let $T \sim W_p(n^{-1}D_\lambda, n)$ and $D_\lambda = \text{diag}(\lambda_1, \ldots, \lambda_p)$ with $\lambda_1 > \lambda_2 > \cdots > \lambda_p > 0$. Then the asymptotic density function of eigenvalues $l_1, \ldots, l_p$ of $T$ is given by*

$$c_p(n)|D_\lambda|^{-\frac{1}{2}(n-p-1)}|D_l|^{\frac{1}{2}(n-p-1)}\left(\frac{\alpha(D_l)}{\alpha(D_\lambda)}\right)^{1/2} \text{etr}\left(-\tfrac{1}{2}nD_\lambda^{-1}D_l\right)\left(\sum_{j=0}^{\infty}\frac{a_j}{n^j}\right),$$

*where* $D_l = \text{diag}(l_1, \ldots, l_p)$, $\alpha(D_l) = \prod_{i=1}^{p-1}\prod_{j=i+1}^{p}(l_i - l_j)$, $a_0 = 1$, $a_1 = \frac{1}{2}\sum_{i=1}^{p-1}\sum_{j=i+1}^{p}[\lambda_i\lambda_j/(\lambda_i - \lambda_j)(l_i - l_j)]$, *and* $c_p(n) = (\frac{1}{2}n)^{\frac{1}{2}pn - \frac{1}{4}p(p-1)}\{\prod_{j=1}^{p}\Gamma[\frac{1}{2}(n-j+1)]\}^{-1}$.

The proof of this theorem is given at the end of this subsection. First, we shall give some of its consequences. Transforming $l_i$ to $x_i$ by

$$x_i = \left(\tfrac{1}{2}n\right)^{1/2}\left(l_i\lambda_i^{-1} - 1\right) \qquad \text{for} \quad i=1,2,\ldots,p,$$

one can establish

**Theorem 9.4.3** *The joint density function of* $x_i=(\frac{1}{2}n)^{1/2}(l_i\lambda_i^{-1}-1)$ *for* $i=1,2,\ldots,p$, *when* $\lambda_1>\lambda_2>\cdots>\lambda_p>0$, *is asymptotically given by*

$$\prod_{i=1}^{p}\phi_i(x_i)\left[1+\left(\frac{2}{n}\right)^{1/2}\sum_{i=1}^{p}p_{1i}(x_i)\right.$$

$$+\frac{2}{n}\left[\sum_{i=1}^{p}p_{2i}(x_i)+\sum_{i=1}^{p-1}\sum_{j=i+1}^{p}p_{1i}(x_i)p_{1j}(x_j)+\frac{1}{2}\sum_{i=1}^{p-1}\sum_{j=i+1}^{p}\frac{x_ix_j\lambda_i\lambda_j}{(\lambda_i-\lambda_j)^2}\right]$$

$$\left.+O(n^{-3/2})\right],$$

*where* $\phi(\cdot)$ *denotes the standard normal density function,*

$$p_{1i}(x)=\tfrac{1}{6}\left[2H_3(x)+3A_iH_1(x)\right],\qquad A_i=\sum_{\substack{j=1\\(j\neq i)}}^{p}{}'\,\frac{\lambda_j}{\lambda_i-\lambda_j},$$

$$p_{2i}(x)=\tfrac{1}{72}\left[4H_6(x)+18H_4(x)+12A_iH_4(x)-18B_iH_2(x)+9A_i^2H_2(x)\right],$$

$$B_i=\sum_{\substack{j=1\\(j\neq i)}}^{p}{}'\,\frac{\lambda_j^2}{(\lambda_i-\lambda_j)^2},$$

*and* $H_j(x)$ *is the Hermite polynomial of degree* $j$.

For the details, one can refer to Muirhead and Chikuse (1975). The following corollary gives the marginal density of $x_i$.

**Corollary 9.4.4** *The asymptotic marginal density function of* $x_i=(\frac{1}{2}n)^{1/2}(l_i\lambda_i^{-1}-1)$ *is given by*

$$\phi(x_i)\left[1+\left(\frac{2}{n}\right)^{1/2}p_{1i}(x_i)+\frac{2}{n}p_{2i}(x_i)+O(n^{-3/2})\right].$$

*Further, if*

$$\left(\tfrac{1}{2}n\right)^{1/2}\frac{l_i\lambda_i^{-1}-1-A_i\lambda_i/n}{\lambda(1-B_i/n)^{1/2}}=\tilde{x}_i,$$

*then* $\tilde{x}_i$ *is asymptotically normal with* $E(\tilde{x}_i)=O(n^{-3/2})$ *and* $V(\tilde{x}_i)=1+O(n^{-2})$.

**Corollary 9.4.5** *The first two asymptotic moments of $l_1,\dots,l_p$ are*

$$E(l_i)=\lambda_i+A_i\lambda_i n^{-1}+O(n^{-2}),\qquad V(l_i)=2\lambda_i^2 n^{-1}-2\lambda_i^2 B_i n^{-2}+O(n^{-3}),$$

*and*

$$\operatorname{Cov}(l_i,l_j)=2\frac{\lambda_i\lambda_j}{(\lambda_i-\lambda_j)^2}n^{-2}+O(n^{-3}).$$

*The third and fourth approximate cumulants of $l_i$ are*

$$\kappa_3(l_i)=8\lambda_i^3 n^{-2}+O(n^{-3})\quad\text{and}\quad\kappa_4(l_i)=48\lambda_i^4 n^{-3}+O(n^{-4}).$$

This corollary can be obtained from the results of Section 9.4.2. Now we shall give a detailed proof of Theorem 9.4.2.

PROOF OF THEOREM 9.4.2. Let us use the transformations

$$T=ED_lE',\qquad I+E=2\left(I-\tfrac{1}{2}n^{-1/2}Q\right)^{-1},\tag{9.4.34}$$

where $E$ is an orthogonal matrix with positive diagonal elements and $|I+E|\neq 0$, $D_l=\operatorname{diag}(l_1,\dots,l_p)$, $l_1>l_2>\dots>l_p>0$, and $Q$ is a skew-symmetric matrix. For the Jacobian of the transformation, we use the results of Chapter 1. The differential of $E$ gives

$$dE=n^{-1/2}\left(I-\tfrac{1}{2}n^{-1/2}Q\right)^{-1}dQ\left(I-\tfrac{1}{2}n^{-1/2}Q\right)^{-1}$$

and

$$E'dE=n^{-1/2}\left[\left(I-\tfrac{1}{2}n^{-1/2}Q\right)^{-1}\right]'dQ\left(I-\tfrac{1}{2}n^{-1/2}Q\right)^{-1}.$$

Hence, using Theorems 1.11.2 and 1.11.4, the Jacobian of the transformations is

$$\begin{aligned}J(S\to Q,D_l)&=J(S\to E,D_l)J(E\to E'dE)J(E'dE\to dQ)\\&=\alpha(D_l)|I+(4n)^{-1}QQ'|^{-\frac{1}{2}(p-1)}n^{-\frac{1}{4}p(p-1)},\end{aligned}\tag{9.4.35}$$

where $\alpha(D_l)$ is defined in Theorem 9.4.2. Hence, the joint density function of $D_l$ and $Q$ is

$$f(D_l)f_1(Q)f_2(Q)f_3(Q),\tag{9.4.36}$$

where

(9.4.37) $$f(D_l)=c_p(n)|D_lD_\lambda^{-1}|^{\frac{1}{2}(n-p-1)}[\alpha(D_l)/\alpha(D_\lambda)]^{-1}\operatorname{etr}(-\tfrac{1}{2}nD_lD_\lambda^{-1}),$$

(9.4.38) $$f_1(Q)=\prod_{i=1}^{p-1}\prod_{j=i+1}^{p}\left[(2\pi)^{-1/2}g_{ij}^{1/2}\exp(-\tfrac{1}{2}g_{ij}q_{ij}^2)\right],$$

(9.4.39) $$f_2(Q)=\exp\left(\tfrac{1}{2}n\operatorname{tr}D_lD_\lambda^{-1}+\frac{1}{2}\sum_{i=1}^{p-1}\sum_{j=i+1}^{p}g_{ij}q_{ij}^2-\tfrac{1}{2}n\operatorname{tr}D_\lambda^{-1}ED_lE'\right),$$

(9.4.40) $$f_3(Q)=|I+(4n)^{-1}QQ'|^{-\frac{1}{2}(p-1)},$$

and

(9.4.41) $$g_{ij}=\frac{(l_i-l_j)(\lambda_i-\lambda_j)}{\lambda_i\lambda_j}\qquad\text{for } i\neq j,\quad i,j=1,2,\ldots,p.$$

The problem is to integrate out the $q_{ij}$'s. We have to expand $(I-\frac{1}{2}n^{-1/2}Q)^{-1}$ in an infinite series, and this can be done if $\max_i|\operatorname{ch}_i(Q)|>2n^{1/2}$ has nearly zero probability with respect to the density given by Eq. (9.4.38), where $\operatorname{ch}_i(Q)$ is the $i$th characteristic root of $Q$. To see this, we observe that

(9.4.42) $$\left(\operatorname*{Max}_i|\operatorname{ch}_i(Q)|\right)^2\leqslant\operatorname{ch}_1(QQ')\leqslant\operatorname{tr}QQ'=2\sum_{i=1}^{p-1}\sum_{j=i+1}^{p}q_{ij}^2$$
$$\leqslant 2\sum_{i=1}^{p-1}\sum_{j=i+1}^{p}\frac{q_{ij}^2g_{ij}}{c},$$

where $\operatorname{ch}_1(QQ')$ is the largest eigenvalue of $QQ'$ and $c=\operatorname{Min}_{i,j}g_{ij}$. With respect to the density (9.4.38),

(9.4.43) $$\sum_{i=1}^{p-1}\sum_{j=i+1}^{p}g_{ij}q_{ij}^2\sim\chi_r^2,$$

where

(9.4.44) $$r=\tfrac{1}{2}p(p-1).$$

Let $t=[\frac{1}{2}r]=$ the greatest integer contained in $[0,\frac{1}{2}r]$. Then, if $x\geqslant 1$

(9.4.45) $$\frac{x^{\frac{1}{2}r-1}}{\Gamma(\frac{1}{2}r+1)}\leqslant\frac{x^t}{t!}<e^x.$$

Hence

$$P\left\{\operatorname*{Max}_i |\mathrm{ch}_i(Q)|>2n^{1/2}\right\}\leqslant P\{\chi_r^2\geqslant 2cn\} \tag{9.4.46}$$
$$=\Gamma^{-1}\left(\tfrac{1}{2}r\right)\int_{cn}^{\infty} y^{\frac{1}{2}r-1}e^{-y}\,dy$$
$$\leqslant \frac{(cn)^{\frac{1}{2}r}}{\Gamma\left(\frac{1}{2}r\right)}\int_1^{\infty} x^{\frac{1}{2}r-1}e^{-cnx}\,dx$$
$$\leqslant \left(\tfrac{1}{2}re\right)(cn)^{\frac{1}{2}r}\left[e^{-cn}/(cn-1)\right],$$

which can be made less than $\epsilon$, however small, by choosing $n$ sufficiently large. On account of Eq. (9.4.46), we can write

$$f_3(Q)=\operatorname{etr}\left(\sum_{j=1}^{\infty}(p-1)(2^{2j+1}n)^{-1}Q^{2j}\right), \tag{9.4.47}$$

and using $E=I+\sum_{j=1}^{\infty}(\frac{1}{2})^{j-1}n^{-\frac{1}{2}j}Q^j$,

$$\ln f_2(Q)=\tfrac{1}{2}n\operatorname{tr}D_\lambda^{-1}D_l+\tfrac{1}{2}\left(\operatorname{tr}D_\lambda^{-1}D_lQ^2-\operatorname{tr}D_\lambda^{-1}QD_lQ\right) \tag{9.4.48}$$
$$-\tfrac{1}{2}n\operatorname{tr}\left[D_\lambda^{-1}\left(I+\sum_{j=0}^{\infty}\left(\tfrac{1}{2}\right)^{j-1}n^{-\frac{1}{2}j}Q^j\right)\right.$$
$$\left.\times D_l\left(I-\sum_{j=1}^{\infty}\left(-\tfrac{1}{2}\right)^{j-1}n^{-\frac{1}{2}j}Q^j\right)\right]$$
$$=\sum_{j=1}^{\infty}n^{-\frac{1}{2}j}\{Q^{j+2}\},$$

where

$$\{Q^{j+2}\}=\left(\tfrac{1}{2}\right)^{2m+1}\sum_{\alpha=0}^{m}(-1)^{\alpha}\operatorname{tr}D_\lambda^{-1}Q^{2m-\alpha+2}D_lQ^{\alpha+1} \tag{9.4.49a}$$
$$\text{if } j=2m+1,$$

and

$$\{Q^{j+2}\}=\left(\tfrac{1}{2}\right)^{2m}\sum_{\alpha=0}^{m-1}(-1)^{\alpha}\operatorname{tr}D_\lambda^{-1}Q^{2m-\alpha+1}D_lQ^{\alpha+1} \tag{9.4.49b}$$
$$+(-1)^m\left(\tfrac{1}{2}\right)^{2m+1}\operatorname{tr}D_\lambda^{-1}Q^{m+1}D_lQ^{m+1}$$
$$-\left(\tfrac{1}{2}\right)^{2m+1}\operatorname{tr}D_\lambda^{-1}D_lQ^{2m+2}\qquad\text{if } j=2m.$$

Let

$$\mathfrak{D} = \{ q_{ij} : -\infty < q_{ij} < \infty \}$$

and

$$\{Q^{j+2}\}_1 = \begin{cases} \{Q^{j+2}\} & \text{if } j = 2m+1 \\ \{Q^{j+2}\} + (p-1)\left(\tfrac{1}{2}\right)^{2m+1} \operatorname{tr} Q^{2m} & \text{if } j = 2m. \end{cases}$$

Notice that $\{Q^{2m+3}\}_1$ is an odd function of $Q$ and $\{Q^{2m+2}\}_1$ is an even function of $Q$, and this shows that

$$\prod_j \{Q^{j+2}\}_1^{\alpha_j} \quad \text{is an odd function of} \quad Q \quad \text{if} \quad \Sigma j\alpha_j \quad \text{is odd}$$

and

$$\int_{\mathfrak{D}} \cdots \int f_1(Q)\left( \prod_j \{q^{j+2}\}_1^{\alpha_j} \right) dQ = 0 \quad \text{if} \quad \Sigma j\alpha_j \quad \text{is odd.}$$

Hence, integration of $Q$ from Eq. (9.4.36) gives

$$\text{(9.4.50)} \quad \int_{\mathfrak{D}} \cdots \int f_i(Q) \exp\left( \sum_{j=1}^{\infty} n^{-\frac{1}{2}j} \{Q^{j+2}\}_1 \right) dQ = 1 + \sum_{j=1}^{\infty} n^{-j} a_j.$$

This gives the asymptotic density function of $D_l$ as mentioned in Theorem 9.4.2, except that we need to obtain $a_1$ explicitly. This is given by

$$\text{(9.4.51)} \quad a_1 = E\left( \{Q^4\}_1 + \tfrac{1}{2}\{Q^3\}_1^2 \right) = \int_{\mathfrak{D}} f_1(Q)\left( \{Q^4\}_1 + \tfrac{1}{2}\{Q^3\}_1^2 \right) dQ.$$

We observe the following:

$$2\{Q^3\}_1 = \operatorname{tr} D_\lambda^{-1} Q^2 D_l Q = \sum_{i<j<k} h(i,j,k) q_{ij} q_{jk} q_{ki},$$

$$\begin{aligned} 8\{Q^4\}_1 &= -\operatorname{tr} D_\lambda^{-1} D_l Q^4 + (p-1)\operatorname{tr} Q^2 - \operatorname{tr} D_\lambda^{-1} Q^2 D_l Q^2 + 2 \operatorname{tr} D_\lambda^{-1} Q^3 D_l Q \\ &= -2(p-1)\sum_{i<j} q_{ij}^2 + \sum_{\substack{i \neq k \\ j \neq \alpha}} f(i,j,k) q_{ij} q_{jk} q_{\alpha k} q_{i\alpha} \\ &\quad - \sum_{\substack{i,j,k \\ (i \neq j \neq k)}} q_{ij}^2 q_{jk}^2 [ f(i,j,k) + f(j,i,j) ] \\ &\quad - \sum_{i \neq j} q_{ij}^4 f(i,j,i), \end{aligned}$$

where

$$f(i,j,k)=\frac{1}{\lambda_i}(l_i-2l_j+l_k),$$

$$h(i,j,k)=\frac{1}{\lambda_i}(l_k-l_j)+\frac{1}{\lambda_j}(l_i-l_k)+\frac{1}{\lambda_k}(l_j-l_i).$$

It can be shown that

$$[h(i,j,k)]^2=g_{ij}^2+g_{ik}^2+g_{jk}^2-2(g_{ij}g_{ik}+g_{ij}g_{jk}+g_{ik}g_{jk}).$$

Hence,

$$(9.4.52)\quad 4E(\{Q^3\}_1^2)=\sum_{i<j<k}(g_{ij}g_{jk}g_{ik})^{-1}\times(g_{ij}^2+g_{ik}^2+g_{jk}^2-2g_{ij}g_{ik}-2g_{ij}g_{jk}-2g_{ik}g_{jk})$$
$$=-2(p-2)\sum_{i<j}(g_{ij})^{-1}+\sum_{i<j<k}\frac{g_{ij}^2+g_{ik}^2+g_{jk}^2}{g_{ij}g_{jk}g_{ik}},$$

and

$$(9.4.53)\quad 8E(\{Q^4\}_1)=-2(p-1)\sum_{i<j}(g_{ij})^{-1}-\sum_{i\neq j\neq k}(g_{ij}g_{jk})^{-1}[f(i,j,k)+f(j,i,j)]-3\sum_{i\neq j}(g_{ij})^{-2}f(i,j,i).$$

We have

$$(9.4.53\text{a})\quad \sum_{i\neq j}g_{ij}^{-2}f(i,j,i)=-2\sum_{i<j}(g_{ij})^{-1},$$

and

$$(9.4.53\text{b})\quad \sum_{i\neq j\neq k}(g_{ij}g_{jk})^{-1}[f(i,j,k)+f(j,i,j)]$$
$$=\sum_{i\neq j\neq k}(g_{ij}g_{jk})^{1}f_1(i,j,k)$$
$$=\sum_{i<j<k}\{(g_{ij}g_{jk})^{-1}[f_1(i,j,k)+f_1(k,j,i)]+(g_{ij}g_{ik})^{-1}[f_1(k,i,j)+f_1(j,i,k)]+(g_{ik}g_{jk})^{-1}[f_1(j,k,i)+f_1(i,k,j)]\}$$
$$=\sum_{i<j<k}[(g_{ij}g_{jk})^{-1}(g_{ik}-2g_{ij}-2g_{jk})+(g_{ij}g_{ik})^{-1}(g_{jk}-2g_{ij}-2g_{ik})+(g_{ik}g_{jk})^{-1}(g_{ij}-2g_{ik}-2g_{jk})]$$
$$=-4(p-2)\sum_{i<j}(g_{ij})^{-1}+\sum_{i<j<k}\frac{g_{ij}^2+g_{ik}^2+g_{jk}^2}{g_{ij}g_{jk}g_{ik}}.$$

Using Eqs. (9.4.52) and (9.4.53) in Eq. (9.4.51) and after some simplifications, we get

$$a_1 = \tfrac{1}{2} \sum_{i<j} (g_{ij})^{-1}, \qquad \square$$

as required in Theorem 9.4.2.

A proof of the above type was given by Khatri and Srivastava (1978). These results have been extended to the case when $\Sigma$ has multiple roots: see Khatri and Srivastava (1978), Sugiuara (1976), Chikuse (1976), Constantine and Muirhead (1976), James (1969), and Chattopadhyay and Pillai (1973).

### 9.4.4 *Variances and Covariances of Unordered Characteristic Roots when the Population Roots Are Equal (Girshick)*

Let $l_1, l_2, \ldots, l_p$ denote the *unordered* characteristic roots of $S = n^{-1}V$, where $V \sim W_p(I, n)$; without loss of generality we have taken the population covariance matrix as $I$ rather than $\lambda I$. Then

$$E(s_{ij}s_{km}) = \delta_{ij}\delta_{km} + \frac{1}{n}(\delta_{ik}\delta_{jm} + \delta_{im}\delta_{jk}), \tag{9.4.54}$$

where $S = (s_{ij})$ and $\delta_{pq}$ is the Kronecker delta. From the results on the trace and characteristic roots of a matrix, we get

$$\operatorname{tr}_1 S = \sum_{i=1}^{p} s_{ii} = \sum_{i=1}^{p} l_i, \tag{9.4.55}$$

$$\operatorname{tr}_2 S = \sum_{i<j} (s_{ii}s_{jj} - s_{ij}^2) = \sum_{i<j} l_i l_j.$$

Hence

$$\left(\sum s_{ii}\right)^2 = \sum s_{ii}^2 + 2\sum_{i<j} s_{ii}s_{jj} = \sum l_i^2 + 2\sum_{i<j} l_i l_j. \tag{9.4.56}$$

This gives

$$\sum s_{ii}^2 + 2\sum_{i<j} s_{ij}^2 = \sum l_i^2. \tag{9.4.57}$$

Hence

$$E(l_i) = 1 \tag{9.4.58}$$

and

$$E\left(\sum l_i^2\right) = E\left(\sum s_{ii}^2 + 2\sum_{i<j} s_{ij}^2\right). \tag{9.4.59}$$

The second equation gives

(9.4.60)
$$pE(l_i^2) = pEs_{ii}^2 + p(p-1)E(s_{ij}^2)$$
$$= p\left(1 + \frac{2}{n}\right) + p\frac{p-1}{n}.$$

Hence

(9.4.61)
$$E(l_i^2) = 1 + \frac{p+1}{n}.$$

Thus from Eqs. (9.4.58) and (9.4.61), we get

(9.4.62)
$$\mathrm{Var}(l_i) = \frac{p+1}{n}.$$

It is left as an exercise for the reader to show that

(9.4.63)
$$\mathrm{Cov}(l_i, l_j) = -\frac{1}{n}, \qquad i \neq j.$$

## 9.5 Inference on Characteristic Roots

Consider a situation in which all measurements are made in the same units and by the same kind of measuring device. If the errors of measurements in all variables are mutually independent (having the same variance) and are independent of the true measurements, then the covariance matrix of the observed variables can be expressed as $\Sigma = \Psi + \lambda I$, where $\Psi$ is the covariance matrix of the true measurements and $\lambda$ is the variance of error. Thus, if $\Psi$ is a matrix of rank $k$ only, then the last $q \equiv p-k$ roots of $\Sigma$ are equal and equal to $\lambda$. Based on a sample of size $N$, the likelihood-ratio test for the equality of the last $p-q$ roots to $\lambda$ can easily be derived and is given by

(9.5.1)
$$Q = \left[ \frac{\prod_{j=k+1}^{p} l_j}{\left( \sum_{j=k+1}^{p} \frac{l_j}{q} \right)^q} \right]^{\frac{1}{2}N}, \qquad q = p-k,$$

where $l_1 > l_2 > \cdots > l_p$ are the roots of the sample covariance matrix $S$ given by

(9.5.2)
$$nS = \sum_{i=1}^{N} (\mathbf{x}_i - \bar{\mathbf{x}})(\mathbf{x}_i - \bar{\mathbf{x}})', \qquad n = N-1.$$

In fact, one arrives at the same kind of test criterion for testing the equality of any subset of the roots. The criterion is a monotonic function of the geometric mean of the relevant roots divided by the arithmetic mean. We shall now derive the asymptotic distribution of $-2(n/N)\ln Q$ under the null hypothesis.

Under the null hypothesis, $-2(n/N)\ln Q$ is asymptotically equivalent to (see Theorem 9.4.1)

$$(9.5.3)\quad -n\ln\prod_{j=k}^{p} l_j + nq\ln\sum_{j=k+1}^{p}\frac{l_j}{q}$$

$$= -n\sum_{j=k+1}^{p}\ln(\lambda+n^{-1/2}h_j) + nq\ln\sum_{j=k+1}^{p}\frac{\lambda+n^{-1/2}h_j}{q}$$

$$= n\left[-\sum_{j=k+1}^{p}\ln\left(1+\frac{h_j}{n^{1/2}\lambda}\right)+q\ln\left(1+\frac{\Sigma h_j}{n^{1/2}q\lambda}\right)\right]$$

$$= n\left[-\sum_j\left(\frac{h_j}{n^{1/2}\lambda}-\frac{h_j^2}{2n\lambda^2}+\cdots\right)+q\left(\frac{\Sigma h_j}{n^{1/2}q\lambda}-\frac{(\Sigma h_j)^2}{2nq^2\lambda^2}+\cdots\right)\right]$$

$$= n\left[\sum_{j=k+1}^{p}\frac{h_j^2}{2n\lambda^2}-\frac{\left(\sum_{j=k}^{p}h_j\right)^2}{2nq\lambda^2}+O(n^{-3/2})\right]$$

$$= \left(\tfrac{1}{2}\lambda^{-2}\right)\left[\sum_{j=k+1}^{p}h_j^2-q^{-1}\left(\sum_{j=k+1}^{p}h_j\right)^2\right]+O(n^{-1/2}),$$

where $q=p-k$ and the asymptotic joint density of $h_{k+1},\ldots,h_p$ is given by Eq. (9.4.2) with $\lambda_i=\lambda$ and $q_i=p-k=q$. If $U=(u_{ij})$ is a $q\times q$ symmetric matrix and the $u_{ij}$'s are independent variates such that $u_{ii}\sim N(0,2\lambda^2)$ and $u_{ij}(i<j)\sim N(0,\lambda^2)$, then the distribution of the characteristic roots of $U$ is the same as that of $h_{k+1},\ldots,h_p$. Hence, we can replace $\Sigma h_j^2$ by $\operatorname{tr}U^2$ and $\Sigma h_j$ by $\operatorname{tr}U$ in Eq. (9.5.3) and write

$$(9.5.4)\quad -2\frac{n}{N}\ln Q=\sum_{i<j}\frac{u_{ij}^2}{\lambda^2}+\left[\Sigma u_{ii}^2-q^{-1}(\Sigma u_{ii})^2\right](2\lambda^2)^{-1}+O(n^{-1/2}).$$

Since $\Sigma_{i<j}u_{ij}^2/\lambda^2$ is asymptotically $\chi^2$ with $\frac{1}{2}(p-k)(p-k-1)$ d.f. and is distributed independently of

$$(9.5.5)\quad \frac{\Sigma u_{ii}^2-(p-k)^{-1}(\Sigma u_{ii})^2}{2\lambda^2}\sim\chi^2_{p-k-1},$$

we find that under the null hypothesis, $-2\ln Q$ is asymptotically distributed as $\chi^2$ with $\frac{1}{2}(p-k)(p-k+1)-1=\frac{1}{2}(p-k-1)(p-k+2)$ d.f.

As an improvement on the above asymptotic distribution, Lawley [see also James (1969) and Fujikoshi (1978)] has shown (by using his results as in Subsection 9.4.2) that

$$(9.5.6)\qquad -\frac{2}{N}\left\{n-k-\tfrac{1}{6}\left[2(p-k)+1+2(p-k)^{-1}\right]+\hat{\lambda}^2\sum_{j=1}^{k}\frac{1}{\left(l_r-\hat{\lambda}\right)^2}\right\}\ln Q$$

is asymptotically $\chi^2$ with $\frac{1}{2}(p-k)(p-k+1)-1$ d.f., where $\hat{\lambda}=q^{-1}\sum_{i=k+1}^{p}l_i$.

## 9.6 Asymptotic Test for a Given Principal Component (Anderson)

Let $\Sigma=\Gamma D_\lambda \Gamma'$, where $\Gamma=(\gamma_1,\ldots,\gamma_p)$ and $D_\lambda=\operatorname{diag}(\lambda_1,\ldots,\lambda_p)$ with $\lambda_1>\lambda_2\geqslant\lambda_3\geqslant\cdots\geqslant\lambda_p>0$. Suppose we wish to test the hypothesis $H:\gamma_1=\mathbf{a}$, where $\mathbf{a}$ is specified, against the alternative $A:\gamma_1\neq\mathbf{a}$. We shall now obtain a test for this problem on the basis of asymptotic results given in Section 9.4. Let

$$(9.6.1)\qquad \mathbf{y}=n^{1/2}(\mathbf{c}_1-\gamma_1),$$

where $\mathbf{c}_1$ is the characteristic vector corresponding to the largest root of the sample covariance matrix based on $n+1$ observations from $N_p(\boldsymbol{\mu},\Sigma)$. Then $\mathbf{y}$ has a limiting normal distribution with mean $\mathbf{0}$ and covariance matrix

$$(9.6.2)\qquad \sum_{j=2}^{p}\frac{\lambda_1\lambda_j}{(\lambda_1-\lambda_j)^2}\gamma_j\gamma_j'\equiv\Gamma_2\Delta^2\Gamma_2',$$

where

$$(9.6.3)\qquad \Gamma_2=(\gamma_2,\ldots,\gamma_p),$$

$$\Delta^2=\operatorname{diag}\left(\lambda_1\lambda_2(\lambda_1-\lambda_2)^{-2},\lambda_1\lambda_3(\lambda_1-\lambda_3)^{-2},\ldots,\lambda_1\lambda_p(\lambda_1-\lambda_p)^{-2}\right).$$

Hence, $\mathbf{z}=\Delta^{-1}\Gamma_2'\mathbf{y}$ has a limiting normal distribution with mean $\mathbf{0}$ and covariance matrix $I_{p-1}$. Consequently,

$$(9.6.4)\qquad \mathbf{z}'\mathbf{z}=\mathbf{y}'\Gamma_2\Delta^{-2}\Gamma_2'\mathbf{y}$$

has a limiting chi-square distribution with $p-1$ d.f. Since

$$(9.6.5)\qquad \Sigma^{-1}=\Gamma D^{-1}\Gamma'=\sum_{j=1}^{p}\lambda_j^{-1}\gamma_j\gamma_j',$$

$$\Gamma\Gamma'=\sum_{j=1}^{p}\gamma_j\gamma_j'=I,$$

$$\Sigma=\Gamma D_\lambda\Gamma'=\sum_{j=1}^{p}\lambda_j\gamma_j\gamma_j',$$

the matrix of the quadratic form in $\mathbf{y}$ in Eq. (9.6.4) is

$$(9.6.6)\quad \Gamma_2\Delta^{-2}\Gamma_2'=\Gamma_2\begin{bmatrix}\frac{\lambda_1}{\lambda_2}-2+\frac{\lambda_2}{\lambda_1} & & 0\\ & \ddots & \\ 0 & & \frac{\lambda_1}{\lambda_p}-2+\frac{\lambda_p}{\lambda_1}\end{bmatrix}\Gamma_2'$$

$$=\lambda_1\Gamma_2\begin{bmatrix}\lambda_2^{-1} & & 0\\ & \ddots & \\ 0 & & \lambda_p^{-1}\end{bmatrix}\Gamma_2'-2\Gamma_2\Gamma_2'+\lambda_1^{-1}\Gamma_2\begin{bmatrix}\lambda_2 & & 0\\ & \ddots & \\ 0 & & \lambda_p\end{bmatrix}\Gamma_2'$$

$$=\lambda_1\sum_{j=2}^{p}\lambda_j^{-1}\gamma_j\gamma_j'-2I+2\gamma_1\gamma_1'+\lambda_1^{-1}\sum_{j=2}^{p}\lambda_j\gamma_j\gamma_j'$$

$$=\lambda_1\Sigma^{-1}-2I+\lambda_1^{-1}\Sigma.$$

Thus, asymptotically

$$(9.6.7)\qquad n(\mathbf{c}_1-\gamma_1)'(\lambda_1\Sigma^{-1}-2I+\lambda_1^{-1}\Sigma)(\mathbf{c}_1-\gamma_1)$$

has a $\chi^2$ distribution with $p-1$ d.f.

Thus if $\Sigma$ is not known, we can estimate $\Sigma$, $\Sigma^{-1}$, and $\lambda_1$ consistently by $S$, $S^{-1}$, and $l_1$, respectively ($l_1=\mathbf{c}_1'S\mathbf{c}_1$), and use the test criterion based on the statistic

$$(9.6.8)\qquad n(\mathbf{c}_1-\mathbf{a})'\left(l_1S^{-1}-2I+l_1^{-1}S\right)(\mathbf{c}_1-\mathbf{a})$$
$$=n\left(l_1\mathbf{a}'S^{-1}\mathbf{a}+l_1^{-1}\mathbf{a}'S^{-1}\mathbf{a}-2\right),$$

which has a limiting $\chi^2$ distribution with $p-1$ d.f., for testing the hypothesis $H:\gamma_1=\mathbf{a}$ against the alternative $A:\gamma_1\neq\mathbf{a}$. The right side of Eq. (9.6.8) is obtained from the fact that

$$(9.6.9)\qquad S\mathbf{c}_1=l_1\mathbf{c}_1,\qquad S^{-1}\mathbf{c}_1=l_1^{-1}\mathbf{c}_1,$$
$$\left(l_1S^{-1}-2I+l_1^{-1}S\right)\mathbf{c}_1=\mathbf{0}.$$

## 9.7 Exact Test for a Given Principal Component

We consider the testing of the hypothesis $H$: ($\mathbf{a}'\mathbf{x}$ is a principal component) against the alternative $A$: ($\mathbf{a}'\mathbf{x}$ is not a principal component). Let $(\mathbf{a},B_1)=B$ be an orthogonal matrix. Then under $H$,

$$B'\Sigma B=\begin{pmatrix}\delta & \mathbf{0}_{1,p-1}\\ \mathbf{0}_{p-1,1} & \Sigma_1\end{pmatrix},$$

and the problem of testing $H$ against $A$ reduces to that of testing whether the multiple correlation between $\mathbf{a}'\mathbf{x}$ and the set $B_1'\mathbf{x}$ is zero or not. Using the multiple-correlation theory, the exact test for $H$ against $A$ is based on the statistic

$$F_{p-1,n-p+1}=(n-p+1)(p-1)^{-}(\mathbf{a}'(\mathbf{z}'S^{-1}\mathbf{a})(\mathbf{a}'S\mathbf{a})-1],$$

which is distributed as $F$ with $p-1$ and $n-p+1$ d.f. under $H$. The above test has been derived by Mallows (1960) by using likelihood-ratio test procedure. For generalizations to more than one principal component, one may refer to Mallows (1960).

Kshirsagar (1961) has given an exact test procedure for testing $H$ against $A$ when there is only one nonisotropic principal component. The nonisotropic principal component is $\mathbf{a}'\mathbf{x}$, and there is only one such component iff the structure of the covariance matrix $\Sigma$ is

$$\Sigma=\delta I+(\delta_1-\delta)\mathbf{a}\mathbf{a}'.$$

If $\delta$ is known, then Kshirsagar (1961) proposes the test based on

$$\chi_d^2=\left[(\mathbf{a}'S^2\mathbf{a})(\mathbf{a}'S\mathbf{a})^{-1}-\mathbf{a}'S\mathbf{a}\right]n/\delta$$

for testing $H$ against $A$ under the assumption that there is only one nonisotropic principal component, and $\chi_d^2$ is distributed as chi square with $p-1$ d.f. under $H$. If $\delta$ is not known, he proposes to use the $F$ test based on the statistic

$$F_{p-1,(n-1)(p-1)}=\frac{\delta(n-1)\chi_d^2}{n(\operatorname{tr}S-\mathbf{a}'S\mathbf{a})},$$

which is distributed as $F$ with $p-1$ and $(n-1)(p-1)$ d.f. under $H$ and under the assumption of only one nonisotropic principal component. This type of result is generalized by Kshirsagar and Gupta (1965) to more than one nonisotropic principal component. The noncentral distribution of $\chi_d^2$ is given by Kshirsagar (1966).

## Problems

**9.1.** Show that Eqs. (9.2.8) and (9.2.9) are minimized at $T$ given by Eq. (9.2.12).

**9.2.** Suppose we wish to approximate the $p$-component random vector $\mathbf{x}$ with covariance matrix $\Sigma$ by a linear form $B\mathbf{y}+\mathbf{c}$, where $B:p\times q$, $\mathbf{y}:q\times 1$, and $\mathbf{c}:p\times 1$, such that the roots of the matrix

$$E(\mathbf{x}-B\mathbf{y}-\mathbf{c})(\mathbf{x}-B\mathbf{y}-\mathbf{c})'$$

is minimized. Show that the optimum choice of $B$, $\mathbf{y}$, and $\mathbf{c}$ is given by

$$B=(\gamma_1,\ldots,\gamma_q),\qquad \mathbf{y}=B'\mathbf{x},\quad\text{and}\quad \mathbf{c}=\mathbf{0},$$

where

$$\Gamma=(\gamma_1,\ldots,\gamma_p) \quad \text{and} \quad \Gamma\Sigma\Gamma'=\operatorname{diag}(\lambda_1,\ldots,\lambda_p).$$

**9.3.** Obtain the likelihood-ratio test for testing the hypothesis that the last $q=p-k$ roots of $\Sigma$ are equal to a *specified* value $\lambda_0$. Derive its asymptotic distribution.

**9.4.** Let $\Sigma=\Gamma D_\delta \Gamma'$, where $\Gamma=(\gamma_1,\ldots,\gamma_p)$ and $D_\delta=\operatorname{diag}(\delta_1,\delta_2,\ldots,\delta_p)$. Let $\delta_1>\delta_2>\cdots>\delta_p>0$. Derive the likelihood-ratio test for the following hypotheses:

(i) $H_1:\gamma_i=\gamma_i^0$, $i=1,2,\ldots,k\leqslant p$, $\gamma_1^0$ specified, against $A_1=\gamma_i\neq\gamma_i^0$.

(ii) $H_2:\gamma_i=\gamma_i^0$ and $\delta_i=\delta_i^0$, $i=1,2,\ldots,k\leqslant p$, $\gamma_i^0$ and $\delta_i^0$ specified, against $A_2\neq H_2$.

Show that when $H_1$ is true, $-2\ln\Lambda_1$ is asymptotically $\chi_r^2$ with $r=k(p-1)-\frac{1}{2}k(k-1)$ d.f. Similarly, show that $-2\ln\Lambda_2$ is asymptotically $\chi_{r+k}^2$. Here $\Lambda_1$ and $\Lambda_2$ are the likelihood-ratio criteria for parts (i) and (ii) respectively.

**9.5.** Prove Eq. (9.4.63).

# 10

# Monotonicity and Unbiasedness of Some Power Functions

## 10.1 Introduction

For a particular test of hypothesis, there exists generally a large class of possible test criteria. This class of tests may be reduced by imposing the conditions of invariance and unbiasedness. An additional property that may be required for a test is that the corresponding power function of the test be a monotonically nondecreasing function of the maximal invariant parameters under the alternative hypothesis. This monotonicity property ensures the unbiasedness of a test. In this chapter, the monotonicity of the power function of several test criteria is shown; in some cases only unbiasedness is available. For convenience, we shall consider only the canonical form of the tests.

## 10.2 Some Probability Inequalities

In this section we shall derive some inequalities for certain integrals useful in showing the monotonicity of power functions. First, we give a definition.

**Definition 10.2.1** For $\mathbf{x} \in R^n$, let $f(\mathbf{x}) \geq 0$ be a function such that the set

$$K_u = \{\mathbf{x} : f(\mathbf{x}) > u\}$$

is convex for every $0 < u < \infty$. Then the function $f$ is said to be *unimodal*.

For example, if $f(\mathbf{x})=c\exp(-\mathbf{x}'\Sigma^{-1}\mathbf{x})$, $\Sigma>0$, $c>0$, then the set $K_u=\{\mathbf{x}: f(\mathbf{x})\geqslant u\}$ for $0<u<\infty$ is convex. If $u<c$, the set $K_u$ is the null set, which is considered as convex (as well as concave).

**Theorem 10.2.1 (Anderson 1955)** *Let $E$ be a convex set in $R^n$, symmetric about the origin. For $\mathbf{x}\in R^n$, let $f(\mathbf{x})\geqslant 0$ be a function such that*

(i) $f(\mathbf{x})=f(-\mathbf{x})$,
(ii) $\int_E f(\mathbf{x})\,d\mathbf{x}<\infty$,
(iii) $f(\mathbf{x})$ *is unimodal.*

*Then for every $\mathbf{y}\in R^n$, and for $0\leqslant k\leqslant 1$,*

$$\int_E f(\mathbf{x}+\mathbf{y})\,d\mathbf{x}\geqslant\int_E f(\mathbf{x}+k\mathbf{y})\,d\mathbf{x}.$$

First we prove the following lemma.

**Lemma 10.2.1** *Let $V\{\ \}$ denote the volume of a set,* i.e., *$V(A)=\int_A d\mathbf{x}$. Let $E$ and $K$ be convex sets in $n$ space symmetric about the origin. Denote by $E+\mathbf{y}$ the set $E$ translated by the vector $\mathbf{y}$. Then for $0\leqslant k\leqslant 1$, $\mathbf{y}\in R^n$,*

$$V\{(E+k\mathbf{y})\cap K\}\geqslant V\{(E+\mathbf{y})\cap K\}.$$

PROOF. For arbitrary sets $A$, $B$ in $R^n$ let $\alpha A+(1-\alpha)B$ be the set of all points of the form $\alpha\mathbf{a}+(1-\alpha)\mathbf{b}$ where $\mathbf{a}\in A$ and $\mathbf{b}\in B$, $0\leqslant\alpha\leqslant 1$. Let $\alpha=\frac{1}{2}(k+1)$. Then

$$(E+k\mathbf{y})\cap K\supset\alpha\left[(E+\mathbf{y})\cap K\right]+(1-\alpha)\left[(E-\mathbf{y})\cap K\right].$$

For if $\mathbf{x}\in\alpha[(E+\mathbf{y})\cap K]+(1-\alpha)[(E-\mathbf{y})\cap K]$, then $\mathbf{x}$ can be represented as $\mathbf{x}=\alpha(\mathbf{z}_1+\mathbf{y})+(1-\alpha)(\mathbf{z}_2-\mathbf{y})=\alpha\mathbf{z}_1+(1-\alpha)\mathbf{z}_2+(2\alpha-1)\mathbf{y}$, where $\mathbf{z}_1,\mathbf{z}_2\in E$. But $E$ is convex and $2\alpha-1=k$; therefore, $\mathbf{x}\in E+k\mathbf{y}$. Alternatively $\mathbf{x}$ can be represented as $\alpha\mathbf{u}_1+(1-\alpha)\mathbf{u}_2$ where $\mathbf{u}_1,\mathbf{u}_2\in K$. But $K$ is convex; therefore $\mathbf{x}\in K$. Hence $\mathbf{x}\in(E+k\mathbf{y})\cap K$. Now as $E$ and $K$ are symmetric about the origin, $V\{(E+\mathbf{y})\cap K\}=V\{(E-\mathbf{y})\cap K\}$. Therefore we have that

$$\begin{aligned}V\{(E+k\mathbf{y})\cap K\}&\geqslant V\{\alpha\left[(E+\mathbf{y})\cap K\right]+(1-\alpha)\left[(E-\mathbf{y})\cap K\right]\}\\&\geqslant\alpha V\{(E+\mathbf{y})\cap K\}+(1-\alpha)V\{(E-\mathbf{y})\cap K\}\\&=V\{(E+\mathbf{y})\cap K\},\end{aligned}$$

where the second step comes from the Brunn–Minkowski inequality [see Bonnesen and Fenchel (1948)]. □

PROOF OF THEOREM 10.2.1. Let $H_1(u) = V\{E \cap K_u\}$, $H(u) = V\{(E + k\mathbf{y}) \cap K_u\}$, and $H^*(u) = V\{(E+\mathbf{y}) \cap K_u\}$. Also let $I\{\ \}$ denote the indicator function of a set. Now

$$\infty > \int_E f(\mathbf{x})\,d\mathbf{x} = \int_E \int_0^{f(\mathbf{x})} du\,d\mathbf{x}.$$

Applying Fubini's theorem, we get

$$\int_E f(\mathbf{x})\,d\mathbf{x} = \int_0^\infty \int_E I\{\mathbf{x} : f(\mathbf{x}) \geqslant u\}\,d\mathbf{x}\,du = \int_0^\infty H_1(u)\,du.$$

From Lemma 10.2.1, $H_1(u) \geqslant H(u) \geqslant H^*(u)$; therefore, as $H_1$, $H$, $H^*$ are positive functions,

$$\int_0^\infty H(u)\,du \quad \text{and} \quad \int_0^\infty H^*(u)\,du$$

exist, that is, are finite. But

$$\int_0^\infty H(u)\,du = \int_0^\infty \int_{E+k\mathbf{y}} I\{\mathbf{x} : f(\mathbf{x}) \geqslant u\}\,d\mathbf{x}\,du$$

$$= \int_0^\infty \int_E I\{\mathbf{x} : f(\mathbf{x} + k\mathbf{y}) \geqslant u\}\,d\mathbf{x}\,du$$

$$= \int_E \int_0^{f(\mathbf{x}+k\mathbf{y})} du\,d\mathbf{x} = \int_E f(\mathbf{x} + k\mathbf{y})\,d\mathbf{x}.$$

Similarly $\int_0^\infty H^*(u)\,du = \int_E f(\mathbf{x}+\mathbf{y})\,d\mathbf{x}$. Therefore, as $0 \leqslant H^*(u) \leqslant H(u)$, we have that

$$\int_E f(\mathbf{x} + k\mathbf{y})\,d\mathbf{x} \geqslant \int_E f(\mathbf{x}+\mathbf{y})\,d\mathbf{x}. \qquad \square$$

**Lemma 10.2.2** *Let* $\mathbf{x} \in R$, *and let* $A$ *be a* $n \times n$ p.s.d. *symmetric matrix. Then the region defined by*

$$E = \{\mathbf{x} \in R^n : \mathbf{x}'A\mathbf{x} \leqslant c\}, \qquad 0 \leqslant c \leqslant \infty,$$

*is a convex region.*

PROOF. Suppose $\mathbf{x}, \mathbf{y} \in E$. Then $\mathbf{z} = \alpha\mathbf{x} + (1-\alpha)\mathbf{y}$, $0 \leqslant \alpha \leqslant 1$, satisfies the following relation:

$$\mathbf{z}'A\mathbf{z} = \alpha^2\mathbf{x}'A\mathbf{x} + (1-\alpha)^2\mathbf{y}'A\mathbf{y} + 2\alpha(1-\alpha)\mathbf{x}'A\mathbf{y}.$$

By applying the Cauchy–Schwartz inequality to the last terms we obtain that

$$\mathbf{z}'A\mathbf{z} \leqslant \alpha^2\mathbf{x}'A\mathbf{x} + (1-\alpha)^2\mathbf{y}'A\mathbf{y} + 2\alpha(1-\alpha)(\mathbf{x}'A\mathbf{x}\mathbf{y}'A\mathbf{y})^{1/2}.$$

As $\mathbf{x},\mathbf{y}\in E$, we have $\mathbf{x}'A\mathbf{x}\leqslant c$ and $\mathbf{y}'A\mathbf{y}\leqslant c$; therefore

$$\mathbf{z}'A\mathbf{z}\leqslant\left[\alpha+(1-\alpha)\right]^2c=c.$$

Hence $\mathbf{z}\in E$. □

**Corollary 10.2.1** *For* $\mathbf{x}$ *a* $p$ *vector and* $\Sigma$ *a* $p\times p$ p.d. *symmetric matrix, the function* $f(\mathbf{x})=(2\pi)^{-p/2}|\Sigma|^{-1/2}\exp(-\frac{1}{2}\mathbf{x}'\Sigma^{-1}\mathbf{x})$ *is a unimodal function (see Definition* 10.2.1*).*

**Corollary 10.2.2** *For* $\mathbf{x}\in R^p$, $A$ p.s.d. *and* $g\geqslant 0$ *a monotonically decreasing function on* $R$, *the function* $f(\mathbf{x})=g(\mathbf{x}'A\mathbf{x})$ *is a unimodal function.*

PROOF. $f(\mathbf{x})$ is unimodal iff $\{\mathbf{x}:f(\mathbf{x})\geqslant u\}=K_u$ is convex for every $u$, $0<u<\infty$. Now

$$K_u=\{\mathbf{x}:f(\mathbf{x})\geqslant u\}=\{\mathbf{x}:\mathbf{x}'A\mathbf{x}\leqslant g^{-1}(u)\}$$

is, from Lemma 10.2.2, a convex set. □

**Theorem 10.2.2** *Let* $\mathbf{x}_1,\ldots,\mathbf{x}_s$, $Y$ *be mutually independent. Suppose* $\mathbf{x}_i\sim N_p(k_i\boldsymbol{\mu}_i,\Sigma_i)$, $i=1,\ldots,s$, *where* $\boldsymbol{\mu}_i$ *is a* $p$ *vector and* $0\leqslant k_i<\infty$. *If* $\omega$ *is a convex set, symmetric about zero in the sample space for each* $\mathbf{x}_i$ *given* $\mathbf{x}_j$ $(j\neq i)$ *and* $Y$, *then* $P(\omega)$ *decreases in each* $k_i$.

PROOF. For arbitrary $i$ let

$$R_i=\{\mathbf{x}_i:(\mathbf{x}_1,\ldots,\mathbf{x}_{i-1},\mathbf{x}_i,\mathbf{x}_{i+1},\ldots,\mathbf{x}_s,Y)\in\omega \text{ given } \mathbf{x}_j\ (j\neq i) \text{ and } Y\}.$$

Let $f_i(\mathbf{x}_i)$ be the density $N_p(\mathbf{0},\Sigma_i)$. From Theorem 10.2.1,

$$\int_{R_i}f_i(\mathbf{x}_i+k_i\boldsymbol{\mu}_i)\,d\mathbf{x}_i\geqslant\int_{R_i}f_i(\mathbf{x}_i+\tilde{k}_i\boldsymbol{\mu}_i)\,d\mathbf{x}_i\qquad\text{for}\quad k_i\leqslant\tilde{k}_i.$$

Multiplying both sides by the joint marginal distribution of $\mathbf{x}_j$ $(j\neq i)$ and $Y$, and integrating over $\omega$, we obtain

$$P(\omega|k_1,\ldots,k_{i-1},k_i,k_{i+1},\ldots,k_s)\geqslant P(\omega|k_1,\ldots,k_{i-1},\tilde{k}_i,k_{i+1},\ldots,k_s).$$ □

## 10.3 Monotonicity of Some Tests for MANOVA

From Chapter 6 we have, in the canonical form, $Y_1\sim N_{p,m_1}(\mu_1,\Sigma,I)$, $Y_2\sim N_{p,m_2}(\mu_2,\Sigma,I)$, and $Y_3\sim N_{p,n}(0,\Sigma,I)$, and we wish to test the hypothesis

(10.3.1) $$H:\mu_2=0\quad\text{vs}\quad A:\mu_2\neq 0.$$

The problem is invariant under transformations of the form

$$Y_1 \to CY_1 + D, \qquad (10.3.2) \qquad Y_2 \to CY_2, \qquad Y_3 \to CY_3,$$

where $C$ is a $p\times p$ nonsingular matrix and $D$ is a $p\times m_1$ matrix. We shall therefore restrict our attention to invariant tests only. Invariant tests are based on functions of the maximal invariants $l_1 \geqslant \dots \geqslant l_t$, where $t=\min(p,m_2)$, the roots of $(Y_3Y_3')^{-1}Y_2Y_2'$. These roots depend only on the maximal invariants of the parameter space, that is, on $\gamma_1 \geqslant \dots \geqslant \gamma_c$, the nonzero roots of $\mu_2\mu_2'\Sigma^{-1}$, $c \leqslant \mathrm{Min}(p,m_2)$. In Subsection 6.3.11, the following four invariant tests were proposed:

(i) The likelihood-ratio test, with acceptance region

$$\prod_{i=1}^{t}(1+l_i) \leqslant c_1, \qquad c_1 \geqslant 1.$$

(ii) The Lawley–Hotelling trace test, with acceptance region

$$\sum_{i=1}^{t} l_i \leqslant c_2, \qquad c_2 \geqslant 0.$$

(iii) Roy's maximum-root test, with acceptance region

$$l_1 \leqslant c_3, \qquad c_3 \geqslant 0.$$

(iv) The Pillai–Nanda trace test, with acceptance region

$$\sum_{i=1}^{t} l_i(1+l_i)^{-1} \leqslant c_4, \qquad 0 \leqslant c_4 \leqslant \mathrm{Min}(p,m_2).$$

Tests (i)–(iv) will be shown to have power functions (for $0\leqslant c_4 \leqslant 1$) monotonically nondecreasing coordinatewise in each of the parameters $\gamma_1,\dots,\gamma_c$ [see Das Gupta, Anderson, and Mudholkar (1964) or J. N. Srivastava (1964) for (i)–(iii)]. From the invariance of the test statistics we shall assume that $\Sigma = I$ and $\mu_2$ is of the form

$$\mu_2 = \begin{pmatrix} D_\gamma^{1/2} & 0 \\ 0 & 0 \end{pmatrix}.$$

where $D_\gamma$ is the diagonal matrix with $\gamma_1,\dots,\gamma_c$ on the diagonal. First, we prove

**Theorem 10.3.1** *Let $\mathbf{y}_1,\dots,\mathbf{y}_{m_2}$ be the columns of $Y_2$, defined above. If the acceptance region $\omega$ of an invariant test is convex in each $\mathbf{y}_i$ for fixed $\mathbf{y}_j$ $(j\neq i)$ and $Y_3$, then the power of the test increases in each $\gamma_i$.*

PROOF. Let $\mathbf{e}_i$ be the $p$ vector with 1 in the $i$th position and 0 elsewhere, $i=1,\ldots,p$. Then for arbitrary $j$, $\mathbf{y}_j \sim N_p(\gamma_j^{1/2}\mathbf{e}_j, I)$. As the convex acceptance region $\omega$ depends only on the roots of $(Y_2Y_2')(Y_3Y_3')^{-1}$, the region is invariant under the transformation $\mathbf{y}_j \to -\mathbf{y}_j$. Hence from Theorem 10.2.2, $P(\omega)$ decreases in each $\gamma_j^{1/2}$; hence, the power of the test is nondecreasing in each $\gamma_j$. □

**Corollary 10.3.1** *If the acceptance region of an invariant test is convex in $Y_2$ for each fixed $Y_3$, then the power of the test increases monotonically in each $\gamma_i$.*

We have shown that any invariant test with a convex acceptance region in the columns of $Y_2$ has a monotone increasing power function. All that remains to show is that tests based on the statistics (i)–(iv) have a convex acceptance region.

**Theorem 10.3.2** *The acceptance regions* (i)–(iii) *are convex. The acceptance region* (iv) *is convex if* $0 \leqslant c_4 \leqslant 1$.

PROOF. Let $\mathbf{y}_j$ be the $j$th column of $Y_2$, and let

$$A = \sum_{\substack{i \neq j \\ i=1}}^{m_2} \mathbf{y}_i\mathbf{y}_i' \quad \text{and} \quad B = Y_3Y_3'.$$

Then for fixed $A$ and $B$ the acceptance region in (i) becomes

$$\left\{\mathbf{y}_j : 1 + \mathbf{y}_j'(A+B)^{-1}\mathbf{y}_j \leqslant c_1|B|/|A+B|\right\}. \tag{10.3.3}$$

It follows from Lemma 10.2.2 that this region is convex in $\mathbf{y}_j$.

The convexity of the acceptance regions of the Lawley–Hotelling trace test and Roy's maximum-root test are left as exercises to the readers in Problems 10.3–10.5. The fourth test mentioned, the Pillai–Nanda trace test, has the acceptance region given by

$$\left\{(Y_2, Y_3) : \operatorname{tr}\left[Y_2Y_2'(Y_2Y_2' + Y_3Y_3')^{-1}\right] \leqslant c_4\right\}. \tag{10.3.4}$$

We shall now show that if $0 \leqslant c_4 \leqslant 1$, the above region (10.3.4) is convex in the columns of $Y_2$. Let

$$Y_2 \equiv (\mathbf{y}, Z), \qquad Y_2Y_2' \equiv \mathbf{y}\mathbf{y}' + A, \quad \text{and} \quad Y_3Y_3' \equiv B. \tag{10.3.5}$$

For fixed $Z$ and $Y_3$, the region becomes

$$\left\{\mathbf{y} : \operatorname{tr}(\mathbf{y}\mathbf{y}' + A)(\mathbf{y}\mathbf{y}' + A + B)^{-1} \leqslant c_4\right\}. \tag{10.3.6}$$

Thus, we need to show the convexity of the above region in $\mathbf{y}$. Since

$$(A+B+\mathbf{y}\mathbf{y}')^{-1}=(A+B)^{-1}-\frac{(A+B)^{-1}\mathbf{y}\mathbf{y}'(A+B)^{-1}}{1+\mathbf{y}'(A+B)^{-1}\mathbf{y}},$$

(10.3.7)
$$\begin{aligned}\operatorname{tr}(\mathbf{y}\mathbf{y}'+A)(\mathbf{y}\mathbf{y}'+A+B)^{-1} &= p-\operatorname{tr}B(\mathbf{y}\mathbf{y}'+A+B)^{-1}\\ &= p-\operatorname{tr}(A+B)^{-1}B+\frac{\mathbf{y}'(A+B)^{-1}B(A+B)^{-1}\mathbf{y}}{1+\mathbf{y}'(A+B)^{-1}\mathbf{y}}.\end{aligned}$$

The above region is convex in $\mathbf{y}$ iff the region

(10.3.8)
$$\begin{aligned}\mathbf{y}'(A+B)^{-1/2}\{&(A+B)^{-1/2}B(A+B)^{-1/2}\\ &-[\operatorname{tr}(A+B)^{-1}B-p+c_4]I\}(A+B)^{-1/2}\mathbf{y}\\ &\leqslant \operatorname{tr}(A+B)^{-1}B-p+c_4\end{aligned}$$

is convex in $\mathbf{y}$.

From Eqs. (10.3.6) and (10.3.7), the right side of the inequality (10.3.8) is positive ($\geqslant 0$). Hence from Lemma 10.2.2, the region (10.3.6) is convex in $\mathbf{y}$ if

(10.3.9) $$(A+B)^{-1/2}B(A+B)^{-1/2}-[\operatorname{tr}(A+B)^{-1}B-p+c_4]I\geqslant 0.$$

Let $1\geqslant\theta_1\geqslant\ldots\geqslant\theta_p\geqslant 0$ be the roots of $(A+B)^{-1}B$. Then the region (10.3.6) is convex if

(10.3.10) $$D_\theta-I\left(\sum_{i=1}^{p}\theta_i\right)+(p-c_4)I\geqslant 0,$$

where $D_\theta=\operatorname{diag}(\theta_1,\ldots,\theta_p$—that is, if

(10.3.11) $$-D_\mu+(p-c_4)I\geqslant 0,$$

where $D_\mu=\operatorname{diag}(\mu_1,\ldots,\mu_p)$ with $\mu_i=\Sigma\theta_j$; the $\Sigma$ denotes the summation over all $j$ except $j\neq i$. But $\mu_i\leqslant p-1$. Hence for $0\leqslant c_4\leqslant 1$, Eq. (10.3.11) holds. □

### *10.3.1 Bibliographical Note*

Roy and Mikhail (1961) were the first to give the monotonicity result of Roy's maximum-root test. Their proof, however, was incomplete, and before a completed proof could be published (made available to us), Das Gupta, Anderson, and Mudholkar (1964) and J. N. Srivastava (1964) obtained independently not only the monotonicity of Roy's test but also of the likelihood-ratio test and Lawley's test. The monotonicity of Pillai—Nanda's test was given by Perlman (1974) for $0\leqslant c_4\leqslant 1$. When $c_4$ does not belong to the above range, it is not even known if the test is unbiased, contrary to the general belief.

## 10.4 Monotonicity of Tests for Covariance

In this section the monotonicity of various tests for covariance is presented. In some cases only unbiasedness is known.

### *10.4.1 Monotonicity of the Modified Likelihood-Ratio Test for $\Sigma = I$*

Let $\mathbf{y} \sim N_p(\boldsymbol{\mu}, \Sigma)$ and $V \sim W_p(\Sigma, n)$ be independently distributed. The critical region of the likelihood-ratio test for $H: \Sigma = I$ against the alternative $A: \Sigma \neq I$ is given by

$$\omega = \left\{ V : V > 0 \text{ and } |V|^{\frac{1}{2}N} \operatorname{etr}\left(-\tfrac{1}{2}V\right) \leq k \right\}, \tag{10.4.1}$$

where $N = n+1$. When $p=1$, it follows from Lehmann (1959, p. 165) or from Problem 10.1 that this test is not even unbiased, while the modified likelihood-ratio test in which $N$ is replaced by $n = N-1$, the number of d.f. associated with $V$, is not only unbiased but UMPU.

Thus when $p=1$, the monotonicity result is a trivial consequence. For completeness, however, we prove this result here in Theorem 10.4.1. The critical region in this case is given by

$$\omega^* = \left\{ v : v > 0 \text{ and } v^{n/2} e^{-\frac{1}{2}v} \leq k \right\}, \tag{10.4.2}$$

where $v/\sigma^2$ has a $\chi^2$ distribution with $n$ d.f.

**Theorem 10.4.1.** *Let $s$ be a random variable such that the distribution of $s/\sigma^2$ is $\chi^2$ with $n$ d.f. Let*

$$\beta(\sigma^2) = P_{\sigma^2}\left\{ s : s > 0 \text{ and } s^{n/2} e^{-\frac{1}{2}s} \leq k \right\}. \tag{10.4.3}$$

*Then*

$$\frac{d\beta(\sigma^2)}{d\sigma^2} \gtreqless 0 \quad \text{according as} \quad \sigma^2 \gtreqless 1.$$

PROOF. Since the equation $s^{\frac{1}{2}n} e^{-\frac{1}{2}s} = k$ has exactly two solutions, $s = c_1$ and $s = c_2$ ($c_1 < c_2$), where $c_1$ and $c_2$ are given by

$$c_1^{n/2} e^{-\frac{1}{2}c_1} = c_2^{n/2} e^{-\frac{1}{2}c_2} = k, \tag{10.4.4}$$

we get

$$\begin{aligned}
\beta(\sigma^2) &= c_{1n} \int_0^{c_1} \left(\frac{s}{\sigma^2}\right)^{\frac{1}{2}n-1} \exp\left(-\frac{s}{2\sigma^2}\right) d\left(\frac{s}{\sigma^2}\right) \\
&\quad + c_{1n} \int_{c_2}^{\infty} \left(\frac{s}{\sigma^2}\right)^{\frac{1}{2}n-1} \exp\left(-\frac{s}{2\sigma^2}\right) d\left(\frac{s}{\sigma^2}\right) \\
&= c_{1n} \int_0^{c_1/\sigma^2} u^{\frac{1}{2}n-1} e^{-\frac{1}{2}u}\, du + c_{1n} \int_{c_2/\sigma^2}^{\infty} u^{\frac{1}{2}n-1} e^{-\frac{1}{2}u}\, du,
\end{aligned}$$

where $c_{1n}^{-1}=2^{\frac{1}{2}n}\Gamma(\frac{1}{2}n)$. Hence

$$\frac{d\beta(\sigma^2)}{d\sigma^2}=c_{1n}\left\{\left(\frac{c_1}{\sigma^2}\right)^{\frac{1}{2}n-1}\exp\left[-\frac{1}{2}\frac{c_1}{\sigma^2}\right]\left(-\frac{c_1}{\sigma^4}\right)\right.$$
$$\left.-\left(\frac{c_2}{\sigma^2}\right)^{\frac{1}{2}n-1}\exp\left[-\frac{1}{2}\frac{c_2}{\sigma^2}\right]\left(-\frac{c_2}{\sigma^4}\right)\right\}$$
$$=c_{1n}(\sigma^2)^{-\frac{1}{2}n-1}\left[c_2^{\frac{1}{2}n}\exp\left(-\frac{c_2}{2\sigma^2}\right)-c_1^{n/2}\exp\left(-\frac{c_1}{2\sigma^2}\right)\right]$$
$$=c_{1n}(\sigma^2)^{-\frac{1}{2}n-1}\left[c_1^{\frac{1}{2}n}\exp\left(\frac{-c_2}{2\sigma^2}\right)\right]$$
$$\times\left[\exp\tfrac{1}{2}(c_2-c_1)-\exp\tfrac{1}{2}\sigma^{-2}(c_2-c_1)\right]$$

from Eq. (10.4.4). Thus if $\sigma^2>1$, then $d\beta(\sigma^2)/d\sigma^2>0$, and if $\sigma^2<1$, then $d\beta(\sigma^2)/d\sigma^2<0$. □

**Theorem 10.4.2** *Let* $\mathbf{y}\sim N_p(\boldsymbol{\mu},\Sigma)$ *and* $V\sim W_p(\Sigma,n)$ *be independently distributed. Then the power of the modified likelihood-ratio test (replacing N by n) for testing the hypothesis* $H:\Sigma=I$ vs $A:\Sigma\neq I$ *increases monotonically as the absolute deviation of each characteristic root of* $\Sigma$ *from* 1 *increases.*

PROOF. The critical region of the modified likelihood-ratio test (see Chapter 7) is given by

$$\omega=\left\{V:V>0\text{ and }|V|^{\frac{1}{2}n}\operatorname{etr}\left(-\tfrac{1}{2}V\right)\leqslant k\right\},\tag{10.4.5}$$

where $k$ is determined by the size of the test. Note that

$$|V|^m\operatorname{etr}\left(-\tfrac{1}{2}V\right)=|V|^m\left(\prod_{j=1}^{p}v_{jj}^m\right)^{-1}\prod_{j=1}^{p}\left[v_{jj}^m\exp\left(-\tfrac{1}{2}v_{jj}\right)\right]$$
$$\equiv|R|^m\prod_{j=1}^{p}\left[v_{jj}^m\exp\left(-\tfrac{1}{2}v_{jj}\right)\right],$$

where $V=(v_{ij})$ and $R=(r_{ij})$ with $r_{ij}=v_{ij}/(v_{ii}v_{jj})^{1/2}$. Since the problem and the critical region $\omega$ in Eq. (10.4.5) remain invariant under the group of orthogonal transformations, we may assume without any loss of generality that $\Sigma\equiv\Delta=\operatorname{diag}(\lambda_1,\ldots,\lambda_p)$. From the results of Chapter 3 it is known that when $\Sigma=\Delta$, $R$ and $v_{jj}$ $(j=1,2,\ldots,p)$ are mutually independently distributed. Thus $P_\Delta(\omega)$

$$=C(p,n)\int_{R>0}|R|^{\frac{1}{2}(n-p-1)}dR\int_{\omega_R}|\Delta|^{-\frac{1}{2}n}\prod_{j=1}^{p}\left[v_{jj}^{\frac{1}{2}(n-p-1)}\exp\left(-\frac{1}{2}\frac{v_{jj}}{\lambda_j}\right)\right]dv_{jj}$$
$$=C(p,n)\int_{R>0}|R|^{\frac{1}{2}(n-p-1)}\beta(\lambda_1,\ldots,\lambda_p|R)\,dR,$$

where

$$\omega_R = \left\{ V : \prod_{j=1}^{p} \left[ v_{jj}^{\frac{1}{2}n} \exp\left(-\tfrac{1}{2}v_{jj}\right)\right] \leqslant k|R|^{-\frac{1}{2}n}|R \right\},$$

$$\beta(\lambda_1,\ldots,\lambda_p|R) = P\left[\omega_{11},\ldots,\omega_{pp} \in \omega_R|R\right]$$

and $C(p,n)$ is the constant of the Wishart distribution. By considering $v_{22},\ldots,v_{pp}$ fixed, it can be shown from Theorem 10.4.1 that $\beta(\lambda_1,\ldots,\lambda_p|R)$ increases monotonically as $|\lambda_1 - 1| > 0$. The result now follows by repeating the above argument for other variables. □

This result was first proved by Nagao (1967) and independently by Das Gupta (1969).

### *10.4.2 Monotonicity of the Likelihood-Ratio Tests for Sphericity and Homogeneity of Variances*

Let $S \sim W_p(\Sigma, n)$. Then the critical region for testing the hypothesis $H:\Sigma = \sigma^2 I$, $\sigma^2$ unknown, against the alternative $A:\Sigma \neq \sigma^2 I$ is given by

$$B = \left\{ S : |S|^n (\operatorname{tr} S)^{-np} \leqslant c \right\}, \tag{10.4.6}$$

where $c$ is determined by the size of the test, and $N = n+1$. The problem of testing $H:\Sigma = \sigma^2 I$ against $A:\Sigma > 0$ remains invariant under the transformation $S \to b^2\Gamma S\Gamma'$, where $\Gamma'\Gamma = I$, and $b$ is a scalar. Thus, if $\sigma_1 \geqslant \sigma_2 \geqslant \cdots \geqslant \sigma_p$ denote the ordered roots of $\Sigma$, the problem becomes that of testing $H:\sigma_1 = \sigma_2 = \cdots = \sigma_p$ against $A:\sigma_i \neq \sigma_j$ for at least one pair $(i,j)$, $i \neq j$, $i,j = 1,2,\ldots,p$. The power of the test will depend on the $p-1$ ratios of these roots, say $\delta_j = \sigma_j/\sigma_{j+1}$, $j = 1,2,\ldots,p-1$, $\delta_j \geqslant 1$. These $p-1$ ratios form a set of maximal invariants. It is shown in Theorem 10.4.4 that the power of the test in (10.4.6) is a monotonically nondecreasing function of $\delta_k$, while the remaining $p-2$ parameters $\delta_i = \sigma_i/\sigma_{i+1}$, $i = 1,2,\ldots,p-1$, $i \neq k$, are held fixed. This will be shown to be a special case of the monotonicity of the modified likelihood-ratio test for the homogeneity of variances. First we state this result for two populations:

**Lemma 10.4.1** *Let $x$ have the density given by*

$$\left[B(f_1,f_2)\right]^{-1}\lambda^{f_1}x^{f_1-1}(1+\lambda x)^{-f_1-f_2} \qquad \text{for} \quad 0 < x < \infty,$$

*where $0 < \lambda, f_1, f_2 < \infty$. Then, for any constant $0 < c < 1$, $g(\lambda) = P(x^{f_1}(1+x)^{-f_1-f_2} \leqslant c)$ increases as $\lambda$ increases or decreases from 1.*

The proof follows from the UMPU property of the $F$ test for the equality of two variances. Alternatively, it can be proved on the lines of Theorem 10.4.1.

Next, we consider the monotonicity of the power function of the modified likelihood-ratio test for the homogeneity of $p$ variances. Let $s_j$, $j=1,\dots,p$, be independently distributed as $\sigma_j \chi^2_{n_j}$. The likelihood-ratio criterion (hereafter referred to as LRC) for testing

$$H_0:\sigma_1=\dots=\sigma_p=\sigma \text{ (say)} \quad \text{vs} \quad H_1:\sigma_i\neq\sigma_j \quad \text{for some} \quad i\neq j, \quad i,j=1,\dots,p, \tag{10.4.7}$$

where $\sigma$ is unknown, is as follows: Reject $H_0$ if

$$\frac{\prod_{j=1}^{p} s_j^{n_j}}{\left(\sum_{j=1}^{p} s_j\right)^{N(p)}} \leqslant c, \tag{10.4.8}$$

where

$$N(r)=\sum_{j=1}^{r} n_j \tag{10.4.9}$$

and $c$ is so chosen that the error of the first kind (size of the test) is at a specified level. Without loss of generality we shall assume that the parameters $\sigma_i$ are ordered as $\sigma_1 \geqslant \dots \geqslant \sigma_p$. The power of this test is shown below in Theorem 10.4.3 to be a monotone nondecreasing function of $\delta_j=\sigma_j/\sigma_{j+1}$, $j=1,2,\dots,p-1$, $\delta_j \geqslant 1$.

It is to be noted that the above test is the *modified* LRC (Bartlett 1937) in which $N_i$, the sample size from the $i$th population, is replaced by $n_i$, the number of d.f. associated with $s_i$, for testing the homogeneity of variances of $p$ normal populations. It may be mentioned that the *unmodified* LRC is not even unbiased unless the sample sizes are equal [see Sugiura and Nagao (1968)].

**Theorem 10.4.3** *Let $s_j$ be independently distributed as $\frac{1}{2}\sigma_j\chi^2_{n_j}$, $j=1,\dots,p$. Suppose also that $\sigma_1 \geqslant \dots \geqslant \sigma_p$. Let*

$$A=A(s_1,\dots,s_p) = \left\{ (s_1,\dots,s_p): \left(\prod_{j=1}^{p} s_j^{n_j}\right)\left(\sum_{j=1}^{p} s_j\right)^{-N(p)} \leqslant c \right\}, \tag{10.4.10}$$

*where we define $N(r)=\sum_{j=1}^{r} n_j$, and $c$ is an arbitrary constant. Let $P(A)=P((s_1,\dots,s_p)\in A)$. Then for any $k$ $(1 \leqslant k \leqslant p-1)$, $P(A)$ is a nondecreasing function of $\delta_k=\sigma_k/\sigma_{k+1}$, while the remaining $p-2$ parameters $\delta_i=\sigma_i/\sigma_{i+1}$, $i=1,\dots,p-1$, $i\neq k$, are held fixed.*

PROOF. As the region $A$ is invariant under scale transformations of the $s_i$'s, we consider the following transformation:

$$x_j = s_j / s_k, \qquad j = 1, 2, \dots, p, \quad j \neq k. \tag{10.4.11}$$

Then, integrating over $s_k$ we obtain the joint density function of $x_1, \dots, x_{k-1}, x_{k+1}, \dots, x_p$ as

$$c_1 \left[ \prod_{\substack{j=1 \\ j \neq k}}^{p} \left( \frac{\sigma_k}{\sigma_j} \right)^{f_j} x_j^{f_j - 1} \right] \left[ 1 + \sum_{\substack{i=1 \\ i \neq k}}^{p} \frac{\sigma_k x_i}{\sigma_i} \right]^{-f_0}, \tag{10.4.12}$$

where $f_j = \frac{1}{2} n_j, j = 1, \dots, p, f_0 = \Sigma_{j=1}^{p} f_j$, and $c_1 = [\Pi_{j=1}^{p} \Gamma(f_j)] / \Gamma(f_0)$.

Now consider a transformation of the following kind:

$$x_i = x_{k+1} u_i, \qquad i = k+2, \dots, p, \tag{10.4.13}$$

$$x_{k+1} = u_{k+1} \frac{1 + \sum_{j=1}^{k-1} x_j}{1 + \sum_{i=k+2}^{p} u_i}. \tag{10.4.14}$$

The joint pdf of $x_1, \dots, x_{k-1}, u_{k+1}, \dots, u_p$ is given by

$$\begin{aligned} & c_1 \left( \prod_{j=1}^{k-1} \lambda_j^{f_j} x_j^{f_j - 1} \right) \left( \prod_{i=k+2}^{p} \lambda_i^{f_i} u_i^{f_i - 1} \right) (\delta_k \gamma)^{f_{(k+1)}} \\ & \times u_{k+1}^{f_{(k+1)} - 1} (1 + \delta_k \gamma u_{k+1})^{-f_0} \left( 1 + \sum_{i=k+2}^{p} \gamma_i u_i \right)^{-f_{(k+1)}} \\ & \times \left( 1 + \sum_{j=1}^{k-1} \lambda_j x_j \right)^{-\Sigma_{j=1}^{k} f_j}, \end{aligned} \tag{10.4.15}$$

where $f_{(k+1)} = \Sigma_{i=k+1}^{p} f_i$, $\lambda_j = \sigma_k / \sigma_j$ for $1 \leqslant j \leqslant k$, $\lambda_i = \sigma_{k+1} / \sigma_i$ for $k+2 \leqslant i \leqslant p$, $\delta_k = \sigma_k / \sigma_{k+1}$, and

$$\gamma = \left( 1 + \sum_{j=1}^{k-1} x_j \right) \left( 1 + \sum_{i=k+2}^{p} \lambda_i u_i \right) \left( 1 + \sum_{j=1}^{k-1} \lambda_j x_j \right)^{-1} \left( 1 + \sum_{i=k+2}^{p} u_i \right)^{-1}. \tag{10.4.16}$$

Hence the conditional pdf of $u_{k+1}$ given $x_1, \dots, x_{k-1}, u_{k+2}, \dots, u_p$ is given by

$$\text{const} (\delta_k \gamma)^{f(k+1)} u_{k+1}^{f_{(k+1)} - 1} (1 + \gamma \delta_k u_{k+1})^{-f_0},$$

and the conditional region for $u_{k+1}$ is

$$(10.4.17)\qquad \omega_1 : u_{k+1}^{f_{(k+1)}}(1+u_{k+1})^{-f_0} \leqslant \frac{c\left(1+\sum_{i=k+2}^{p} u_i\right)^{f_{(k+1)}}\left(1+\sum_{j=1}^{k-1} x_j\right)^{\sum_{j=1}^{k} f_j}}{\left(\prod_{j=1}^{k-1} x_j^{f_j}\right)\left(\prod_{i=k+2}^{p} u_i^{f_i}\right)}.$$

Since the parameter point is $\sigma_1 \geqslant \ldots \geqslant \sigma_k \geqslant \sigma_{k+1} \geqslant \ldots \geqslant \sigma_p$, we obtain that $\lambda_1 \leqslant \ldots \leqslant \lambda_{k-1} \leqslant 1 \leqslant \lambda_{k+2} \leqslant \ldots \leqslant \lambda_p$, and hence that $\gamma \geqslant 1$. Therefore, from the properties of the $F$ test given in Lemma 10.4.1,

$$(10.4.18)\qquad P(u_{k+1} \in \omega_1 | x_1, \ldots, x_{k-1}, u_{k+2}, \ldots, u_p)$$

increases as $\delta_k$ increases from 1 with $\lambda_i$, $i=1,\ldots,p-1$, $i \neq k$, kept fixed, or equivalently with $\delta_i = \sigma_i / \sigma_{i+1}$, $i=1,\ldots,p-1$, $i \neq k$, kept fixed. Hence, on averaging with respect to the conditional variables, we obtain the required result. □

**Corollary 10.4.1** *Suppose that for any test the conditional acceptance region given $x_1,\ldots,x_{k-1},u_{k+2},\ldots,u_p$ is of the form $a \leqslant u_{k+1} \leqslant b$. If $a^{f_{(k+1)}}(1+a)^{-f_0} \geqslant b^{f_{(k+1)}}(1+b)^{-f_0}$, then the power of the test is monotone increasing in $\delta_k = \sigma_k/\sigma_{k+1}$ for fixed $\delta_i = \sigma_i/\sigma_{i+1}$, $i=1,\ldots,p-1$, $i \neq k$.*

We shall now use the above results to show the monotonicity of the sphericity test. Let $S$ be distributed as $W_p(\frac{1}{2}\Sigma, n)$. The density of $S$ is given by

$$\pi^{-\frac{1}{4}p(p-1)} \prod_{j=1}^{p} \Gamma^{-1}\left(\frac{n-j+1}{2}\right) |\Sigma|^{-\frac{1}{2}n} |S|^{\frac{1}{2}[n-(p+1)]} \exp(-\operatorname{tr}\Sigma^{-1}S).$$

Define $P(B) = P(S \in B)$, where $B = \{S : |S|^n (\operatorname{tr} S)^{-pn} \leqslant c\}$. As $B$ is invariant under orthogonal transformations, we shall assume $\Sigma$ to be diagonal $(\sigma_1,\ldots,\sigma_p)$ with $\sigma_1 \geqslant \ldots \geqslant \sigma_p$. Therefore

$$(10.4.19)\qquad P(B) = \int_B \pi^{-\frac{1}{4}p(p-1)} \prod_{j=1}^{p} \Gamma^{-1}\left[\tfrac{1}{2}(n-j+1)\right] |S|^{\frac{1}{2}n - \frac{1}{2}(p+1)} \times \prod_{j=1}^{p} \sigma_j^{-\frac{1}{2}n} \exp\left(-\sum_{j=1}^{p} s_{jj}\sigma_j^{-1}\right) dS.$$

By making the transformation

$$(10.4.20)\qquad S = \operatorname{diag}(s_{11}^{1/2},\ldots,s_{pp}^{1/2})\, R \,\operatorname{diag}(s_{11}^{1/2},\ldots,s_{pp}^{1/2}),$$

we obtain

(10.4.21) $P(B)$

$$= \int_R \pi^{-\frac{1}{4}p(p-1)}\left(\prod_{j=1}^{p}\Gamma^{-1}\left[\tfrac{1}{2}(n-j+1)\right]\right)\Gamma^p\left(\tfrac{1}{2}n\right)|R|^{\frac{1}{2}n-\frac{1}{2}(p+1)}$$

$$\times\int_B \prod_{j=1}^{p}\sigma_j^{-\frac{1}{2}n}\Gamma^{-p}\left(\tfrac{1}{2}n\right)\prod_{j=1}^{p}s_{jj}^{\frac{1}{2}n-1}\exp\left(-\sum_{j=1}^{p}s_{jj}\sigma_j^{-1}\right)\prod_{j=1}^{p}ds_{jj}\,dR,$$

where

$$B \equiv B(s_{jj}, j=1,\ldots,p|R) = \left\{(s_{11},\ldots,s_{pp}):\left(\prod_{j=1}^{p}s_{jj}^{n}\right)\left(\sum_{j=1}^{p}s_{jj}\right)^{-np}\leqslant c|R|^{-n}\right\}.$$

By simply applying Theorem 10.4.3 to the inner integral we obtain the result that $P(B)$ is a monotone nondecreasing function of each $\delta_k = \sigma_k/\sigma_{k+1}$, when other $\delta_i$ $(i=1,2,\ldots,p-1)$ $i\neq k$ are held fixed.

**Theorem 10.4.4** *The power function of the sphericity test* (10.4.6) *is a monotone nondecreasing function of* $\delta_k=\sigma_k/\sigma_{k+1}$ *while the remaining* $p-2$ *parameters* $\delta_i=\sigma_i/\sigma_{i+1}$, $i=1,2,\ldots,p-1$, $i\neq k$, *are held fixed.*

The complex sphericity test has exactly the same monotonicity properties as the real test. The proof is identical except that $S$ and hence $R$ are Hermitian matrices. These results are due to Carter and Srivastava (1977).

For the case of equal sample sizes Cohen and Strawderman (1971) have shown the unbiasedness of a large class of tests for the homogeneity of variances.

### *10.4.3 Unbiasedness of the Likelihood-Ratio Test for* $\boldsymbol{\mu}=\mathbf{0}$ *and* $\Sigma = I$

The unbiasedness of the likelihood-ratio test for $\boldsymbol{\mu}=\mathbf{0}$ and $\Sigma=I$ was proved by Sugiura and Nagao (1968) and independently by Das Gupta (1969). The result presented here is due to Das Gupta. The monotonicity result is not yet available.

**Theorem 10.4.5** *Let* $\mathbf{y}\sim N_p(\boldsymbol{\mu},\Sigma)$ *and* $V\sim W_p(\Sigma,n)$ *be independently distributed. Then the likelihood-ratio test for the hypothesis* $H:\boldsymbol{\mu}=\mathbf{0}$ *and* $\Sigma=I$ *against the alternative* $A\neq H$ *is unbiased.*

PROOF. The critical region of the likelihood-ratio test is given by

(10.4.21a) $$\omega=\left\{(V,\mathbf{y}):V>0 \text{ and } |V|^{N/2}\operatorname{etr}\left[-\tfrac{1}{2}(V+\mathbf{y}\mathbf{y}')\right]\leqslant k\right\}.$$

Thus the unbiasedness can be shown if we can establish the following two inequalities:

(i) $P(\bar{\omega}|\boldsymbol{\mu}=\mathbf{0},\Sigma=I_p)>P(\bar{\omega}|\boldsymbol{\mu}=\mathbf{0},\Sigma)$ for any positive definite matrix $\Sigma\neq I_p$.

(ii) $P(\bar{\omega}|\boldsymbol{\mu}=\mathbf{0},\Sigma)>P(\bar{\omega}|\boldsymbol{\mu},\Sigma)$ for any $\boldsymbol{\mu}\neq 0$, where $\bar{\omega}$ denotes the complement of $\omega$.

To prove inequality (i) we may assume without loss of generality that $\Sigma$ is diagonal. Let

(10.4.22)
$$A=V+\mathbf{y}\mathbf{y}'.$$

Then

(10.4.23)
$$|V|^{N/2}\operatorname{etr}\left(-\tfrac{1}{2}A\right)=\left[|V|/|A|\right]^{N/2}\left[|A|^{N/2}\operatorname{etr}\left(-\tfrac{1}{2}A\right)\right].$$

When $\boldsymbol{\mu}=\mathbf{0}$, the distribution of the first factor in the above expression is free from any parameters and is independent of the second factor. The inequality thus follows from Theorem 10.4.2. The second inequality follows from the results of Section 10.3, since the region $\omega$ is convex for fixed $V$. □

### *10.4.4 Some Monotonicity and Unbiasedness Results for the Modified Likelihood-Ratio Test for $\Sigma_1=\Sigma_2$*

Let $\mathbf{y}_1\sim N_p(\boldsymbol{\mu}_1,\Sigma_1)$, $V_1\sim W_p(\Sigma_1,n_1)$, $\mathbf{y}_2\sim N_p(\boldsymbol{\mu}_2,\Sigma_2)$, and $V_2\sim W_p(\Sigma_2,n_2)$ be independently distributed. Then the likelihood-ratio test for the problem of testing the hypothesis $H:\Sigma_1=\Sigma_2$ against the alternative $A:\Sigma_1\neq\Sigma_2$ where $\boldsymbol{\mu}_1$ and $\boldsymbol{\mu}_2$ are unspecified has the acceptance region

(10.4.24)
$$\omega'=\left\{(V_1,V_2):V_1>0,\ V_2>0,\ |V_1|^{\frac{1}{2}N_1}|V_2|^{\frac{1}{2}N_2}|V_1+V_2|^{-\frac{1}{2}(N_1+N_2)}\geqslant c_\alpha\right\},$$

where $N_i=n_i+1$. For $p=1$, Brown (1939) has shown that this acceptance region gives an unbiased test iff $N_1=N_2$. Also, it can be obtained from Lehmann (1959, p. 170) that the uniformly most powerful unbiased test has the acceptance region $\omega$ obtained from $\omega'$ by replacing $N_i$ with $n_i$; this is called the modified likelihood-ratio test and was proposed by Bartlett (1937).

We shall first establish

**Theorem 10.4.6** *Let the density function of a random* p.d. *matrix $Q$ be*

$$h(Q|A)=\left[B_p(f_1,f_2)\right]^{-1}|A|^{f_1}|Q|^{f_1-\frac{1}{2}(p+1)}|I+AQ|^{-f_1-f_2}$$

*for* $Q > 0$, $f_1 > 0$, *and* $f_2 > 0$, *where*

$$A = \operatorname{diag}(a_1 I_{p_1}, a_2 I_{p_2}, \ldots, a_r I_{p_r}), \quad a_1 > a_2 > \ldots > a_r > 0, \quad \sum_{i=1}^{r} p_i = p,$$

$$B_p(f_1, f_2) = \frac{\Gamma_p(f_1)\Gamma_p(f_2)}{\Gamma_p(f_1 + f_2)}$$

*and*

$$\Gamma_p(n) = \pi^{\frac{1}{4}p(p-1)} \prod_{i=1}^{p} \Gamma\left(n - \frac{i-1}{2}\right).$$

*Let* $g(a_1, a_2, \ldots, a_r) = P(|Q|^{f_1}|I+Q|^{-f_1-f_2} \leqslant c)$ *for any* $0 < c < 1$. *Then, for* $a_1 \geqslant \tilde{a}_1 \geqslant \operatorname{Max}(1, a_2)$,

$$g(a_1, a_2, \ldots, a_r) \geqslant g(\tilde{a}_1, a_2, \ldots, a_r),$$

*while for* $a_r \leqslant \tilde{a}_r \leqslant \operatorname{Min}(1, a_{r-1})$,

$$g(a_1, a_2, \ldots, a_r) \geqslant g(a_1, a_2, \ldots, a_{r-1}, \tilde{a}_r).$$

In order to prove Theorem 10.4.6, we need

**Lemma 10.4.2** *Let* $D = \operatorname{diag}(I_{p_1}, 0)$ *be a* $p \times p$ *diagonal matrix whose first* $p_1$ *diagonal elements are one and the rest zeros. Then*

$$|I+AQ|^{-1}\frac{\partial}{\partial a_1}|I+AQ| = \operatorname{tr} Q(I+AQ)^{-1}D$$

$$= \frac{p_1 - \operatorname{tr}(I+AQ)^{-1}D}{a_1}.$$

PROOF. By definition

$$\frac{\partial}{\partial a_1}|I+AQ| = \lim_{\varepsilon \downarrow 0} \frac{|I+(A+\varepsilon D)Q| - |I+AQ|}{\varepsilon}$$

$$= \lim_{\varepsilon \downarrow 0} |I+AQ| \frac{|I+\varepsilon DQ(I+AQ)^{-1}| - 1}{\varepsilon}$$

$$= \lim_{\varepsilon \downarrow 0} |I+AQ| \frac{1 + \varepsilon \operatorname{tr} DQ(I+AQ)^{-1} + O(\varepsilon^2) - 1}{\varepsilon}$$

$$= |I+AQ| \operatorname{tr} DQ(I+AQ)^{-1}$$

$$= |I+AQ| \operatorname{tr} DA^{-1}AQ(I+AQ)^{-1}$$

$$= |I+AQ| \operatorname{tr} DA^{-1}(I+AQ-I)(I+AQ)^{-1}$$

$$= |I+AQ|\left[\operatorname{tr} DA^{-1} - \operatorname{tr} DA^{-1}(I+AQ)^{-1}\right]$$

$$= |I+AQ|\left[\frac{p_1}{a_1} - \frac{1}{a_1}\operatorname{tr} D(I+AQ)^{-1}\right]. \qquad \square$$

PROOF OF THEOREM 10.4.6. Let $g \equiv g(a_1,\ldots,a_r)$ and

$$\mathfrak{D} = \{Q : |Q|^{f_1}|I+Q|^{-f_1-f_2} \leqslant c\}$$

Then from Lemma 10.4.2, we get

$$\frac{\partial g}{\partial a_1} = \frac{p_1 f_1}{a_1} g - (f_1+f_2)\frac{p_1}{a_1} \times \int_{\mathfrak{D}} h(Q|A)\left[1 - p_1^{-1}\operatorname{tr}(I+AQ)^{-1}D\right] dQ. \tag{10.4.25}$$

Let

$$(I+AQ)^{-1} = (b_{ij}).$$

Then it is easy to verify that

$$\int_{\mathfrak{D}} b_{ii} h(Q|A)\, dQ = \int_{\mathfrak{D}} b_{11} h(Q|A)\, dQ, \qquad i = 1,2,\ldots,p_1.$$

Hence Eq. (10.4.25) can be written as

$$\frac{\partial g}{\partial a_1} = \frac{p_1 f_1}{a_1} g - \frac{p_1}{a_1}(f_1+f_2)\int_{\mathfrak{D}} (1-b_{11}) h(Q|A)\, dQ. \tag{10.4.26}$$

Let

$$Q = \begin{pmatrix} q_{11} & \mathbf{q}' \\ \mathbf{q} & Q_{11} \end{pmatrix}, \qquad A = \begin{pmatrix} a_1 & \mathbf{0}' \\ \mathbf{0} & A_1 \end{pmatrix},$$

$$P_1 = Q_{11}^{-1} - (I+Q_{11})^{-1} = (Q_{11} + Q_{11}^2)^{-1},$$

$$P_2 = Q_{11}^{-1} - (A_1^{-1} + Q_{11})^{-1} = (Q_{11} + Q_{11} A Q_{11})^{-1}.$$

The transformations

$$u = q_{11} - \mathbf{q}'(I+Q_{11})^{-1}\mathbf{q} \quad \text{and} \quad \mathbf{q} = u^{1/2}\mathbf{x}$$

give the following results:

$$J \equiv J(q_{11}, \mathbf{q} \to u, \mathbf{x}) = J(q_{11} \to u) J(\mathbf{q} \to \mathbf{x}) = u^{\frac{1}{2}(p-1)}$$

$$|Q| = |Q_{11}|(q_{11} - \mathbf{q}' Q_{11}^{-1} \mathbf{q}) = |Q_{11}| u (1 - \mathbf{x}' P_1 \mathbf{x}),$$

$$|I_p + AQ| = |I_{p-1} + A_1 Q_{11}|(1 + a_1 \delta u),$$

$$1 - b_{11} = \frac{a_1 \delta u}{1 + a_1 \delta u},$$

and

$$\mathfrak{D} = \mathfrak{D}_1 \cap \mathfrak{D}_2,$$

where

$$\mathfrak{D}_1 : \{ Q_{11} > 0, \mathbf{x} : \mathbf{x}' P_1 \mathbf{x} < 1 \},$$

$$\mathfrak{D}_2 : \{ u : u^{f_1}(1+u)^{-f_1-f_2} \leq c_0 \},$$

(10.4.27)
$$c_0 = c_0(Q_{11}, \mathbf{x}) = \frac{c|I + Q_{11}|^{f_1+f_2}}{|Q_{11}|^{f_1}(1-\mathbf{x}'P_1\mathbf{x})^{f_1}},$$

and

(10.4.28)
$$\delta = 1 + \mathbf{x}'(P_2 - P_1)\mathbf{x}.$$

With the above notations, we have

$$a_1^{-1}(1-b_{11})Jh(Q|A) = h_1(Q_{11}, \mathbf{x}|A_1)\frac{(a_u\delta)^{f_1}u^{f_1}}{(1+a_1\delta u)^{f_1+f_2+1}},$$

$$h_1(Q_{11}, \mathbf{x}|A_1) = \frac{|A_1|^{f_1}|Q_{11}|^{f_1 - \frac{1}{2}(p+1)}(1-\mathbf{x}'P_1\mathbf{x})^{f_1 - \frac{1}{2}(f+1)}}{B_p(f_1, f_2)|I_{p-1} + A_1 Q_{11}|^{f_1+f_2}\delta^{f_1-1}},$$

and hence Eq. (10.4.26) can be written as

(10.4.29)
$$\frac{\partial g}{\partial a_1} = p_1 \int_{\mathfrak{D}_1} h_1(Q_{11}, \mathbf{x}|A_1)\, dQ_{11}\, d\mathbf{x} \int_{\mathfrak{D}_2} \frac{\partial}{\partial u}\left[\frac{(a_1\delta)^{f_1-1}u^{f_1}}{(1+a_1\delta u)^{f_1+f_2}}\right] du$$

$$= p_1 \int_{\mathfrak{D}_1} h_1(Q_{11}, \mathbf{x}|A_1)(a_1\delta)^{f_1-1}\left[\frac{u^{f_1}}{(1+a_1\delta u)^{f_1+f_2}}\right]_{\mathfrak{D}_2} dQ_{11}\, d\mathbf{x}.$$

Now, there are exactly two solutions $c_1$ and $c_2$, $c_2 > c_1$, of the equation

$$u^{f_1}(1+u)^{-f_1-f_2} = c_0$$

and then,

$$\mathfrak{D}_2 = \{0 \leq u \leq c_1\} \cup \{c_2 \leq u < \infty\}.$$

Therefore,

(10.4.30)
$$\left[\frac{u^{f_1}}{(1+a_1\delta u)^{f_1+f_2}}\right]_{\mathfrak{D}_2}$$

$$= c_1^{f_1}(1+a_1\delta c_1)^{-f_1-f_2} - c_2^{f_1}(1+a_1\delta c_2)^{-f_1-f_2}$$

$$= c_2^{f_1}(1+a_1\delta c_2)^{-f_1-f_2}\left[\left(\frac{(1+c_1)(1+a_1\delta c_2)}{(1+c_2)(1+a_1\delta c_1)}\right)^{f_1+f_2} - 1\right],$$

since $c_1^{f_1}=c_2^{f_2}[(1+c_1)/(1+c_2)]^{f_1+f_2}$. Now, since

$$\begin{aligned} a_1\delta &= a_1 + a_1\mathbf{x}'\left[(Q_{11}+Q_{11}A_1Q_{11})^{-1}-(Q_{11}+Q_{11}^2)^{-1}\right]\mathbf{x} \\ &= 1+(a_1-1)(1-\mathbf{x}'P_1\mathbf{x}) \\ &\quad +\mathbf{x}'\left[a_1(Q_{11}+Q_{11}A_1Q_{11})^{-1}-(Q_{11}+Q_{11}^2)^{-1}\right]\mathbf{x}, \end{aligned}$$

and since

$$\mathbf{x}'P_1\mathbf{x}\leqslant 1 \quad \text{and} \quad a_1(Q_{11}+Q_{11}A_1Q_{11})^{-1}-(Q_{11}+Q_{11}^2)^{-1}\geqslant 0,$$

we get

$$(10.4.31) \qquad a_1\delta > 1 \qquad \text{if} \quad a_1\geqslant 1.$$

Using Eqs. (10.4.31) and (10.4.30), we find that

$$\left[\frac{u^{f_1}}{(1+a_1\delta u)^{f_1+f_2}}\right]_{\mathfrak{D}_2} > 0 \qquad \text{if} \quad a_1\geqslant 1,$$

and consequently Eq. (10.4.29) gives

$$(10.4.32) \qquad \frac{\partial g}{\partial a_1}\geqslant 0 \qquad \text{if} \quad a_1\geqslant 1 \text{ and } a_1>a_2>\cdots>a_r.$$

Similarly, it can be established that

$$(10.4.33) \qquad \frac{\partial g}{\partial a_r}\leqslant 0 \qquad \text{if} \quad a_r\leqslant 1 \text{ and } a_1>a_2>\cdots>a_r.$$

These show that $g$ increases if $a_1$ increases from 1 and $a_r$ decreases from 1. This proves the theorem. □

**Theorem 10.4.7** *Let $\mathbf{y}_1\sim N_p(\boldsymbol{\mu}_1,\Sigma_1)$, $V_1\sim W_p(\Sigma_1,n_1)$, $\mathbf{y}_2\sim N_p(\boldsymbol{\mu}_2,\Sigma_2)$, and $V_2\sim W_p(\Sigma_2,n_2)$ be independently distributed, and let the distinct nonzero characteristic roots of $\Sigma_1\Sigma_2^{-1}$ be $\lambda_1>\lambda_2>\cdots>\lambda_r>0$. Then, the power of the modified likelihood-ratio test for $H:\Sigma_1=\Sigma_2$ against $A:\Sigma_1\neq\Sigma_2$ increases if $\lambda_1$ increases from 1 and $\lambda_r$ decreases from 1.*

PROOF. The rejection region is given by

$$\omega=\left\{(V_1,V_2): V_1>0,\, V_2>0,\, |V|^{\frac{1}{2}n_1}|V_2|^{\frac{1}{2}n_2}|V_1+V_2|^{-\frac{1}{2}(n_1+n_2)}\leqslant c\right\}.$$

Since the problem and the rejection region are invariant under the group of nonsingular linear transformations $(V_1,V_2)\to(AV_1A',AV_2A')$, we assume without any loss of generality $\Sigma_1=I$ and $\Sigma_2=\Lambda=\text{diag}(\lambda_1 I_{p_1},\ldots,\lambda_r I_{p_r})$, $p_1+p_2+\cdots+p_r=p$. Then let $Q=V_2^{-1/2}V_1V_2^{-1/2}$. We now

where $f(U,\Sigma,m)$ denotes the pdf of $U \sim W_p(\Sigma,m)$. Let

(10.4.40) $$A(W)=\{W:(\text{ch roots of } W)\in\omega\}.$$

Then

(10.4.41) $$P_\Delta(\omega)$$
$$=\int_{V_2>0}|V_2|^{\frac{1}{2}(p+1)}dV_2\int_{A(W)}f(V_2,I,n_2)f\left(W,V_2^{-1/2}\Delta V_2^{-1/2},n_1\right)dW.$$

Since the region $A(W)$ is invariant under the transformations $W\to\Gamma W\Gamma'$, $\Gamma$ orthogonal, we may assume $V_2^{-1/2}\Delta V_2^{-1/2}$ is a diagonal matrix $D$ of the characteristic roots of $V_2^{-1/2}\Delta V_2^{-1/2}$. Let

(10.4.42) $$B(Y|D)\equiv B(Y|\Delta,V_2)=\{Y:(\text{ch roots of } D^{1/2}YD^{1/2})\in\omega\}.$$

Then $\{V_2>0\}\cap A(W)=\{V_2>0\}\cap\{B(Y|\Delta,V_2)\}$, and by making the transformation $Y=D^{-1/2}WD^{-1/2}$, we get

(10.4.43) $$P_\Delta(\omega)=\int_{V_2>0}f(V_2,I,n_2)\,dV_2\int_{B(Y|\Delta,V_2)}f(Y,I,n_1)\,dY.$$

Let $\Delta_1$ be a diagonal matrix such that $\Delta_1-\Delta$ is positive semidefinite. Let $D$ and $D_1$ be the diagonal matrices corresponding to the characteristic roots of $V_2^{-1/2}\Delta V_2^{-1/2}$ and $V_2^{-1/2}\Delta_1 V_2^{-1/2}$, respectively. Then $D_1-D$ is also positive (Problem 1.18).

Since the region $\omega$ satisfies Assumption A, it follows that

(10.4.44) $$B(Y|D)\supset B(Y|D_1).$$

Hence

(10.4.45) $$P_\Delta(\omega)\geqslant P_{\Delta_1}(\omega).$$ □

**Corollary 10.4.3** *Suppose an invariant test has an acceptance region such that if* $(l_1,\ldots,l_p)$ *is in the region, then so is* $(\tilde{l}_1,\tilde{l}_2,\ldots,\tilde{l}_2)$ *for* $\tilde{l}_i\leqslant l_i$. *Then the power of the test is a monotonically increasing function of each* $\gamma_i$.

**Corollary 10.4.4** *If* $g(l_1,\ldots,l_p)$ *is monotonically increasing in each of its arguments, a test with acceptance region* $g(l_1,\ldots,l_p)\leqslant k$ *has a monotonically increasing function in each* $\gamma_i$.

Thus any test with acceptance region $\Sigma_{i=1}^p d_iT_i\leqslant a$, where $d_i\geqslant 0$ and $T_i$ is the sum of all different products of $l_1,\ldots,l_p$ taken $i$ at a time, will have the monotonicity property. Special cases of this region are

$$\left(\frac{n_2}{n_1}\right)^p\prod_{i=1}^p l_i=\frac{|S_1|}{|S_2|}\leqslant a,\qquad \frac{n_2}{n_1}\sum_{i=1}^p l_i=\operatorname{tr}S_1S_2^{-1}\leqslant a,$$

where $S_1$ and $S_2$ are sample covariances $(n_i^{-1}V_i)$. Tests with acceptance regions of the type $\Sigma_{i,j=1}^{p} a_{ij} W_{ij} \leqslant \mu$, where $a_{ij} \geqslant 0$ and where $W_{ij} = T_i / T_j$ $(i > j)$, are also special cases of Corollary 10.4.2. Thus the two tests proposed by Roy (1957), having acceptance regions $l_1 \leqslant a_1$ and $l_p \geqslant a_p$, respectively, will have the monotonicity property.

**Corollary 10.4.5** *For testing $\Sigma = I$ (or $\Sigma = \Sigma_0$) in the one-sample problem, it follows from Eqs.* (10.4.41) *and* (10.4.43) *that, given any invariant test with acceptance region such that if $(l_1, \ldots, l_p)$ is in the region, so also is $(\tilde{l}_1, \ldots, \tilde{l}_p)$ with $\tilde{l}_i \leqslant l_i$, the power function of that test is an increasing function of each characteristic root of $\Sigma$.*

The above results are due to Anderson and Das Gupta (1964a).

## 10.5 Monotonicity of Some Tests for the Independence of Two Vectors

Let $\mathbf{z} \sim N_{(p+q)}(\boldsymbol{\mu}, \Sigma)$ and $V \sim W_{(p+q)}(\Sigma, n)$ be independently distributed where $\Sigma$ and $V$ are partitioned into $p$ and $q$ rows and columns as

$$\Sigma = \begin{pmatrix} \Sigma_{11} & \Sigma_{12} \\ \Sigma_{12}' & \Sigma_{22} \end{pmatrix} \quad \text{and} \quad V = \begin{pmatrix} V_{11} & V_{12} \\ V_{12}' & V_{22} \end{pmatrix}.$$

Without loss of generality we assume that $p \leqslant q$. Consider the problem of testing the hypothesis

$$H : \Sigma_{12} = 0 \qquad \text{against} \quad A : \Sigma_{12} \neq 0.$$

In order to study the monotonicity property of some invariant tests, we first obtain canonical reductions.

### *10.5.1 Canonical Reduction and Invariance*

The problem of testing $H$ against $A$ remains invariant under transformations

$$(10.5.1) \qquad V \to \begin{pmatrix} B_1 & 0 \\ 0 & B_2 \end{pmatrix} V \begin{pmatrix} B_1' & 0 \\ 0 & B_2' \end{pmatrix}, \qquad \mathbf{z} \to \begin{pmatrix} B_1 & 0 \\ 0 & B_2 \end{pmatrix} \mathbf{z} + \mathbf{b},$$

where $B_1$ and $B_2$ are nonsingular matrices of order $p$ and $q$ respectively. Under these transformations the matrix $\Sigma$ can be assumed to be of the form

$$(10.5.2) \qquad \begin{array}{c} \begin{matrix} p & p & q-p \end{matrix} \\ \begin{bmatrix} I & P & 0 \\ P & I & 0 \\ 0 & 0 & I \end{bmatrix} \begin{matrix} p \\ p \\ q-p \end{matrix} \end{array},$$

where $P$ is the $p \times p$ diagonal matrix with elements $\rho_1, \ldots, \rho_p$ arranged in decreasing order down the diagonal. The $\rho_i$ are the *canonical correlations*. They are easily seen to satisfy the determinantal equation

$$|\Sigma_{12}\Sigma_{22}^{-1}\Sigma_{12}' - \rho^2\Sigma_{11}| = 0. \tag{10.5.3}$$

Any function of the elements of $\Sigma$ that is invariant under the transformations (10.5.1) must be a function of the $\rho_i$, so that Eq. (10.5.2) is the *canonical form* of $\Sigma$ under such transformations. Thus the problem of testing $H$ reduces to that of testing the hypothesis that $\rho_1 = \cdots = \rho_p = 0$.

A test procedure that is invariant under the transformations (10.5.1) depends only on the characteristic roots $\lambda_1 \geqslant \lambda_2 \geqslant \cdots \geqslant \lambda_p$ of $V_{11}^{-1}V_{12}V_{22}^{-1}V_{12}'$. Under the transformation (10.5.1), we shall assume without loss of generality that $\Sigma$ is as mentioned in Eq. (10.5.2). Then, by Theorem 3.3.5, $V_{1\cdot2} = V_{11} - V_{12}V_{22}^{-1}V_{12}'$ and $(V_{12}, V_{22})$ are independently distributed, $V_{1\cdot2} \sim W_p(I - P^2, n-q)$, $V_{22} \sim W_q(I, n)$; and given $V_{22}$, $V_{12}V_{22}^{-1/2} \sim N_{p,q}((P,0)V_{22}^{1/2}, I - P^2, I)$, where $V_{22}^{1/2}$ is a symmetric square root of $V_{22}$. Since $\lambda_1, \lambda_2, \ldots, \lambda_p$ are invariant under the transformation $V_{11} \to CV_{11}C'$ and $V_{12} \to CV_{12}V_{22}^{-1/2}E$, where $C$ is nonsingular and $E$ is orthogonal, we choose $C$ and $E$ such that (see Chapter 1)

$$C(I-P)C' = I \quad \text{and} \quad C(P,0)V_{22}^{1/2}E = (D_\gamma, 0),$$

where $D_\gamma$ is a diagonal matrix with diagonal elements $\gamma_i$ where the $\gamma_i^2$'s are the characteristic roots of $C(P,0)V_{22}(P,0)'C'$ or of $V_{22}(P,0)'(I - P^2)^{-1}(P,0)$ or of

$$V_{22}\begin{pmatrix} D_\theta & 0 \\ 0 & 0 \end{pmatrix}$$

with $D_\theta = P(I-P^2)^{-1}P$ and $\theta_i = \rho_i^2/(1-\rho_i^2)$. Hence, given $V_{22}$, $W = CV_{1\cdot2}C'$ and $Y = CV_{12}V_{22}^{-\frac{1}{2}}E$ are independently distributed, $W \sim W_p(I, n-q)$ and $Y \sim N_{p,q}((D_\gamma, 0), I_p, I_q)$. Further, $\lambda_1, \lambda_2, \ldots, \lambda_p$ are the characteristic roots of $(W + YY')^{-1}YY'$. Thus, as shown in the next subsection, the problem reduces to that of the MANOVA model of Section 10.3, when $V_{22}$ is considered as fixed.

### *10.5.2 Monotonicity of Some Invariant Tests*

Let $\mathrm{ch}_i(\cdot)$ denote the $i$th largest characteristic root of a matrix. Then, if $V_{22} = TT'$ and

$$B = \begin{pmatrix} D_\theta & 0 \\ 0 & 0 \end{pmatrix},$$

we have

$$\mathrm{ch}_i(B\mathbf{T}\mathbf{T}') = \mathrm{ch}_i(\mathbf{T}'B\mathbf{T}), \qquad i = 1, 2, \ldots, p.$$

Let $B_1$ be the matrix obtained by changing the nonzero elements of $B$ from $\theta_i$ to $\tilde{\theta}_i$, where $\tilde{\theta}_i \geqslant \theta_i$ $(i=1,2,\ldots,p)$. Then $\mathbf{T}'B_1\mathbf{T}-\mathbf{T}'B\mathbf{T}$ is a positive semidefinite matrix and

$$\mathrm{ch}_i(\mathbf{T}'B_1\mathbf{T}) \geqslant \mathrm{ch}_i(\mathbf{T}'B\mathbf{T}).$$

Thus, for fixed $Y$, the problem is reduced to the MANOVA model of Section 10.3. Hence if we consider an acceptance region that is convex in each column vector of $Y$ for each set of fixed $W$ and fixed values of the other column vectors of $Y$, then it follows from Corollary 10.3.1 that the power of the test increases monotonically in $\theta_i$ and hence in each $\rho_i$. Thus we get

**Theorem 10.5.1** *An invariant test for which the acceptance region is convex in each column vector of $Y$ for each set of fixed $W$ and fixed values of the other column vectors of $Y$ has a power function that is monotonically increasing in each $\rho_i$.*

Thus from Section 10.3, we get the following

**Theorem 10.5.2** *The power function of the likelihood-ratio test increases monotonically as each $\rho_i$ increases.*

It follows from Problems 10.3–10.5 that Roy's maximum-root test and any test having the acceptance region $\Sigma_{k=1}^{b} a_k w_k \leqslant \mu$, $a_k \geqslant 0$ have the monotonicity property, where $w_k$ is the sum of all different products of $d_i \equiv (1-\lambda_i)^{-1}$ taken $k$ at a time $(k=1,2,\ldots,p)$.

These results are due to Anderson and Das Gupta (1964b).

## Problems

**10.1** Let $\beta(\sigma^2)=P[S^r\exp(-\frac{1}{2}S)\leqslant k,\ r>0]$, where $S/\sigma^2$ has a $\chi^2$ distribution with $n$ d.f. Show that

$$\frac{d\beta}{d\sigma^2} \gtreqless 0 \quad \text{according as} \quad \sigma^2 \gtreqless \frac{2r}{n}.$$

Using this result, show that the likelihood-ratio test for testing $\sigma^2=1$ vs $\sigma^2\neq 1$, based on a sample of size $N$ from $N(\mu,\sigma^2)$, is biased.

**10.2** Let $y_1 \sim N_p(\mu_1,\Sigma_1)$, $y_2 \sim N_p(\mu_2,\Sigma_2)$, $V_1 \sim W_p(\Sigma_1,n_1)$, and $V_2 \sim W_p(\Sigma_2,n_2)$ be independently distributed. By considering a special case of $p=1$, show that the likelihood-ratio test is unbiased iff $n_1=n_2$.

**10.3** Show that for any p.s.d. matrix $B:n\times n$, the region

$$E=\{A:\mathrm{ch}_1(AA'B)\leqslant \mu\}$$

is convex in $A:n\times m$, where $\text{ch}_1(D)$ denotes the largest characteristic root of the matrix $D$. Hence, show that Roy's maximum-root test proposed in Section 10.3 has a power function that is monotonically increasing in each $\gamma_i$.

**10.4** Consider a matrix $A=(\mathbf{a}_1,\ldots,\mathbf{a}_n):m\times n$, where the $\mathbf{a}_j$'s are the column vectors of $A$. Define $W_k(A)$ as the sum of all $k$-row principal minors of $AA'+I_m$, or equivalently as the sum of all different products of the roots of $AA'+I_m$ taken $k$ at a time. Then show that for any $j$ and $k$ ($j=1,\ldots,n$, $k=1,\ldots,m$) and for $\mathbf{a}_i$ fixed ($i\neq j$), $W_k(A)$ is a positive definite quadratic form in $\mathbf{a}_j$ plus a constant.

**10.5** Let $\lambda_1>\cdots>\lambda_p$ be the roots of $(Y_3Y_3')^{-1}Y_2Y_2'$. Let $d_i=1+\lambda_i$, $i=1,2,\ldots,p$, and let $W_k$ be the sum of all different products of $d_1,\ldots,d_p$ taken $k$ at a time ($k=1,2,\ldots,p$). Using the results in Problem 10.4, show that any invariant test having the acceptance region $\Sigma_{k=1}^{p}a_kW_k\leqslant\mu$ ($a_k\geqslant 0$) has a power function that is monotonically increasing in each $\gamma_i$.

**10.6** For the growth-curve model of Chapter 6, obtain the monotonicity result of the likelihood-ratio test for $\xi=0$.

# Bibliography


Afifi, A. A., and Elashoff, E. M. (1966) Missing observations in multivariate statistics I. Review of the literature. *J. Amer. Statist. Assoc.* 61, 595–604.

Ali, M., Fraser, D. A. S., and Lee, Y. S. (1970) Distribution of the correlation matrix. *J. Statist. Res.* 4, 1–15.

Amir, M., and Ali, R. (1956) Extreme properties of eigenvalues of Hermitian transformation and singular values of the sum and product of linear transformations. *J. Duke Math.* 23, 463–476.

Anderson, G. A. (1965) An asymptotic expansion for the distribution of the latent roots of the estimated covariance matrix. *Ann. Math. Statist.* 36, 1153–1173.

Anderson, T. W. (1951) Classification by multivariate analysis. *Psychometrika* 16, 31–50.

——— (1955) The integral of a symmetric unimodal function over a symmetric convex set and some probability inequalities. *Proc. Am. Math. Soc.* 6, 170–176.

——— (1957) Maximum likelihood estimates for a multivariate normal distribution when some observations are missing. *J. Amer. Statist. Assoc.* 52, 200–203.

——— (1958) *Introduction to Multivariate Statistical Analysis*. Wiley, New York.

——— (1963) Asymptotic theory for principal component analysis. *Ann. Math. Statist.* 34, 122–148.

——— (1965) Some optimum confidence bounds for roots of determinantal equations. *Ann. Math. Statist.* 36, 468–488.

——— (1973) An asymptotic expansion of the distribution of the studentized classification statistic. *Ann. Statist.* 1, 964–972.

Anderson, T. W., and Bahadur, R. R. (1962) Classification into two multivariate normal distributions with different covariance matrices. *Ann. Math. Statist.* 420–431.

Anderson, T. W., and Das Gupta, S. (1964a) A monotonicity property of the power functions of some tests of the equality of two covariance matrices. *Ann. Math. Statist.* 35, 1059–1063.

——— (1964b) Monotonicity of the power functions of some tests of independence between two sets of variates. *Ann. Math. Statist.* 35, 206–208.

Anderson, T. W., and Girshick, M. A. (1944) Some extensions of the Wishart distribution. *Ann. Math. Statist.* 15, 345–357.

Banerjee, K. S., and Marcus, L. F. (1965) Bounds in a minimax classification procedure, *Biometrika* 52, 653–654.

Barnard, M. M. (1935) The secular variations of skull characters in four series of Egyptian skulls. *Ann. Eugen.* 6, 352–371.

Barnes, E. W. (1899) The theory of gamma functions. *Messeng. Math.* 29, 64–129.

Bartlett, M. S. (1937) Properties of sufficiency and statistical tests. *Proc. Camb. Phil. Soc. A* 160, 268–282.
——— (1938) Further aspects of the theory of multiple regression. *Proc. Camb. Phil. Soc.* 34, 33–40.
——— (1950) Tests of significance in factor analysis. *British J. Psych. (Statist. Sec.)* 3, 77–85.
——— (1951a) The effect of standardization on a $\chi^2$-approximation in factor analysis. *Biometrika* 38, 337–344.
——— (1951b) A further note on tests of significance in factor analysis. *British J. Psych. (Statist. Sec.)* 4, 1–2.
——— (1954) A note on the multiplying factors for various $\chi^2$-approximations. *J. Roy. Statist. Soc. B* 16, 296–298.
Basu, D. (1955) On statistics independent of a complete sufficient statistic. *Sankhyā* 15, 377–380.
Basu, D., and Khatri, C. G. (1969) On some characterizations of statistics. *Sankhyā A* 31, 1–10.
Bellman, R. (1960) *Introduction to Matrix Analysis.* McGraw-Hill, New York.
Bennett, B. M. (1951) Note on a solution of the generalized Behrens–Fisher problem. *Ann. Inst. Statist. Math.* 2, 87.
Bhargava, R. P. (1962) Multivariates tests of hypothesis with incomplete data. *Technical Report No. 3,* Stanford University.
Bhargava, R. P., and Srivastava, M. S. (1973) On Tukey's confidence intervals for the contrasts in the means of the intraclass correlation model. *J. Roy. Statist. Soc.* 35, 147–152.
Bingham, C. (1972) An asymptotic expansion for the distribution of the eigenvalues of a 3 by 3 Wishart matrix. *Ann. Math. Statist.* 43, 1498–1506.
Bonnesen, T., and Fenchel, W. (1948) *Theorie der konvexen Korper.* Chelsea, New York.
Bose, R. C., and Roy, S. N. (1938) The distribution of the studentized $D^2$-statistic. *Sankhyā* 4, 19–38.
Bowker, A. H. (1960) A representation of Hotelling's $T^2$ and Anderson's classification statistic $W$ in terms of simple statistics. No. 12 in *Contributions to Probability and Statistics, Essays in Honour of H. Hotelling.* Stanford University Press.
Bowker, A. H., and Sitgreaves, R. (1961) An asymptotic expansion for the distribution function of the $W$-classification statistic. In *Studies in Item Analysis and Prediction* (H. Solomon, ed.). Stanford University Press.
Box, G. E. P. (1949) A general distribution theory for a class of likelihood criteria. *Biometrika* 36, 317–346.
Broffit, J., and Williams, J. S. (1973) Minimum variance estimators for misclassification probabilities in discriminant analysis. *J. Multiv. Anal.* 3, 311–327.
Brown, G. W. (1939) On the power of the $L_1$-test for equality of several variances. *Ann. Math. Statist.* 10, 119–128.
Cacoullos, T. (1965a) Comparing Mahalanobis distance I: Comparing distance between $k$ normal populations and another unknown. *Sankhyā A* 27, 1–22.
——— (1965b) Comparing Mahalanobis distance II: Bayes procedures when the mean vectors are unknown. *Sankhyā A* 27, 23–32.

——— (1972) *Discriminant Analysis and Its Applications.* Academic Press, New York.

Cartan, H. (1963) *Elementary Theory of Analytic Functions of One or Several Complex Variables.* Addison-Wesley, Reading, Mass.

Carter, E. M. (1975) Characterization and testing problems in the complex Wishart distribution. Unpublished Ph.D. thesis, University of Toronto.

Carter, E. M., Khatri, C. G., and Srivastava, M. S. (1979) The effect of inequality of variances on the $t$-test. *Sankhyā B* (to appear).

Carter, E. M., and Srivastava, M. S. (1976) Asymptotic non-null distribution for the locally most powerful invariant test. Unpublished.

——— (1977) Monotonicity of the power functions of the modified likelihood ratio criterion for the homogeneity of variances and of the sphericity test. *J. Multiv. Anal.* 7, 229–233.

Chattopadhyay, A. K., and Pillai, K. C. S. (1973) Asymptotic expansions for the distribution of characteristic roots when the parameter matrix has several multiple roots. In *Multivariate Analysis III* (P. R. Krishnaiah ed.). Academic Press, New York.

Chikuse, Y. (1976) Asymptotic Distributions of the latent roots of the covariance matrix with multiple population roots. *J. Multiv. Anal.* 6, 237–249.

Clunier-Ross, C. W., and Riffenburgh, R. H. (1960) Geometry and linear discrimination. *Biometrika* 47, 185–189.

Cochran, W. G. (1934) The distribution of quadratic forms in a normal system with applications to the analysis of variance. *Proc. Camb. Phil. Soc.* 30, 178–191.

Cohen, A., and Strawderman, W. E. (1971) Unbiasedness of tests for homogeneity of variance. *Ann. Math. Statist.* 42, 355–360.

Constantine, A. G., and Muirhead, R. J. (1976) Asymptotic expansions for distributions of latent roots in multivariate analysis. *J. Multiv. Anal.* 6, 369–391.

Consul, P. C. (1967a) On the exact distributions of likelihood-ratio criteria for testing independence of sets of variates under the null hypothesis. *Ann. Math. Statist.* 38, 1160–1169.

——— (1967b) On the exact distributions of the criterion $W$ for testing spherecity in a $p$-variate normal distribution. *Ann. Math. Statist.* 38, 1170–1174.

——— (1969) The exact distribution of the likelihood criteria for different hypotheses. In *Multivariate Analysis II* (P. R. Krishnaiah, ed.). Academic Press, New York, pp. 171–181.

Craig, A. T. (1943) Note on the independence of certain quadratic forms. *Ann. Math. Statist.* 14, 195.

Cramèr, H. (1937) Random variables and probability distributions. *Cambridge Tracts in Mathematics, No. 36.* Cambridge University Press.

——— (1946) *Mathematical Methods of Statistics.* Princeton University Press.

Dantzig, G. B. (1940) On the non-existence of tests of 'Student's' hypothesis having power functions independent of $\sigma$. *Ann. Math. Statist.* 11, 186.

Darroh, J. N. (1965) An optimum property of principal components. *Ann. Math. Statist.* 36, 1579.

Das Gupta, S. (1964) Nonparametric classification rules. *Sankhyā A* 26, 25–30.

——— (1965) Optimum classification rules for classification into two multivariate normal populations. *Ann. Math. Statist.* 36, 1174–1184. Correction (1970), 326.

——— (1969) Properties of power functions of some tests concerning dispersion matrices of multivariate normal distributions. *Ann. Math. Statist.* 40, 697–701.

——— (1971) Nonsingularity of the sample covariance matrix. *Sankhyā A* 33, 475–478.

——— (1974) Probability inequalities and errors in classification. *Ann. Statist.* 2, 751–762.

Das Gupta, S., Anderson, T. W., and Mudholkar, G. S. (1964) Monotonicity of power functions of some tests of the multivariate linear hypothesis. *Ann. Math. Statist.* 35, 200.

David, F. N. (1938) *Tables of the Correlation Coefficient*. Cambridge University Press.

Davis, A. W. (1970a) Exact distribution of Hotelling's generalized $T_0^2$. *Biometrika* 57, 187–191.

——— (1970b) On the null distribution of sum of the roots of a multivariate beta distribution. *Ann. Math. Statist.* 41, 1557–1562.

Deemer, W. L., and Olkin, I. (1951) The Jacobians of certain matrix transformations useful in multivariate analysis. Based on lectures of P. L. Hsu at the University of North Carolina. *Biometrika* 38, 345–367.

Dempster, A. P. (1964) Tests for the equality of two covariance matrices in relation to a best linear discriminator analysis. *Ann. Math. Statist.* 35, 191–199.

Drygas, H. (1977) Best quadratic unbiased estimation in variance component models, *Math. Operationsf. u. Statist. Ser. Statist.* 8(2), 211–231.

Dunn, O. J. (1958) Estimation of the means of dependent variables. *Ann. Math. Statist.* 29, 1095–1111.

Dwyer, P. S. (1967) Some applications of matrix derivatives in multivariate analysis. *J. Am. Statist. Assoc.* 62, 607–625.

Dwyer, P. S., and MacPhail, M. S. (1948) Symbolic matrix derivatives. *Ann. Math. Statist.* 19, 517–534.

Dykstra, R. L. (1970) Establishing the positive definiteness of the sample covariance matrix. *Ann. Math. Statist.* 41, 2153–2154.

Eaton, M. L., and Perlman, M. D. (1973) The nonsingularity of generalized sample covariance matrices. *Ann. Statist.* 1, 710–717.

——— (1974) A monotonicity property of the power functions of some invariant tests for MANOVA. *Ann. Statist.* 2, 1022–1028.

Elfving, G. (1947) A simple method of deducing certain distributions connected with multivariate sampling. *Skand. Actuarietidskr.* 30, 56.

Ellison, B. E. (1962) A classification problem in which information about alternative distribution is based on samples. *Ann. Math. Statist.* 33, 213–223.

Eredelyi, A. (1953) *Higher Transcendental Function*. McGraw-Hill, New York.

Fisher, R. A. (1915) Frequency distribution of the values of the correlation coefficient in samples from an indefinitely large population. *Biometrika* 10, 507–521.

——— (1928) The general sampling distribution of the multiple correlation coefficient. *Proc. Roy. Soc. A* 121, 654–673.

——— (1936) The use of multiple measurement in taxonomic problems. *Ann. Eugen.* 7, 179–188.

——— (1939) The sampling distribution of some statistics obtained from nonlinear equations. *Ann. Eugen.* 9, 238–249.

Fraser, D. A. S. (1976) *Probability and Statistics: Theory and Application.* Duxbury Press, North Scituate, Massachusetts.

Frisch, R. (1929) Correlation and scatter in statistical variables. *Nordic Statist. J.* 8, 36–102.

Fujikoshi, Y. (1973) Monotonicity of the power functions of some tests in general MANOVA models. *Ann. Statist.* 1, 388–391.

——— (1978) Asymptotic expansions of the distribution of the likelihood ratio statistic for the equality of the $q$ smallest latent roots of a covariance matrix. (Unpublished).

Gajjar, A. V. (1967) Limiting distributions of certain transformations of multiple correlation coefficients. *Metron* 26, 189–193.

Ghosh, J. K. (1969) Only linear transformations preserve normality. *Sankhyā A* 31, 309–312.

Ghurye, S. G., and Olkin, I. (1962) A characterization of the multivariate normal distribution. *Ann. Math. Statist.* 33, 533–541.

Giri, N. (1964) On the likelihood ratio test of a normal multivariate testing problem. *Ann. Math. Statist.* 35, 181–190. Correction. **35**, 1388.

——— (1965) On the complex analogues of $T^2$ and $R^2$ tests. *Ann. Math. Statist.* 36, 664–670.

——— (1968) On tests of the equality of two covariance matrices. *Ann. Math. Statist.* 39, 275–277.

Giri, N., Kiefer, J., and Stein, C. (1963) Minimax character of Hotelling's $T^2$-test in the simplest case. *Ann. Math. Statist.* 34, 1524.

Girshick, M. A. (1939) On the sampling theory of roots of determinantal equations. *Ann. Math. Statist.* 10, 203–224.

Goodman, N. R. (1963) Statistical analysis based on a certain multivariate complex Gaussian distribution. *Ann. Math. Statist.* 34, 152–176.

Graybill, F. A. (1969) *Introduction to Matrices with Applications in Statistics.* Wadsworth, Belmont, Calif.

Graybill, F. A., and Milliken, G. A. (1969) Quadratic forms and idempotent matrices with random elements. *Ann. Math. Statist.* 40, 1430–1438.

Gupta, R. P. (1967) Latent roots and vectors of a Wishart matrix. *Ann. Inst. Statist. Math.* 19, 157–165.

Gurland, J. (1955) Distribution of definite and of indefinite quadratic forms. *Ann. Math. Statist.* 26, 122–127.

——— (1968) A relatively simple form of the distribution of the multiple correlation coefficient. *J. Roy. Statist. Soc. B* 30, 276–283.

Gurland, J., and Milton, R. (1970) Further consideration of the distribution of the multiple correlation coefficient. *J. Roy. Statist. Soc. B* 32, 381–394.

Halmos, P. R. (1950) *Measure Theory.* D. Van Nostrand Co., Princeton, N.J.

Hamedani, G. G., and Tata, M. N. (1975) On the determination of the bivariate normal distribution from distributions of linear combinations of the variables. *Am. Math. Monthly* 82, 913–915.

Hardy, G. E., Littlewood, J. E., and Polya, G. (1964) *Inequalities.* Cambridge University Press.

Harition, G. (1972) Multivariate mixture model. Unpublished Ph.D. thesis. University of Toronto.

Harley, B. L. (1954) A note on the probability integral of the correlation coefficient. *Biometrika* 41, 278–280.

——— (1956) Some properties of an angular transformation for the correlation coefficient. *Biometrika* 43, 219–224.

Hartley, H. O., and Hocking, R. R. (1971) The analysis of incomplete data. *Biometrics* 27, 783–823.

Heck, D. L. (1960) Charts of some upper percentage points of the distribution of the largest characteristic root. *Am. Math. Statist.* 31, 625–642.

Hogg, R. V. (1963) On the independence of certain Wishart variables. *Ann. Math. Statist.* 34, 935–939.

Hotelling, H. (1933) Analysis of a complex of statistical variables into principal components. *J. Educ. Psych.* 24, 417–441, 498–520.

——— (1944) Note on a matrix theorem of A. T. Carig. *Ann. Math. Statist.* 15, 427–429.

——— (1947) Multivariate quality control illustrated by the air testing of sample bombsights. In *Techniques of Statistical Analysis.* (Eisenhart, C., Hastay, M. W., and Wallis, W. A., eds.) McGraw-Hill, New York, pp. 113–184.

——— (1953) New light on the correlation coefficient and its transform. *J. Roy. Statist. Soc. B* 15, 193–225.

Hsu, P. L. (1938) Notes on Hotelling's generalized $T^2$. *Ann. Math. Statist.* 9, 231–243.

——— (1939) On the distribution of the roots of certain determinantal equations. *Ann. Eugen.* 9, 250–258.

——— (1942) On the limiting distribution of the canonical correlations. *Biometrika* 32, 38.

Ingham, A. E. (1933) An integral that occurs in statistics. *Proc. Camb. Phil. Soc.* 29, 271.

Jack, H. (1965) Jacobians of transformations involving orthogonal matrices. *Proc. Roy. Soc. Edinburgh*, 47, 81–103.

Jack, H., and Macbeath, A. M. (1959) The value of a certain set of matrices. *Proc. Camb. Phil. Soc.* 55, 213–223.

James, A. T. (1964) Distributions of matrix variates and latent roots derived from normal samples. *Ann. Math. Statist.* 35, 475–501.

——— (1966) Inference on latent roots by calculation of hypergeometric functions of matrix arguments. In *Multivariate Analysis I* (P. R. Krishnaiah, ed.). Academic Press, New York.

——— (1969) Tests of equality of latent roots of the covariance matrix. In *Multivariate Analysis II* (P. R. Krishnaiah, ed.). Academic Press, New York.

James, G. S. (1952) Note on a theorem of Cochran. *Proc. Camb. Phil. Soc.* 48, 443–446.

John, S. (1961) Errors in discrimination. *Ann. Math. Statist.* 32, 1125–1144.
——— (1971) Some optimal multivariate tests. *Biometrika* 38, 123–127.
——— (1972) The distribution of a statistic used for testing sphericity of normal distributions. *Biometrika* 39, 169–174.
Johnson, N. L., and Kotz, S. (1967a, b) *Continuous Univariate Distributions,* 2 volumes. Houghton Mifflin, Boston.
Kagan, A., Linnik, Yu. V., and Rao, C. R. (1972) *Characterization Problems of Mathematical Statistics* (in Russian). Academy Nank, Moscow; English Ed. (1973), Wiley, New York.
Khatri, C. G. (1959a) On the mutual independence of certain statistics. *Ann. Math. Statist.* 30, 1258–1262.
——— (1959b) On the conditions for the forms of the type $XAX'$ to be distributed independently or to obey Wishart distribution. *Bull. Calcutta Statist. Assoc.* 8, 162–168.
——— (1960) On certain problems in multivariate analysis: multicollinearity of means and power series distributions. Unpublished Ph.D. thesis, M. S. University of Baroda, India.
——— (1961) A simplified approach to the derivation of the theorems of the rank of a matrix. *J. M. S. University of Baroda* 10, 1–5.
——— (1962a) A necessary and sufficient condition on the equality of a number of nonzero characteristic roots and a rank of a matrix. *Vidya* 5, 1–4.
——— (1962b) Conditions for Wishartness and independence of second degree polynomials in a normal vector. *Ann. Math. Statist.* 33, 1002–1007.
——— (1963a) Further contributions of Wishartness and independence of second degree polynomials in normal vectors. *J. Indian Statist. Assoc.* 1, 61–70.
——— (1963b) Wishart distribution. Queries and answers section, *J. Indian Statist. Assoc.* 1, 30.
——— (1965a) Classical statistical analysis based on a certain multivariate complex Gaussian distribution. *Ann. Math. Statist.* 36, 98–114.
——— (1965b) A note on the confidence bounds for the characteristic roots of dispersion matrices of normal variates. *Ann. Inst. Statist. Math. (Tokyo)* 17, 175–183.
——— (1965c) A test for reality of a covariance matrix in a certain complex Gaussian distribution. *Ann. Math. Statist.* 36, 115–119.
——— (1966a) A note on a MANOVA model applied to problems in growth curve. *Ann. Inst. Statist. Math.* 18, 75–86.
——— (1966b) A note on a large sample distribution of a transformed multiple correlation coefficient. *Ann. Inst. Statist. Math.* 18, 375–380.
——— (1967) On certain inequalities for normal distributions and their applications to simultaneous confidence bounds. *Ann. Math. Statist.* 38, 1853–1867.
——— (1968a) Some results for the singular normal multivariate regression models. *Sankhyā A* 30, 267–280.
——— (1968b) A note on exact moments of arc sine correlation with the help of characteristic function. *Ann. Inst. Statist. Math.* 20, 143–149.

——— (1970) A note on Mitra's paper "A density free approach to the matrix variate beta distributions." *Sankhyā A* 32, 311–317.

——— (1971a) *Mathematics of matrices* (in Gujarati). Gujarat University Press, Ahmedabad, India.

——— (1971b) Series representations of distributions of quadratic form in normal vectors and generalized variance. *J. Multiv. Anal.* 1, 199–214.

——— (1973) Testing some covariance structures under a growth curve model. *J. Multiv. Anal.* 3.

——— (1975) A characterization property of a normal distribution. *Gujarat Statist. Rev.* 2, 24–27.

——— (1976a) A note on an inequality for a multivariate normal distribution. *Gujarat Statist. Rev.* 3, 1–12.

——— (1976b) A note on multiple and canonical correlation for a singular covariance matrix. *Psychometrika* 41, 465–470.

——— (1978a) Minimum variance quadratic unbiased estimate of variance under ANOVA model. *Gujarat Statist. Rev.* 5, 33–41.

——— (1978b) Some optimization problems with applications to canonical correlations and sphericity tests. *J. Multiv. Anal.* 8, 453–467.

Khatri, C. G. and Bhargava, R. P. (1975) The distribution of product of independent beta random variables with application to multivariate analysis. Unpublished.

Khatri, C. G., and Mitra, S. K. (1976) Hermitian and nonnegative definite solutions of linear matrix equations, *SIAM J. Appl. Math.* 31, 579–585.

Khatri, C. G., and Pillai, K. C. S. (1965) Some results on the noncentral multivariate beta distribution and moments of traces of two matrices. *Ann. Math. Statist.* 36, 1511–1520.

——— (1966) On the moments of the trace of a matrix and approximation to its noncentral distribution. *Ann. Math. Statist.* 37, 1312–1318.

——— (1967) On the moments of traces of two matrices in multivariate analysis. *Ann. Inst. Statist. Math.* 19, 143–156.

Khatri, C. G., and Ramachandran, K. V. (1958) Certain multivariate distribution problems I (Wishart's distribution). *J. M. S. University of Baroda* 7, 79–82.

Khatri, C. G., and Rao, C. R. (1972) Functional equations and characterization of probability laws through linear functions of random variables. *J. Multiv. Anal.* 2, 162–173.

——— (1976) Characterization of multivariate normality, I, through independence of some statistics. *J. Multiv. Anal.* 6, 81–94.

Khatri, C. G., and Srivastava, M. S. (1971) On exact non-null distributions of likelihood ratio criteria for sphericity test and equality of two covariance matrices. *Sankhyā A* 33, 201–206.

——— (1974a) On the likelihood ratio test for covariance matrix in growth curve model. In *Applied Statistics* (R. P. Gupta, ed.), Proc. Conf. Dalhousie University, Halifax, May 2–4, North-Holland Publishing Co.

——— (1974b) Asymptotic expansions of the non-null distributions of likelihood ratio criteria for covariance matrices. *Ann. Statist.* 2, 109–117.

——— (1974c) Asymptotic expansions of the non-null distributions of the likelihood ratio criteria for covariance matrices II. Proc. Carleton University, Ottawa. *Metron* 36, 55–71 (1976).

——— (1978) Asymptotic expansion for distributions of characteristic roots of covariance matrices. *South African Statist. J.* 161–186.

Kiefer, J., and Schwartz, R. (1965) Admissible Bayes character of $T^2 - R^2$ and other fully invariant tests for classical multivariate normal problems. *Ann. Math. Statist.* 36, 747–770.

Kleffe, J. (1977) Simultaneous estimation of expectation and covariance matrix in linear models. Notes on lectures given at the Indian Statist. Inst., New Delhi. Unpublished.

——— (1978) On Hsu's theorem in multivariate regression. Unpublished.

Kotz, S., Johnson, N. L., and Boyd, D. W. (1967a) Series representations of distributions of quadratic forms in normal variables I. Central Case. *Ann. Math. Statist.* 38, 823–837.

——— (1967b) Series representations of distributions of quadratic forms in normal variables II. Noncentral Case. *Ann. Math. Statist.* 38, 838–848.

Kramer, H. C. (1963) Tables for constructing confidence limits on multiple correlation coefficient. *J. Am. Statist. Assoc.* 58, 1082–1085.

Krishnaiah, P. R. (1976) Some recent developments on complex multivariate distributions. *J. Multiv. Anal.* 6 (1), 1–30.

——— (1978) Some recent developments on real multivariate distributions. In *Developments in Statistics I.* Academic Press, New York.

Krishnaiah, P. R. and Pathak, P. K. (1967) Tests for the equality of covariance matrices under the intraclass correlation model. *Ann. Math. Statist.* 38, 1286–1288.

Krishnaiah, P. R., and Schuurmann, F. J. (1974) On the exact distributions of the ratios of the extreme roots of the real and complex Wishart matrices. Technical Report No. ARL TR 74–0118, Aerospace Research Laboratory, Wright-Patterson Air Force Base, Ohio.

Kshirsagar, A. M. (1959) Bartlett decomposition and Wishart distribution. *Ann. Math. Statist.* 30, 239.

——— (1961) The goodness of fit of a single (non-isotropic) hypothetical principal component. *Biometrika* 48, 397.

——— (1966) The non-null distribution of a direction statistic in principal component analysis. *Biometrika* 53, 590.

Kshirsagar, A. M., and Gupta, R. P. (1965) The goodness of fit of two or more principal components. *Ann. Inst. Statist. Math.* 17, 347.

Kullback, S. (1959) *Information Theory and Statistics*. Wiley, New York.

Laha, R. G. (1956) On the stochastic independence of two second degree polynomial statistics in normally distributed variates. *Ann. Math. Statist.* 27, 790–796.

Lancaster, H. O. (1954) Traces and cumulants of quadratic forms in normal variables. *J. Roy. Statist. Soc. B* 16, 247–254.

——— (1967) *The Chi-Square Distribution*. Wiley, New York.

Lawley, D. N. (1938) A generalization of Fisher's *z* test. *Biometrika* 30, 180–187.

——— (1953) A modified method of estimation in factor analysis and some large

sample results. *In Uppsala Symposium in Psychological Factor Analysis, 17–19 March 1953*. Almqvist and Wicksell, Uppsala, pp. 33–42.
——— (1956a) Tests of significance for the latent roots of covariance and correlation matrices. *Biometrika* 43, 128–136.
——— (1956b) A general method for approximating to the distributions of likelihood ratio criteria. *Biometrika* 43, 295–303.
——— (1963) On testing a set of correlation coefficients for equality. *Ann. Math. Statist*. 34, 149–151.
Lawley, D. N., and Maxwell, A. E. (1963) *Factor Analysis as a Statistical Method*. Butterworth, London.
Lee, Y. S. (1971) Some results on the sampling distribution of the multiple correlation coefficient. *J. Roy. Statist. Soc. B* 33, 117–130.
——— (1972a) Tables of upper percentage points of the multiple correlation coefficient. *Biometrika* 59, 175–189.
——— (1972b) Some results on the distribution of Wilk's likelihood-ratio criterion. *Biometrika* 59, 649–664.
Lehmann, E. L. (1959) Testing Statistical Hypothesis. John Wiley and Sons, New York.
Leung, C. Y. (1977) Discriminant Analysis and testing problems based on a general regression model. Unpublished Ph.D. Thesis, University of Toronto.
Madow, W. G. (1938) Contributions to the theory of multivariate statistical analysis. *Trans. Am. Math. Soc*. 44, 454.
Mahalanobis, P. C. (1936) On the generalized distance in statistics. *Proc. Natl. Inst. Sci. India* 12, 49–55.
Mahalanobis, P. C., Bose, R. C., and Roy, S. N. (1937) Normalization of statistical variates and the use of rectangular coordinates in the theory of sampling distribution. *Sankhyā* 3, 1.
Mallows, C. L. (1960) Latent vectors of random symmetric matrices. *Biometrika* 48, 133–149.
Maraglia, G., and Styan, G. P. H. (1974) Rank conditions for generalized inverses of partitioned matrices. *Sankhyā A* 36, 437–442.
Marcus, M., and Minc, H. (1964) *A Survey of Matrix Theory and Matrix Inequalities*. Allyn and Bacon.
Marden, J., and Perlman, M. D. (1977) Invariant tests for means with covariates. *Technical Report No. 45*, Department of Statistics, University of Chicago.
Mase, S. (1977) Some theorems of normality preserving transformations. *Sankhyā A* 39, 186–190.
Mathai, A. M., and Rathie, P. N. (1970) The exact distribution for the sphericity test. *J. Statist. Res*. 4, 140–159.
——— (1971) The exact distribution of Wilks' criterion. *Ann. Math. Statist*. 42, 1010–1019.
Mauchly, J. W. (1940) Significance tests for sphericity of a normal *n*-variate distribution. *Ann. Math. Statist*. 11, 204–209.
Mauldon, J. G. (1955) Pivotal quantities for Wishart's and related distributions and a paradox in fiducial theory. *J. Roy. Statist. Soc. B* 17, 79.

Milliken, G. A. (1971) New criteria for estimability for linear models. *Ann. Math. Statist.* 42, 1588–1594.
Mitra, S. K. (1969) Some characteristics and non-characteristic properties of the Wishart distributions. *Sankhyā A* 31, 19–22.
——— (1970) A density free approach to the matrix variate beta distribution. *Sankhyā A* 32, 81–88.
——— (1971) Another look at Rao's MINQUE of variance components. *Bull. Inst. Intern. Statist.* 44, 279–283.
——— (1973) Unified least squares approach to linear estimation in a general Gauss–Markov model. *SIAM J. Appl. Math.* 25, 671–680.
——— (1976) Non-existence of uniformly minimum variance unbiased estimators: some examples. *Gujarat Statist. Rev.* 3, 27–30.
Mitra, S. K., and Moore, B. J. (1973) Gauss–Markov estimation with an incorrect dispersion matrix. *Sankhyā A* 35, 139–152.
Mitra, S. K., and Rao, C. R. (1968) Some results in estimation and tests of linear hypotheses under the Gauss–Markoff model. *Sankhyā A* 30, 281–290.
Moore, E. H. (1935) *General Analysis*. Am. Phil. Soc., Philadelphia.
Morrow, D. J. (1948) On the distribution of the sums of the characteristic roots of a determinantal equation. (Abstract.) *Bull. Am. Math. Soc.* 54, 75.
Mudholkar, G. S. (1965) A class of tests with monotone power functions for two problems in multivariate statistical analysis. *Ann. Math. Statist.* 36, 1794–1801.
——— (1966) The integral of an invariant unimodal function over an invariant convex set—an inequality and applications. *Proc. Am. Math. Soc.* 17, 1327–1333.
Muirhead, R. J., and Chikuse, Y. (1975) Asymptotic expansions for the joint and marginal distribution of the latent roots of the covariance matrix. *Ann. Statist.* 3, 1011–1017.
Nagao, H. (1967) Monotonicity of the modified likelihood ratio test for a covariance matrix. *J. Sci. Hiroshima University A1* 31, 147–150.
——— (1970) Asymptotic expansions of some test criteria for homogeneity of variances and covariance matrices from normal populations. *J. Sci. Hiroshima University A1* 34, 153–247.
——— (1973a) On some test criteria for covariance matrix. *Ann. Math. Statist.* 1, 700–709.
——— (1973b) Asymptotic expansions of the distribution of Bartlett's test and sphericity test under the local alternatives. *Ann. Inst. Statist. Math.* 25, 407–422.
Nagarsenker, B. N., and Pillai, K. C. S. (1973) The distribution of the sphericity test criterion. *J. Multiv. Anal.* 3, 226–235.
Nanda, D. N. (1948a) Distribution of a root of a determinantal equation. *Ann. Math. Statist.* 19, 47–57.
——— (1948b) Limiting distribution of a root of a determinantal equation. *Ann. Math. Statist.* 19, 340–350.
——— (1950) Distribution of the sum of roots of a determinantal equation under a certain condition. *Ann. Math. Statist.* 21, 432–439.

Narian, R. D. (1948) A new approach to the sampling distributions of the multivariate normal theory. *J. Indian Soc. Agric. Statist.* 1, 59.
Ogawa, J. (1950) On the independence of quadratic forms in a noncentral normal system. *Ann. Inst. Math. Statist.* 2, 151–159.
——— (1953) On the sampling distributions of classical statistics in multivariate analysis. *Osaka Math. J.* 5, 13.
Okamoto, M. (1963) An asymptotic expansion for the distribution of the linear discriminant function. *Ann. Math. Statist.* 34, 1286–1301. Corrections: 39, 1358–1359.
——— (1969) Optimality of principal components. In *Multivariate Analysis II* (P. R. Krishnaiah, ed.) Academic Press, New York.
——— (1973) Distinctness of the eigenvalues of a quadratic form in multivariate sample. *Ann. Statist.* 1, 763–765.
Okamoto, M., and Kanazawa, M. (1968) Minimization of eigenvalues of matrix and optimality of principal components. *Ann. Math. Statist.* 30, 859.
Olkin, I., and Press, S. J. (1969) Testing and Estimation for a circular stationary model. *Ann. Math. Statist.* 40, 1358–1373.
Olkin, I., and Roy, S. N. (1954) On multivariate distribution theory. *Ann. Math. Statist.* 25, 329.
Olkin, I., and Rubin, H. (1962) A characterization of the Wishart distribution. *Ann. Math. Statist.* 33, 1483.
Olkin, I., and Sampson, A. R. (1972) Jacobians of matrix transformations and induced functional equations. *Linear Algebra and Its Applications* 5, 257–276.
Olkin, I., and Shrikhande, S. S. (1970) An extension of Wilks' test for the equality of means. *Ann. Math. Statist.* 41, 683–687.
Penrose, R. (1955) A generalized inverse for matrices. *Proc. Camb. Phil. Soc.* 51, 406–413.
Perlman, M. D. (1974) On the monotonicity of the power function of tests based on traces of multivariate beta matrices. *J. Multi. Anal.* 4, 22–30.
——— (1978) Unbiasedness of the likelihood-ratio tests for equality of several convariance matrices and equality of several multivariate normal populations (Unpublished).
Pillai, K. C. S. (1955) Some new test criteria in multivariate analysis. *Ann. Math. Statist.* 26, 117–121.
——— (1967) Upper percentage points of the largest root of a matrix in multivariate analysis. *Biometrika* 54, 189–194.
Pillai, K. C. S., and Gupta, A. K. (1969) On the exact distribution of Wilks' criterion. *Biometrika* 56, 109.
Pitman, E. J. G. (1939) Tests of hypotheses concerning location and scale parameters. *Biometrika* 31, 200–215.
Potthoff, R. F., and Roy, S. N. (1964) A generalized multivariate analysis of variance model useful especially for growth curve problems. *Biometrika* 51, 313.
Prekopa, A. (1973) On logarithmically concave measures and functions. *Acta. Sci. Math.* 34, 335–343.

Ramachandran, K. V. (1958) A test of variances. *J. Am. Statist. Assoc.* 53, 741–748.
Rao, C. R. (1948) Tests of significance in multivariate analysis. *Biometrika* 35, 58–79.
——— (1949) On some problems arising out of discrimination with multiple characters. *Sankhyā* 9, 343–366.
——— (1950) A note on the distribution of $D_{p+q}^2 - D_p^2$ and some computation aspects of $D^2$ statistic and discriminant function. *Sankhyā* 10, 257–268.
——— (1952) *Advanced Statistical Methods in Biometric Research.* Wiley, New York.
——— (1954) A general theory of discrimination when the information about alternative population distributions is based on samples. *Ann. Math. Statist.* 25, 651–670.
——— (1959) Some problems involving linear hypothesis in multivariate analysis. *Biometrika* 46, 49–58.
——— (1962) A note on a generalized inverse of a matrix with applications to problems in mathematical statistics. *J. Roy. Statist. Soc. B* 24, 152–158.
——— (1964) The use and interpretation of principal component analysis in applied research. *Sankhyā A* 26, 329–358.
——— (1966a) Characterization of the distribution of random variables in linear structural relations. *Sankhyā A* 28, 251–260.
——— (1966b) Covariance adjustment and related problems in multivariate analysis. In *Multivariate Analysis I* (P. R. Krishnaiah, ed.) Academic Press, New York, pp. 87–103.
——— (1967) Least squares theory using an estimated dispersion matrix and its application to measurement of signals. *Proc. Fifth Berkeley Sym.* 1, 355–372.
——— (1969) Some characterizations of the multivariate normal distributions. In *Multivariate Analysis II* (P. R. Krishnaiah, ed.). Academic Press, New York, pp. 321–328.
——— (1970) Estimation of heteroscedastie variances in a linear model. *J. Am. Statist. Assoc.* 65, 161–172.
——— (1971) Estimation of variance and covariance components—MINQUE Theory. *J. Multiv. Anal.* 1, 257–275.
——— (1972) Estimation of variance and covariance components in linear models. *J. Am. Statist. Assoc.* 67, 112–115.
——— (1973a) *Linear Statistical Inference and Its Applications*, 2nd ed. Wiley, New York.
——— (1973b) Representation of best linear unbiased estimators in the Gauss–Markoff model with a singular dispersion matrix. *J. Multiv. Anal.* 3, 276–292.
——— (1974) Some problems in the characterization of the multivariate normal distribution. Linnik Memorial Lecture. In *Statistical Distribution in Scientific Work*, Volume 3, *Characterizations and Applications* (G. P. Patil et al., eds.). D. Reidel Publishing Co., Dordrecht, Holland, and Boston, U.S.A., pp. 1–14.
Rao, C. R., and Mitra, S. K. (1971) *Generalized Inverses of Matrices and Its Applications*. Wiley, New York.
Rasch, G. (1948) A functional equation for Wishart's distribution. *Ann. Math. Statist.* 19, 262.

Roy, J. (1958) Step-down procedure in multivariate analysis. *Ann. Math. Statist.* 29, 1177–1187.

Roy, S. N. (1939) A note on the distribution of the studentized $D^2$-statistic. *Sankhyā* 4, 373–380.

——— (1953) On a heuristic method of test construction and its use in multivariate analysis. *Ann. Math. Statist.* 24, 220–238.

——— (1957) *Some Aspects of Multivariate Analysis*. Wiley, New York.

Roy, S. N., and Bargmann, R. E. (1958) Tests of multiple independence and the associated confidence bounds. *Ann. Math. Statist.* 29, 491–503.

Roy, S. N., and Mikhail, N. (1961) On the monotonic character of the power functions of two multivariate tests. *Ann. Math. Statist.* 32, 1145.

Roy, S. N., and Roy, J. (1959) A note on a class of problems in "normal" multivariate analysis of variance. *Ann. Math. Statist.* 30, 577–581.

Ruben, H. (1962) Probability content of regions under spherical normal distributions, IV: the distribution of homogeneous and non-homogeneous quadratic functions of normal variables. *Ann. Math. Statist.* 33, 542–570.

——— (1963) A new result on the distribution of quadratic forms. *Ann. Math. Statist.* 34, 1582–1584.

Salaevski, O. V. (1968) Minimax character of Hotelling's $I^2$ Test. *Sov. Math. Dokl.* 9, 733–735.

Saxena, A. K. (1966) On the complex analogue of Hotelling's $T^2$ for two populations. *J. Indian Statist. Assoc.* 4.

Schatzoff, M. (1966) Exact distribution of Wilks' likelihood-ratio criterion. *Biometrika* 53, 347.

Scheffé, H. (1943) On solution of the Behrens–Fisher problem based on the $t$-distribution. *Ann. Math. Statist.* 14, 35–44.

Scott, A. (1967) A note on conservative confidence regions for the mean of a multivariate normal mean. *Ann. Math. Statist.* 38, 278–380.

Seely, J. (1970) Vector spaces and unbiased estimation—application to the mixed linear model. *Ann. Math. Statist.* 41, 1735–1748.

Selliah, J. B. (1964) Estimation and testing problems in a Wishart distribution. *Technical Report No. 10*, Stanford University.

Sen, A. K., and Srivastava, M. S. (1973) On multivariate tests for detecting change in mean. *Sankhyā A* 35, 173–185.

——— (1975a) On tests for detecting change in mean. *Ann. Statist.* 3, 98–108.

——— (1975b) On tests for detecting change in mean when variance is unknown. *Ann. Inst. Statist. Math. Tokyo* 27, 479–486.

——— (1975c) Some one-sided tests for change in level. *Technometrics* 17, 61–64.

Shah, B. K. (1963) Distributions of definite and of indefinite quadratic forms from a noncentral normal distribution. *Ann. Math. Statist.* 34, 186–190.

Shah, B. K., and Khatri, C. G. (1961) Distribution of a definite quadratic form for noncentral normal variates. *Ann. Math. Statist.* 32, 883–887.

Sidak, Z. (1967a) Rectangular confidence regions for the means of multivariate normal distributions. *J. Am. Statist. Assoc.* 62, 626–633.

——— (1967b) Rectangular confidence regions for the means of multivariate normal distributions, II. *Statist. Lab. Publ. No. 7*, Michigan State University.

——— (1968) On multivariate normal probabilities of rectangles: their dependence on correlations. *Ann. Math. Statist.* 39, 1425–1434.

——— (1971) On probabilities of rectangles in multivariate Student distributions: their dependence on correlations. *Ann. Math. Statist.* 42, 169–175.

——— (1975) A note on C. G. Khatri's and A. Scott's papers on multivariate normal distributions. *Ann. Inst. Statist. Math. Tokyo* 27, 181–184.

Simaika, J. B. (1941) An optimum property of two statistical tests. *Biometrika* 32, 70.

Sitgreaves, R. (1952) On the distribution of two random matrices used in classification procedures. *Ann. Math. Statist.* 23, 263–270.

Smith, C. A. B. (1947) Some examples of discrimination. *Ann. Eugen.* 13, 272–282.

Srivastava, J. N. (1964) On the monotonicity property of the three main tests for multivariate analysis of variance. *J. Roy. Statist. Soc. B* 26, 77–81.

Srivastava, J. N., and Zaatar, N. K. (1972) On the maximum likelihood classification rule for incomplete multivariate sample and its admissibility. *J. Multiv. Analy.* 2, 115–126.

Srivastava, M. S. (1964) Optimum procedures for classification and related problems. *Technical Report No. 11*, Stanford University.

——— (1965a) On the complex Wishart distribution. *Ann. Math. Statist.* 36, 313–315.

——— (1965b) Some tests for the intraclass correlation model. *Ann. Math. Statist.* 36, 1802–1806.

——— (1966) On a multivariate slippage problem I. *Ann. Inst. Statist. Math.* 18, 299–305.

——— (1967a) On fixed width confidence bounds for regression parameters and mean vector. *J. Roy. Statist. Soc. B* 29, 132–140.

——— (1967b) Classification into multivariate normal populations when the population means are linearly restricted. *Ann. Inst. Statist. Math.* 19, 473–478.

——— (1967c) Comparing distances between multivariate populations—the problem of minimum distance. *Ann. Math. Statist.* 38, 550–556.

——— (1968) On the distribution of a multiple correlation matrix, noncentral multivariate beta distributions. *Ann. Math. Statist.* 39, 227–232. Correction: 39, 1350.

——— (1970a) On a sequential analogue of the Behrens–Fisher problem. *J. Roy. Statist. Soc. B* 32 (1), 144–148.

——— (1970b) On a class of non-parametric tests for multivariate regression parameters. *J. Statist. Res.* 4 (1), 25–31.

——— (1970c) On a class of non-parametric tests for regression parameters. *J. Statist. Res.* 4 (2), 117–132.

——— (1971) On fixed width confidence bounds for regression parameters. *Ann. Math. Statist.* 42, 1403–1411.

——— (1972a) Some sequential procedures for ranking multivariate populations. *Ann. Inst. Statist. Math. Tokyo* 24 (3), 455–464.

——— (1972b) Asymptotically most powerful rank tests for regression parameters in MANOVA. *Ann. Inst. Statist. Math. Tokyo* 24 (2), 285–297.

——— (1973a) On a class of nonparametric tests for independence—bivariate case. *Can. Math. Bull.* 16 (3), 337–342.

——— (1973b) Evaluation of misclassification errors. *Can. J. Statist.* 1, 35–50.

——— (1973c) A sequential approach to classification. *J. Multiv. Anal.* 3, 173–183.

——— (1975a) On a class of rank score tests for censored data. *Ann. Inst. Statist. Math. Tokyo* 25 (3), 69–78.

——— (1975b) On a class of rank score tests for censored data. *Ann. Inst. Statist. Math. Tokyo* 25 (3), 69–78.

Srivastava, M. S., Khatri, C. G., and Carter, E. M. (1978) On some monotonicity property of the modified likelihood ratio test for the equality of two covariances. *J. Multiv. Anal.* 8, 262–267.

Srivastava, M. S., and Saleh, A. K. M. E. (1970) On a class of non-parametric estimates for regression parameters. *J. Statist. Res.* 4 (2), 133–139.

Srivastava, M. S., and Selliah, J. B. (1976) On errors of misclassification. Unpublished.

Srivastava, M. S., and Taneja, V. S. (1973a) Some sequential procedures for a mutlivariate slippage problem. *Metron* 31, 1–10.

——— (1973b) Some sequential procedures for ranking multivariate populations. *Ann. Inst. Statist. Math. Tokyo* 24 (3), 455–464.

Stein, C. (1956) The admissibility of Hotelling's $T^2$-test. *Ann. Math. Statist.* 27, 616.

——— (1969) Multivariate analysis I. *Technical Report No.* 42, Stanford University.

Subramaniam, K., and Subramaniam, K. (1973) On the distribution of $(D_{p+q}^2 - D_q^2)$ statistics: percentage points and the power of the test. *Sankhyā B* 35, 51–78.

Sugiura, N. (1969) Asymptotic expansions of the distributions of the likelihood-ratio criteria for covariance matrix. *Ann. Math. Statist.* 40, 2051–2063.

——— (1973) Asymptotic non-null distributions of the likelihood ratio criteria for covariance matrix under local alternatives. *Ann. Statist.* 1, 718–728.

——— (1976) Asymptotic expansions of the distributions of the latent roots and the latent vector of the Wishart and multivariate *F* matrices. *J. Multiv. Anal.* 6, 500–525.

Sugiura, N., and Nagao, H. (1968) Unbiasedness of some test criteria for the equality of two covariance matrices. *Ann. Math. Statist.* 39, 1686–1692.

Sugiyama, T. (1966) On the distribution of the largest root and the corresponding latent vectors for principal component analysis. *Ann. Math. Statist.* 37, 995.

Sverdrup, E. (1947) Derivation of the Wishart distribution of the second order sample moments by straightforward integration of a multiple integral. *Skand. Aktuarietidskr.* 30, 151.

Tang, P. C. (1938) The power function of the analysis of variance tests with tables and illustrations of their use. *Statist. Res. Mem.* 2, 126–157.

Tiku, M. L. (1975) *Selected Tables in Mathematical Statistics*, Volume 2. Edited by Inst. Math. Statist. Amer. Math. Soc., Providence, Rhode Island.

Tracy, D. S., and Dwyer, P. S. (1969) Multivariate maxima and minima with matrix derivatives. *J. Amer. Statist. Assoc.* 64, 1576–1594.

Trawinski, Irene Monohan, and Bargmann, R. E. (1964) Maximum likelihood

Estimation with incomplete multivariate data. *Ann. Math. Statist.* 35, 647–657.

Venables, W. (1973) Computation of the null distribution of the largest or smallest latent note of a beta matrix. *J. Multiv. Anal.* 3, 125–131.

Wald, A. (1944) On a statistical problem arising in the classification of an individual into one of two groups, *Ann. Math. Statist.* 15, 145–162.

Wald A., and Brookner, R. J. (1941) On the distribution of Wilks' statistic for testing independence of several groups of variates. *Ann. Math. Statist.* 12, 137–152.

Watson, G. W. (1964) A note on maximum likelihood, *Sankhyā A* 26, 303–304.

Welch, B. L. (1939) Note on discriminant function. *Biometrika* 31, 218–220.

Wijsman, R. A. (1957) Random orthogonal transformations and their use in some classical distribution problems in multivariate analysis. *Ann. Math. Statist.* 28, 415–423.

Wilks, S. S. (1932) Certain generalizations in the analysis of variance, *Biometrika* 24, 471–494.

——— (1935) On the independence of $k$ sets of normally distributed statistical variables. *Econometrika* 3, 309–326.

——— (1946) Sample criteria for testing equality of means, equality of variances, and equality of covariances in a normal multivariate distribution. *Ann. Math. Statist.* 17, 257–281.

Williams, E. J. (1970) Comparing means of correlated variables. *Biometrika* 57, 459–461.

Wishart, J. (1928) Proofs of the distribution law of the second order moment statistics. *Biometrika* 35, 55.

Yao, Ying (1965) An approximate degrees of freedom solution to the multivariate Behrens–Fisher problem. *Biometrika* 52, 139–147.

Zyskind, G., and Martin F. (1969) On best linear estimation and a general Gauss–Markov theorem in linear models with arbitrary non-negative covariance structure. *SIAM J. Appl. Math.* 17, 1190–1202.

# Author Index

# Subject Index

# D

## S



## T



## U



## V



## W